QUANTUM SYSTEMS IN CHEMISTRY AND PHYSICS. TRENDS IN METHODS AND APPLICATIONS

TOPICS IN MOLECULAR ORGANIZATION AND ENGINEERING

Volume 16

The titles published in this series are listed at the end of this volume.

QUANTUM SYSTEMS IN CHEMISTRY AND PHYSICS. TRENDS IN METHODS AND APPLICATIONS

edited by

Roy McWeeny, Jean Maruani, Yves G. Smeyers

and

Stephen Wilson

KLUWER ACADEMIC PUBLISHERS

DORDRECHT / BOSTON / LONDON

Library of Congress Cataloging-in-Publication Data

A C.I.P. Catalogue record for this book is available from the Library of Congress.

ISBN 0-7923-4699-8

Published by Kluwer Academic Publishers,
P.O. Box 17, 3300 AA Dordrecht, The Netherlands.

Sold and distributed in the U.S.A. and Canada
by Kluwer Academic Publishers,
101 Philip Drive, Norwell, MA 02061, U.S.A.

In all other countries, sold and distributed
by Kluwer Academic Publishers,
P.O. Box 322, 3300 AH Dordrecht, The Netherlands.

Printed on acid-free paper

Table of Contents *

Preface vii

* Introduction by ROY McWEENY / The European Workshop "Quantum Systems in Chemistry and Physics" 1

I) TRENDS IN METHODS

* R. McWEENY / Separability in Quantum Mechanics 7

* T. GRABO, T. KREIBICH and E.K.U. GROSS / Optimized Effective Potential for Atoms and Molecules 27

* I. MARTIN / The Relativistic Quantum Defect Orbital Method and Some of its Applications 51

* T. THORSTEINSSON, D.L. COOPER, J. GERRATT and M. RAIMONDI / A New Approach for Valence-bond Calculations: CASVB 67

* M.A.C. NASCIMENTO / Studies on Chemical Structure, Spectroscopy and Electron Scattering Using Generalized Multistructural Wavefunctions 87

II) TRENDS IN APPLICATIONS

* S.G. CHRISTOV / The Characteristic (Crossover) Temperature in the Theory of Thermally Activated Tunneling Processes 109

* L. MIHAILOV, M. KIRTCHEVA and A. MANOV / An Improved method of Relaxation Rate Calculation in Double-well Potential Systems 149

* H. CHOJNACKI / Correlation Effects in the Double Proton Transfer of the Formic Acid Dimer 161

* I. ROSENBLUM, E.I. DASHEVSKAYA, E.E. NIKITIN and I. OREF / Effect of the Vibrational/Rotational Energy Partitioning on the Energy Transfer in Atom-triatomic Molecule Collisions 169

* Contributions indicated with an asterisk also appear in the journal *Molecular Engineering*, Volume 7, Nos. 1–2.

* E. BUONOMO and F.A. GIANTURCO / Fragmentation of Ar_3^+: the Role of Rotational and Vibrational Predissociation Dynamics 185

* G. DELGADO-BARRIO, A. GARCIA-VELA, C. GARCIA-RIZO, M.I. HERNANDEZ and P. VILLAREAL / Sampling the Initial Conditions for Quasi-classical Trajectory Studies of Vibrational Predissociation Dynamics 219

* C. PISANI and S. CASASSA / Energy Estimates for Local Chemical Processes in Condensed Matter 231

* A. MARKOVITS, J. AHDJOUDJ and C. MINOT / Theoretical Study of the TiO_2 and MgO Surface Acidity and the Adsorption of Acids and Bases 245

* M. GEORGIEV and L. MIHAILOV / Optical Spectra of High-temperature Superconductors 263

III) ADDITIONAL CONTRIBUTIONS

J. MASIK and I. HUBAC / Multireference Brillouin–Wigner Coupled-Cluster Theory: Hilbert Space Approach 283

L.S. GEORGIEV, YA.I. DELCHEV, R.L. PAVLOV and J. MARUANI / Core-Valence Separation for an Open-shell Atom in the LST-DFT Scheme 309

D. MONCRIEFF and S. WILSON / On the Accuracy of the Algebraic Approximation in Molecular Electronic Structure Studies: Matrix Valence-bond Calculations for the Hydrogen Molecule Ground State 323

C. AMOVILLI / On the Evaluation of the Matrix Elements Between Non-Orthogonal Slater Determinants 343

R. MOCCIA, S. MENGALI and R. MONTUORO / The Calculation of Some Processes Involving Transitions in the Continuum 349

M. IVANOVITCH, M. GEORGIEV, M. MLADENOVA, G. BALDACCHINI, R.M. MONTEREALI, U.M. GRASSANO and A. SCACCO / Revisiting the Off-center Impurity Problem: Reorientational Rates of Hindered Rotators 373

Subject Index 397

Preface

This volume contains a representative selection of the papers presented at the European Workshop on 'Quantum Systems in Chemistry and Physics', held at San Miniato (PI), Italy, 14–17 April, 1996. The meeting was sponsored by the European Union, as a part of the COST initiative for stimulating cooperation in science and technology, within the states of the EU and those of Central and Eastern Europe. Generous support was also offered by the University of Pisa, the Italian National Research Council (through the Istituto di Chimica Quantistica e Energetica Molecolare, Pisa), and the Unione Industriale Pisana. The Cassa di Risparmio di San Miniato provided a delightful venue for the meeting at their Study Centre, 'i Cappuccini', an ex-convent set amidts the hills of Tuscany. To all who helped to make the Workshop a success, including those who gave so freely of their time and energy, the organizers are indeed grateful.

The purpose of the Workshop was to bring together chemists and physicists with a common field of interest – the quantum-mechanical many-body problem – and to encourage collaboration at the fundamental level of innovative theory and conceptual development. For this reason, many spectacular applications of currently available computing power were scarcely mentioned. The emphasis of the meeting will be clear from the titles of the main sessions, as follows:

(1) Density matrices and density functionals – which provide a common language, bridging areas of activity that range from nuclear shell models to solid-state theory. (2) Relativistic formulations – now recognized as essential in the consideration of molecules containing even moderately heavy atoms. (3) Electron correlation, CI and many-body methods – an area of undiminished interest and importance, in which there is constant progress. (4) Valence theory and the description of chemical bonds – where the newer theoretical techniques find an inexhaustible field of application. (5) Nuclear motion, vibronic effects and flexible molecules – where the problems of 'separation' of electronic and nuclear motion, and of their interaction, remain severe. (6) Response theory and molecular properties – a rapidly developing area in which there is a particularly strong interplay between theory and experiment, involving the calculation and use of propagators. (7) Response theory and molecular spectra – where the time-dependent perturbation, with which the system is 'probed', becomes the radiation field. (8) Condensed matter, crystal and clusters, surfaces and interfaces – in which the number of interacting particles becomes immense and the new methodologies developed for atoms and molecules must be replaced by other techniques, often with a more 'macroscopic' and semi-empirical orientation. (9) Chemical reactions – a closing (double-) session, concerned with what many would call the 'real' problems of chemistry – how do molecules react

and what are the dynamical processes involved? This session was planned as a special celebration of the work of the distinguished Bulgarian scientist, Academician Stefan Christov, on the occasion of his 85th birthday; this pioneering work on the formulation and use of the adiabatic hypothesis extends back to the time of Ehrenfest and his later work on reaction-rate theory remains of lasting value. Much of the work in this session was devoted to the identification of new horizons and techniques, often involving a blend of both classical and quantum dynamics.

The participants in this European Workshop came from many different countries and many different disciplines and backgrounds; their interactions were scientifically productive and personally rewarding. We left San Miniato feeling refreshed and enriched, and looking forward to future contacts.

THE EDITORS

Molecular Engineering **7**: 1–6, 1997.

Introduction

The European Workshop 'Quantum Systems in Chemistry and Physics'

Theoprop and the Workshop

ROY McWEENY
Chairman of the Organizing Committee.

Abstract. The scope of Theoprop* – theories of the properties, structure and dynamics of many-body quantum systems – is very broad, ranging from nuclear shell models to atoms, molecules, the solid state and the dynamics of chemical reactions; but the project has a certain unity and logical structure – and the Organizing Committee has tried to echo that logical structure in the Workshop itself. There are 10 Sessions, to which experts from many countries have been invited: here I simply want to reflect on our intentions in planning the programme.

Key words: Density matrices, density functionals, relativistic formulations, valence theory, nuclear motion, vibronic effects, flexible molecules, response theory, condensed matter, chemical reactions.

Density Matrices and Density Functionals

It was clear to us at the outset that density matrix- and density functional theory (DMT and DFT) could underpin the whole project and provide a unifying theme; both are direct descendents of the *reduced density matrices*, introduced by Husimi in 1940 and carried into quantum chemistry in the mid-fifties.[1] Briefly, as you will all know, the reduced density matrices are derived from the wavefunction Ψ of an N-electron system by integration over 'redundant' variables to give the one- and two-body densities (ρ and π) which are sufficient for describing a system with only one- and two-body terms in the Hamiltonian; and which also provide a transparent physical interpretation of many properties in terms of concepts such as electron

* Theoprop was used as the short title of a collaborative research project proposed in 1993 by Professor Stefan Christov (Bulgarian Academy of Sciences) and partly supported by the European Union as part of its COST initiative. The Western European Member States involved were France, Italy, Spain and the United Kingdom and the present European Workshop was a direct outcome of the resultant collaborative activity.

[1] For early work in this field see Husimi (1940) *Proc. Phys.-Math. Soc. Jpn.* **22**, 264; Löwdin (1955) *Phys. Rev.* **97**, 1474; McWeeny (1955) *Proc. Roy. Soc. Lond.* **A232**, 114; McWeeny (1960) *Rev. Mod. Phys.* **32**, 335.

density, spin density, current density – all of which can be easily visualized in three-dimensional space. DMT is not committed to orbital approximations and therefore stands outside all the intricacies of actual wave-function calculation, working in terms of ρ and π or (for a spinless Hamiltonian) the electron density P and the pair function Π – obtained by spin integration of ρ and π. But how do we know that any functions we might use could actually be derived from an N-electron wavefunction? This is the famous (and still unsolved) N-representability problem. Without imposing the implied constraints, there is no energy upper bound and all conventional methods of calculation must be abandoned. This problem appears to be circumvented in DFT, where – according to the Kohn–Hohenberg theorem – the *exact* energy can be expressed as a universal *functional* of the electron density P *alone*. By using a density based on the orbitals of a 'model' system it possible to make progress; but the connection between the model system and the actual system remains obscure and in practice DFT, however successful, still appears to contain empirical elements.

Much work in this area is in progress,[2] in Theoprop and elsewhere, and in Session 1 of the Workshop we shall find these ideas carried forward in various new directions.

Relativistic Formulations

A constant worry in calculations on many-electron systems is the importance of relativistic effects: it is well known that even for second- and third-row atoms such effects can become alarmingly significant. One of the aims of Theoprop was to investigate the systematic generalization of DMT and DFT to admit relativistic corrections. Naturally, the starting point is the Dirac equation, in which the usual two-component spin-orbitals.

$$\psi = \phi_\alpha \alpha + \phi_\beta \beta$$

are replaced by corresponding four-component quantities

$$\psi = \phi_1 e_1 + \phi_2 e_2 + \phi_3 e_3 + \phi_4 e_4.$$

The solutions are of two types, with negative- and positive-energy eigenvalues, respectively, one pair of components being in each case very small. Dirac's interpretation of the first type of solutions was that, using an independent-particle model, they would hold a filled 'sea' of 'negative energy electrons'; solutions of the second type would hold the 'ordinary' positive-energy electrons. A sufficient amount of

[2] For bibliography and references see, in particular, Davidson (1978) *Reduced Density Matrices for Quantum Chemistry*, Academic Press; Parr and Yang (1989) *Density Functional Theory of Atoms and Molecules*, Oxford University Press, Oxford; Kryachko and Ludeña (1990) *Energy Density Functional Theory of Many-Electron System*, Kluwer, Dordrecht; Petkov and Stoitsov (1991) *Nuclear Density Functional Theory*, Oxford University Press, Oxford.

energy would serve to knock a particle out of a negative energy state, leaving a 'hole' which would behave like a particle of opposite charge and producing an ordinary electron in a positive energy state. Thus, an electron-positron pair would be produced. Of course, Dirac's interpretation is inconsistent: the Dirac equation refers to a single particle and we cannot apply it directly to a many-electron system with strong inter-particle interactions. But a certain rationalization is possible: a system with two groups of non-interacting electrons could be described by an antisymmetrized product of two 'strong-orthogonal' functions,[3] Φ_0 for the negative energy 'sea' and Φ for N positive energy electrons. To refine this model, and admit particle interactions, one can use a standard CI procedure: $\Psi = \hat{A}[\Phi_0\Phi]$ can be improved by adding terms describing 'virtual excitations' from the occupied states in Ψ to a complementary set of empty solutions of the Dirac equation. As long as we exclude negative-to-positive energy excitations (i.e. use a no virtual (electron-positron) pair approximation), we can calculate Φ (which describes N ordinary electrons) *just as if the electrons were alone*, the CI with positive-to-positive energy excitations leading in principle to a fully-correlated wavefunction.

The possibility of proceeding in this way depends only on the mutual orthogonality of the negative and positive energy solutions of the Dirac equation. Of course, we are still using four-component spin-orbitals and much generalization needs to be done. For example, the one-electron density matrix (in fact an operator kernel).

$$\rho(\mathbf{x};\mathbf{x}') = P_{\alpha\alpha}(\mathbf{r};\mathbf{r}')\alpha(s)\alpha^*(s') + P_{\alpha\beta}(\mathbf{r};\mathbf{r}')\alpha(s)\beta^*(s') + \cdots,$$

which has four components, has to be replaced by a 16-component quantity in which, instead of the Pauli spin dyads, we shall have Dirac dyads such as $e_1(s)e_1^*(s')$. It is perfectly possible to do this: there are two triads of Dirac operator products that have properties isomorphic with those of the Pauli spin operators and have the es as eigenvectors; and to maintain the parallel with the non-relativistic theory it is only necessary to introduce an extra 'spin variable', using σ, τ in place of s. These and related problems are under active investigation in Theoprop.

A final problem to consider is whether a four-component formulation is essential or not. As we all know, by 'absorbing' the two small components into the two large components (essentially by a 'partitioning' technique) it is possible to derive a 'Dirac–Pauli' Hamiltonian, which is the usual non-relativistic Hamiltonian with additional 'small terms' – defined up to a certain order in the inverse velocity of light. The definition of these small terms is a matter of recent controversy – of which we shall hear much more in Session 2.

Electron correlation. CI and many-body methods

The usual 'electron correlation problem' remains of course one of the great challenges, at either the non-relativistic or relativistic level, in all calculations on many-

[3] For a full account of antisymmetrized products of this kind, see the last reference in footnote 1.

electron systems – a problem that is confronted in Session 3. With wave-functions constructed from discrete basis sets (the usual 'algebraic approximation'), the approach is in principle simple and direct and leads to large systems of secular equations to determine the coefficients in a CI expansion. Unfortunately, correlation depends on the cusp behaviour of the pair function $\Pi(\mathbf{r}_1, \mathbf{r}_2)$ as two electrons approach; and to describe this cusp well the expansion is so long that the secular equations become gigantic. They may be solved directly only by using sophisticated methods of matrix element evaluation and iterative methods of evaluating the coefficients without ever writing down the full matrices – which are much too big to go in the computer. The alternative is to use perturbation methods (usually with diagrammatic techniques) in which the coefficients in the CI expansion are not evaluated directly but are approximated order-by-order, using only limited classes of diagrams. Expert assessments of progress in this field are included in Session 3.

Valence Theory: The Description of Chemical Bonds

In Session 4 we turn from mathematical techniques to chemical applications – to the discussion of chemical bonding and molecular electronic structure. In recent years it has become evident that the IPM approach, which leads naturally to MO theory, is not well adapted to the discussion of many areas of chemistry – in which the making and breaking of chemical bonds play a central role. As a consequence, the old valence bond approach, after several decades of neglect, has been resurrected in various *ab initio* forms. This Session is characterized by the interplay of the alternative approaches (MO-based and VB-based), with their interrelationship, and with the possibility of exploiting the best features of both.

Nuclear Motion, Vibronic Effects and Flexible Molecules

The systems we deal with contain heavy nuclei as well as electrons; and nuclear motion, even if slow compared with that of the electrons, cannot be neglected. There arises the problem of how to 'separate' in some sense electronic and nuclear motion, possibly dealing with the latter in a semi-classical way. Soon after Ehrenfest, Christov became one of the great pioneers in the formulation and use of this 'adiabatic' hypothesis (later elaborated by Born and Oppenheimer); and one of his recent papers[4] contains an authoritative and illuminating review of the adiabatic approximation. Session 5 opens with the general mathematical problem of separating nuclear and electronic motion, where fundamental and conceptual difficulties still remain, and proceeds to the study of some implications of nuclear motion – in particular to vibronic effects and to molecular flexibility and its consequences.

---

[4] Critical Reviews in Solid State and Materials Sciences **1990**, 337.

Response Theory: Molecular Properties

What is the connection between the largely mathematical study of many-body systems, on the one hand, and the experimental study of physical and chemical systems, on the other hand? Certainly, the total energy of an electronic system is one of its least important properties – for it is never directly observed. All experimental observations are concerned more with small energy changes, such as excitation energies, dissociation energies, or the response of the system to some external perturbation (the observer and his 'probes').

Session 6 opens with the general theory of response and of the propagation of perturbations to which an electronic system may be subjected. The effects are typically (i) the 'polarization' of the system (e.g. by application of fields) without adding or subtracting electrons; and (ii) ionization and attachment processes, in which the number of electrons will change. Both are studied in 'propagator theory' – which is very closely related to density matrix theory – effects of type (i) being determined by the polarization propagator and those of type (ii) by the electron propagator. The propagators are complex quantities and, if the perturbation is infinitesimal and oscillating, poles will occur at certain values of the frequency (resonances): these values give directly transition, ionization and attachment energies, while the residues at the poles give information on oscillator strengths and cross-sections for the corresponding processes. The Session includes various applications of these beautiful techniques.

Response Theory: Molecular Spectra

Session 7 is concerned more explicitly with response to oscillating perturbations, namely to the radiation field. In this case, there are particular problems, notably how to ensure the gauge invariance of the results and how to take into account correctly the presence of the electronic continuum. After a general introduction to radiation-molecule interactions, both these problems are addressed. The session also includes applications to fields of experimental importance, such as the interpretation of Auger spectra.

Condensed Matter: Crystals and Clusters, Surfaces and Interfaces

In earlier sessions, the systems considered have usually contained relatively few particles; and the methodology discussed, sophisticated as it may have been, could be realistically applied to such small systems. In dealing with the solid state, however, quite new problems arise as a result of the immense number of particles to be included. The application of the standard techniques of ab initio calculation to crystals is consequently precluded – with one exception: at the Hartree–Fock level crystal symmetry allows one to reduce the problem to manageable dimensions by introducing a 'repeating unit' with periodic boundary conditions. The self-consistent field method can then be generalized, the (density-dependent) effective

Hamiltonian being obtained in every iteration by integrating over an occupied Brillouin-zone in k-space (i.e., over a set of Bloch-type crystal orbitals) – instead of summing density matrix contributions over a discrete set of molecular orbitals. This method is now standard and widely employed in energy-band calculations. Much work in progress at present is devoted to its detailed exploitation.

To go beyond the Hartree–Fock approximation is exceedingly difficult for real crystals, and there has thus been a growing interest in the use of clusters, containing only a small number of atoms, which may be expected to simulate crystal properties as the number is increased – and for which more complete calculations are feasible.

On the other hand, many important crystal properties lie well beyond the scope of *ab initio* calculation, typically involving nuclear motion and electron-phonon interaction. In such cases it is inevitable that one must fall back on relatively primitive models.

Session 8 contains a small selection of papers showing that, however insurmountable the problems may seem, impressive progress can still be made.

Chemical Reactions

The last session, in fact a double session, is in a certain sense a culmination of the Workshop: it is dedicated to the dynamics of chemical reactions, the problem with which most of chemistry is concerned, and it is a special celebration of the pioneering work of Stefan Christov – some of whose major contributions to science were in this field – on the occasion of his 85th birthday.

The session opens with an appreciation of Stefan Christov, prepared by Raymond Daudel, another of the great pioneers of quantum chemistry, and continues with two main themes: first the use of semi-classical models, in which chemical reactions (most commonly in the liquid phase) are treated using a blend of classical and quantum mechanical methods; and secondly the attempts now being made to combine the study of electronic and nuclear motions in a single, wholly quantum mechanical scheme. The first approach is by now familiar, the reacting systems being discussed using potential energy surfaces calculated using quantum mechanics, but with the liquid environment regarded as a continuous dielectric – to which classical physics may be applied: the second, at present applicable only on a severely limited time scale, brings us to the limits of what can be done by present-day quantum dynamics. Perhaps the future will depend in no small degree on the development of computing machines at the molecular level.

Molecular Engineering **7**: 7–26, 1997.

Separability in Quantum Mechanics

ROY McWEENY
Dipartimento di Chimica e Chimica Industriale, Via Risorgimento 35, 56100 Pisa, Italy.

(Received and accepted: 17 June 1997)

Abstract. The idea of 'separability' is concerned with the quantum mechanical description of systems in which it is possible to recognize experimentally the presence of certain 'structural units', each with a 'personal' identity. The aim of this paper is to account for the striking success of this model and to develop the mathematical tools for its refinement and exploitation. At the root of the concept of separability lies the possibility of factorizing the wave-function for the whole system into a product of factors, each of which refers to a constituent part – a certain 'group' of electrons. In this 'group function' approximation the wavefunction for the whole system is written as an antisymmetrized product of strong-orthogonal group functions, one for each subsystem. Although the resultant ansatz is well known, its deeper implications and full potential have never been fully explored. Progress has been made in several directions: these include the separation of charge, current and spin densities into subsystem contributions, which provides a basis for the discussion of 'additivity rules'; an extension of the formalism to admit relativistic effects, using both 2-component and 4-component (Dirac-type) spin-orbitals; and a practical implementation of a variational procedure which leads to an optimal separation of given form, whilst preserving strong-orthogonality of the subsystems. To illustrate the approach, the dissociation curve for the molecule LiCl has been calculated using a 'two-group' approximation in which 4 'valence electrons' (a bond pair and an electron pair) are separated from the remaining 16 'core electrons'. The bond-breaking process is well described and there appears to be much potential for using this approach in the study of chemical reactions.

Key words: Separability, wavefunction factorization, additivity rules, LiCl dissociation, group functions.

1. Introduction

What do we mean when we say that a molecule consists of bonded atoms; or that liquid water consists of H_2O molecules in motion; or that a crystal consists of atoms or molecules located at points on a lattice? We mean, it can be argued, that certain structural units, such as atoms or molecules, can be recognized *experimentally* as distinct components of a more extended physical system. Indeed, these units are 'visible' to us in the pictures provided by X-ray spectroscopy; and their presence can also be inferred using a vast range of experimental techniques. It is the task of *theory* to account for the success of this model, in which the matter we observe is 'mentally separated' into interacting units, each with its own identity, by providing a quantum mechanical foundation for this separation. The separation is ubiquitous: it is carried to deeper levels, of course, whenever one refers to 'atomic inner shells'; or to 'valence electrons' or to 'chemical bonds' and 'lone pairs'; or to 'sigma electrons' and 'pi electrons'.

At the root of the concept of separability, which can have only an approximate status – because in reality there can be no sharp boundaries between different parts of a quantum system, lies the possibility of representing the wavefunction for the whole system as a product of factors, each of which refers to a constituent part; for the probability of finding two subsystems, A and B, simultaneously in well-defined conditions will be a *product*, $P = P_A P_B$, when and only when $\Psi = \Psi_A \Psi_B$. In this idealized situation, the subsystems would be statistically independent: in reality, the behaviour of A and B will be at least weakly correlated and the wavefunction will depart from product form. There are two obstacles to the formulation of a theory of separability:

(i) the fact that, even when the subsystems are *spatially* well separated, this cannot be true of their electrons – which are in principle shared by all subsystems and are subject to conditions imposed by the Pauli principle; and

(ii) the fact that, in constructing a wavefunction even for *one* subsystem from a complete set of 1-electron functions, the *same* set must serve for *all* subsystems – there can be no separation of basis functions.

In short, neither electrons nor basis functions can be separated into groups associated uniquely with the various subsystems; and it is for this reason that separability necessarily involves approximations. The best that can be done is to minimize the defects associated with the use of a wavefunction of 'separable' form, in which each subsystem is represented by one factor in a product. To this end, application of an antisymmetrizer will ensure that the Pauli principle is satisfied and that no particular group of electrons 'belongs' to any specific subsystem; but the possibility of separating the *basis* into subsets which can each describe one subsystem with high accuracy raises delicate questions of overcompleteness. In order to avoid overcompleteness problems it is necessary to *truncate* the function space associated with each group of particles, to eliminate excessive overlap between the functions of different groups. Ideally, the truncated sets describing different groups should be mutually orthogonal and in this case the subsystem wavefunctions are said to be *strong orthogonal*. The total wavefunction is then written in the form

$$\begin{aligned}\Psi_{Aa,Bb,\ldots}(\mathbf{x}_1, \mathbf{x}_2, \ldots, \mathbf{x}_N) &= M\hat{A}[\Phi_{Aa}(\mathbf{x}_1, \ldots, \mathbf{x}_{N_A}) \times \\ &\quad \times \Phi_{Bb}(\mathbf{x}_{N_A+1}, \ldots, \mathbf{x}_{N_A+N_B}) \ldots], \qquad (1)\end{aligned}$$

as first proposed by Parr and co-workers [1]. Here $\hat{A}$ is the antisymmetrizer for all particles and (1) defines a 'generalized product function' [2] – the analogue of a Slater determinant in which each spin-orbital is replaced by a many-electron function for a particular electron group. In (1), M is merely a normalizing factor, spins are included in the space-spin coordinates $\mathbf{x}_i = \mathbf{r}_i s_i$, and indices $a, b, \ldots$ indicate a selection of 'states' for the individual groups. More generally, a type of configuration interaction (CI) may be admitted by using instead of (1) the wavefunction

$$\Psi(\mathbf{x}_1, \mathbf{x}_2, \ldots, \mathbf{x}_N) = \sum_{a',b',\ldots} C_{a'b',\ldots} \Psi_{Aa',Bb',\ldots}, \tag{2}$$

which is the most general form of group function approximation: the CI then extends over any number of local 'excited' states for any number of groups. An *exact* wavefunction cannot in general be represented even in the form (2), because it does not contain terms corresponding to *electron transfer* between different groups: but it is capable of describing both *intra*-group electron correlation and 'long-range' inter-group correlation (which results, for example, in the dispersion interactions between distant molecules).

For present purposes, it will be sufficient to consider the 'one-configuration' group function product (1) and to suppress the state labels $(a, b, \ldots)$ for the separate groups. Before exploring the implications of this approximation it is useful to introduce the electron density functions, and more generally density *matrices*, which will give a physically transparent interpretation of separability.

2. Density Functions

The electron density is only one of a whole series of distribution functions referring to one electron, a pair of electrons, or a cluster of n electrons, simultaneously at given points in space. As is well known (see for example [3]), the most general definition of the electron density is

$$P(\mathbf{r}) = N \int \Psi(\mathbf{x}, \mathbf{x}_2, \ldots, \mathbf{x}_N) \Psi^*(\mathbf{x}, \mathbf{x}_2, \ldots, \mathbf{x}_N) \, \mathrm{d}s \mathrm{d}\mathbf{x}_2 \ldots \mathrm{d}\mathbf{x}_N \tag{3}$$

– which, more correctly, is the probability/unit volume of finding an electron (no matter which) at point $\mathbf{r}$. In this definition, the wavefunction (in principle exact) is assumed antisymmetric in the space-spin variables $(\mathbf{x}_i = \mathbf{r}_i s_i)$ and all reference to spin is eliminated in the integrations.

The quantity $P(\mathbf{r})$ is in fact the 'diagonal element', $P(\mathbf{r}) = P(\mathbf{r}; \mathbf{r})$, of the (spinless) 1-electron density *matrix* $P(\mathbf{r}; \mathbf{r}')$, in which the $\mathbf{r}$ of Ψ^* in (3) is replaced by a second variable $\mathbf{r}'$. More generally, $P(\mathbf{r}; \mathbf{r}')$ is derived from a density matrix $\rho(\mathbf{x}; \mathbf{x}')$ *with spin included*:

$$\rho(\mathbf{x}; \mathbf{x}') = N \int \Psi(\mathbf{x}, \mathbf{x}_2, \ldots, \mathbf{x}_N) \Psi^*(\mathbf{x}', \mathbf{x}_2, \ldots, \mathbf{x}_N) \, \mathrm{d}\mathbf{x}_2, \ldots, \mathrm{d}\mathbf{x}_N \tag{4}$$

from which, identifying s', s and integrating,

$$P(\mathbf{r}; \mathbf{r}') = \int_{s'=s} \rho(\mathbf{x}; \mathbf{x}') \, \mathrm{d}s. \tag{5}$$

A basic property of the density matrix is, of course, that the expectation value of any symmetric sum of 1-electron operators (e.g. the potential and kinetic energy terms in the Hamiltonian) can be written directly in terms of the *one*-electron quantity. Thus,

$$\langle V\rangle = \int V(\mathbf{r})P(\mathbf{r})\,\mathrm{d}\mathbf{r}, \qquad \langle T\rangle = \int_{\mathbf{r}'=\mathbf{r}} \frac{\hat{p}^2}{2m}P(\mathbf{r};\mathbf{r}')\,\mathrm{d}\mathbf{r}, \tag{6}$$

where the off-diagonal quantity is required only for the *kinetic* energy. The potential energy is thus a very simple *functional* of the electron density, with an obviously 'classical' interpretation – a result of special importance for 'density functional theory' (DFT) [4, 5] whose aim is to express all properties of the system in a similar way; but the kinetic energy is a functional of the density *matrix* and has so far resisted all attempts to express it in terms of the electron density alone.[1] This difficulty is fundamental, arising from the non-commuting nature of position and momentum operators, and in most practical applications of DFT it is now customary to express $\langle T\rangle$ as in (6) i.e. as a functional of the density *matrix* rather than the density.

Since $P(\mathbf{r};\mathbf{r}')$ arises by spin integration in (5), it is important to know the general form of $\rho(\mathbf{x};\mathbf{x}')$. Evidently,

$$\begin{aligned}\rho(\mathbf{x};\mathbf{x}') \;=\; & P_{\alpha\alpha}(\mathbf{r};\mathbf{r}')\alpha(s)\alpha^*(s') + P_{\alpha\beta}(\mathbf{r};\mathbf{r}')\alpha(s)\beta^*(s') + \\ & + P_{\beta\alpha}(\mathbf{r};\mathbf{r}')\beta(s)\alpha^*(s') + P_{\beta\beta}(\mathbf{r};\mathbf{r}')\beta(s)\beta^*(s'),\end{aligned} \tag{7}$$

this being the integral-operator form of the density *operator* $\hat{\rho}$, whose effect on a space-spin function is given by $\hat{\rho}\psi(\mathbf{x}) = \int \rho(\mathbf{x};\mathbf{x}')\psi(\mathbf{x}')\,\mathrm{d}\mathbf{x}'$. For any state of definite spin, however, only the $\alpha\alpha$ and $\beta\beta$ components are non-zero. On using (5),

$$P(\mathbf{r};\mathbf{r}') = P_{\alpha\alpha}(\mathbf{r};\mathbf{r}') + P_{\beta\beta}(\mathbf{r};\mathbf{r}') \tag{8}$$

and on using the analogue of (6), with spin included, to evaluate the expectation value of the total spin z component, we obtain

$$\langle \hat{S}_z\rangle = \int_{\mathbf{x}'=\mathbf{x}} \hat{S}_z\rho(\mathbf{x};\mathbf{x}')\,\mathrm{d}\mathbf{x} = \int Q_z(\mathbf{r};\mathbf{r})\,\mathrm{d}\mathbf{r}, \tag{9}$$

where, generally,

$$Q_z(\mathbf{r};\mathbf{r}') = \tfrac{1}{2}[P_{\alpha\alpha}(\mathbf{r};\mathbf{r}') - P_{\beta\beta}(\mathbf{r};\mathbf{r}')] \tag{10}$$

and has a diagonal element[2] which is the density of spin angular momentum around the z axis. Thus

$$P(\mathbf{r}) = P_\alpha(\mathbf{r}) + P_\beta(\mathbf{r}), \tag{11a}$$

[1] Except in the trivial case of a free-electron gas.

[2] In referring to diagonal elements we drop both the second variable and the corresponding spin index. The spin density is often normalized to unity, by dividing by M.

$$Q_z(\mathbf{r}) = \tfrac{1}{2}[P_\alpha(\mathbf{r}) + P_\beta(\mathbf{r})], \tag{11b}$$

are, respectively, the electron density and the *spin density*. It must be emphasized, however, that Q_z is only one component of a *vector* density,[3] the field components $Q_x(\mathbf{r})$, $Q_y(\mathbf{r})$ being defined in strict analogy with (9). Attention is usually focused on Q_z (z being the quantization axis) only because for a pure spin state (quantum numbers S, M) the components transverse to the axis are everywhere zero. For such a state, $Q_x = Q_y = 0$ and $Q_z = (M/S)Q_z^{\mathrm{st}}$, where Q_z^{st} denotes the density for a standard state – the 'top' state ($M = S$) of the multiplet. This is not the case in the presence of spin-orbit coupling (see, for example, [6, 7]).

The density matrix $\rho(\mathbf{x};\mathbf{x}')$ defines another spatial density of great importance – the *current density* $\mathbf{J}(\mathbf{r})$ which is again a vector density. The components $J_x(\mathbf{r})$, $J_y(\mathbf{r})$, $J_z(\mathbf{r})$ determine the flux of electron (probability) density out of the volume element $\mathrm{d}\mathbf{r}$ at any point $\mathbf{r}$ in space. Since the flux is zero everywhere for a state described by a *real* wavefunction, it is desirable in discussing this density to introduce in the Hamiltonian a magnetic field (however weak) in order to define the quantization axis and to introduce an imaginary component in Ψ. The presence of such a field requires only that the usual kinetic energy term in the Hamiltonian, namely $(1/2m)\Sigma_i\hat{p}^2(i)$, be replaced by

$$\hat{T} = (1/2m)\Sigma_i\hat{\pi}^2(i), \quad \hat{\pi}_\lambda = \hat{P}_\lambda + eA_\lambda \quad (\lambda = x, y, z). \tag{12}$$

Here $\hat{\pi}$ is the 'gauge invariant momentum' and contains the vector potential $\mathbf{A}$ of the field: $\hat{\pi}/m$ is in fact the operator associated with the electron *velocity*. The components of the *current density matrix* may be defined most symmetrically as

$$J_\lambda(\mathbf{r}) = m^{-1}[\tfrac{1}{2}(\pi_\lambda + \pi_\lambda'^\dagger)P(\mathbf{r};\mathbf{r}')]_{\mathbf{r}'=\mathbf{r}}, \tag{13}$$

where the adjoint operator $\pi_\lambda'^\dagger$ works on the *primed* variables in the density matrix.

The diagonal element $J_\lambda(\mathbf{r})$, has all the properties of a *local* flux density at point $\mathbf{r}$, as defined in classical physics. In particular, for a wavefunction satisfying the time-dependent Schrödinger equation, it is easy to derive a conservation equation for the electron density:

$$\operatorname{div}\mathbf{J}(\mathbf{r}) = \frac{-\partial P(\mathbf{r})}{\partial t} \tag{14}$$

– the rate of decrease of the probability of finding an electron in volume element $\mathrm{d}\mathbf{r}$ at point $\mathbf{r}$ is the divergence of the current density at that point. It follows that for a system in a stationary state the currents must be *circulating* ($\operatorname{div}\mathbf{J} = 0$); and that these induced currents (created by the external field) will produce a secondary magnetic field. The components of this field, at point $\mathbf{r}_0$, are found to be

$$B_\lambda^{\mathrm{ind}}(\mathbf{r}_0) = \left(\frac{\mu_0}{4\pi}\right)\int \frac{[(\mathbf{r} - \mathbf{r}_0) \times -e\mathbf{J}(\mathbf{r})]_\lambda}{|\mathbf{r} - \mathbf{r}_0|^3}\mathrm{d}\mathbf{r} \tag{15}$$

[3] More correctly a *pseudo*vector density, being an angular momentum per unit volume.

– just as if (remembering the Biot–Savart law) the electron distribution were a conducting medium containing electric currents of density $-e\mathbf{J}(\mathbf{r})$. Another interesting property of the current density is that the total electronic energy of an arbitrary system in the presence of any external field whatever, with vector potential $\mathbf{A}$, may be written [8]

$$E = E_0 - \int \mathbf{A} \cdot [-e\mathbf{J}_0(\mathbf{r})]\, d\mathbf{r} - \frac{1}{2}\int \mathbf{A} \cdot [-e\mathbf{J}_{\mathrm{ind}}(\mathbf{r})]\, d\mathbf{r}. \tag{16}$$

Here E_0 and $\mathbf{J}_0$ are the energy and current density in the zero-field limit ($\mathbf{J}_0$ vanishing except for systems of high symmetry in states with unquenched angular momentum, e.g. paramagnetic atoms), while $\mathbf{J}_{\mathrm{ind}}$ is the field-linear part of the current density and therefore describes the currents *induced* by the applied field. The two field-dependent terms in (16) represent, respectively, the potential energy of a system of 'permanent' electric currents, of density $-e\mathbf{J}_0(\mathbf{r})$, and that of a system of induced currents, of density $-e\mathbf{J}_{\mathrm{ind}}(\mathbf{r})$, in a conducting medium in the presence of a field – as would be calculated from classical physics.

The density functions introduced above are important in the 'quantum fluid dynamics' (QFD) interpretation of quantum mechanics (for a review, see Ref. [9]) and provide a basis for the semiclassical interpretation of a wide range of electronic properties, as reviewed elsewhere [10]. It must be emphasized that (apart from the use of a non-relativistic Born–Oppenheimer Hamiltonian) *no approximations need be made*: the density functions are all rigorously derivable, in principle, from an exact wavefunction containing no orbital approximations. To give only one example, (15) allows one to calculate the secondary magnetic field at a nucleus, due to induced currents in the electron distribution, and hence the NMR shielding tensor.

Up to this point only *one*-electron density functions have been considered, but of course electron interactions are also present in the usual Hamiltonian $\hat{H}$ and the expectation value of the interaction energy, or any other *two*-electron quantity, depends on a two-electron density. The analogue of $\rho(\mathbf{x};\mathbf{x}')$ is the *pair* density matrix $\pi(\mathbf{x}_1,\mathbf{x}_2;\mathbf{x}_1',\mathbf{x}_2')$; while that of $P(\mathbf{r};\mathbf{r}')$ is $\Pi(\mathbf{r}_1,\mathbf{r}_2;\mathbf{r}_1',\mathbf{r}_2')$. The electron interaction energy depends in fact[4] only on the diagonal element $\Pi(\mathbf{r}_1,\mathbf{r}_2)$, which is just like the pair function in the classical theory of fluids and determines the probability of two particles being present simultaneously in volume elements at points $\mathbf{r}_1$ and $\mathbf{r}_2$. In terms of this function, the electron interaction energy assumes the form

$$V_{\mathrm{ee}} = \frac{1}{2}\int g(1,2)\Pi(\mathbf{r}_1,\mathbf{r}_2)\, d\mathbf{r}_1\, d\mathbf{r}_2, \tag{17}$$

a result which, again, is valid for any system at any level of approximation.

[4] In the usual case of spin-independent coulombic interactions.

It is now time to turn to the separability concept, knowing that many of the most important properties of a physical system – in particular those associated with applied electric and magnetic fields (used in experimentally 'probing' the system) – are expressible in terms of *one*-electron densities.

3. Separability of the Density Functions

The theory of generalized product wavefunctions constructed from strong orthogonal groups was developed many years ago [2, 3]. It will be sufficient to summarize the most important results: these relate the density functions of the whole system to those of its strong-othogonal parts. Thus, using

$$\begin{aligned}\Psi(\mathbf{x}_1, \mathbf{x}_2, \ldots, \mathbf{x}_N) &= M\hat{A}[\Phi_A(\mathbf{x}_1, \ldots, \mathbf{x}_{N_A}) \times \\ &\quad \times \Phi_B(\mathbf{x}_{N_A+1}, \ldots, \mathbf{x}_{N_A+N_b}) \ldots].\end{aligned} \tag{18}$$

The first two results are that

$$\rho(\mathbf{x}_1; \mathbf{x}'_1) = \sum_R \rho_R(\mathbf{x}_1; \mathbf{x}'_1), \tag{19}$$

for the one-body density, while

$$\begin{aligned}\pi(\mathbf{x}_1, \mathbf{x}_2; \mathbf{x}'_1, \mathbf{x}'_2) &= \sum_R \pi_R(\mathbf{x}_1, \mathbf{x}_2; \mathbf{x}'_1, \mathbf{x}'_2) + \\ &\quad + \frac{1}{2} \sum_{R,S}{}' [\rho_R(\mathbf{x}_1; \mathbf{x}'_1)\rho_S(\mathbf{x}_2; \mathbf{x}'_2) + \\ &\quad + [\rho_R(\mathbf{x}_2; \mathbf{x}'_1)\rho_S(\mathbf{x}_1; \mathbf{x}'_2)],\end{aligned} \tag{20}$$

for the corresponding two-body density. In dealing with quantities whose associated operators do not act on spin variables, we may use (5) and its two-electron analogue to derive, first

$$P(\mathbf{r}_1; \mathbf{r}'_1) = \sum_R P_R(\mathbf{r}_1; \mathbf{r}'_1), \tag{21}$$

and, provided the individual group functions correspond to spin zero,

$$\begin{aligned}\Pi(\mathbf{r}_1, \mathbf{r}_2; \mathbf{r}'_1, \mathbf{r}'_2) &= \sum_R \Pi_R(\mathbf{r}_1, \mathbf{r}_2; \mathbf{r}'_1, \mathbf{r}'_2) + \\ &\quad + \frac{1}{2} \sum_{R,S}{}' [P_R(\mathbf{r}_1; \mathbf{r}'_1)P_S(\mathbf{r}_2; \mathbf{r}'_2) + \\ &\quad + \tfrac{1}{2} P_R(\mathbf{r}_2; \mathbf{r}'_1)P_S(\mathbf{r}_1; \mathbf{r}'_2)].\end{aligned} \tag{22}$$

It is hard to exaggerate the importance of these results. The one-body density for the whole system, with or without inclusion of spin, is a sum of the densities of the constituent parts. Thus, the electron density is (using $\mathbf{r}$ as variable)

$$P(\mathbf{r}) = \sum_R P_R(\mathbf{r}) \tag{23}$$

– a simple superposition of the electron densities of the separate groups. And by use of (13) it follows that the total *current* density may also be expressed as a sum of strictly additive contributions from the separate electron groups. Observable molecular properties that depend only on these densities (including molecular electric moments, electric and magnetic fields due to charge and current distributions within the molecule, magnetic susceptibility, and many other properties) must consequently each be expressible as a sum of contributions from separate groups.

This is not to say, of course, that the subsystems are *independent*: the presence, outside any given subsystem, of other nuclei and electron groups may provide a considerable perturbation of individual group properties. But provided such effects are recognized by defining subsystems *in the molecule* (and not in isolation), then the concept of separability is valid and useful. The possibility of doing so depends on setting up an *effective* Hamiltonian for each subsystem in the field due to the others and then performing a variational calculation to establish the optimum forms of all electron groups. In early work (e.g. [11]), such variational calculations were subject to constraints imposed by an *a priori* partitioning of a basis set for the whole system into subsets associated with its various parts. Ideally, the partitioning should arise as a consequence of optimization of the whole wavefunction: this possibility has been fully explored [12] and viable procedures have been formulated for achieving this end. In this way it is possible to eliminate the major defects of group function theory in its original form.

At this point it is worth putting on record the variational principle to be applied in optimizing a single antisymmetrized product of strong-othogonal group functions. This is obtained [2, 3] by writing the total electronic energy as the sum of the terms $\langle T \rangle$, $\langle V \rangle$, and V_{ee}, given in Equations (6) and (17), and substituting the expressions (19) and (20). The energy then assumes the form

$$E = \sum_R E_R + \frac{1}{2} \sum_{R,S}{}' E_{RS} = E_R^{\text{eff}} + \sum_{S(\neq R)} E_S, \tag{24}$$

in which all the interaction terms have been associated with one particular group R to give an 'effective' energy for that group in the field provided by all others. The wavefunction for the whole system may then be optimized by variation of each group function Φ_R, subject to orthogonality constraints, by minimizing each E_R^{eff} in turn – the rest of the expression (24) being independent of Φ_R.

The explicit form of the group energy E_R^{eff} is found to be simply

$$
\begin{aligned}
E_R^{\text{eff}} &= \langle \hat{H}_R^{\text{eff}} \rangle \\
&= \left\langle \left| \sum_{i=1}^{N_R} \hat{h}_R^{\text{eff}}(i) + \frac{1}{2} \sum_{i,j=1}^{N_{R'}} g(i,j) \right| \right\rangle ,
\end{aligned}
\tag{25}
$$

which refers explicitly *only to the N_R electrons of group R*. In this way the optimization of the wavefunction for a large number of particles (N) is reduced to a number of single-group optimizations, each involving only N_R electrons at a time – clearly a great simplification of the original problem. The effective one-electron Hamiltonian $\hat{h}_R^{\text{eff}}$ is obtained by adding other-group Coulomb-exchange terms, thus:

$$
\hat{h}_R^{\text{eff}}(i) = \hat{h}(i) + \sum_{S(\neq R)} [\hat{J}_S(i) - \hat{K}_S(i)], \tag{26}
$$

where the J and K operators have the general effect

$$
\begin{aligned}
\hat{J}_S \psi(\mathbf{x}_1) &= [\textstyle\int g(1,1')\rho_S(\mathbf{x}_1';\mathbf{x}_1')\,\mathrm{d}\mathbf{x}_1']\psi(\mathbf{x}_1) \\
\hat{K}_S \psi(\mathbf{x}_1) &= \textstyle\int g(1,1')\rho_S(\mathbf{x}_1;\mathbf{x}_1')\psi(\mathbf{x}_1')\,\mathrm{d}\mathbf{x}_1'.
\end{aligned}
\tag{27}
$$

In the preceding equations, $\psi(\mathbf{x})$ indicates a general function of space-spin variables ($\mathbf{x} = \mathbf{r}, s$): when the operators do not involve spin, the integrations over spin may be completed immediately to give similar results with ρ_S replaced by P_S and (in the usual case of a group with total spin zero) with a factor $\frac{1}{2}$ attached to the exchange term. The term in square brackets in (26) then yields the potential energy of an electron at point $\mathbf{r}_i$ in the field of the electron distribution, of density $P_S(\mathbf{r}_1')$, of group S: the exchange term, involving an *off*-diagonal density matrix element, is usually relatively small.

The practical details of optimization will not be discussed here: the important points are that (i) instead of making an *a priori* partitioning of the one-electron basis set into disjoint orthogonal sets for the various electron groups, a single 'global' set must be used; and (ii) projection operators must be introduced in order to ensure that the variational improvement of any group R will conserve strong orthogonality with all other groups. The optimization is essentially an iterative process, because after revising all groups the effective field experienced by any given group must also be revised. In a later section one example of this procedure will be presented.

4. Relativistic Considerations

For systems containing heavy atoms, it is now well recognized that the use of some form of relativistic quantum mechanics is mandatory: instead of the non-relativistic Schrödinger Hamiltonian one should employ a Hamiltonian that satisfies the requirements of relativity theory. For a many-electron system, no completely satisfactory theory of that kind exists: it is usual to construct an approximate Hamiltonian in which each one-electron term takes the form of a Dirac Hamiltonian and electron interaction terms are added in a somewhat *ad hoc* manner. No attempt will be made here to discuss the relativistic form of the Hamiltonian: but it is nevertheless interesting to enquire whether the density functions used so far, and in particular the separability concept, might be incorporated into a relativistic theory. For present purposes the main question is how to build up a parallel interpretative scheme in which two-component spin-orbitals, $\psi = \phi^\alpha \alpha + \phi^\beta \beta$, used as building blocks in the construction of non-relativistic wavefunctions, are replaced by the four-component spinors that occur as solutions of the Dirac equation. Instead of a complete set of spin-orbitals, it will be necessary to employ a complete set of 'Dirac spin-orbitals' which may be written in the form

$$\psi = \phi^{(1)} e_1 + \phi^{(2)} e_2 + \phi^{(3)} e_3 + \phi^{(4)} e_4, \tag{28}$$

so as to preserve the analogy with conventional Schrödinger theory. The Dirac 'spin space 'is carried by the four basis vectors e_μ, $\mu = 1, , \ldots, 4$, which provide a representation of the Dirac operators. Little need be said here about the Dirac equation itself; only that the solutions, of type (28), fall into sets corresponding to negative energy and positive energy (separated by a gap of $2m_0c^2$) and that all are orthogonal with the metric $\langle e_j | e_k \rangle = \delta_{jk}$. The conventional interpretation of the fact that only the positive-energy solutions describe an electron is that all the negative-energy states are already full, providing an unobservable 'sea' into which a positive-energy electron cannot fall without violating the Pauli principle. A *positron* is observed when an electron in the 'sea' is given enough energy to reach a positive-energy state, leaving behind a 'hole' which behaves like a positive charge.

The unsatisfactory nature of the Dirac equation and its interpretation has been nicely put by Messiah [13]: "By postulating the occupation of the quasi-totality of negative-energy states *the theory ceases to be a one-particle theory*, even when it sets out to describe a single electron. The formalism of the Dirac theory ... is therefore insufficient for describing such a situation ... [and] having defined the 'vacuum' [or 'sea'] as composed of an infinite number of electrons, it is inconsistent to assume that these electrons do not interact." Despite these defects of the approach, and the even worse defects of the many-electron generalizations proposed by Breit and others, there seems to be little doubt that many-electron *wavefunctions*, and the density functions to which they lead, can be built up from antisymmetrized products of spin-orbitals of four-component form. With this assumption, it is evident that a filled negative-energy sea, together with *any number N* of electrons allocated to the positive-energy spin-orbitals, may be regarded as a system comprising two strong-orthogonal subsystems. The group-function formalism can be taken over in its entireity, the 'upper' and 'lower' groups remaining strong orthogonal as long as any

CI admitted does not correspond to electron transfer between them (i.e. electron-positron pair production); in this way correlation within the N-electron subsystem can be admitted, in principle, by conventional CI methods. Evidently, this approach provides a rationalization of the now widely used 'no-pair' approximation (e.g. [14, 15]). It remains only to discuss the generalization of the density functions, in particular the one-electron densities, when the wavefunction is constructed using four-component spinors of type (28).

Let us focus on the N-electron group, whose wavefunction is built from determinants of four-component spin-orbitals. The one-body density will then always appear in the general form

$$\rho(\mathbf{x};\mathbf{x}') = \sum_{i,j} \rho_{ij}\psi_i(\mathbf{x})\psi_j^*(\mathbf{x}'), \tag{29}$$

in which

$$\psi_i(\mathbf{x}) = \sum_{\mu} \phi_i^{\mu}(\mathbf{r})e_{\mu}(\eta) \tag{30}$$

and a 'spin variable' η has been introduced to maintain the analogy between the Pauli spin functions $\alpha(s), \beta(s)$ and the 'Dirac' spin functions which provide a basis for the usual matrix representation of the Dirac equation. The quantities ρ_{ij} in (29) are the elements of the *matrix* associated with the one-body density *operator* $\hat{\rho}$, in the discrete basis $\{\psi_i\}$; and the effect of this operator on an arbitrary $\psi(\mathbf{x})$ is to produce

$$\hat{\rho}\psi(\mathbf{x}) = \int \rho(\mathbf{x};\mathbf{x}')\psi(\mathbf{x}')\,\mathrm{d}\mathbf{x}', \tag{31}$$

where the integration runs over both space and spin variables. It is also well known that a unitary transformation of spin-orbitals can bring ρ to diagonal form, in which case (29) is replaced by

$$\rho(\mathbf{x};\mathbf{x}') = \sum_{k} n_k\psi_k(\mathbf{x})\psi_k^*(\mathbf{x}') \ : \tag{32}$$

this is referred to as the 'natural' expansion of the density, ψ_k being a 'natural' spin-orbital and n_k its occupation number.

Although it will not be necessary to discuss the Dirac equation in detail, the known properties of the solutions will provide useful guidelines for studying systems with $N > 1$. We therefore start from the observations that (i) the spin-orbitals describing electrons have two large components and two small; and (ii) the one-electron (Dirac) equation contains operators $\alpha_1, \ldots, \alpha_4$, which work on the basis vectors $e_1, \ldots, e_4$ just as the Pauli spin operators of non-relativistic theory work on the spin eigenvectors α, β. The Dirac equation has, in fact, the form

$$(c\alpha\cdot\pi + V + m_0c^2\alpha_4)\psi = i\hbar\partial\psi/\partial t \tag{33}$$

where the operator scalar product $\alpha \cdot \pi = \sum_{\mu=1}^{3} \alpha_\mu \pi_\mu$ and the operators must have the properties

$$\alpha_\mu^2 = 1, \quad (\alpha_\mu \alpha_\nu + \alpha_\nu \alpha_\mu) = 0 \quad \mu \neq \nu. \tag{34}$$

It is usual to pass directly to a matrix representation in which a 4×4 matrix is associated with each α. To maintain the parallel with Pauli theory, however, a different approach will be used here.

Let us look for a representation of the 'Dirac space' as a Kronecker product of two spaces of Pauli type, each providing a representation of an operator set with the same properties as the Pauli spin operators. Two such sets can easily be found: they are

$$\hat{S}_1 = i\alpha_2\alpha_3, \quad \hat{S}_2 = i\alpha_3\alpha_1, \quad \hat{S}_3 = i\alpha_1\alpha_2 \tag{35}$$

and

$$\hat{T}_1 = i\alpha_4\alpha_5, \quad \hat{T}_2 = i\alpha_5, \quad \hat{T}_3 = i\alpha_4 \tag{36}$$

where $\alpha_5 = \alpha_1\alpha_2\alpha_3\alpha_4$. The 'spin operators' in the first set will correspond to twice the usual Pauli operators (with $x, y, z \to 1, 2, 3$). To exploit this parallel let us introduce basis vectors u, u_2 and v_1, v_2 to carry the two representations, supposing

$$\hat{S}_3(u_1 \ u_2) = (u_1 \ u_2)\begin{pmatrix} 1 & 0 \\ 0 & -1 \end{pmatrix} \quad \hat{T}_3(v_1 \ v_2) = (v_1 \ v_2)\begin{pmatrix} 1 & 0 \\ 0 & -1 \end{pmatrix}. \tag{37}$$

It can now be verified that the Dirac basis which carries the standard matrix representation of the αs can be expressed as a Kronecker product:

$$(e_1 \ e_2 \ e_3 \ e_4) = (u_1v_1 \ u_2v_1 \ u_1v_2 \ u_2v_2). \tag{34}$$

From (37) it follows that the basis vectors $e_1, \ldots, e_4$ are simultaneous eigenvectors of $\hat{S}_3$ and $\hat{T}_3$ with eigenvalues (+1, +1), (−1, +1), (+1, −1), (−1, −1), respectively.

The one-body density matrix in Pauli approximation is usually written in terms of space and spin 'variables', so that Equation (4) assumes the form

$$\begin{aligned} \rho(\mathbf{x};\mathbf{x}') = {} & P_{\alpha\alpha}(\mathbf{r};\mathbf{r}')\alpha(s)\alpha^*(s') + P_{\alpha\beta}(\mathbf{r};\mathbf{r}')\alpha(s)\beta(s') + \\ & P_{\beta\alpha}(\mathbf{r};\mathbf{r}')\beta(s)\alpha^*(s') + P_{\beta\beta}(\mathbf{r};\mathbf{r}')\beta(s)\beta^*(s'). \end{aligned} \tag{39}$$

The spin variable s is merely a convenient formal device which allows us to treat coordinate space and spin space in a similar way, avoiding mixtures of integration and summation. In fact, (39) is an expanded representation of the density *operator* in (31). The factor $\alpha(s)\alpha^*(s')$, for example, represents the spin dyad $\alpha\alpha^*$, which operates on a general spin-orbital to project out the α-part: thus, $\alpha\alpha^*(\phi^\alpha\alpha + \phi^\beta\beta)$

$= \phi^\alpha\alpha(\alpha^*\alpha) + \phi^\beta\alpha(\alpha^*\beta) = \phi^\alpha\alpha$, where the symbolic scalar products such as $\alpha^*\beta$ are interpreted in the continuous representation as $\alpha^*\beta = \langle\alpha|\beta\rangle = \int \alpha^*(s)\beta(s)\,ds$. When s and s' in (39) are identified and a spin integration is performed the result in (5) is obtained in the form $P(\mathbf{r};\mathbf{r}') = P_{\alpha\alpha}(\mathbf{r};\mathbf{r}') + P_{\beta\beta}(\mathbf{r};\mathbf{r}')$, as a sum of 'up-spin' and 'down-spin' components.

Such considerations may be extended immediately to relativistic density operators: for a single Dirac particle there will be *two* spin variables, such that $u_i \to u_i(s)$, $v_i \to v_i(t)$, and the density operator

$$\rho = \psi\psi^\dagger = \sum_{\mu\nu} \phi_\mu\phi_\nu^* e_\mu e_\nu^* \tag{40}$$

will yield the kernel (density matrix)

$$\rho(\mathbf{r}, s, t\,:\,\mathbf{r}', s', t') = P_{11}(\mathbf{r};\mathbf{r}')u_1v_1u_1^*v_1^* + \cdots + P_{44}(\mathbf{r};\mathbf{r}')u_2v_2u_2^*v_2^*, \tag{41}$$

which is analogous to (39) – but now there will be 16 terms. When the non-spatial variables are removed by integration, in the usual way, the *spinless* density matrix appears in the form

$$P(\mathbf{r};\mathbf{r}') = \sum_\mu P_{\mu\mu}(\mathbf{r};\mathbf{r}'), \tag{42}$$

where the label μ (a compound index for u- and v-factors) corresponds to the Dirac basis labelling in (38) and only the terms with $\mu = \nu$ survive the spin integrations. The electron density, obtained by putting $\mathbf{r}' = \mathbf{r}$ in (42), thus contains only four terms. For the single Dirac particle

$$P(\mathbf{r}) = [\phi_1(\mathbf{r})\phi_1^*(\mathbf{r}) + \phi_2(\mathbf{r})\phi_2^*(\mathbf{r})] + [\phi_3(\mathbf{r})\phi_3^*(\mathbf{r}) + \phi_4(\mathbf{r})\phi_4^*(\mathbf{r})], \tag{43}$$

where the large and small components appear, respectively, in the first and second square brackets.[5] The second term, absent in a Pauli-type approximation, contains the correction arising from use of a four-component Dirac spinor: it is of order $(2m_0c)^{-2}$, except at singularities in the potential, and is thus normally negligible. The components of a current density and a spin density may also be derived from (41); but at this point we may proceed at once to the case of N Dirac particles, with a completely general CI wavefunction built up from a basis of four-component spin-orbitals.

For a general multiconfiguration wavefunction, the density matrix will arise from the expansions (29) and (30); and if we choose the natural expansion (32) it will take the form (including spin variables)

[5] For the positive-energy solutions of the Dirac equation, the convention adopted here is that the 'large components' correspond to μ = 1, 2 and the 'small components' to μ = 3, 4. Some authors, notably Slater, use the reverse convention.

$$\rho(\mathbf{x};\mathbf{x}') = \sum_k n_k \psi_k(\mathbf{x})\psi_k^*(\mathbf{x}')$$
$$= \sum_k n_k \sum_{\mu\nu} \phi_k^\mu(\mathbf{r})\phi_k^{\nu*}(\mathbf{r}')e_\mu(\eta)e_\nu^*(\eta'), \tag{44}$$

where η stands for the *pair* of spin variables (s,t). On putting $\eta' = \eta$ and integrating over spins, the spinless density matrix becomes

$$P(\mathbf{r};\mathbf{r}') = \sum_\mu P_{\mu\mu}(\mathbf{r};\mathbf{r}') \tag{45a}$$

where

$$P_{\mu\mu}(\mathbf{r};\mathbf{r}') = \sum_k n_k \phi_k^\mu(\mathbf{r})\phi_k^{\mu*}(\mathbf{r}'). \tag{45b}$$

As in (42), only four terms survive the spin integrations: the first pair will be large, referring to the Pauli-like components, the second pair small.

According to (13), the current density should also follow from (45a,b): but for $N = 1$ the usual relativistic derivation, which we indicate first, is somewhat different. From the Dirac equation, one obtains a current density by appeal to a conservation equation for the electron density $\rho = \psi\psi^*$. *Provided ψ satisfies the Dirac equation*, the components of the current are found to be

$$J_\mu^D(\mathbf{r}) = \psi^\dagger(c\alpha_\mu)\psi, \tag{46}$$

in this context α_μ is the *matrix* associated with the operator; ψ is used to denote the column of spatial components ϕ_1, ϕ_2, ϕ_3, ϕ_4 and $\psi^\dagger$ is obtained by transposition and starring. By introducing explicitly the Dirac Hamiltonian, and expressing small components in terms of large, one shows that the 'Dirac current' (46) contains a 'conventional' part (Section 2) together with a small contribution which is the curl of a *magnetization density* associated with the spin. The spin therefore enters 'automatically'.

The Dirac current with its spin contribution, is *not* obtained when we proceed from the general four-component density (45a,b): instead we find (putting $\mathbf{r}' = \mathbf{r}$)

$$J_x(\mathbf{r}) = J_x^L(\mathbf{r}) + J_x^S(\mathbf{r}), \tag{47}$$

(with similar results for J_y, J_z). Here 'L' and 'S' indicate the terms deriving from the Large and Small parts of (45a) (μ = 1, 2 and μ = 3, 4, respectively, and $J_x^S(\mathbf{r})$ is of order $(2m_0c)^{-2}$ relative to $J_x^L(\mathbf{r})$ – one order of magnitude smaller than the spin contribution to the Dirac current.

Let us now turn to the 'conventional' spin density, defined in the non-relativistic case through (9). According to this equation, the spin density should also follow from the density matrix (44) which includes the spin variables. As in (45a,b), $Q_x(\mathbf{x};\mathbf{x}')$ will be a sum of terms from the various spinor components, summed over all spin-orbitals in the natural expansion. A typical term will be

$$Q_x^k(\mathbf{r};\mathbf{r}') = \int \left[\sum_{\mu\nu} \hat{S}_x \phi_k^{\mu}(\mathbf{r}) e_\mu(\eta) \phi_k^{\nu *}(\mathbf{r}') e_\nu^*(\eta') \right]_{\eta'=\eta} \mathrm{d}\eta, \tag{48}$$

where the operator works on the unprimed variables. On expressing the Dirac basis vectors in the Kronecker product form (41) it follows that

$$\begin{aligned} \hat{S}_x(e_1, e_2, e_3, e_4) &= \hat{S}_x(u_1v_1, u_2v_1, u_1v_2, u_2v_2) \\ &= (u_2v_1, u_1v_1, u_2v_2, u_1v_2) \\ &= (e_2, e_1, e_4, e_3), \end{aligned}$$

from which it follows at once that

$$Q_x(\mathbf{r};\mathbf{r}') = Q_x^L(\mathbf{r};\mathbf{r}') + Q_x^S(\mathbf{r};\mathbf{r}'), \tag{49}$$

where the Large and Small parts are given by

$$Q_x^L(\mathbf{r};\mathbf{r}') = \sum_k n_k[\phi_k^2(\mathbf{r})\phi_k^{1*}(\mathbf{r}') + \phi_k^1(\mathbf{r})\phi_k^{2*}(\mathbf{r}')], \tag{50a}$$

$$Q_x^S(\mathbf{r};\mathbf{r}') = \sum_k n_k[\phi_k^4(\mathbf{r})\phi_k^{3*}(\mathbf{r}') + \phi_k^3(\mathbf{r})\phi_k^{4*}(\mathbf{r}')], \tag{50b}$$

There are corresponding results for Q_y, Q_z and again the 'L' and 'S' parts differ by two orders of magnitude, the large parts reproducing the densities obtained in the usual 2-component Pauli approximation (cf. (9) *et seq.*), the small parts giving the relativistic corrections.

In summary, although (to quote Slater [16]) "... the Dirac wavefunction has the capability of giving complete information about the orientation of the spin in space ...", the same is true for wavefunctions built from either two-component or four-component spin-orbitals, for either 1 or N electrons, and there is no need to invoke explicitly the Dirac equation or the expansion of small components in terms of large ones. All the densities defined in the conventional (two-component) approach follow also, from the usual density matrix definitions, at the 'fully relativistic' (four-component) level; and the possibility of 'separating' a large system into weakly interacting subsysystems is not affected by relativistic generalizations.

An attractive way of dealing with large molecules containing heavy atoms, for which relativistic effects must be included, might thus be to use the group function formalism with a fully relativistic treatment only of the heavy-atom group.

5. Optimization of the Wavefunction: An Example

It remains to be verified that the constrained optimization of a wavefunct ion ansatz of type (1) presents no special problems. To this end it will be sufficient to consider a two-group system, at the non-relativistic level, the two groups comprising the core and 'valence' electrons of a molecule undergoing dissociation: in this case the bond breaking affects mainly the valence electrons, but the core function also responds quite strongly to the changes occurring.

5.1. DISSOCIATION OF LiCl

Conventionally, the two atoms are regarded as possessing inert-gas cores, Li[$1s^2$] and Cl[$1s^2 2s^2 2p^6$], the remaining electrons belonging to the valence shell: the elementary textbook description assigns two electrons to a bond formed by overlap of the Li $2s$ orbital with a Cl $3p$ orbital ($3p_\sigma$, say), the remainder forming lone pairs Cl[$3s^2$ $3p_\pi^4$]. To study the bond-breaking process, it is convenient to adopt a VB description in which there are two electron pairs of sigma symmetry, the lone pair Cl[$3s^2$] and the bond pair Li($2s$)—Cl($3p_\sigma$): the remaining electrons, whose orbitals are less strongly affected by bonding, will form an extended core. The orbitals used in describing the 16 core electrons and four valence electrons will not, of course, be simple AOs; they will be general linear combinations of a 'global' set of basis functions and their forms will be determined automatically in the constrained optimization that keeps the two groups strong orthogonal.

The global basis χ used in the test calculations is of simple contracted Gaussian form:

Cl (on left):	$s1\ s2\ s3\ s4\ x1\ y1\ z1\ x2\ y2\ z2\ x3\ y3\ z3$
Li (on right):	$S1\ S2\ S3\ X1\ Y1\ Z1\ X2\ Y2\ Z2.$

$s1$ and $S1$ are six-term approximations to Cl $1s$ and Li $1s$, respectively; the other functions are used in building the 'split-valence' approximations to the $2s$, $3s$ and $2p$, $3p$ orbitals.

First approximations to the orbitals ϕ fall into three sets, as follows:

Core orbitals:	Cl($1s$) Cl($2s$) Cl($2px$) Cl($2py$) Cl($2pz$) Li($1s$) Cl($3px$) Cl($3py$)
Valence orbitals:	Cl($3s$) Cl($3pz$) Li($2s$)
Virtual orbitals:	Cl($3s'$) Li($2s'$) Cl($3pz'$) Li($2pz$) Li($2pz'$) Cl($3pz'$) Cl($3py'$) Li($2px$) Li($2py$) Li($2px'$) Li($2py'$).

The 'virtual' orbitals are arbitrary as long as all three sets span the whole 22-dimensional space; the notation is self-explanatory, the primes simply indicating more diffuse 'partners' of the functions occurring previously. When the orbitals are written in the matrix form $\phi = \chi\mathbf{T}$ the square matrix $\mathbf{T}$ thus takes the form

$$\mathbf{T} = (\mathbf{T}_{\text{cor}}|\mathbf{T}_{\text{val}}|\mathbf{T}_{\text{vir}})$$

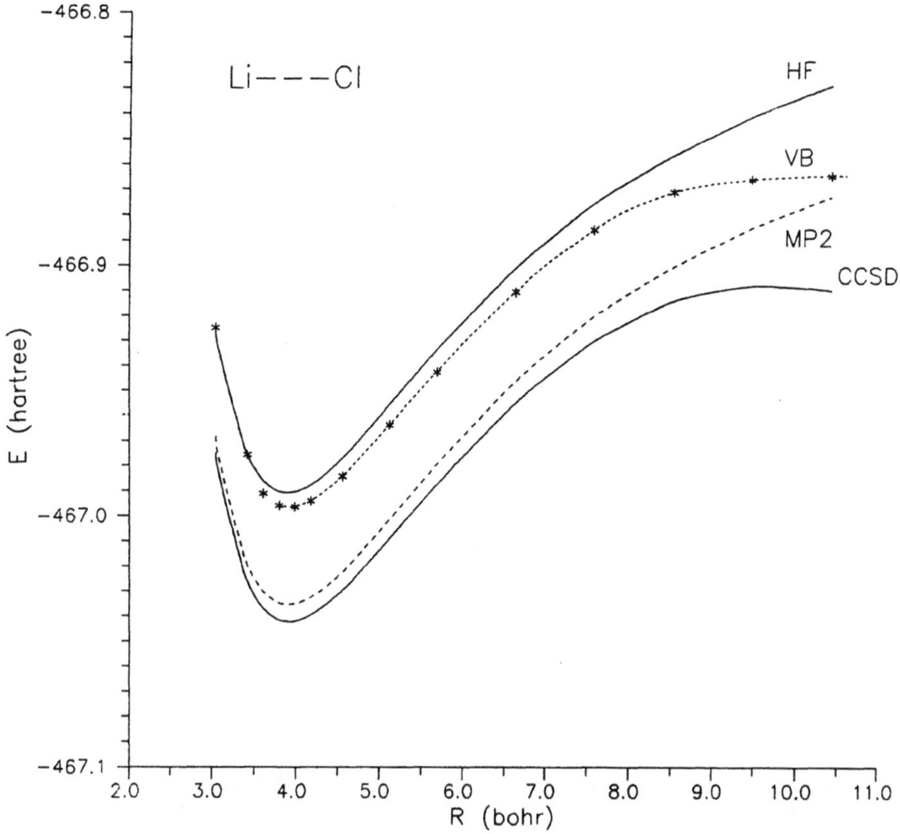

Figure 1. Potential energy curves for LiCl. HF = Hartree–Fock (restricted); MP2 = Hartree–Fock with 2nd order MP correction; CCSD = coupled-cluster with singles and doubles (based on UHF); VB = Valence Bond (this paper); points indicated refer to a CAS-SCF calculation with the same partitioning (the two curves coincide).

where the eight columns of $\mathbf{T}_{\rm cor}$ define the doubly occupied core orbitals; the three columns of $\mathbf{T}_{\rm val}$ define the sigma lone pair and the two orbitals of the bond pair; and the 11 columns of $\mathbf{T}_{\rm vir}$ represent the virtual set.

The energy and one-electron density matrix for the single-structure VB function are obtained using a spin-free method described elsewhere [17]: corresponding quantities for the core, regarded as a closed-shell subsystem with 16 electrons occupying eight orbitals are obtained by standard SCF methods. The interaction terms require only knowledge of the one-electron dms of the two groups, and these are upgraded as iteration proceeds. Optimization of the VB function is performed using a simple Powell-type procedure [18] which does not require the evaluation of energy derivatives. Strong orthogonality of the two groups is imposed and maintained throughout the iteration by adjusting the matrix $\mathbf{T}$ at the start of every

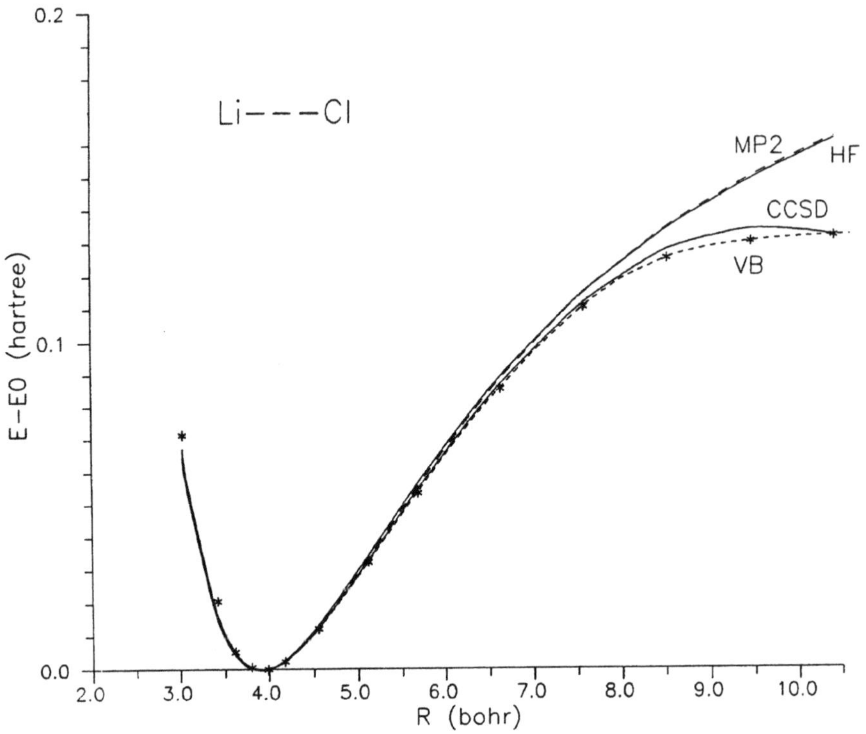

Figure 2. Potential energy curves for LiCl. (See Figure 1. All curves have been shifted to a common energy zero.

iterative cycle: $\mathbf{T}_{val}$ and $\mathbf{T}_{vir}$) are orthogonalized against $\mathbf{T}_{cor}$, by annihilating their overlap with the core space, and $\mathbf{T}_{vir}$) is similarly orthogonalized against the core+valence space. The optimization of the VB function is performed in the valence+virtual space; that of the core function is performed in the core+virtual space. During the VB optimization, the function stays strong-orthogonal to the core; and, during the core optimization, the SCF function stays strong-orthogonal to the VB function. The improved functions lose their strong orthogonality, but are re-orthogonalized before entering the next cycle. In spite of its simplicity, this procedure has been found to give satisfactory convergence.

The PE curve obtained for LiCl, over the range of internuclear distances 3.0 – 13.0 bohr, is shown in Figure 1, where it is compared with the results of various standard calculations, including a CAS-SCF calculation with which it virtually coincides. It is particularly noteworthy that although the *absolute* value of the energy is poor compared with the results of more massive computations the *shape* of the PE curve reproduces faithfully the best reults available. This is clear from Figure 2, where the various curves are shifted to a common zero at the equilibrium distance.

6. Conclusion

The main conclusions to be drawn from the work reported in this paper are:

- Whenever an N-electron system is represented in terms of subsystems $A, B, \ldots$, using the generalized product ansatz (1), the entire wavefunction can be optimized by repeated optimization of the subsystem wavefunctions
- The subsystem wavefunction Φ_R refers to only N_R electrons, described by an 'effective' N_R-electron Hamiltonian which depends on the dm's of all subsystems, and can be optimized without *a priori* partitioning of the global basis set
- The resultant total wavefunction provides an optimum 'separation' of the whole system, many of whose properties are then expressible as sums of subsystem contributions
- Separability can be exploited even with admission of relativistic effects, using a basis of four-component Dirac spin-orbitals, provided the dm formalism is suitably generalized
- The optimization method proposed is computationally viable: a test calculation on the 20-electron system LiCl, regarded as a 'two-group' system with four valence electrons (one lone pair, one bond pair) moving in the field of a 16-electron 'core', yields a PE curve virtually coincident with that obtained from a CAS-SCF calculation the same basis and partitioning.

Acknowledgements

The work reported in this paper formed part of a project supported by the European Union, under the terms of the COST initiative. A preliminary account was presented at the COST Workshop 'Quantum Systems in Chemistry and Physics' (San Miniato, Italy, 14–17 April 1996); support for this meeting is gratefully acknowledged, along with useful comments from Stephen Wilson (Rutherford Appleton Laboratory, UK), Jean Maruani (UPMC, Paris), and colleagues at the Bulgarian Academy of Sciences. I am indebted to Dr. Claudio Amovilli for obtaining the curves used for comparison purposes in the figures.

References

1. R. G. Parr, F. O. Ellison and P. G. Lykos: *J. Chem. Phys.* **24**, 1106 (1956).
2. R. McWeeny: *Proc. Roy. Soc. Lond.* **A253** 242 (1959).
3. R. McWeeny: *Rev. Mod. Phys.* **32**, 335 (1960).
4. R. G. Parr and W. Yang: *Density Functional Theory of Atoms and Molecules*, Oxford: OUP (1989).
5. E. S. Kryachko and E. V. Ludena: *Energy Density Functional Theory of Many-Electron Systems*, Dordrecht: Kluwer (1990).
6. R. McWeeny: *Spins in Chemistry*, New York: Brooklyn Polytechnic Press (1969).
7. R. McWeeny: *Phil. Mag.* **B69**, 727 (1994).
8. R. McWeeny: *Proc. Ind. Acad. Sci.* **96**, 263 (1986).

9. B. M. Deb and S. K. Ghosh: in *The Single-Particle Density Physics and Chemistry* N. H. March and B. M. Deb (eds.), London: Academic (1989).
10. R. McWeeny: *Methods of Molecular Quantum Mechanics*, 2nd ed. London: Academic (1992) – see especially Ch. 9.
11. M. Klessinger and R. McWeeny: *J. Chem. Phys.* **42**, 3343 (1965).
12. R. McWeeny, to be published.
13. A. Messiah: *Quantum Mechanics*, Vol. II, Amsterdam: North-Holland (1962).
14. B. A. Hess: *Phys. Rev.* **A33**, 3742 (1986).
15. E. van Lenthe, E. J. Baerends, and J. G. Snijders: *J. Chem. Phys.* **99**, 4597 (1993).
16. J. C. Slater: *Quantum Theory of Atomic Structure*, Vol. II, New York: McGraw-Hill (1960).
17. R. McWeeny: *Int. J. Quantum Chem.* **34**, 25 (1988).
18. W. H. Press, B. P. Flannery, S. A. Teukolsky, and W. T. Vetterling: *Numerical Recipes*, Cambridge: CUP (1989).

Molecular Engineering **7**: 27–50, 1997.

Optimized Effective Potential for Atoms and Molecules

T. GRABO, T. KREIBICH and E. K. U. GROSS
Institut für Theoretische Physik, Universität Würzburg, Am Hubland, D-97074 Würzburg, Germany.

(Received: 6 July 1997; in final form: 16 September 1997)

Abstract. We describe the optimized effective potential method of density functional theory and the semi-analytical approximation due to Krieger, Li and Iafrate. Results for atomic and molecular systems including correlation contributions are presented and compared with conventional Kohn–Sham methods. The combination of the exact exchange energy functional with the correlation energy functional of Colle and Salvetti works extremely well for atomic systems, while further improvement is required for molecular systems.

Key words: Optimized effective potential, density functional theory, exchange functional, correlation functionals.

1. Introduction

Density functional heory (DFT) is a powerful quantum mechanical method for calculating the electronic structure of atoms, molecules and solids [1–3]. The success of DFT hinges on the availability of good approximations for the total-energy functional. In this article we shall describe a particular approach to the construction of such approximations which involves explicitly orbital-dependent functionals. In order to describe the nature of this approach we will first briefly review the foundations of DFT.

We are concerned with Coulomb systems described by Hamiltonians of the type

$$\hat{H} = \hat{T} + \hat{W}_{\text{Clb}} + \hat{V} \tag{1}$$

where (atomic units are used throughout this article)

$$\hat{T} = \sum_{i=1}^{N} \left(-\frac{1}{2} \nabla_i^2 \right) \tag{2}$$

denotes the kinetic energy operator,

$$\hat{W}_{\text{Clb}} = \frac{1}{2} \sum_{\substack{i,j=1 \\ i \neq j}}^{N} \frac{1}{|\mathbf{r}_i - \mathbf{r}_j|} \tag{3}$$

represents the Coulomb interaction between the particles, and

$$\hat{V} = \sum_{i=1}^{N} v(\mathbf{r}_i) \tag{4}$$

contains all external potentials of the system, typically the Coulomb potentials of the nuclei.

Modern DFT is based on the celebrated theorem of Hohenberg and Kohn (HK) [4] which may be summarized by the following three statements:

1. The ground-state density ρ uniquely determines the external potential $v = v[\rho]$ as well as the ground-state wave function $\Psi[\rho]$. As a consequence, any observable of a static many-particle system is a functional of its ground-state density.
2. The total-energy functional
$$E_{v_0}[\rho] := \langle \Psi[\rho] | \hat{T} + \hat{W}_{\mathrm{Clb}} + \hat{V}_0 | \Psi[\rho] \rangle \tag{5}$$
of a particular physical system characterized by the external potential v_0 is equal to the exact ground-state energy E_0 if and only if the exact ground-state density ρ_0 is inserted. For all other densities $\rho \neq \rho_0$ the inequality
$$E_0 < E_{v_0}[\rho] \tag{6}$$
holds. Consequently, the exact ground-state density ρ_0 and the exact ground-state energy E_0 can be determined by solving the Euler–Lagrange equation
$$\frac{\delta}{\delta \rho(\mathbf{r})} E_{v_0}[\rho] = 0. \tag{7}$$
3. The functional
$$F[\rho] := \langle \Psi[\rho] | \hat{T} + \hat{W}_{\mathrm{Clb}} | \Psi[\rho] \rangle \tag{8}$$
is universal in the sense that it is independent of the external potential v_0 of the particular system considered, i.e. it is of the same functional form for all systems with a fixed particle-particle interaction ($\hat{W}_{\mathrm{Clb}}$ in our case).

The proof of the HK theorem does not depend on the particular form of the particle–particle interaction. It is valid for *any* given particle–particle interaction $\hat{W}$, in particular also for $\hat{W} \equiv 0$, i.e. for non-interacting systems described by Hamiltonians of the form

$$\hat{H}_S = \hat{T} + \hat{V}_S. \tag{9}$$

Hence the potential $V_S(\mathbf{r})$ is uniquely determined by the ground-state density:

$$V_S(\mathbf{r}) = V_S[\rho](\mathbf{r}). \tag{10}$$

As a consequence, all single-particle orbitals satisfying the Schrödinger equation

$$\left(-\frac{\nabla^2}{2} + V_S[\rho](\mathbf{r})\right) \varphi_j(\mathbf{r}) = \varepsilon_j \varphi_j(\mathbf{r}) \tag{11}$$

are functionals of the density as well:

$$\varphi_j(\mathbf{r}) = \varphi_j[\rho](\mathbf{r}). \tag{12}$$

The HK total-energy functional of non-interacting particles is given by

$$E_S[\rho] = T_S[\rho] + \int d^3r \rho(\mathbf{r}) V_S(\mathbf{r}) \tag{13}$$

where $T_S[\rho]$ is the kinetic-energy functional of non-interacting particles:

$$T_S[\rho] = \sum_{\substack{i=1 \\ \text{lowest } \varepsilon_i}}^{N} \int d^3r \varphi_i^*[\rho](\mathbf{r}) \left(-\frac{\nabla^2}{2}\right) \varphi_i[\rho](\mathbf{r}). \tag{14}$$

We emphasize that the quantity (14) really represents a *functional of the density*: functional means that we can assign a unique number $T_S[\rho]$ to any function $\rho(\mathbf{r})$. This is done by first calculating that actual potential $V_S(\mathbf{r})$ which uniquely corresponds to $\rho(\mathbf{r})$. Several numerical schemes have been devised to achieve this task [5–10]. Then we take this potential, solve the Schrödinger Equation (11) with it to obtain a set of orbitals $\{\varphi_j(\mathbf{r})\}$ and use those to calculate the number T_S by evaluating the right-hand side of Equation (14). As a matter of fact, by the same chain of arguments, *any orbital functional is an (implicit) functional of the density*, provided the orbitals come from a local, i.e. multiplicative potential.

Returning to the *interacting* system of interest we now define the so-called exchange-correlation (xc) energy functional by

$$E_{\mathrm{xc}}[\rho] := F[\rho] - \frac{1}{2}\int d^3r \int d^3r' \frac{\rho(\mathbf{r})\rho(\mathbf{r}')}{|\mathbf{r}-\mathbf{r}'|} - T_S[\rho]. \tag{15}$$

The HK total-energy functional (5) can then be written as

$$\begin{aligned} E_{v_0}[\rho] &= T_S[\rho] + \int d^3r \rho(\mathbf{r}) v_0(\mathbf{r}) + \\ &+ \frac{1}{2}\int d^3r \int d^3r' \frac{\rho(\mathbf{r})\rho(\mathbf{r}')}{|\mathbf{r}-\mathbf{r}'|} + E_{\mathrm{xc}}[\rho]. \end{aligned} \tag{16}$$

In historical retrospect we may identify three generations of density functional schemes which may be classified according to the level of approximations used for the universal functionals $T_S[\rho]$ and $E_{\mathrm{xc}}[\rho]$.

In what we call the *first generation of DFT, explicitly* density-dependent functionals are used to approximate both $T_S[\rho]$ and $E_{\rm xc}[\rho]$. For example, the simplest and historically first approximation of this kind is the Thomas–Fermi model, where $E_{\rm xc}[\rho]$ is neglected completely and $T_S[\rho]$ is approximated by

$$T_S^{\rm TF}[\rho] = \frac{3}{10}(3\pi^2)^{2/3} \int \mathrm{d}^3 r \rho(\mathbf{r})^{5/3} \tag{17}$$

yielding

$$\begin{aligned} E_{v_0}^{\rm TF}[\rho] &= \frac{3}{10}(3\pi^2)^{2/3} \int \mathrm{d}^3 r \rho(\mathbf{r})^{5/3} + \int \mathrm{d}^3 r v_0(\mathbf{r})\rho(\mathbf{r}) \\ &+ \frac{1}{2}\int \mathrm{d}^3 r \int \mathrm{d}^3 r' \frac{\rho(\mathbf{r})\rho(\mathbf{r}')}{|\mathbf{r}-\mathbf{r}'|} \end{aligned} \tag{18}$$

as approximate expression for the total-energy functional. For functionals of this type the HK variational principle (7) can be used directly, leading to equations of the Thomas–Fermi type. As these equations only contain one basic variable, namely the density $\rho(\mathbf{r})$ of the system, they are readily solved numerically. The results obtained in this way, however, are generally of moderate quality.

The *second generation* of DFT employs the *exact* functional (14) for the non-interacting kinetic energy and an approximate density functional for the xc energy:

$$\begin{aligned} E_{v_0}^{\rm KS}[\rho] &= T_S^{\rm exact}[\rho] + \int \mathrm{d}^3 r v_0(\mathbf{r})\rho(\mathbf{r}) + \\ &+ \frac{1}{2}\int \mathrm{d}^3 r \int \mathrm{d}^3 r' \frac{\rho(\mathbf{r})\rho(\mathbf{r}')}{|\mathbf{r}-\mathbf{r}'|} + E_{\rm xc}[\rho]. \end{aligned} \tag{19}$$

This total-energy expression leads to the Kohn–Sham (KS) version of DFT [11] as will be shown in the following. Plugging Equation (19) into the variational principle (7) yields

$$0 = \frac{\delta T_S^{\rm exact}[\rho]}{\delta\rho(\mathbf{r})} + v_0(\mathbf{r}) + \int \mathrm{d}^3 r' \frac{\rho(\mathbf{r}')}{|\mathbf{r}-\mathbf{r}'|} + \frac{\delta E_{\rm xc}[\rho]}{\delta\rho(\mathbf{r})}. \tag{20}$$

The variation of the non-interacting kinetic energy functional is given by

$$\begin{aligned} \delta T_S^{\rm exact}[\rho] &= \delta \sum_{i=1}^{N} \left\langle \varphi_i[\rho] \left| -\frac{\nabla^2}{2} \right| \varphi_i[\rho] \right\rangle \\ &= \delta \left[\sum_{i=1}^{N} \varepsilon_i[\rho] - \int \mathrm{d}^3 r' V_S[\rho](\mathbf{r}')\rho(\mathbf{r}') \right] \end{aligned} \tag{21}$$

where the one-particle Equation (11) has been used. Since the HK theorem ensures the one-to-one correspondence between the density and the one-particle potential, a variation $\delta\rho$ of the former will result in a variation δV_S of the latter. Therefore, the variation of the one-particle energies ε_i can be calculated using first-order perturbation theory yielding

$$\delta\varepsilon_i = \langle\varphi_i[\rho]|\delta V_S[\rho]|\varphi_i[\rho]\rangle. \tag{22}$$

Using this result in (21) gives

$$\delta T_S^{\text{exact}}[\rho] = -\int d^3r' V_S[\rho](\mathbf{r}')\delta\rho(\mathbf{r}') \tag{23}$$

which, combined with Equation (20) leads to

$$V_S[\rho](\mathbf{r}) = v_0(\mathbf{r}) + \int d^3r' \frac{\rho(\mathbf{r}')}{|\mathbf{r}-\mathbf{r}'|} + V_{\text{xc}}[\rho](\mathbf{r}) \tag{24}$$

where we have defined the xc potential as

$$V_{\text{xc}}[\rho](\mathbf{r}) := \frac{\delta E_{\text{xc}}[\rho]}{\delta\rho(\mathbf{r})}. \tag{25}$$

Being the HK variational equation of the *interacting* system, Equation (20) determines the exact ground-state density of the interacting system. Since Equation (24), on the other hand, is equivalent to Equation (20), the density

$$\rho(\mathbf{r}) = \sum_{\substack{i=1 \\ N \text{ lowest } \varepsilon_i}}^{N} |\varphi_i(\mathbf{r})|^2 \tag{26}$$

resulting from the solution of the Schrödinger Equation (11) with the potential (24) must be identical with the ground-state density of the *interacting* system of interest. Equations (11), (24), (25) and (26) are known as Kohn–Sham equations. In practice, these equations have to be solved self-consistently employing approximate but explicitly density-dependent functionals for $E_{\text{xc}}[\rho]$. The resulting scheme is still easy to solve numerically and – especially for sophisticated density-gradient-dependent approximations of $E_{\text{xc}}[\rho]$ – gives excellent results for a wide range of atomic, molecular and solid-state systems [2,12].

Finally, in the *third generation of DFT*, in addition to the *exact* expression for T_S one also employs the *exact* expression for the exchange energy given by

$$E_{\text{x}}^{\text{exact}}[\rho] = -\frac{1}{2}\sum_{\sigma=\uparrow,\downarrow}\sum_{j,k=1}^{N_\sigma}\int d^3r \int d^3r' \frac{\varphi_{j\sigma}^*(\mathbf{r})\varphi_{k\sigma}^*(\mathbf{r}')\varphi_{k\sigma}(\mathbf{r})\varphi_{j\sigma}(\mathbf{r}')}{|\mathbf{r}-\mathbf{r}'|}. \tag{27}$$

Only the correlation part of $E_{xc}[\rho]$ needs to be approximated in this approach. In contrast to the conventional second-generation KS scheme, the third generation allows for the treatment of explicitly orbital-dependent functionals for E_c as well, giving more flexibility in the construction of such approximations.

The central equation in the third generation of DFT is still the KS Equation (11). The difference between the second and third generations lies in the level of approximation to the xc-energy. As a consequence of the orbital dependence of E_{xc} in the third generation of DFT the calculation of $V_{xc}[\rho](\mathbf{r})$ from Equation (25) is somewhat more complicated. A detailed derivation will be given in the following section for the spin-dependent version of DFT. The result is an integral equation determining the xc potential. This integral equation, known as the optimized effective potential (OEP) equation, is very hard to solve numerically. To avoid a full-scale numerical solution, Krieger, Li and Iafrate (KLI) [13–21] have devised a semi-analytical scheme for solving the OEP integral equation approximately. This scheme is described in the subsequent section. After that, some rigorous properties of the OEP and KLI solutions will be briefly discussed and finally numerical results for atomic and molecular systems will be presented.

We finally mention that a time-dependent generalization of the OEP has recently been developed [22] to deal with explicitly time-dependent situations such as atoms in strong laser pulses [23]. In the linear-response regime this method has led to a rather successful procedure [24] to calculate excitation energies from the poles of the frequency-dependent density response. Time-dependent applications of this kind will not be discussed in the present article. The interested reader is referred to a recent review of time-dependent DFT [25].

2. The OEP Method

2.1. DERIVATION OF THE OEP EQUATIONS

We are going to derive the OEP equations for the spin-dependent version of DFT [26, 27], where the basic variables are the spin-up and spin-down densities $\rho_\uparrow(\mathbf{r})$ and $\rho_\downarrow(\mathbf{r})$, respectively. They are obtained by self-consistently solving the single-particle Schrödinger equations

$$\left(-\frac{\nabla^2}{2} + V_{S\sigma}[\rho_\uparrow, \rho_\downarrow](\mathbf{r})\right) \varphi_{j\sigma}(\mathbf{r}) = \varepsilon_{j\sigma}\varphi_{j\sigma}(\mathbf{r}) \quad j = 1, \ldots, N_\sigma$$

$$\sigma = \uparrow, \downarrow \tag{28}$$

where

$$\rho_\sigma(\mathbf{r}) = \sum_{i=1}^{N_\sigma} |\varphi_{i\sigma}(\mathbf{r})|^2. \tag{29}$$

For convenience we shall assume in the following that infinitesimal symmetry-breaking terms have been added to the external potential to remove any possible degeneracies.

The KS orbitals can then be labeled such that

$$\varepsilon_{1\sigma} < \varepsilon_{2\sigma} < \ldots < \varepsilon_{N_\sigma\sigma} < \varepsilon_{(N_\sigma+1)\sigma} < \ldots . \tag{30}$$

The Kohn–Sham potentials $V_{S\sigma}(\mathbf{r})$ may be written in the usual way as

$$V_{S\sigma}(\mathbf{r}) = v_0(\mathbf{r}) + \int \mathrm{d}^3r' \frac{\rho(\mathbf{r}')}{|\mathbf{r}-\mathbf{r}'|} + V_{\mathrm{xc}\sigma}(\mathbf{r}), \tag{31}$$

$$\rho(\mathbf{r}) = \sum_{\sigma=\uparrow,\downarrow} \rho_\sigma(\mathbf{r}) \tag{32}$$

where

$$V_{\mathrm{xc}\sigma}(\mathbf{r}) := \frac{\delta E_{\mathrm{xc}}[\rho_\uparrow, \rho_\downarrow]}{\delta \rho_\sigma(\mathbf{r})}. \tag{33}$$

The starting point of the OEP method is the total-energy functional

$$\begin{aligned} E_{v_0}^{\mathrm{OEP}}[\rho_\uparrow, \rho_\downarrow] = & \sum_{\sigma=\uparrow,\downarrow} \sum_{i=1}^{N_\sigma} \int \mathrm{d}^3r \varphi_{i\sigma}^*(\mathbf{r}) \left(-\frac{1}{2}\nabla^2\right) \varphi_{i\sigma}(\mathbf{r}) + \\ & + \int \mathrm{d}^3r v_0(\mathbf{r})\rho(\mathbf{r}) + \\ & + \frac{1}{2} \int \mathrm{d}^3r \int \mathrm{d}^3r' \frac{\rho(\mathbf{r})\rho(\mathbf{r}')}{|\mathbf{r}-\mathbf{r}'|} + \\ & + E_{\mathrm{xc}}^{\mathrm{OEP}}[\{\varphi_{j\tau}\}] \end{aligned} \tag{34}$$

where, in contrast to ordinary spin DFT, the xc energy is an *explicit* (approximate) functional of spin orbitals and therefore only an *implicit* functional of the spin densities $\rho_\uparrow$ and $\rho_\downarrow$. In order to calculate the xc potentials from Equation (33) we use the chain rule for functional derivatives to obtain

$$\begin{aligned} V_{\mathrm{xc}\sigma}^{\mathrm{OEP}}(\mathbf{r}) &= \frac{\delta E_{\mathrm{xc}}^{\mathrm{OEP}}[\{\varphi_{j\tau}\}]}{\delta \rho_\sigma(\mathbf{r})} \\ &= \sum_{\alpha=\uparrow,\downarrow} \sum_{i=1}^{N_\alpha} \int \mathrm{d}^3r' \frac{\delta E_{\mathrm{xc}}^{\mathrm{OEP}}[\{\varphi_{j\tau}\}]}{\delta \varphi_{i\alpha}(\mathbf{r}')} \frac{\delta \varphi_{i\alpha}(\mathbf{r}')}{\delta \rho_\sigma(\mathbf{r})} + \mathrm{c.c.} \end{aligned} \tag{35}$$

and, by applying the functional chain rule once more,

$$V_{\mathrm{xc}\sigma}^{\mathrm{OEP}}(\mathbf{r}) = \sum_{\alpha=\uparrow,\downarrow}\sum_{\beta=\uparrow,\downarrow}\sum_{i=1}^{N_\alpha}\int \mathrm{d}^3r' \int \mathrm{d}^3r'' \times$$
$$\times \left(\frac{\delta E_{\mathrm{xc}}^{\mathrm{OEP}}[\{\varphi_{j\tau}\}]}{\delta\varphi_{i\alpha}(\mathbf{r}')} \frac{\delta\varphi_{i\alpha}(\mathbf{r}')}{\delta V_{S\beta}(\mathbf{r}'')} + \mathrm{c.c.} \right) \frac{\delta V_{S\beta}(\mathbf{r}'')}{\delta\rho_\sigma(\mathbf{r})}. \tag{36}$$

The last term on the right-hand side is readily identified with the inverse $\chi_S^{-1}(\mathbf{r},\mathbf{r}')$ of the density response function of a system of non-interacting particles

$$\chi_{S\alpha,\beta}(\mathbf{r},\mathbf{r}') := \frac{\delta\rho_\alpha(\mathbf{r})}{\delta V_{S\beta}(\mathbf{r}')}. \tag{37}$$

This quantity is diagonal with respect to the spin variables so that Equation (36) reduces to

$$V_{\mathrm{xc}\sigma}^{\mathrm{OEP}}(\mathbf{r}) = \sum_{\alpha=\uparrow,\downarrow}\sum_{i=1}^{N_\alpha}\int \mathrm{d}^3r' \int \mathrm{d}^3r'' \times$$
$$\times \left(\frac{\delta E_{\mathrm{xc}}^{\mathrm{OEP}}[\{\varphi_{j\tau}\}]}{\delta\varphi_{i\alpha}(\mathbf{r}')} \frac{\delta\varphi_{i\alpha}(\mathbf{r}')}{\delta V_{S\sigma}(\mathbf{r}'')} + \mathrm{c.c.} \right) \chi_{S\sigma}^{-1}(\mathbf{r}'',\mathbf{r}). \tag{38}$$

Acting with the response operator (37) on both sides of Equation (38) one obtains

$$\int \mathrm{d}^3r' V_{\mathrm{xc}\sigma}^{\mathrm{OEP}}(\mathbf{r}')\chi_{S\sigma}(\mathbf{r}',\mathbf{r}) = \sum_{\alpha=\uparrow,\downarrow}\sum_{i=1}^{N_\alpha}\int \mathrm{d}^3r' \frac{\delta E_{\mathrm{xc}}^{\mathrm{OEP}}[\{\varphi_{j\tau}\}]}{\delta\varphi_{i\alpha}(\mathbf{r}')} \frac{\delta\varphi_{i\alpha}(\mathbf{r}')}{\delta V_{S\sigma}(\mathbf{r})} + \mathrm{c.c.} \tag{39}$$

Finally, the second functional derivative on the right-hand side of Equation (39) is calculated using first-order perturbation theory. This yields

$$\frac{\delta\varphi_{i\alpha}(\mathbf{r}')}{\delta V_{S\sigma}(\mathbf{r})} = \delta_{\alpha,\sigma} \sum_{\substack{k=1\\k\neq i}}^{\infty} \frac{\varphi_{k\sigma}(\mathbf{r}')\varphi_{k\sigma}^*(\mathbf{r})}{\varepsilon_{i\sigma} - \varepsilon_{k\sigma}} \varphi_{i\sigma}(\mathbf{r}). \tag{40}$$

Using this equation, the response function

$$\chi_{S\alpha,\beta}(\mathbf{r},\mathbf{r}') = \frac{\delta}{\delta V_{S\beta}(\mathbf{r}')} \left(\sum_{i=1}^{N_\alpha} \varphi_{i\alpha}^*(\mathbf{r})\varphi_{i\alpha}(\mathbf{r}) \right) \tag{41}$$

is readily expressed in terms of the orbitals as

$$\chi_{S\sigma}(\mathbf{r},\mathbf{r}') = \sum_{i=1}^{N_\sigma} \sum_{\substack{k=1\\k\neq i}}^{\infty} \frac{\varphi_{i\sigma}^*(\mathbf{r})\varphi_{k\sigma}(\mathbf{r})\varphi_{k\sigma}^*(\mathbf{r}')\varphi_{i\sigma}(\mathbf{r}')}{\varepsilon_{i\sigma} - \varepsilon_{k\sigma}} + \mathrm{c.c.} \tag{42}$$

Inserting (40) and (42) in Equation (39), we obtain the standard form of the OEP integral equation:

$$\sum_{i=1}^{N_\sigma} \int \mathrm{d}^3r' (V_{\mathrm{xc}\sigma}^{\mathrm{OEP}}(\mathbf{r}') - u_{\mathrm{xc}i\sigma}(\mathbf{r}')) G_{Si\sigma}(\mathbf{r}', \mathbf{r}) \varphi_{i\sigma}(\mathbf{r}) \varphi_{i\sigma}^*(\mathbf{r}') + \mathrm{c.c.} = 0 \quad (43)$$

where

$$u_{\mathrm{xc}i\sigma}(\mathbf{r}) := \frac{1}{\varphi_{i\sigma}^*(\mathbf{r})} \frac{\delta E_{\mathrm{xc}}^{\mathrm{OEP}}[\{\varphi_{j\tau}\}]}{\delta \varphi_{i\sigma}(\mathbf{r})} \quad (44)$$

and

$$G_{Si\sigma}(\mathbf{r}, \mathbf{r}') := \sum_{\substack{k=1 \\ k \neq i}}^{\infty} \frac{\varphi_{k\sigma}(\mathbf{r}) \varphi_{k\sigma}^*(\mathbf{r}')}{\varepsilon_{i\sigma} - \varepsilon_{k\sigma}}. \quad (45)$$

The derivation of the OEP integral Equation (43) described here was first given by Görling and Levy [28, 29]. It is important to note that the same expression results [15, 30–32] if one demands that the local one-particle potential appearing in Equation (28) be the *optimized* one yielding orbitals minimizing the total-energy functional (34), i.e. that

$$\left. \frac{\delta E_{v_0}^{\mathrm{OEP}}}{\delta V_{S\sigma}(\mathbf{r})} \right|_{V_{S\sigma} = V_\sigma^{\mathrm{OEP}}} = 0. \quad (46)$$

This equation is the historical origin [30] of the name *optimized effective potential*. As was first pointed out by Perdew and co-workers [33, 34], Equation (46) is equivalent to the HK variational principle. This is most easily seen by applying the functional chain rule to Equation (46) yielding

$$0 = \frac{\delta E_{v_0}^{\mathrm{OEP}}}{\delta V_{S\sigma}(\mathbf{r})} = \sum_\alpha \int \mathrm{d}^3r' \frac{\delta E_{v_0}^{\mathrm{OEP}}}{\delta \rho_\alpha(\mathbf{r}')} \frac{\delta \rho_\alpha(\mathbf{r}')}{\delta V_{S\sigma}(\mathbf{r})}. \quad (47)$$

Once again, the last term on the right-hand side of Equation (47) can be identified with the KS response function (37). Hence, acting with the inverse response operator on Equation (47) leads to the HK variational principle

$$0 = \frac{\delta E_{v_0}^{\mathrm{OEP}}}{\delta \rho_\sigma(\mathbf{r})}. \quad (48)$$

2.2. APPROXIMATION OF KRIEGER, LI AND IAFRATE

In order to use the OEP method derived in the last section we have to solve Equation (43) for the xc potential $V_{\mathrm{xc}\sigma}^{\mathrm{OEP}}$. Unfortunately, there is no known analytic solution for

$V_{xc\sigma}^{OEP}$ depending explicitly on the set of single-particle orbitals $\{\varphi_{i\sigma}\}$. Furthermore, Equation (43) is not in a form that allows for successive approximations of the xc potential. Thus, we need to solve the full integral equation numerically, which is a rather demanding task and has been achieved so far only for systems of high symmetry such as spherical atoms [15, 18, 19, 31, 35] and for solids within the linear muffin tin orbitals atomic sphere approximation [36–38].

However, Krieger, Li and Iafrate [13–21] recently proposed a transformation of Equation (43) that leads to an alternative but still exact form of the OEP equation which lends itself as a starting point for a highly accurate approximation of the OEP potential. Defining

$$\begin{aligned}\psi_{i\sigma}^*(\mathbf{r}) &:= \sum_{\substack{k=1\\k\neq i}}^{\infty} \frac{\int d^3r' \varphi_{i\sigma}^*(\mathbf{r}')(V_{xc\sigma}^{OEP}(\mathbf{r}') - u_{xci\sigma}(\mathbf{r}'))\varphi_{k\sigma}(\mathbf{r}')}{\varepsilon_{i\sigma} - \varepsilon_{k\sigma}} \varphi_{k\sigma}^*(\mathbf{r}) \\ &= \int d^3r' \varphi_{i\sigma}^*(\mathbf{r}')(V_{xc\sigma}^{OEP}(\mathbf{r}') - u_{xci\sigma}(\mathbf{r}'))G_{Si\sigma}(\mathbf{r}', \mathbf{r}),\end{aligned} \tag{49}$$

the OEP integral Equation (43) can be rewritten as

$$\sum_{i=1}^{N_\sigma} \psi_{i\sigma}^*(\mathbf{r})\varphi_{i\sigma}(\mathbf{r}) + \text{c.c.} = 0. \tag{50}$$

Since the KS orbitals $\{\varphi_{i\sigma}\}$ span an orthonormal set we readily conclude from equation (49) that the function $\psi_{i\sigma}(\mathbf{r})$ is orthogonal to $\varphi_{i\sigma}(\mathbf{r})$:

$$\int d^3r \psi_{i\sigma}^*(\mathbf{r})\varphi_{i\sigma}(\mathbf{r}) = 0. \tag{51}$$

The quantity $G_{Si\sigma}(\mathbf{r}', \mathbf{r})$ given by Equation (45) is the Green's function of the KS equation projected onto the subspace orthogonal to $\varphi_{i\sigma}(\mathbf{r})$, i.e., it satisfies the equation

$$(\hat{h}_{S\sigma}(\mathbf{r}) - \varepsilon_{i\sigma})G_{Si\sigma}(\mathbf{r}', \mathbf{r}) = -[\delta(\mathbf{r}' - \mathbf{r}) - \varphi_{i\sigma}(\mathbf{r}')\varphi_{i\sigma}^*(\mathbf{r})] \tag{52}$$

where $\hat{h}_{S\sigma}(\mathbf{r})$ is a shorthand notation for the KS Hamiltonian

$$\hat{h}_{S\sigma}(\mathbf{r}) := -\frac{\nabla^2}{2} + V_{S\sigma}[\rho_\uparrow, \rho_\downarrow](\mathbf{r}). \tag{53}$$

Using Equation (52), we can act with the operator $(\hat{h}_{S\sigma} - \varepsilon_{i\sigma})$ on Equation (49), leading to

$$(\hat{h}_{S\sigma}(\mathbf{r}) - \varepsilon_{i\sigma})\psi_{i\sigma}^*(\mathbf{r}) = -[V_{xc\sigma}^{OEP}(\mathbf{r}) - u_{xci\sigma}(\mathbf{r}) - (\bar{V}_{xci\sigma} - \bar{u}_{xci\sigma})]\varphi_{i\sigma}^*(\mathbf{r}) \tag{54}$$

where $\bar{V}_{\mathrm{xc}i\sigma}$ denotes the average of $V_{\mathrm{xc}\sigma}(\mathbf{r})$ with respect to the ith orbital, i.e.

$$\bar{V}_{\mathrm{xc}i\sigma} := \int \mathrm{d}^3 r \varphi^*_{i\sigma}(\mathbf{r}) V^{\mathrm{OEP}}_{\mathrm{xc}\sigma}(\mathbf{r}) \varphi_{i\sigma}(\mathbf{r}) \tag{55}$$

and

$$\bar{u}_{\mathrm{xc}i\sigma} := \int \mathrm{d}^3 r \varphi^*_{i\sigma}(\mathbf{r}) u_{\mathrm{xc}i\sigma}(\mathbf{r}) \varphi_{i\sigma}(\mathbf{r}). \tag{56}$$

Using the KS Equation (28), we can solve Equation (54) for $V_{S\sigma}(\mathbf{r})\psi^*_{i\sigma}(\mathbf{r})$:

$$\begin{aligned} V_{S\sigma}(\mathbf{r})\psi^*_{i\sigma}(\mathbf{r}) &= -[V^{\mathrm{OEP}}_{\mathrm{xc}\sigma}(\mathbf{r}) - u_{\mathrm{xc}i\sigma}(\mathbf{r}) - (\bar{V}_{\mathrm{xc}i\sigma} - \bar{u}_{\mathrm{xc}i\sigma})]\varphi^*_{i\sigma}(\mathbf{r}) + \\ &\quad + \left(\frac{\nabla^2}{2} + \varepsilon_{i\sigma}\right)\psi^*_{i\sigma}(\mathbf{r}). \end{aligned} \tag{57}$$

The differential Equation (54) has the structure of a KS equation with an additional inhomogeneity term. Equation (54) plus the boundary condition that $\psi^*_{i\sigma}(\mathbf{r})$ tends to zero as $r \to \infty$ uniquely determines $\psi^*_{i\sigma}(\mathbf{r})$. We can prove this statement by assuming that there are two independent solutions $\psi^*_{i\sigma,1}(\mathbf{r})$ and $\psi^*_{i\sigma,2}(\mathbf{r})$ of Equation (54). Then the difference between these two solutions, $\Psi^*_{i\sigma}(\mathbf{r}) := \psi^*_{i\sigma,1}(\mathbf{r}) - \psi^*_{i\sigma,2}(\mathbf{r})$, satisfies the homogeneous KS equation

$$(\hat{h}_{S\sigma} - \varepsilon_{i\sigma})\Psi^*_{i\sigma}(\mathbf{r}) = 0, \tag{58}$$

which has a unique solution

$$\Psi^*_{i\sigma}(\mathbf{r}) = \varphi^*_{i\sigma}(\mathbf{r}), \tag{59}$$

if the above boundary condition is fulfilled. However, this solution leads to a contradiction with the orthogonality relation (51) so that $\Psi^*_{i\sigma}(\mathbf{r})$ can only be the trivial solution of Equation (58),

$$\Psi^*_{i\sigma}(\mathbf{r}) \equiv 0. \tag{60}$$

This completes the proof.

At this point it is useful to attach some physical meaning to the quantity $\psi_{i\sigma}$: from Equation (49) it is obvious that $\psi_{i\sigma}$ is the usual first-order shift in the wavefunction caused by the perturbing potential $\delta V_{i\sigma} = V^{\mathrm{OEP}}_{\mathrm{xc}\sigma} - u_{\mathrm{xc}i\sigma}$. This fact also motivates the boundary condition assumed above. In x-only theory, $u_{\mathrm{x}i\sigma}$ is the local, orbital-dependent HF exchange potential so that $\psi_{i\sigma}$ is the first order shift of the KS wavefunction towards the HF wavefunction. One has to realize, however, that the first-order change of the orbital dependent potential $u_{\mathrm{x}i\sigma}[\{\varphi_{i\sigma}\}]$ has been neglected. This change can be expected to be small compared to $\delta V_{i\sigma}$ [20].

Having found an equation satisfied by $\psi^*_{i\sigma}(\mathbf{r})$ in which $V^{\mathrm{OEP}}_{\mathrm{xc}\sigma}(\mathbf{r})$ shows up explicitly, we may use this identity to further transform the OEP Equation (50). To this end we multiply Equation (50) by the KS potential $V_{S\sigma}(\mathbf{r})$

$$\sum_{i=1}^{N_\sigma} V_{S\sigma}(\mathbf{r})\psi^*_{i\sigma}(\mathbf{r})\varphi_{i\sigma}(\mathbf{r}) + \text{c.c.} = 0 \tag{61}$$

and employ Equation (57) to obtain

$$\begin{aligned} 0 = \sum_{i=1}^{N_\sigma} \Big\{ & [V^{\mathrm{OEP}}_{\mathrm{xc}\sigma}(\mathbf{r}) - u_{\mathrm{xc}i\sigma}(\mathbf{r}) - (\bar{V}_{\mathrm{xc}i\sigma} - \bar{u}_{\mathrm{xc}i\sigma})]\varphi^*_{i\sigma}(\mathbf{r}) - \\ & - \left(\frac{\nabla^2}{2} + \varepsilon_{i\sigma}\right)\psi^*_{i\sigma}(\mathbf{r})\Big\}\varphi_{i\sigma}(\mathbf{r}) + \text{c.c.} \end{aligned} \tag{62}$$

Solving this equation for $V^{\mathrm{OEP}}_{\mathrm{xc}\sigma}$ yields

$$\begin{aligned} V^{\mathrm{OEP}}_{\mathrm{xc}\sigma}(\mathbf{r}) = & \frac{1}{2\rho_\sigma(\mathbf{r})}\sum_{i=1}^{N_\sigma}\Big\{ |\varphi_{i\sigma}(\mathbf{r})|^2[u_{\mathrm{xc}i\sigma}(\mathbf{r}) + (\bar{V}_{\mathrm{xc}i\sigma} - \bar{u}_{\mathrm{xc}i\sigma})] + \\ & + \left(\frac{\nabla^2}{2}\psi^*_{i\sigma}(\mathbf{r}) + \varepsilon_{i\sigma}\psi^*_{i\sigma}(\mathbf{r})\right)\varphi_{i\sigma}(\mathbf{r})\Big\} + \text{c.c.} \end{aligned} \tag{63}$$

The second term in the curly brackets may be rewritten by using the KS and the OEP equation again, leading to

$$\begin{aligned} & \sum_{i=1}^{N_\sigma}\left(\frac{\nabla^2}{2}\psi^*_{i\sigma}(\mathbf{r}) + \varepsilon_{i\sigma}\psi^*_{i\sigma}(\mathbf{r})\right)\varphi_{i\sigma}(\mathbf{r}) + \text{c.c.} \\ & = \sum_{i=1}^{N_\sigma}\left[\left(\frac{\nabla^2}{2}\psi^*_{i\sigma}(\mathbf{r})\right)\varphi_{i\sigma}(\mathbf{r}) - \psi^*_{i\sigma}(\mathbf{r})\left(\frac{\nabla^2}{2}\varphi_{i\sigma}(\mathbf{r})\right)\right] + \text{c.c.} \\ & = -\sum_{i=1}^{N_\sigma}\nabla\cdot(\psi^*_{i\sigma}(\mathbf{r})\nabla\varphi_{i\sigma}(\mathbf{r})) + \text{c.c.} \end{aligned} \tag{64}$$

In this way Equation (63) may be written as

$$V^{\mathrm{OEP}}_{\mathrm{xc}\sigma}(\mathbf{r}) = \frac{1}{2\rho_\sigma(\mathbf{r})}\sum_{i=1}^{N_\sigma}|\varphi_{i\sigma}(\mathbf{r})|^2[v_{\mathrm{xc}i\sigma}(\mathbf{r}) + (\bar{V}_{\mathrm{xc}i\sigma} - \bar{u}_{\mathrm{xc}i\sigma})] + \text{c.c.} \tag{65}$$

with

$$v_{\mathrm{xc}i\sigma}(\mathbf{r}) = u_{\mathrm{xc}i\sigma}(\mathbf{r}) - \frac{1}{|\varphi_{i\sigma}(\mathbf{r})|^2}\nabla \cdot (\psi_{i\sigma}^*(\mathbf{r})\nabla\varphi_{i\sigma}(\mathbf{r})). \tag{66}$$

Equation (65) is an exact transformation of the original OEP integral Equation (43). The advantage of Equation (65), although still being an integral equation, lies in the fact that it may serve as a starting point for constructing approximations of $V_{\mathrm{xc}\sigma}^{\mathrm{OEP}}$: We only need to approximate $\psi_{i\sigma}^*$ in Equation (66) by a suitable functional of the orbitals.

The simplest possible approximation is obtained by completely neglecting the terms involving $\psi_{i\sigma}^*$, i.e. by replacing $v_{\mathrm{xc}i\sigma}$ by $u_{\mathrm{xc}i\sigma}$. At first sight, this approximation might appear rather crude. It can be interpreted, however, as a mean-field approximation in the sense that the neglected terms averaged over the ground-state spin density $\rho_\sigma(\mathbf{r})$ vanish. To demonstrate this, we investigate the quantity

$$\int \mathrm{d}^3 r \nabla \cdot \frac{1}{2}\sum_{i=1}^{N_\sigma}(\psi_{i\sigma}^*(\mathbf{r})\nabla\varphi_{i\sigma}(\mathbf{r})) \tag{67}$$

which amounts to the difference between the exact $V_{\mathrm{xc}\sigma}^{\mathrm{OEP}}(\mathbf{r})$ and the approximated xc potential averaged over $\rho_\sigma(\mathbf{r})$. By virtue of the divergence theorem, the integral can be transformed to a surface integral. The latter vanishes because $\psi_{i\sigma}^*(\mathbf{r})$ and $\varphi_{i\sigma}(\mathbf{r})$ decrease exponentially for $r \to \infty$ [15]. Hence, the neglected terms have zero average value. The resulting equation, known as the KLI approximation, is given by

$$V_{\mathrm{xc}\sigma}^{\mathrm{KLI}}(\mathbf{r}) = \frac{1}{2\rho_\sigma(\mathbf{r})}\sum_{i=1}^{N_\sigma}|\varphi_{i\sigma}(\mathbf{r})|^2[u_{\mathrm{xc}i\sigma}(\mathbf{r}) + (\bar{V}_{\mathrm{xc}i\sigma}^{\mathrm{KLI}} - \bar{u}_{\mathrm{xc}i\sigma})] + \mathrm{c.c.} \tag{68}$$

which has proved to be an excellent approximation to the full xc potential $V_{\mathrm{xc}\sigma}^{\mathrm{OEP}}(\mathbf{r})$ [15, 18, 19]. We immediately recognize that this form is very similar to the Slater potential.

In contrast to the full OEP Equation (43), the KLI equation, still being an integral equation, can be solved explicitly in terms of the orbitals $\{\varphi_{i\sigma}\}$: multiplying Equation (68) by $|\varphi_{j\sigma}(\mathbf{r})|^2$ and integrating over space yields

$$\bar{V}_{\mathrm{xc}j\sigma}^{\mathrm{KLI}} = \bar{V}_{\mathrm{xc}j\sigma}^{S} + \sum_{i=1}^{N_\sigma - 1} M_{ji\sigma}\left(\bar{V}_{\mathrm{xc}i\sigma}^{\mathrm{KLI}} - \frac{1}{2}(\bar{u}_{\mathrm{xc}i\sigma} + \bar{u}_{\mathrm{xc}i\sigma}^*)\right), \tag{69}$$

where

$$\bar{V}_{\mathrm{xc}j\sigma}^{S} := \int \mathrm{d}^3 r \frac{|\varphi_{j\sigma}(\mathbf{r})|^2}{\rho_\sigma(\mathbf{r})}\sum_{i=1}^{N_\sigma}|\varphi_{i\sigma}(\mathbf{r})|^2\frac{1}{2}(u_{\mathrm{xc}i\sigma}(\mathbf{r}) + u_{\mathrm{xc}i\sigma}^*(\mathbf{r})) \tag{70}$$

and

$$M_{ji\sigma} := \int d^3r \frac{|\varphi_{j\sigma}(\mathbf{r})|^2 |\varphi_{i\sigma}(\mathbf{r})|^2}{\rho_\sigma(\mathbf{r})}. \tag{71}$$

The term corresponding to the highest occupied orbital $\varphi_{N_\sigma \sigma}$ has been excluded from the sum in Equation (69) because $\bar{V}_{\mathrm{xc}N_\sigma\sigma} = \bar{u}_{\mathrm{xc}N_\sigma\sigma}$ [15]. The remaining unknown constants $(\bar{V}^{\mathrm{KLI}}_{\mathrm{xc}i\sigma} - \frac{1}{2}(\bar{u}_{\mathrm{xc}i\sigma} + \bar{u}^*_{\mathrm{xc}i\sigma}))$ are determined by the linear equation

$$\sum_{i=1}^{N_\sigma - 1} (\delta_{ji} - M_{ji\sigma}) \left(\bar{V}^{\mathrm{KLI}}_{\mathrm{xc}i\sigma} - \frac{1}{2}(\bar{u}_{\mathrm{xc}i\sigma} + \bar{u}^*_{\mathrm{xc}i\sigma}) \right)$$

$$= \left(\bar{V}^S_{\mathrm{xc}j\sigma} - \frac{1}{2}(\bar{u}_{\mathrm{xc}j\sigma} + \bar{u}^*_{\mathrm{xc}j\sigma}) \right), \tag{72}$$

with $j = 1, \ldots, N_\sigma - 1$. Solving Equation (72) and substituting the result into Equation (68), we obtain an explicitly orbital dependent functional.

We note that the KLI Equation (68) can also be obtained by a less rigorous derivation, namely by approximating the energy denominator in the Green's function (52) by a single constant as was first suggested by Sharp and Horton [30] and further elaborated by Krieger, Li, and Iafrate [13, 15].

Numerical calculations [15, 18, 19] for atomic systems neglecting correlation effects have shown that the KLI approximation gives excellent results which deviate only by a few ppm from the much more involved exact solutions of the full OEP integral Equation (43). Results for diatomic molecules [39] appear to be of similar quality. Examples will be presented in Section 3.

2.3. RIGOROUS PROPERTIES OF THE OEP AND KLI POTENTIALS

To conclude this section we will describe (without proof) some exact properties of the OEP method and the KLI approximation.

- *Asymptotics*: for finite systems, both the full OEP and the KLI potential fall off as $-1/r$ for $r \to \infty$ [15] if the exact expression (27) for the exchange-energy functional is employed.
- *Freedom of self-interaction*: if the employed xc-energy functional cancels the self-interaction of the Hartree term, this property is preserved by the KLI approximation. Thus x-only OEP and x-only KLI schemes are self-interaction free. It has to be noted, however, that the inclusion of an LDA-correlation-energy functional might introduce a self-interaction error again.
- *Derivative discontinuities*: an important property of the *exact* xc energy is that it exhibits derivative discontinuities as a function of particle number N at integer values of N. This has important consequences for the values of band gaps in insulators and semiconductors (for a detailed description see e.g. [2]):

Table I. Various self-consistently calculated x-only results for the Ar atom. All values in atomic units

	OEP	KLI	B88	xPW91	xLDA
E_{TOT}	−526.8122	−526.8105	−526.7998	−526.7710	−524.5174
ε_{1s}	−114.4524	−114.4279	−114.1890	−114.1887	−113.7159
ε_{2s}	−11.1534	−11.1820	−10.7911	−10.7932	−10.7299
ε_{2p}	−8.7339	−8.7911	−8.4107	−8.4141	−8.3782
ε_{3s}	−1.0993	−1.0942	−0.8459	−0.8481	−0.8328
ε_{3p}	−0.5908	−0.5893	−0.3418	−0.3441	−0.3338
ε_{4s}	−0.1607	−0.1616	−0.0102	−0.0122	−0.0014
$\langle r^2 \rangle$	1.4465	1.4467	1.4791	1.4876	1.4889
$\langle r^{-1} \rangle$	3.8736	3.8738	3.8731	3.8729	3.8648
$\rho(0)$	3839.7	3832.6	3847.3	3847.0	3818.7

The correct value E^g of the gap is obtained by adding the discontinuity Δ_{xc} of the xc potential to the KS gap, i.e., $E^g = E^g_{\mathrm{KS}} + \Delta_{\mathrm{xc}}$. Neither the LDA nor GGAs reproduce this discontinuity. To date, the OEP and the KLI potential are the only known approximations of $V_{\mathrm{xc}}(\mathbf{r})$ that reproduce this property of the exact xc potential [15].

3. Selected Results

3.1. ATOMIC SYSTEMS

We begin with a comparison of x-only results. In an x-only world, the OEP represents the *exact* KS potential of DFT and can therefore serve as a standard to compare approximations with.

In Table I we show as a typical example various results for the argon atom obtained with different x-only methods. Besides the exact OEP and KLI methods employing the exact exchange energy functional (27) described above, we also list results from conventional KS-DFT obtained with the x-energy-functional approximations due to Becke (B88) [40], Perdew and Wang (PW91) [41] and from the well-known x-only LDA approximation. The KLI results given in the second column of Table I clearly demonstrate the high quality of the KLI approximation as all results differ only slightly from the exact OEP ones. For all standard DFT methods, the disagreement is much more pronounced, especially for the highest occupied orbital energies and even more so for the unoccupied ones.

Extensive calculations for atoms [32, 39, 42] have shown that a suitable correlation-energy functional to be combined with the exact exchange energy functional in the KLI scheme is the one developed by Colle and Salvetti (CS) [43, 44]. It is given by

$$
\begin{aligned}
E_c^{\mathrm{CS}}&[\{\varphi_{i\sigma}\}] \\
&= -ab\int \mathrm{d}^3 r\gamma(\mathbf{r})\xi(\mathbf{r})\times \\
&\quad\times\left[\sum_\sigma \rho_\sigma(\mathbf{r})\sum_i|\nabla\varphi_{i\sigma}(\mathbf{r})|^2-\frac{1}{4}|\nabla\rho(\mathbf{r})|^2-\right. \\
&\quad\left.-\frac{1}{4}\sum_\sigma\rho_\sigma(\mathbf{r})\Delta\rho_\sigma(\mathbf{r})+\frac{1}{4}\rho(\mathbf{r})\Delta\rho(\mathbf{r})\right]- \\
&\quad -a\int \mathrm{d}^3 r\gamma(\mathbf{r})\frac{\rho(\mathbf{r})}{\eta(\mathbf{r})}, \qquad (73)
\end{aligned}
$$

where

$$\gamma(\mathbf{r}) = 4\frac{\rho_\uparrow(\mathbf{r})\rho_\downarrow(\mathbf{r})}{\rho(\mathbf{r})^2}, \qquad (74)$$

$$\eta(\mathbf{r}) = 1 + d\rho(\mathbf{r})^{-1/3}, \qquad (75)$$

$$\xi(\mathbf{r}) = \frac{\rho(\mathbf{r})^{-5/3}\exp[-c\rho(\mathbf{r})^{-1/3}]}{\eta(\mathbf{r})}. \qquad (76)$$

The values of constants a, b, c and d are $a = 0.04918$, $b = 0.132$, $c = 0.2533$, $d = 0.349$.

In Table II we compare total ground state energies of first-row atoms calculated self-consistently with various approximations. The first column, headed KLICS, shows the results from the KLI method employing the exact exchange energy functional (27) plus the CS-correlation energy functional, while the next columns show conventional KS results. The latter were obtained with an LDA functional using the parametrisation of the correlation energy of a homogeneous electron gas by Perdew and Wang [47]; the x-energy functional due to Becke [40] combined with the c-energy functional of Lee, Yang and Parr [48], denoted as BLYP; and the generalized gradient approximation due to Perdew and Wang [41], referred to as PW91. The quantum chemistry values, headed QC, are based on configuration interaction calculations [45]. The exact non-relativistic energies, i.e. the experimental values with relativistic effects subtracted, have been taken from [46]. The mean absolute deviations of the calculated from the exact values, denoted by $\bar{\Delta}$, are about the same for the KLICS and QC approaches, while they are about twice as high for the GGAs and about a factor of 80 higher in the LDA. We emphasize that the numerical effort involved in the KLICS scheme for atoms is only slightly higher than in the LDA and GGA schemes.

Table II. Total absolute ground-state energies for first-row atoms from various self-consistent calculations. Quantum chemistry (QC) values from [45]. $\bar{\Delta}$ denotes the mean absolute deviation from the exact non-relativistic values [46]. All numbers in Hartrees. Taken from [32] and modified.

	KLICS	xcLDA	BLYP	PW91	QC	Exact
He	2.9033	2.8346	2.9071	2.9000	2.9049	2.9037
Li	7.4829	7.3433	7.4827	7.4742	7.4743	7.4781
Be	14.6651	14.4465	14.6615	14.6479	14.6657	14.6674
B	24.6564	24.3525	24.6458	24.6299	24.6515	24.6539
C	37.8490	37.4683	37.8430	37.8265	37.8421	37.8450
N	54.5905	54.1344	54.5932	54.5787	54.5854	54.5893
O	75.0717	74.5248	75.0786	75.0543	75.0613	75.067
F	99.7302	99.1112	99.7581	99.7316	99.7268	99.734
Ne	128.9202	128.2299	128.9730	128.9466	128.9277	128.939
$\bar{\Delta}$	0.0047	0.3813	0.0108	0.0114	0.0045	

Apart from total energies, the highest occupied orbital energies, which should be equal to the exact ionization potential in an exact implementation of DFT, are much closer to the experimental ionization potentials in the KLICS scheme than in the conventional KS approaches. This is shown in Table III: all the conventional KS calculations are inadequate, the numbers are off by about 100% due to the wrong asymptotic behaviour of the xc-potentials in these approximations. Only the KLICS scheme results in a potential with the correct $-1/r$ decay for large r.

3.2. RESULTS FOR DIATOMIC MOLECULES

The feasibility of the KLI approximation for more complex systems has been demonstrated recently [39] by x-only calculations on diatomic molecules. Some x-only results for the N_2 molecule are shown in Table IV. Among all the density functional approaches the KLI scheme yields results closest to the HF values. In principle, the density functional results should not be compared with HF but rather with exact x-only OEP values. For molecules, however, the latter are not available yet. From atomic results [18, 19] it is expected that HF and exact x-only OEP values will agree closely for total energies while the orbital energies are expected to be somewhat different.

It remains to examine the effect of correlation contributions. Therefore, we have implemented the CS functional (73) for E_c in our fully numerical basis-set-free code for diatomic molecules, which is based on the Xα program developed by Laaksonen, Sundholm and Pyykkö [50]. The solution of the Kohn–Sham equation and the exchange energy integrals are calculated by means of relaxation techniques

Table III. Ionization potentials from the highest occupied orbital energy of neutral atoms. $\bar{\Delta}$ denotes the mean absolute deviation from the experimental values, taken from [49]. All values in Hartrees. Taken from [32] and modified.

	KLICS	xcLDA	BLYP	PW91	Experiment
He	0.945	0.570	0.585	0.583	0.903
Li	0.200	0.116	0.111	0.119	0.198
Be	0.329	0.206	0.201	0.207	0.343
B	0.328	0.151	0.143	0.149	0.305
C	0.448	0.228	0.218	0.226	0.414
N	0.579	0.309	0.297	0.308	0.534
O	0.559	0.272	0.266	0.267	0.500
F	0.714	0.384	0.376	0.379	0.640
Ne	0.884	0.498	0.491	0.494	0.792
Na	0.189	0.113	0.106	0.113	0.189
Mg	0.273	0.175	0.168	0.174	0.281
Al	0.222	0.111	0.102	0.112	0.220
Si	0.306	0.170	0.160	0.171	0.300
P	0.399	0.231	0.219	0.233	0.385
S	0.404	0.228	0.219	0.222	0.381
Cl	0.506	0.305	0.295	0.301	0.477
Ar	0.619	0.382	0.373	0.380	0.579
$\bar{\Delta}$	0.030	0.176	0.183	0.177	

Table IV. X-only results for N_2 with bond length of 2.07 a.u. HF values from [50]. Q_2^{tot} and Q_4^{tot} denote the total quadrupole and hexadecapole moments, respectively, calculated from the center of mass. All numbers in atomic units. Taken from [39] and modified.

	HF	KLI	xPW91	xLDA
E_{TOT}	−108.9936	−108.9856	−109.0581	−107.7560
$\varepsilon_{1\sigma g}$	−15.6822	−14.3722	−14.0717	−13.8950
$\varepsilon_{1\sigma u}$	−15.6787	−14.3709	−14.0703	−13.8936
$\varepsilon_{2\sigma g}$	−1.4726	−1.3076	−1.0014	−0.9875
$\varepsilon_{2\sigma u}$	−0.7784	−0.7453	−0.4611	−0.4434
$\varepsilon_{3\sigma g}$	−0.6347	−0.6305	−0.3927	−0.3335
$\varepsilon_{1\pi g}$	−0.6152	−0.6818	−0.3478	−0.3887
Q_2^{tot}	−0.9372	−0.9488	−1.1962	−1.1643
Q_4^{tot}	−7.3978	−6.7476	−6.1809	−6.2553
$\langle 1/r \rangle_N$	21.6543	21.6439	21.6921	21.5820

on a two-dimensional grid. For comparison, we have also performed calculations employing the conventional density dependent xcLDA and PW91 functionals for E_{xc}.

Table V. Results for the Neon atom using various DFT approaches. 1D denotes the exact values obtained with our one-dimensional code, 2D the results from our 2D code with a bond distance of 1 a.u. and one nuclear charge set to zero. 1D calculations performed with a grid size of 800 points, 2D calculations with a grid size of 145 × 241 points. All numbers in atomic units.

	KLICS		PW91		xcLDA	
	1D	2D	1D	2D	1D	2D
E_{TOT}	−128.920235	−128.920234	−128.946580	−128.946580	−128.229914	−128.229915
ε_{1s}	−30.841442	−30.841447	−30.507920	−30.507920	−30.305770	−30.305770
ε_{2s}	−1.741044	−1.741044	−1.334970	−1.334970	−1.322601	−1.322601
$\varepsilon_{2p\sigma}$	−0.884057	−0.884057	−0.494228	−0.494228	−0.497847	−0.497847
$\varepsilon_{2p\pi}$	−0.884057	−0.884057	−0.494228	−0.494228	−0.497847	−0.497847
$\langle 1/r \rangle$	3.111736	3.111751	3.110075	3.110091	3.099824	3.099840
$\langle r \rangle$	0.787235	0.787235	0.799878	0.799878	0.801547	0.801547
$\langle r^2 \rangle$	0.931987	0.931987	0.982608	0.982608	0.985199	0.985200

Table VI. Calculated bond lengths of the closed-shell-first-row dimers and hydrides. HF values taken from [51]. Experimental values from [52] except where noted. All values in atomic units.

	KLICS	PW91	xcLDA	HF	Experiment
H_2	1.378	1.414	1.446	1.379	1.401[a]
Li_2	5.086	5.153	5.120	5.304	5.051
Be_2	–	4.588	4.522	–	4.63[b]
C_2	2.306	2.367	2.354	–	2.3481
N_2	1.998	2.079	2.068	2.037	2.074
F_2	2.465	2.669	2.615	2.542	2.6682
LiH	2.971	3.030	3.030	3.092	3.0154
BH	2.274	2.356	2.373	–	2.3289
FH	1.684	1.756	1.761	1.722	1.7325

[a]Exact value from [53].
[b]From [54].

Table VII. Absolute total ground-state energies of the closed-shell-first-row dimers and hydrides calculated at the bond lengths given in Table VI. Estimates for exact values calculated using dissociation energies from Table VIII and non-relativistic, infinite nuclear mass atomic ground-state energies from [55]. All numbers in Hartrees.

	KLICS	PW91	xcLDA	Exact
H_2	1.171444	1.170693	1.137692	1.174448[a]
Li_2	14.9982	14.9819	14.7245	14.9954
Be_2	29.3197[b]	29.3118	28.9136	29.3385
C_2	75.7736	75.8922	75.2041	75.922
N_2	109.4683	109.5449	108.6959	109.5424
F_2	199.4377	199.5699	198.3486	199.5299
LiH	8.0723	8.0625	7.9189	8.0705
BH	25.2857	25.2688	24.9770	25.29
FH	100.4241	100.4715	99.8490	100.4596

[a]Exact value from [53].
[b]Calculated at the experimental bond length of 4.63 a.u.

To demonstrate the accuracy of our implementation, we show results for the neon atom obtained with our molecular code and compare them with the ones from our one-dimensional atomic structure program in Table V. It is obvious from the table that the accuracy of our molecular code is very good, the deviation from the exact results obtained with the atomic code is a few μHartrees at the most.

We have calculated the ground-state properties of the closed-shell-first-row dimers and hydrides in the approximations mentioned above in a fully self-consistent fashion. As our program uses no basis functions, the results are free

Table VIII. Dissociation energies of the closed-shell-first-row dimers and hydrides calculated at the bond lengths given in Table VI. HF values taken from [51]. All numbers in mHartrees.

	KLICS	PW91	xcLDA	HF	Experiment
H_2	171.444	167.665	180.270	121.0	174.475[b]
Li_2	32.4	33.5	37.9	3.5	39.3[c]
Be_2	−10.5[a]	15.9	20.6	–	3.8[c]
C_2	75.6	239.2	267.5	–	232[d]
N_2	287.3	387.5	427.1	167.5	364.0[d]
F_2	−22.7	106.7	126.2	−54.7	62.1[d]
LiH	89.4	86.8	96.9	48.4	92.4[d]
BH	129.3	137.4	145.8	–	135[e]
FH	193.9	238.4	259.1	130.8	225.7[e]

[a]Calculated at the experimental bond length 4.63 a.u.
[b]Exact value from [53].
[c]From [54].
[d]From [52].
[e]From [56].

Table IX. Absolute values for the highest occupied orbital energies of the closed-shell-first-row dimers and hydrides calculated at the bond lengths given in Table VI. Experimental values are the ionization potentials taken from [52]. All numbers in Hartrees.

	KLICS	PW91	xcLDA	Experiment
H_2	0.621563	0.382656	0.373092	0.5669
Li_2	0.1974	0.1187	0.1187	0.18
Be_2	0.2560[a]	0.1678	0.1660	–
C_2	0.4844	0.2942	0.2987	0.4465
N_2	0.6643	0.3804	0.3826	0.5726
F_2	0.6790	0.3512	0.3497	0.5764
LiH	0.3237	0.1621	0.1612	0.283[b]
BH	0.3692	0.2058	0.2041	0.359
FH	0.6803	0.3567	0.3594	0.5894

[a]Calculated at the experimental bond length 4.63 a.u.
[b] From [57].

of basis-set truncation errors. Where available, we have also included HF results which, however, were obtained with conventional codes using basis-set expansions.

In Table VI we display results for the bond lengths. It is apparent that the KLICS scheme results in distances which are generally too short, an effect present in the HF approximation as well. The LDA and PW91 functionals give values which are clearly superior, the latter further reducing the error of the former.

Total absolute ground-state energies calculated at the bond lengths given in Table VI are shown in Table VII. The exact values for the dimers are from [54], for the hydrides they are calculated by the same method using the exact non-relativistic ground state energies for atoms in [55] and the experimental dissociation energies in [52]. For the lighter molecules H_2, Li_2, Be_2, LiH and BH the KLICS and PW91 results are of the same good quality, yielding errors of a few mHartrees. For the heavier molecules, however, the KLICS results are worse. Being the simplest approximation, it is not surprising that the LDA gives values for the total energies which show the largest errors.

Apart form H_2 and LiH, the dissociation energies as obtained within the KLICS approach are disappointing, as may be read in Table VIII. In most cases, the magnitude is considerably underestimated and for Be_2 and F_2 even the wrong sign is obtained. Since the corresponding *atomic* ground-state energies given in the previous subsection are of excellent quality, the error must be due to correlation effects present in molecules only. In particular, the left-right correlation error well-known in HF theory also occurs in DFT when the exact Fock expression (27) for E_x is employed. Apparently, the error is not sufficiently corrected for by the Colle-Salvetti functional. The LDA and PW91 results are clearly much better, the latter reducing the over-binding tendencies of the former.

Despite all of these shortcomings, the asymptotic form of the KLICS xc potential is of much better quality than that of all the conventional xc-functional approximations. In Table IX we list the absolute values of the highest occupied molecular orbital energies. In an *exact* implementation their values should be equal to the ionization potentials of the systems under consideration. It is evident that the conventional KS approaches represented by the LDA and PW91 functionals yield results which are typically 30 to 40 percent too high, while the KLICS values are much closer to the experimental results. As for atomic systems, this fact may be traced back to the correct asymptotic behaviour of the KLICS xc potential for large r.

4. Conclusions

Due to the wealth of exact properties satisfied by the OEP, we argue that this third generation of density functional theory provides a promising basis for further advances. While the results for atoms are very encouraging, the construction of correlation functionals better adapted to correct the left-right deficiency of the exact exchange energy functional is necessary if reliable results for molecular systems are to be obtained.

Acknowledgments

We thank Dr. E. Engel for providing us with his atomic OEP-KS computer code and Dr. D. Sundholm and Professor P. Pyykkö for their two-dimensional Xα code for molecules. Professor J. Perdew supplied us with the PW91 xc subroutine. Discussions with Dr. E. Engel, Dr. D. Sundholm, M. Petersilka and M. Lüders are gratefully acknowledged. This work was supported by the Deutsche Forschungsgemeinschaft.

References

1. E. K. U. Gross and R. M. Dreizler (eds.): *Density Functional Theory*, Vol. 337 of NATO ASI Series B, Plenum Press, New York (1995).
2. R.M. Dreizler and E. K. U. Gross: *Density Functional Theory*, Springer, Berlin (1990).
3. R. G. Parr and W. Yang: *Density-Functional Theory of Atoms and Molecules*, Oxford University Press, New York (1989).
4. P. Hohenberg and W. Kohn: *Phys. Rev.* **136**, B864 (1964).
5. Q. Zhao and R.G. Parr: *Phys. Rev. A* **46**, 2337 (1992).
6. C. J. Umrigar and X. Gonze: in *High Performance Computing and its Application to the Physical Sciences*, D. A. Browne *et al.* (eds.), World Scientific, Singapore (1993).
7. C. J. Umrigar and X. Gonze: *Phys. Rev. A* **50**, 3827 (1994).
8. R. van Leeuven and E.J. Baerends: *Phys. Rev. A* **49**, 2421 (1994).
9. A. Görling and M. Ernzerhof: *Phys. Rev. A* **51**, 4501 (1995).
10. V. E. Ingamells and N. C. Handy: *Chem. Phys. Lett.* **248**, 373 (1996).
11. W. Kohn and L. J. Sham: *Phys. Rev.* **140**, A1133 (1965).
12. J. P. Perdew, J. A. Cheavry, S. H. Vosko, K. A. Jackson, M. A. Pederson, D. J. Singh, and C. Fiolhais: *Phys. Rev. B* **46**, 6671 (1992).
13. J. B. Krieger, Y. Li, and G. J. Iafrate: *Phys. Lett. A* **146**, 256 (1990).
14. J. B. Krieger, Y. Li, M. R. Norman, and G. J. Iafrate: *Phys. Rev. B* **44**, 10437 (1991).
15. J. B. Krieger, Y. Li, G. J. Iafrate: *Phys. Rev. A* **45**, 101 (1992).
16. J. B. Krieger, Y. Li, and G. J. Iafrate: *Phys. Rev. A* **46**, 5453 (1992).
17. J. B. Krieger, Y. Li, and G. J. Iafrate: *Int. J. Quantum Chem.* **41**, 489 (1992).
18. Y. Li, J. B. Krieger, and G. J. Iafrate: *Chem. Phys. Lett.* **191**, 38 (1992).
19. Y. Li, J. B. Krieger, and G. J. Iafrate: *Phys. Rev. A* **47**, 165 (1993).
20. J. B. Krieger, Y. Li, and G. J. Iafrate: in *Density Functional Theory*, R.M. Dreizler and E.K.U. Gross (eds.), Plenum Press, New York (1995).
21. J. B. Krieger, J. Chen, Y. Li, and G. J. Iafrate: *Int. J. Quantum Chem. Symp.* **29**, 79 (1995).
22. C. A. Ullrich, U. J. Gossmann, and E. K. U. Gross: *Phys. Rev. Lett.* **74**, 872 (1995).
23. C. A. Ullirich, S. Erhard, and E. K. U. Gross: in *Superintense Laser Atom Physics IV*, NATO ASI series 3/13, edited by H. G. Muller and M. V. Fedorov, Kluwer, Dordrecht (1996).
24. M. Petersilka, U. J. Gossmann, and E. K. U Gross: *Phys. Rev. Lett.* **76**, 1212 (1996).
25. E. K. U. Gross, J. F. Dobson, and M. Petersilka: in *Topics in Current Chemistry*, R.F. Nalewajski (ed.), Springer, Berlin (1996).
26. U. von Barth and L. Hedin: *J. Phys. C* **5**, 1629 (1972).
27. M. M. Pant and A. K. Rajagopal: *Sol. State Commun.* **10**, 1157 (1972).
28. A. Görling and M. Levy: *Phys. Rev. A* **50**, 196 (1994).
29. A. Görling and M. Levy: *Int. J. Quantum Chem. Symp.* **29**, 93 (1995).
30. R. T. Sharp and G. K. Horton: *Phys. Rev.* **90**, 317 (1953).
31. J. D. Talman and W. F. Shadwick: *Phys. Rev. A* **14**, 36 (1976).
32. T. Grabo and E. K. U. Gross: *Chem. Phys. Lett.* **240**, 141 (1995).
33. V. Sahni, J. Gruenebaum, and J. P. Perdew: *Phys. Rev. B* **26**, 4371 (1982).
34. J. P. Perdew and M. R. Norman: *Phys. Rev. B* **26**, 5445 (1982).
35. E. Engel and S. H. Vosko: *Phys. Rev. A* **47**, 2800 (1993).

36. T. Kotani: *Phys. Rev. B* **50**, 14816 (1994); ibid., **51**, 13903 (1995)(E).
37. T. Kotani: *Phys. Rev. Lett.* **74**, 2989 (1995).
38. T. Kotani and H. Akai: *Phys. Rev. B* **52**, 17153 (1995).
39. T. Grabo and E. K. U. Gross: *Int. J. Quantum Chem.* **64**, 95 (1997).
40. A. D. Becke: *Phys. Rev. A* **38**, 3098 (1988).
41. J. P. Perdew, K. Burke, and Y. Wang: *Phys. Rev. B* **54**, 16533 (1996).
42. E. K. U. Gross, M. Petersilka, and T. Grabo: in *Chemical Applications of Density Functional Theory*, ACS Symposium Series 629, B. B. Laird, R. B. Ross, and T. Ziegler (eds.), American Chemical Society, Washington, DC (1996).
43. R. Colle and D. Salvetti: *Theoret. Chim. Acta* **37**, 329 (1975).
44. R. Colle and D. Salvetti: *Theoret. Chim. Acta* **53**, 55 (1979).
45. J. A. Montgomery, J. W. Ochterski, and G. A. Petersson: *J. Chem. Phys.* **101**, 5900 (1994).
46. E. R. Davidson, S. A. Hagstrom, S. J. Chakravorty, V. M. Umar, and C. Froese Fischer: *Phys. Rev. A* **44**, 7071 (1991).
47. J. P. Perdew and Y. Wang: *Phys. Rev. B* **45**, 13244 (1992).
48. C. Lee, W. Yang, and R. G. Parr: *Phys. Rev. B* **37**, 785 (1988).
49. A. A. Radzig and B. M. Smirnov: *Reference Data on Atoms and Molecules*, Springer, Berlin (1985).
50. L. Laaksonen, P. Pyykkö, and D. Sundholm: *Comp. Phys. Reports* **4**, 313 (1986).
51. B. G. Johnson, P. M. W. Gill, and J. A. Pople: *J. Chem. Phys.* **98**, 5612 (1993).
52. K. P. Huber and G. Herzberg: *Molecular Spectra and Molecular Structure: IV. Constants of Diatomic Molecules*, Van Nostrand Reinhold, New York (1979).
53. W. Kolos and L. Wolniewicz: *J. Chem. Phys.* **49**, 404 (1968).
54. C. Filippi and C. J. Umrigar: *J. Chem. Phys.* **105**, 213 (1996).
55. S. J. Chakravorty, S. R. Gwaltney, E. R. Davidson, F. A. Parpia, and C. Froese Fischer: *Phys. Rev. A* **47**, 3649 (1993).
56. K. A. Peterson, R. A. Kendall, and T. H. Dunning: *J. Chem. Phys.* **99**, 1930 (1993).
57. O. V. Gritsenko, R. van Leeuven, and E. J. Baerends: *J. Chem. Phys.* **104**, 8535 (1996).

Molecular Engineering **7**: 51–65, 1997.

The Relativistic Quantum Defect Orbital Method and some of its Applications

IMMACULADA MARTIN
Departamento de Quimica Fisica, Facultad de Ciencias, 47005 Valladolid, Spain

(Received: 16 December 1996; in final form: 20 January 1997)

Abstract. Justification of a simple and reliable relativistic procedure for the prediction of a large body of transition probability data is made. The main features of the Relativistic Quantum Defect Orbital (RQDO) formalism are described, and several examples of the data yielded by this method are presented in tabular and graph forms.

Key words: Relativistic quantum defect orbital method, ionized systems, astrophysical plasmas, fusion plasmas, transition probabilities, spectral lines.

1. Introduction

The steady pace at which the theoretical determinations of atomic transition probabilities have proceeded during the last two decades has been in part motivated by the fast advances in the experimental techniques which, quite often in turn demand reliable theoretical data as a reference point. Also, the increasing interest in the composition and behaviour of astrophysical, laboratory and fusion plasmas has played an important role in this respect.

Spectral lines from different ionisation stages of various elements (either naturally occurring or deliberately added) in high-temperature Tokamak fusion plasmas may be used for localised diagnostics of such plasmas (Bascheck, 1984). Most of the relevant atomic systems in this field are heavy and/or in high degrees of ionisation, thus requiring the application of relativistic treatments. As regards the diagnostics of astrophysical plasmas, it is widely acknowledged that one must consider hundreds of thousands of atomic energy levels and millions of spectral lines (Seaton, 1994). A number of *ab initio* codes are available, and some of them have proved to yield quite accurate transition probability data (Hibbert, 1992). However, many of the adequate codes in this context are rather time-consuming and often plagued with convergence problems. It is widely recognised (see, e.g., Laughlin, 1992; 1995) that for this type of study the development of methods that combine reliability and simplicity is of a clear convenience. In addition, some simple semiempirical techniques have revealed regularities among members of isoelectronic and homologous sequences of atoms that are not so well predicted by *ab inito* techniques (Curtis, 1987). The validity of these regularities, which have been amply studied by Wiese and Weiss (Wiese and Weiss, 1969; Wiese, 1987) has been confirmed up to very heavy and highly ionised systems, and it offers the great advantage of allowing one to make interpolative or extrapolative predictions. While some complex *ab initio* theoretical methods have become very powerful, they cannot always replace semiempirical data parametrisations. One of the reasons (Curtis, 1987), in addition to the aforementioned regularities, is that an appropriate semiempirical method may contain clues to an improved theoretical understanding of the dynamics of complex atoms.

In the different semiempirical methods that are currently employed, which quite often derive from a modification of a hydrogen-like wave equation, the various interactions are given a different weighting. Deviations from the Coulomb potential, often classified as 'penetration' and 'polarisation' (depending on the degree of overlap between the active and passive electrons, in the context of electronic transitions) may, at least to a reasonable extent, be accounted for by a model Hamiltonian that contains a parameter related to the quantum defect. More specifically, the two above effects seem to be adequately described in the Quantum Defect Orbital method (Simons, 1974; Martin and Simons, 1975; 1976) which has been applied for years as a non-relativistic (QDO) approach, and recently reformulated in the context of the second-order Dirac equation (RQDO) (Martin and Karwowski, 1991; Karwowski and Martin, 1991), and amply used to calculate atomic transition probabilities in the last few years. Regularities along isoelectronic sequences have also been the subject of some of our studies. Some examples of these studies will be given in Section 3. A summary of the RQDO method will be presented in the next section.

2. The Method

Quantum defect theories originate from early work by Bates and Damgaard (1949) and by Ham (1955). Later, they were developed and applied to the study of atomic processes by the groups of Seaton (1966, 1970) and Fano (1970). The relativistic quantum defect method, derived from the four-component Dirac theory, was built up by Johnson and Cheng (1979). It was further generalised and applied, mainly to scattering processes, by Chang (1983), among others. In all these methods the Hamiltonian is asymptotically correct but its approximate eigenfunctions behave correctly only at large radial distances.

Another formulation of the quantum defect theory, known as the Quantum Defect Orbital (QDO) method, was proposed by Simons (1974) and Martin and Simons (1975, 1976). This approach provides exact eigenfunctions of a model Hamiltonian. The resulting orbitals are also valid in the core region retaining approximate core-valence orthogonality. More recently, the QDO method has been used by Kostelecký and Nieto (1985) under the name 'exact quantum defect theory'. A discussion of this subject may be found in Martin and Barrientos (1991) and Kostelecký and Nieto (1991).

The relativistic formulation of the QDO formalism, as proposed by Martin and Karwowski (1991) and Karwowski and Martin (1991) was based on the decoupling of the Dirac second-order equation, and the interpretation of the resulting solutions, that had been carried out by Karwowski and Kobus (1985, 1986). These authors generalised the decoupling procedure of Biedenharn (1962) in such a way that it may be applied to the case of a general atomic spherical potential. They also showed that both the large and small components of the Dirac wavefunction may be recovered without any computational effort.

The Dirac second-order equation for a particle in a stationary potential $V(r)$, after the elimination of spin and angular variables, may be expressed in the form

$$\left[\Omega\mathbf{I} + \frac{1}{r^2}\mathbf{M}\right]\begin{pmatrix} G_k \\ F_k \end{pmatrix} = 0 \tag{1}$$

where

$$\Omega = -\mathrm{d}^2/\mathrm{d}r^2 + k^2/r^2 + 2(V - E) - \alpha^2(V - E)^2 \tag{2}$$

$$\mathbf{M} = \begin{pmatrix} -k & -\alpha r^2\,\mathrm{d}V/\mathrm{d}r \\ \alpha r^2\,\mathrm{d}V/\mathrm{d}r & k \end{pmatrix} \tag{3}$$

$\mathbf{I}$ is the 2×2 unit matrix, E is the difference between the total and the rest energy of the electron, k is related to the total (j) and orbital (l) angular momentum quantum numbers according to $k = \varepsilon(j + 1/2)$, $j = l + \varepsilon/2$, $\varepsilon = \pm 1$. The equations are written in Hartree atomic units, and the other symbols have their usual meaning. In the particular case of a central-field potential that satisfies

$$r^2\,\mathrm{d}V/\mathrm{d}r = Z = \mathrm{const} \tag{4}$$

the elements of $\mathbf{M}$ are independent, and a non-unitary transformation described by the matrix

$$\mathbf{A} = a\begin{pmatrix} 1 & -\alpha Z/(k+s) \\ -\alpha Z/(k+s) & 1 \end{pmatrix} \tag{5}$$

where a is a constant and

$$s = k(1 - \alpha^2 Z^2/k^2)^{1/2} \tag{6}$$

decouples the equations (Barthelat *et al.*, 1980; Karwowski and Kobus, 1985, 1986; Martin and Glauber, 1958). Since

$$\mathbf{A}^{-1}\mathbf{M}\mathbf{A} = \begin{pmatrix} -s & 0 \\ 0 & s \end{pmatrix} \tag{7}$$

the decoupled system of equations reads

$$\left[-\frac{\mathrm{d}^2}{\mathrm{d}r^2}+\frac{s(s\pm 1)}{r^2}-\frac{2Z(1+\alpha^2 E)}{r}\right]\Phi_k^{\pm}=E(2+\alpha^2 E)\Phi_k^{\pm}. \tag{8}$$

Its solutions are related to the large and small components of the Dirac wavefunction in the form

$$\begin{pmatrix}\Phi_k^{+}\\ \Phi_k^{-}\end{pmatrix}=\mathbf{A}^{-1}\begin{pmatrix}G_k\\ F_k\end{pmatrix}. \tag{9}$$

The constant a in Equation (5) is determined from the normalisation conditions

$$\langle G_k \mid G_k\rangle+\langle F_k \mid F_k\rangle=\langle\Phi^- \mid \Phi^-\rangle+\langle\Phi^+ \mid \Phi^+\rangle=1 \tag{10}$$

and is equal to

$$a=\frac{\tilde{n}}{N|s|}\left[\frac{k(k+s)}{2}\right]^{1/2} \tag{11}$$

where

$$\tilde{n}=n-|k|+|s| \tag{12}$$

and

$$N=(n^2+\alpha^2 Z^2)^{1/2} \tag{13}$$

with n being the principal quantum number of the non-relativistic theory.

Let us note that Ω in Equations (1) and (2) does not depend upon the sign of k. (This is because the energy depends only on $|k|$). Therefore, two interpretations may be given to the solutions of Equation (8). First, they give us two components of the relativistic wavefuntion. Second, they give us two solutions (corresponding to k and to $-k$) of a single *scalar* equation for a *scalar* (quasirelativistic (QR)) wavefunction Ψ_k. We can set

$$\Phi_k^{\pm}=N_{\pm}\Psi_{\pm k} \tag{14}$$

with

$$|N_{\pm}|^2=\frac{k\tilde{n}\pm sN}{2k\tilde{n}} \tag{15}$$

chosen so that

$$\langle\Psi_k \mid \Psi_k\rangle=1. \tag{16}$$

Denoting now

$$Z' = Z(1 + \alpha^2 E) \tag{17}$$

and

$$e = E(1 + 1/2\alpha^2 E). \tag{18}$$

Equation (8) may be transformed into

$$\left[-\frac{\mathrm{d}^2}{\mathrm{d}r^2} + \frac{s(s-1)}{r^2} - \frac{2Z'}{r} \right] \Psi_k = 2e\Psi_k. \tag{19}$$

A QR theory is not only considerably simpler than the relativistic one but, first of all, it closely resembles the Schrödinger formulation. As a consequence, one may rather easily implement the QR formalism in the majority of the methods being developed for the non-relativistic theory. The last equation is formally very similar to the radial hydrogenic equation and passes into it in a trivial manner in the nonrelativistic limit of $\alpha \to 0$.

The quantum defect orbitals are solutions of the Schrödinger equation (Simons, 1974; Martin and Simons, 1975, 1976)

$$\left[-\frac{\mathrm{d}^2}{\mathrm{d}r^2} + \frac{\chi(\chi+1)}{r^2} - \frac{Z_{\mathrm{net}}}{r} \right] \Psi_k^{\mathrm{QD}} = 2E^{\mathrm{QD}}\Psi_k^{\mathrm{QD}} \tag{20}$$

where Z_{net} is the nuclear charge on the active electron at large r and

$$\chi = l - \delta + c \tag{21}$$

where δ is the quantum defect and c is an integer chosen to ensure the correct number of nodes and normalization of the radial wavefunction. The eigenvalue E^{QD} in Equation (20) depends only upon the non-integral part of λ and, hence, is independent of c. The quantum defect is empirically obtained from the following equation:

$$E^{\mathrm{QD}} = E^X = -\frac{Z_{\mathrm{net}}^2}{2(n-\delta)^2} \tag{22}$$

where E^X is the experimental energy. A comparison of Equations (20) and (19) demonstrates that the formal mathematical structures of the QDO theory and the scalar relativistic theory are the same. This formal similarity allowed us (Martin and Karwowski, 1991; Karwowski and Martin, 1991) to reinterpret the QDO theory so that it accounts for a major part of relativistic effects.

The relativistic quantum defect orbital (RQDO) equation is written as

$$\left[-\frac{d^2}{dr^2}+\frac{\Lambda(\Lambda+1)}{r^2}-\frac{2Z'_{net}}{r}\right]\Psi_k^{RD}=2e^{RD}\Psi_k^{RD} \tag{23}$$

where

$$\Lambda=\begin{cases} s-1-\delta'+c & \text{if } j=l+1/2 \\ -s-\delta'+c & \text{if } j=l-1/2 \end{cases} \tag{24}$$

and

$$Z'_{net}=Z_{net}(1+\alpha^2E^X). \tag{25}$$

Analogously to the non-relativistic case, the relativistic quantum defect is determined empirically. We have

$$e^{RD}=-\frac{Z_{net}^2}{2(\tilde{n}-\delta')^2} \tag{26}$$

and, taking into account Equations (18), (25) and (26), the value of δ' is obtained from

$$\frac{E^X(1+1/2\alpha^2E^X)}{(1+\alpha^2E^X)^2}=-\frac{Z_{net}^2}{2(\tilde{n}-\delta')^2}. \tag{27}$$

It should be stressed that this formulation is *exact* in the sense that it is equivalent to a four-component formulation based on the standard first-order form of the Dirac equation. All matrix elements, in particular the transition moments, may be expressed in a simple way, using the solutions of the second-order equation. A set of recurrent formulae which are fulfilled by the radial integrals (Kobus *et al.*, 1987) allows the formalism to be very simple and compact. In any case, this theory is quasirelativistic and any comparison with the multi-component relativistic Dirac–Fock formulation should be performed using electron densities rather than wavefunctions. A detailed comparison between the quasi-relativistic and relativistic theories (Barthelat *et al.*, 1980; Karwowski and Kobus, 1985, 1986) has revealed that the former allows one to obtain about 90% of the relativistic correction to energies, transition probabilities and expectation values of powers of r. From our analysis (Karvowski and Martin, 1991) we could see that the relativistic density distribution approximates very well the exact one at large values of r. At small distances the quality of the density deteriorates, as it happens when the nonrelativistic QDO densities are compared with the exact non-relativistic ones (Simons, 1974).

The most important difference between the RQDO and QDO equations is the explicit dependence of the former on the total angular momentum quantum number k. As a consequence, values of the relativistic quantum defects are determined from the fine-structure energies rather than from their centres of gravity. The corresponding relativistic quantum defect orbitals are different for each component of a multiplet and, if $c = 0$, they retain the nodal structure of the large components of the hydrogenic Dirac wavefunction.

In a recent paper Kwato Njock *et al.* (1994) present again a relativistic generalisation of the QDO method. This formulation is equivalent to the previous one (Martin and Karwowski, 1991; Karwowski and Martin, 1991) but it is, in our view, unnecessarily complicated: by using explicitly the four-component formalism the authors loose the formal simplicity of the original theory in which the Biedenharn (1962) representation is employed.

3. The Oscillator Strength and a few Numerical Examples

The dimensionless electric dipole oscillator strength, or f-value, for an nlj–$n'l'j'$ transition is given by the equation (Cowan, 1981)

$$f_{ij} = \frac{2}{3}\frac{l_m}{2j+1}\Delta E(D_{\rm lin})^2\langle nlj|Q(r)|n'l'j'\rangle^2 \tag{28}$$

where l_m is the largest orbital angular momentum quantum number of those involved in the transition, ΔE is the transition energy in Hartrees, and $Q(r)$ is the dipole-length transition operator in a.u. $D_{\rm lin}$ collects the numbers resulting from integrating over the angular part of the wavefunctions in the transition integral. Hence, the integral in the above equation contains the radial parts of the intial and final wavefunctions only.

Quite often, we employ two different forms of the transition operator. The standard dipole-length one: $Q(r) = r$, and a core-polarisation-corrected expression (core-valence correlation effects are included in the QDO and RQDO methods only in an implicit manner) given by (Bielińska-Wąz, 1992):

$$Q(r) = r\left\{1 - \frac{\alpha_c}{r^3}\left[1 - \exp\left(-\frac{r}{r_c}\right)\right]^3\right\} \tag{29}$$

where α_c and r_c are the core dipole polarisability and a cut-off radius, respectively. The last expression offers the advantage of explicitly accounting for core-valence polarisation effects during the electronic transition without any loss of analyticity in the calculation of the transition integral.

We shall next proceed to include a few examples of QDO/RQDO oscillator strengths calculated by us during the last few years, together with some comparative theoretical data that we find to be relevant and the few experimental data available, to our knowledge, in the literature. Some of our examples will be presented in tabular form whilst some others will consist of graphs intended to reflect the systematic trends in the oscillator strength along an isoelectronic sequence to which mention is made in the Introduction.

Table I. Oscillator strengths for $ns\,^2S \rightarrow np\,^2P$ ($n = 5, 6$; $n' = 5$–10) fine structure transitions in Ag^I. All values have been multiplied by 100.

Transition	QDO[a]	RQDO[a]	RQDO[b]	RMP[c]	RHF[d]	RHF+CP[e]	RHF+CP[f]	RHF[g]	Expt.
$5S \rightarrow 5P$									
1/2 → 1/2	22.713	22.402	19.400	32.9	40.3	34.0	21.40	32.9	19.6[h]
									21.5[i]
1/2 → 3/2	45.425	45.714	39.674	67.2	82.2	69.5	44.5	66.6	45.9[h]
									45.0[i]
Multiplet	68.138	68.116	59.074	100.1	122.5	103.5	66.0	99.500	66.5[h]
									58.0[j]
$5S \rightarrow 6P$									
1/2 → 1/2	1.289	1.170	0.865						
1/2 → 3/2	2.577	2.699	2.035						
$5S \rightarrow 7P$									
1/2 → 1/2	0.370	0.332	0.232						
1/2 → 3/2	0.740	0.779	0.560						
$5S \rightarrow 8P$									
1/2 → 1/2	0.160	0.143	0.098						
1/2 → 3/2	0.321	0.339	0.238						
$5S \rightarrow 9P$									
1/2 → 1/2	0.085	0.076	0.051						
1/2 → 3/2	0.169	0.179	0.124						
$5S \rightarrow 10P$									
1/2 → 1/2	0.051	0.045	0.030						
1/2 → 3/2	0.102	0.109	0.075						
$6S \rightarrow 6P$									
1/2 → 1/2	39.480	39.234	38.572						
1/2 → 3/2	78.960	79.156	77.865						
$6S \rightarrow 7P$									
1/2 → 1/2	1.737	1.545	1.448						
1/2 → 3/2	3.474	3.671	3.465						
$6S \rightarrow 8P$									
1/2 → 1/2	0.450	0.389	0.355						
1/2 → 3/2	0.900	0.963	0.886						
$6S \rightarrow 9P$									
1/2 → 1/2	0.188	0.162	0.144						
1/2 → 3/2	0.376	0.403	0.365						
$6S \rightarrow 10P$									
1/2 → 1/2	0.100	0.084	0.074						
1/2 → 3/2	0.201	0.218	0.196						

[a] Martin *et al.* (1995). [b] With explicit inclusion of core polarisation (Martin *et al.*, 1995). [c] Relativistic model potential (Migdalek and Baylis, 1979). [d] RHF, with experimental energies (Migdalek and Baylis, 1979). [e] RHF + core polarisation in the potential and transition operator (Migdalek and Baylis, 1979). [f] RHF + core polarisation in the potential and transition operation (Migdalek and Baylis, 1979). [g] Cheng and Kim (1979). [h] Moise (1966). [i] Lawrence *et al.* (1965). [j] Andersen *et al.* (1970).

Table II. Oscillator strengths for $ns\ ^2S \rightarrow n'p\ ^2P$ ($n = 5, 6$; $n' = 5$–11) fine structure transitions in Cd^{II}. All values have been multiplied by 100.

Transition	QDO[a]	RQDO[a]	RQDO[b]	RMP[c]	RHF[d]	RHF+CP[e]	RHF+CP[f]	RHF[g]	Expt.
$5S \rightarrow 5P$									
1/2 → 1/2	23.789	22.917	20.081	35.1	38.7	35.4	21.8	28.61	21.5[h]
									19.0[i]
									16.020[j]
1/2 → 3/2	47.578	48.333	42.509	73.9	81.5	74.7	46.8	58.00	46.0[h]
									40.0[i]
									40.5[j]
Multiplet	71.366	72.350	62.590	109.0	120.1	110.1	68.6	86.61	67.0[h]
									59.0[i]
									62.0[j]
$5S \rightarrow 6P$									
1/2 → 1/2	0.306	0.217	0.112						
1/2 → 3/2	0.613	0.713	0.425						
$5S \rightarrow 7P$									
1/2 → 1/2	0.065	0.054	0.023						
1/2 → 3/2	0.120	0.141	0.067						
$5S \rightarrow 8P$									
1/2 → 1/2	0.023	0.018	0.006						
1/2 → 3/2	0.046	0.053	0.022						
$5S \rightarrow 9P$									
1/2 → 1/2	0.008	0.007	0.002						
1/2 → 3/2	0.016	0.017	0.005						
$5S \rightarrow 10P$									
1/2 → 1/2	0.002	0.003	0.0001						
1/2 → 3/2	0.004	0.006	0.001						
$5S \rightarrow 11P$									
1/2 → 1/2	0.021	0.002	0.0002						
1/2 → 3/2	0.041	0.077	0.052						
$6S \rightarrow 6P$									
1/2 → 1/2	40.763	39.681	38.551						
1/2 → 3/2	81.527	82.551	80.252						
$6S \rightarrow 7P$									
1/2 → 1/2	0.407	0.343	0.274						
1/2 → 3/2	0.814	0.882	0.722						
$6S \rightarrow 8P$									
1/2 → 1/2	0.056	0.039	0.023						
1/2 → 3/2	0.112	0.132	0.090						
$6S \rightarrow 9P$									
1/2 → 1/2	0.009	0.006	0.002						
1/2 → 3/2	0.018	0.022	0.010						

[a] to [g]: See footnotes to Table I; [h] Andersen *et al.* (1976). [i] Andersen *et al.* (1973). [j] Baumann and Smith (1970).

Table III. Oscillator strengths for the $4s4p\ ^3P_1$–$4s4d\ ^3D_1$ transitions in zinc-like ions.

Ion	RQDO[a]	RQDO[b]	MCDF[c]	HFR[d]
Se^{V}	0.2237	0.2005		
Br^{VI}	0.2386	0.2117		
Kr^{VII}	0.2427	0.2095	0.2434	
Rb^{VIII}	0.2402	0.1980	0.2390	0.2611
Sr^{IX}	0.2435	0.2042	0.2336	0.2528
Y^{X}	0.2392	0.2011	0.2276	0.2448
Zr^{XI}	0.2342	0.1994	0.2214	0.2365
Nb^{XII}	0.2284	0.1953	0.2151	0.2285
Mo^{XIII}	0.2201	0.1891	0.2088	0.2207
Tc^{XIV}	0.2118	0.1853	0.2027	0.2132
Ru^{XV}	0.2033	0.1784	0.1968	0.2065
Rh^{XVI}	0.1951	0.1730	0.1911	0.1999
Pd^{XVII}	0.1872	0.1657	0.1856	0.1936
Ag^{XVIII}	0.1797	0.1600	0.1804	0.1874
Cd^{XIX}	0.1727	0.1549	0.1753	0.1819
In^{XX}	0.1661	0.1514	0.1706	0.1766
Sn^{XXI}	0.1601	0.1490	0.1660	0.1713
Sb^{XXII}	0.1542	0.1422	0.1617	0.1663
Te^{XXIII}	0.1484	0.1354	0.1576	0.1614
I^{XXIV}	0.1430	0.1359	0.1537	0.1566
Xe^{XXV}	0.1379	0.1302	0.1501	0.1517

[a] Martin *et al.* (1994). [b] With polarisation, Martin *et al.* (1994). [c] MCDF, Martin *et al.* (1994). [d] Biémont *et al.* (1989).

Beginning with one-valence, though complex, atomic systems we display in Tables I and II the fine-structure oscillator strengths (as well as a few multiplet values for comparative purposes) for several $ns\ ^2S$–$n'p\ ^2P$ transitions in neutral silver, Ag I, and its isoelectronic cation Mg II, respectively (Martin *et al.*, 1995). A few experimental data have also been included in the tables. These seem to prove the adequacy of our computational procedures, in particular as compared with rather more expensive theoretical methodologies (Migdalek and Baylis, 1979; Cheng and Kim, 1979). If the experimental values are to be trusted, both the QDO and RQDO f-values without explicit core-polarisation correction seem to be superior in quality to those of the relativistic Hartree–Fock (RHF) and a relativistic model potential (RMP) methods, and appear to be in good agreement with the theoretical data supplied by the most complex methodology of those quoted in Tables I and II, i.e., the fully core-polarisation-corrected (both in the potential and in the transition operator) RHF (RHF + CP) calculation by Migdalek and Baylis (1979). In these two elements the influence of the relativistic effects is not very strong

Table IV. A comparison of f-values for the $4s4p\ ^3P$–$4s4d\ ^3D$ multiplet transitions in zinc-like ions.

Ion	RQDO[a]	RQDO[b]	HFR[c]	MP[d]	MCHF[e]	Exp.
Se^{V}	0.899	0.806			1.069	0.69 ± 0.04[f]
						0.42 ± 0.05[g]
Br^{VI}	0.957	0.849			1.058	0.47 ± 0.03[h]
Rb^{VIII}	0.960	0.792	1.038	1.010		
Sr^{IX}	0.973	0.816	1.006	0.983		
Y^{X}	0.954	0.802	0.974	0.936		
Zr^{XI}	0.932	0.794	0.942	0.924		
Nb^{XII}	0.907	0.777	0.913	0.740		
Mo^{XIII}	0.873	0.751	0.884	0.880	0.854	
Tc^{XIV}	0.837	0.732	0.856			
Ru^{XV}	0.803	0.705	0.830			
Rh^{XVI}	0.769	0.688	0.807			
Ag^{XVIII}	0.705	0.628	0.763			
Cd^{XIX}	0.675	0.606	0.743			
In^{XX}	0.648	0.591	0.725			
Sn^{XXI}	0.623	0.580	0.708		0.662	
Sb^{XXII}	0.597	0.551	0.690			
Te^{XXIII}	0.573	0.523	0.673			
I^{XXIV}	0.550	0.519	0.656			
Xe^{XXV}	0.533	0.500	0.639		0.592	

[a] Martin *et al.* (1994). [b] With polarisation, Martin *et al.* (1994). [c] Biémont *et al.* (1989). [d] Victor and Taylor (1983). [e] Biémont and Godefroid (1980). [f] Bahr *et al.* (1982). [g] Sorensen (1982). [h] Pinnington *et al.* (1977).

(Z = 47, 48) although it is noticeable. The use of the core-polarisation corrected transition operator seems to bring the RQDO f-values to a better agreement with the experimental data in Cd^{II}, but in Ag^{I} it might be possible that the use of this operator leads to an overestimation of such effects.

When comparative data is scarce we sometimes perform parallel multiconfiguration Dirac-Fock (MCDF) calculations using the code of Dyall *et al.* (1989), although we sometimes encounter convergence problems. This is the case of a study of triplet-triplet transitions in zinc-like ions, which possess two valence electrons (Martin *et al.*, 1994). Some of the results are shown in Tables III and IV, where only the RQDO and MCDF results of our calculations have been included, together with those provided by Biémont and co-workers (1989) with a method (HFR) that, according to the authors, accounts for the most important relativistic effects as well as for valence correlation by including all the configurations within the $n = 4$ complex. In Table III great similarities are observed between the RQDO (without the explicit polarisation correction) and the HFR set of values. In both tables only multiplet f-values are displayed for comparison with data supplied by

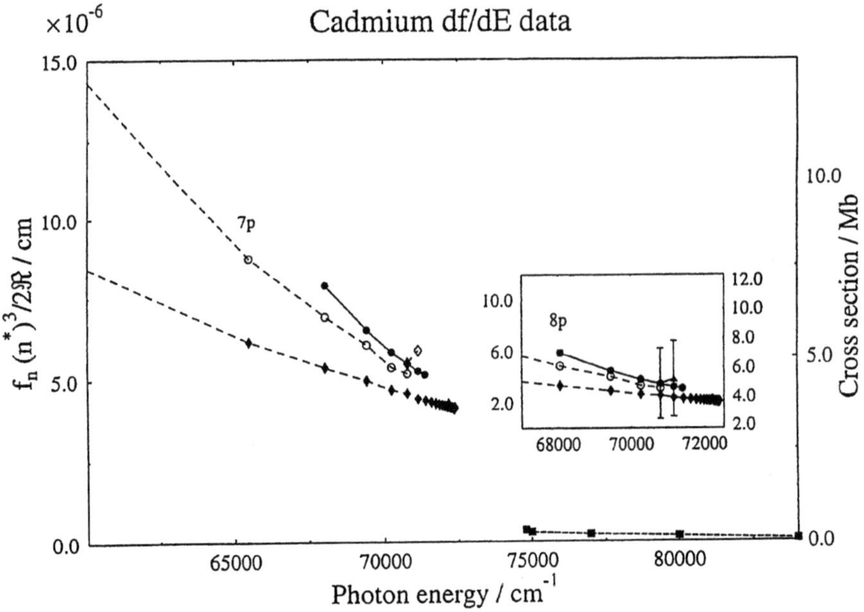

Figure 1. Survey of the relevant experimental and theoretical data of the oscillator strength spectral density of Cadmium. The inset shows calculations and measurements near the ionisation threshold, with error bars for the experimental data only. ●: Experiment (Ahmad *et al.*, 1996). ■: Experiment (Ross and Marr, 1965). ◊: Experiment (Schnitzer and Brehmenburg, 1988). ♦: RQDO (Ahmad *et al.*, 1996). ○: (Zilitis, 1971).

Biémont and co-workers (1989). In Table IV a good general agreement between all the theoretical data is apparent. The f-values of a model potential (MP) calculation by Victor and Taylor (1983), and those of a multiconfiguration Hartree–Fock (MCHF) procedure by Biémont and Godefroid (1980) have also been included. The experimental oscillator strengths are lower in magnitude than the theoretical data. This feature is often encountered when the measurements have been performed with a beam-foil technique from which cascading effects have not been properly corrected.

In Figure 1 we show the RQDO oscillator strength spectral density and photoionisation cross section corresponding to the principal spectral series of neutral cadmium, Cd^{I}, as recently compared with experimental and theoretical data from different sources (Almad *et al.*, 1996). Even though the scale is very narrow (figures are given in units of 10^{-6}) the similarity between the RQDO calculations and the accurate measurements, carried out with magneto-optical rotation spectroscopy in the VUV region, is apparent. From the RQDO results, as well as those of an earlier relativisitc model potential calculation by Zilitis (1971), it has been possible to assess the correctness of the present measurements (Ahmad *et al.*, 1996) versus that of some exisiting cross sections (Ross and Marr, 1965).

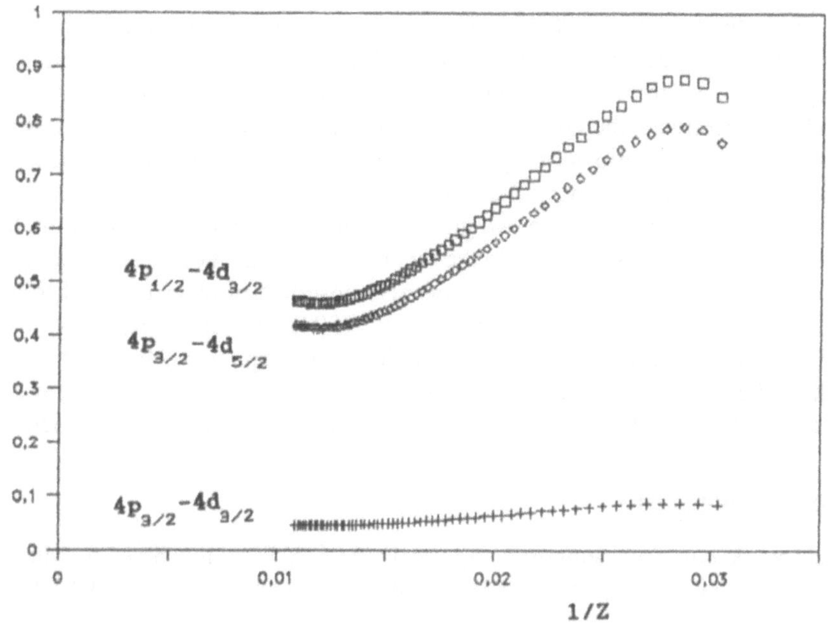

Figure 2. Systematic trends of the QDO oscillator strength for the $4p\,^2P$–$4d\,^2D$ fine-structure transitions in the copper isoelectronic sequence (Lavin *et al.*, 1994).

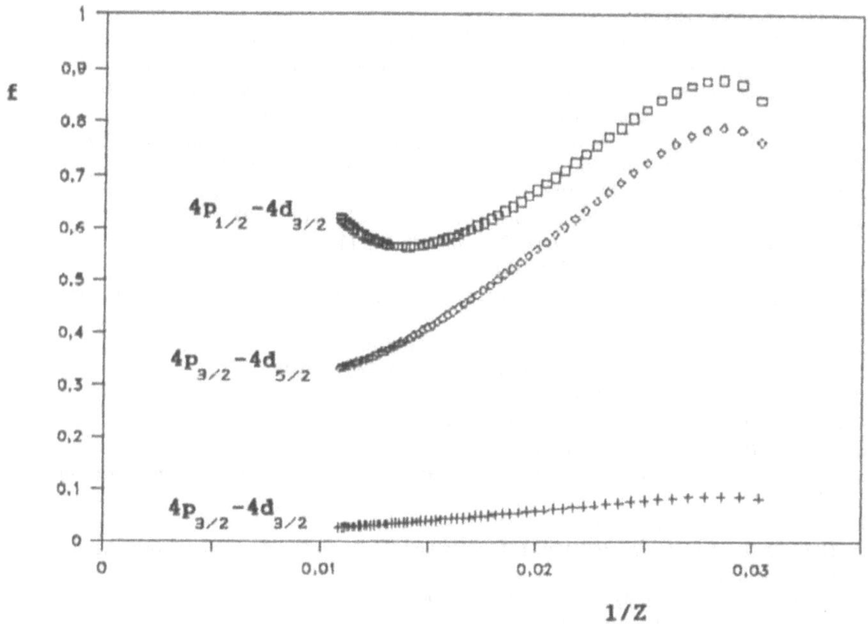

Figure 3. Systematic trends of the RQDO oscillator strength for the $4p\,^2P$–$4d\,^2D$ fine-structure transitions in the copper isoelectronic sequence (Lavin *et al.*, 1994).

As a final example of the adequacy of the RQDO method in this context, and of its clear account for a large part of the relativistic effects, when these are important, e.g., in heavy and highly ionised atoms, we have plotted in Figures 2 and 3, respectively, the QDO and RQDO oscillator strengths for the fine-structure $4p\ ^2P$–$4d\ ^2D$ transitions of a group of ions belonging to the copper isoelectronic sequence, from $Z = 33$ to $Z = 92$, versus the inverse of the atomic number (Lavin *et al.*, 1994). By comparing the two graphs it is apparent that the slope of the 1/2–3/2 and 3/2–5/2 f-values changes, in the relativistic case, at about $Z = 65$, unlike in the non-relativistic case. This feature was also observed in other types of transitions in the Cu isoelectronic series (Lavin *et al.*, 1994), as well as in a previous study of several transitions in the sodium isoelectronic sequence (Martin *et al.*, 1991), where the calculations were carried out to a very high atomic number, and where some of the RQDO curves were proved to agree with available experimental data.

Acknowledgements

This work has been supported by the D.G.I.C.Y.T. of the Spanish Ministry of Education within Projects No. PB91-0207-C02-01 and PB94-1314-C03-03.

References

I. Ahmad, U. Griesmann, I. Martin, W. Dussa, M.A. Baig and J. Hormes: *J. Phys. B.: At. Mol. Opt. Phys.* **29**, 2963 (1996).

T. Andersen, K.A. Jessen and G. Sorensen: *Nucl. Instrum. Methods* **90**, 35 (1970).

T. Andersen, A. Kirkegard, A. Nielsen and G. Sörensen: *Phys. Scr.* **6**, 122 (1972).

T. Andersen, O. Poulsen and P.S. Ramanujan: *J. Quant. Spectrosc. Radiat. Transfer* **16**, 521 (1976).

T. Andersen and G. Sörensen: *J. Quant. Spectrosc. Radiat. Transfer* **13**, 369 (1973).

J.L. Bahr, E.H. Pinnington, J.A. Kernahan and J.A. O'Neill: *Can. J. Phys.* **60**, 1108 (1982).

J.C. Barthelat, M. Pelisier and Ph. Durand: *Phys. Rev. A* **21**, 1773 (1980).

B. Bascheck: *Phys. Scr.* **T8**, 21 (1984).

D.R. Bates and A. Damgaard: *Phil. Trans. Roy. Soc. A* **242**, 101 (1949).

R.R. Baumann and W.H. Smith: *J. Opt. Soc. Am.* **60**, 345 (1970).

L.C. Biedenharn: *Phys. Rev.* **126**, 845 (1962).

D. Bielińska-Wąz: M.Sc. Thesis, Institute of Physics, University Nicholas Copernicus, Torun, Poland (1992).

E. Biémont and M. Godefroid: *Phys. Scr.* **22**, 231 (1980).

E. Biémont, P. Quinet and B.C. Fawcett: *Phys. Scr.* **39**, 562 (1989).

J.J. Chang: *Phys. Rev. A* **28**, 592 (1983).

J. Cheng and Y. Kim: *J. Opt. Soc. Am.* **69**, 125 (1979).

L.J. Curtis: *Phys. Scr.* **35**, 805 (1987).

K.G. Dyall, I.P. Grant, C.T. Johnson, F.A. Parpia and E.P. Plummer: *Comput. Phys. Commun.* **55**, 425 (1989).

U. Fano: *Phys. Rev. A* **2**, 353 (1970).

F.S. Ham: in *Solid State Physics*, Vol. 1, F. Seitz and D. Turnbull (eds.), New York: Academic Press, p. 127 (1955).

A. Hibbert: *Comments At. Mol. Phys.* **27**, 277 (1992).

W.R. Johnson and K.T. Cheng: *J. Phys. B: At. Mol. Phys.* **12**, 863 (1979).

J. Karwowski and J. Kobus: *Int. J. Quant. Chem.* **28**, 741 (1985).

J. Karwowski and J. Kobus: *Int. J. Quant. Chem.* **30**, 809 (1986).

J. Kobus, J. Karwowski, and W. Jaskólski: *J. Phys. A: Math. Gen.* **20**, 3346 (1987).

J. Karwowski and I. Martin: *Phys. Rev. A* **43**, 1539 (1991).
V.A. Kostelecký and M.M. Nieto: *Phys. Rev. A* **32**, 3243 (1985).
V.A. Kostelecký and M.M. Nieto: *Phys. Rev. A* **43**, 4063 (1991).
M.G. Kwato Njock, B. Oumrou, L.C. Owono, J.D. Kenmogne, S.G. Nana and M. Onana Boyomo: *Phys. Letters A* **184**, 352 (1994).
C. Laughlin: *Phys. Scr.* **45**, 238 (1992). *J. Phys. B: At. Mol. Opt. Phys.* **28**, 2787 (1995).
C. Lavin, C. Barrientos and I. Martin: *Int. J. Quant. Chem.* **50**, 411 (1994).
G.M. Lawrence, J.K. Link and R.B. King: *Astrophys. J.* **141**, 293 (1965).
I. Martin, M.A. Almaraz and C. Lavin: *Z. Phys. D* **35**, 239 (1995).
I. Martin and C. Barrientos: *Phys. Rev. A* **43**, 4061 (1991).
I. Martin and J. Karwowski: *J. Phys. B: At. Mol. Opt. Phys.* **24**, 1539 (1991).
I. Martin, J. Karwowski, C. Lavin and G.H.F. Diercksen: *Phys. Scr.* **44**, 567 (1991).
I. Martin and G. Simons: *J. Chem. Phys.* **62**, 4799 (1975).
I. Martin and G. Simons: *Mol. Phys.* **32**, 1017 (1976).
P. Martin, C. Lavin and I. Martin: *Z. Phys. D* **30**, 279 (1994).
P.C. Martin, and R.J. Glauber: *Phys. Rev.* **109**, 1307 (1958).
J. Migdalek and W.E. Baylis: *J. Quant. Spectrosc. Radiat. Transfer* **22**, 113 (1979).
N.L. Moise: *Astrophys. J.* **144**, 774 (1996).
E.H. Pinnington, J.A. Kerhahan and K.E. Donnelly: *J. Opt. Soc. Am.* **67**, 162 (1977).
K.J. Ross and G.V. Marr: *Proc. Phys. Soc.* **85**, 193 (1966).
A. Schitzer and W. Brehmemburg: *Z. Phys. D* **8**, 141 (1988).
M.J. Seaton: *Proc. Phys. Soc.* **88**, 801 (1966).
M.J. Seaton: *Comment. At. Mol. Phys.* **2**, 37 (1970).
M.J. Seaton: in *Advances in Atomic and Molecular Physics*, Vol. 32. D.E. Bates and B. Bederson (eds.), New York: Academic Press, p. 295 (1994).
G. Simons: *J. Chem. Phys.* **60**, 645 (1974).
G. Sörensen: *Phys. Rev. A* **7**, 85 (1973).
G.A. Victor and W.R. Taylor: *At. Data Nucl. Data Tables* **28**, 107 (1983).
W.L. Wiese: *Phys. Scr.* **35**, 846 (1987).
W.L. Wiese and A.W. Weiss: *Phys. Rev.* **175**, 50 (1968).
V.A. Zilitis: *Opt. Spectrosc.* **31**, 86 (1971).

Molecular Engineering **7**: 67–85, 1997.

A New Approach to Valence Bond Calculations: CASVB

THORSTEIN THORSTEINSSON* and DAVID L. COOPER
Department of Chemistry, University of Liverpool, P.O. Box 147, Liverpool L69 7ZD, U.K.

JOSEPH GERRATT
School of Chemistry, University of Bristol, Cantocks Close, Bristol BS8 1TS, U.K.

MARIO RAIMONDI
Dipartimento di Chimica Fisica ed Elettrochimica, Università di Milano, Via Golgi 19, 20133 Milano, Italy.

(Received and accepted: 17 June 1997)

Abstract. We present a new approach to the fully-variational optimization of general types of valence-bond wavefunction. Analogous techniques may be used to generate modern valence-bond representations of CASSCF structure spaces. These various procedures, which we term CASVB, are made possible by a highly efficient algorithm for performing exactly the transformations of full CI spaces that arise from general, non-unitary orbital transformations. As examples, we consider modern valence-bond representations of 'N in N' CASSCF descriptions of methane and diborane, in which the total wavefunction is dominated by covalent structures built from a common product of N *nonorthogonal* orbitals. These two cases highlight some of the possible differences that may arise between representations of CASSCF wavefunctions and the corresponding fully-variational valence bond results.

Key words: Valence-bond wavefunctions, non-unitary orbital transformations, CASVB.

1. Introduction

A widely recognised barrier in the development of valence bond methods is represented by the so-called nonorthogonality problem, which lies mainly in the evaluation of matrix elements involving many-electron functions built from nonorthogonal orbitals. Although recent advances have significantly reduced the computational effort relative to the $N!$ scaling appropriate to some of the earliest efforts, there remain serious difficulties, particularly for schemes that seek to optimize the orbitals. In exact second order methods which employ the full spin space, the very bad scaling with the number of active electrons is associated with the sheer number of elements of the fourth order density matrix, required for the evaluation of the Hessian [1], and with the rapid increase in the number of linearly independent spin eigenfunctions. The currently most efficient methods expand the valence bond

* *Current address:* Department of Chemistry, Chemistry Laboratory IV, Universitetsparken 5, 2100 Copenhagen ø, Denmark.

wavefunction in Slater determinants [1], but a further reduction in computational effort is likely if a basis of projected spin eigenfunctions is used [2].

Due to the severity of these problems, it is natural that many efforts have been directed towards incorporating at least some degree of orthogonality into the orbital set. This is the case, for example, in the widely used GVB approach that employs the so-called 'strong orthogonality' approximation, in which only intra-pair orbital overlaps are retained [3]. Another interesting general approach is to employ biorthogonal orbital sets, in which bra and ket orbitals are not identical [4–7]. These are constructed such that all density matrices coincide with the orthogonal case. In terms of quantitative results, however, biorthogonal methods generally fall somewhat short. This seems to be related mainly to the reduced accuracy of a wavefunction that is optimized using a non-symmetric energy expectation value.

An attractive alternative, which we consider in this work, abandons the direct evaluation of nonorthogonal matrix elements. Instead the valence bond wavefunction is rewritten in terms of orthogonal many-electron functions,

$$\Psi_{\mathrm{VB}} = \sum_I c_I \Phi_I^{\mathrm{VB}} \equiv \sum_I d_I \Phi_I^{\mathrm{MO}}, \tag{1}$$

such that expectation values may be evaluated by use of standard MO procedures [8–11]. We have in this way bypassed the nonorthogonality problem, and the computational effort of our procedure will now be determined instead by

1. the transformation from the $\{\Phi^{\mathrm{VB}}\}$ to the $\{\Phi^{\mathrm{MO}}\}$; and
2. the increased length of the expansion necessary for the $\{\Phi^{\mathrm{MO}}\}$.

The first problem is central and we outline in the next setion our approach to its very efficient solution. Addressing the second point, it is not difficult to realize that the space $\{\Phi^{\mathrm{MO}}\}$ must in general be of full CI type, if no restrictions are to be put on the form of the valence bond orbitals. This is the main reason for a strong link to the CASSCF method [12], which is the most widely used approach employing full CI expansions at present, and prompted us to denote our procedures 'CASVB'. The development of CASSCF methods has come very far and, due to the simplifying properties relating to full CI spaces, extremely long expansions may now be considered.

The structure of this paper is as follows. After reviewing the basis strategy behind the CASVB method in Section 2, in Section 3 we consider modern valence bond representations of CASSCF wavefunctions. Section 4 presents some results for methane and diborane. This is followed by a general discussion in Section 5.

2. The CASVB Method

Current implementations of CASSCF [13, 14] and modern valence bond approaches, such as spin-coupled theory [1, 15], are limited to around 12–14 active electrons if a similar number of active orbitals is used. As such, it is necessary to treat a set of 'core' electrons at a lower level of theory. The computationally simplest procedures, which are found to be satisfactory for most chemical applications, accommodate these in doubly occupied orbitals. We may partition the orbital space as follows:

- Frozen core
- Optimized core
- Active orbitals
- Virtual orbitals.

The frozen core orbitals are not optimized in the calculation, but are instead typically taken from a prior SCF calculation. They may be removed completely from the problem by means of simple modifications of the one-electron integrals [16]. The optimized core orbitals are also doubly occupied in all structures, but they are relaxed during the optimization procedure.

We are concerned with structures written in the form

$$\Phi_I = \mathcal{A}(\Phi^{\text{core}} \times \Phi_I^{\text{act}}), \tag{2}$$

in which Φ^{core} is simply based on a product of $\frac{1}{2}N_{\text{core}}$ doubly occupied orbitals. The Φ_I^{act} describe the N 'active electrons' by varying occupations of the m active orbitals $\{\phi\}$, and incorporates an N-electron spin eigenfunction. It is not in general useful to adapt these structures according to the molecular point group symmetry [11].

Although the number of valence bond expansion functions is likely to be quite small in practical calculations, we allow in the most general case both the $\{\Phi^{\text{VB}}\}$ and the $\{\Phi^{\text{MO}}\}$ to form complete 'N in m' CI spaces. We must therefore address the problem of finding the $N_{\text{CI}} \times N_{\text{CI}}$ transformation, $\mathbf{T}(\mathbf{O})$, of the complete structure space, that arises from a general $m \times m$ orbital transformation $\mathbf{O}$, i.e.

$$\{\phi'\} = \{\phi\}\mathbf{O} \Rightarrow \{\Phi'\} = \{\Phi\}\mathbf{T}(\mathbf{O}). \tag{3}$$

In this notation, the $\{\ \}$ denote row-vectors of the orbitals, ϕ, or many-electron functions, Φ. The effect of $\mathbf{T}(\mathbf{O})$ can be realized in a very efficient manner by writing $\mathbf{O}$ as a product of m^2 simple updates of the form

$$\mathbf{O}_{\mu\nu}(\lambda) : \phi_\nu \to \phi_\nu + \lambda\phi_\mu, \tag{4}$$

i.e.

$$\mathbf{O} = \mathbf{O}_{(\mu\nu)_1}(\lambda_1)\mathbf{O}_{(\mu\nu)_2}(\lambda_2)\cdots\mathbf{O}_{(\mu\nu)_{m^2}}(\lambda_{m^2}). \tag{5}$$

Demonstrating that this can always be achieved is analogous to the observation that *unitary* transformations can always be expressed as products of $\frac{1}{2}m(m-1)$ 2×2 orbital rotations. The advantage of Equation (5) in this context is that the transformation of the structure space corresponding to Equation (4) can be achieved very simply, by application to the structure space of the excitation operators

$$\mathbf{T}(\mathbf{O}_{\mu\nu}(\lambda)) = (\mathbf{1} + \lambda\mathbf{E}^{\alpha}_{\mu\nu})(\mathbf{1} + \lambda\mathbf{E}^{\beta}_{\mu\nu}) = \mathbf{1} + \lambda\mathbf{E}^{(1)}_{\mu\nu} + \lambda^2\mathbf{E}^{(2)}_{\mu\nu}. \tag{6}$$

Here $\mathbf{E}^{\alpha}_{\mu\nu}$ and $\mathbf{E}^{\beta}_{\mu\nu}$ describe excitations of the α and β spin orbital, respectively (appropriate if, for example, Slater determinants are employed), whereas $\mathbf{E}^{(1)}_{\mu\nu}$ and $\mathbf{E}^{(2)}_{\mu\nu}$ describe (spin averaged) single and double excitations. The total structure transformation may then be realized by successive applications of Equation (6), with the m^2 updates in Equations (4, 5), making use of the property of full CI spaces that

$$\mathbf{O} = \mathbf{O}_1\mathbf{O}_2 \Leftrightarrow \mathbf{T}(\mathbf{O}) = \mathbf{T}(\mathbf{O}_1)\mathbf{T}(\mathbf{O}_2). \tag{7}$$

This procedure is central to the viability of the CASVB approach, and so we have included a simple example as an Appendix.

With the transformation implicit in Equation (1) now achieved, we can consider the optimization of the valence bond wavefunction, by minimization of

$$E_{\mathrm{VB}} = \frac{\langle\Psi_{\mathrm{VB}}|\hat{H}|\Psi_{\mathrm{VB}}\rangle}{\langle\Psi_{\mathrm{VB}}|\Psi_{\mathrm{VB}}\rangle}. \tag{8}$$

Both the numerator and denominator in this expression are evaluated by transforming to the basis of orthogonal molecular orbitals, using the identities

$$\mathbf{H}' = \mathbf{T}^{\dagger}(\mathbf{O})\mathbf{H}\mathbf{T}(\mathbf{O}), \tag{9}$$

$$\mathbf{S}' = \mathbf{T}^{\dagger}(\mathbf{O})\mathbf{S}\mathbf{T}(\mathbf{O}), \tag{10}$$

and

$$\mathbf{c}' = \mathbf{T}^{-1}(\mathbf{O})\mathbf{c} = \mathbf{T}(\mathbf{O}^{-1})\mathbf{c}. \tag{11}$$

Here $\mathbf{H}$, $\mathbf{S}$ and $\mathbf{c}$ are the Hamiltonian matrix, overlap matrix and CI vectors appropriate to $\{\Phi\}$, whereas the primed quantities relate to $\{\Phi'\}$. Because the Ψ_{VB}, written in Equation (1) has the form of a CASSCF wavefunction, optimization of the expectation value with respect to the core-active, core-virtual and active-virtual orbital rotations can be done using standard CASSCF procedures. Optimization of the parameters defining the active space is in our procedure achieved using a stabilized Newton–Raphson procedure [17] requiring first and second derivatives. The evaluation of these can again be realized by transforming the first and second order changes in the wavefunction to the MO basis, and from these deriving the first and second order changes in Equation (8). For the first order change of a valence bond CI coefficient (c_I in Equation (1)), the CI vector, $\mathbf{c}$ in Equation

(11) is substituted by $\boldsymbol{\delta}^{(I)}$ defined by $(\boldsymbol{\delta}^{(I)})_J = \delta_{IJ}$. Similarly the first order CI vector corresponding to the simple orbital update given by Equations (4) can be constructed by applying

$$\left[\frac{\partial \mathbf{T}(\mathbf{O}_{\mu\nu}(\lambda))}{\partial \lambda}\right]_{\lambda=0} = \mathbf{E}_{\mu\nu}^{(1)} \tag{12}$$

to the total CI vector (cf Equation (6)). The second derivatives are constructed in an analogous manner, from combinations of first order changes [9, 10].

A fully-variational calculation is hence possible, by alternating optimization of the active and external parameter sets in a two-step procedure. Computationally, this approach has proved competitive with the most efficient of our alternative strategies [1, 18], and an important additional feature is the full utilization of any molecular point group symmetry [11].

3. Interpretation of CASSCF Wavefunctions

As described in the previous section, the CASVB approach can be used for the fully-variational optimization of quite general types of valence bond wavefunction. It is, however, indisputable that the CASSCF solution must be variationally the 'best' for a given number of active orbitals and electrons. As such, any valence bond wavefunction with these values of N and m can formally be derived from the CASSCF wavefunction by imposing the appropriate constraints. Rather than proceeding along such a path, whereby one obtains a less accurate wavefunction, it seems natural to seek instead a compact *representation* of the CASSCF wavefunction, dominated by a small number of valence bond structures. If this can be achieved, then we will have combined the more accurate observables related to the CASSCF wavefunction with the interpretational advantages associated with a valence bond formalism.

The philosophy in this approach is to utilize the invariance of CASSCF wavefunctions under linear transformations of the active orbitals. The CASSCF wavefunction is of course not invariant with respect to transformations of the external parameters (core-active, core-virtual and active-virtual orbital rotations), and so these are kept constant during the optimization. We have considered two basic criteria for defining useful valence bond representations. In the first, we maximize the overlap with the CASSCF wavefunction:

$$S_{\mathrm{VB}} = \frac{\langle \Psi_{\mathrm{CAS}} | \Psi_{\mathrm{VB}} \rangle}{\langle \Psi_{\mathrm{VB}} | \Psi_{\mathrm{VB}} \rangle^{1/2}}. \tag{13}$$

In the second, we minimize the expectation value for the energy, E_{VB} (Equation 8); straightforward minimization will of course be appropriate only for the lowest state of a given symmetry. The quantities (S_{VB} or E_{VB}) are optimized with respect to the orbital parameters that define the transformation $\mathbf{O}$ from $\{\phi^{\mathrm{CAS}}\}$ to $\{\phi^{\mathrm{VB}}\}$.

We may choose to treat the structure coefficients in Equation (1) as further free parameters. Alternatively, they can be derived from the (transformed) CI vector according to

$$\mathbf{c}_{\mathrm{VB}} = \mathbf{P}_{\mathrm{VB}}\mathbf{T}(\mathbf{O}^{-1})\mathbf{c}_{\mathrm{CAS}}, \tag{14}$$

in which $\mathbf{P}_{\mathrm{VB}}$ sets to zero the unwanted structure coefficients.

These considerations lead to four distinct optimization criteria that we investigate in the present work:

CASVB1 Maximize S_{VB} with respect to $\mathbf{O}$ and structure coefficients;

CASVB2 Maximize S_{VB} with respect to $\mathbf{O}$, extracting $\mathbf{c}_{\mathrm{VB}}$ from the transformed CASSCF CI vector as in Equation (14);

CASVB3 Minimize E_{VB} with respect to $\mathbf{O}$ and structure coefficients;

CASVB4 Minimize E_{VB} with respect to $\mathbf{O}$, extracting $\mathbf{c}_{\mathrm{VB}}$ from the transformed CASSCF CI vector as in Equation (14).

A number of applications have now been studied, most of which have shown extraordinarily close agreement between analogous CASSCF interpretations and valence bond results [8–11].

4. Results

All of the CASSCF [13, 14] and CASVB [8–11] calculations described in the present work were carried out using the MOLPRO [19] suite of programs. The spin-coupled calculations were performed using our own code [1]. As an illustration of our procedure, we have chosen two systems, methane and diborane, for which the CASSCF interpretations can differ markedly from the fully-variational spin-coupled descriptions.

4.1. METHANE

In the calculations on CH_4 (T_d symmetry, $r_{\mathrm{CH}} = 2.065$ bohr) correlation-consistent pVTZ basis sets due to Dunning [20] were used for C/H, consisting of $(10s5p2d/5s2p)$ Cartesian Gaussians contracted to $[4s3p2d/3s2p]$. An '8 in 8' CASSCF calculation was performed for the valence electrons, keeping two electrons in the optimized core ($\approx 1s^2$). The correlation energy thus retrieved amounts to 83.0 millihartree (see Table I).

We then carried out the analogous spin-coupled calculation. The two core electrons were described by the core orbital from the CASSCF calculation without further relaxation, in order to facilitate comparison between the two approaches. The spin-coupled solution for methane is well established [21, 22], and has many similarities with Pauling's original description [23] in terms of sp^3 hybrids on carbon overlapping with $1s$ functions on hydrogen to form four equivalent

Table I. Energies for methane. Further details of the various calculations are given in the text.

Calculation	E/hartree	$(E - E_{CAS})$/millihartree
SCF	-40.212541	83.00
SC^{CAS}	-40.275607	19.93
SC	-40.278158	17.38
SC+CI	-40.291257	4.28
CASSCF	-40.295536	(0)

Table II. Symmetry-unique orbital overlaps for methane.

Calculation	$\langle\phi_1\vert\phi_2\rangle$	$\langle\phi_1\vert\phi_3\rangle$	$\langle\phi_1\vert\phi_4\rangle$	$\langle\phi_2\vert\phi_4\rangle$
SC	0.692	0.529	0.096	−0.152
CASVB1	0.201	0.916	−0.062	−0.322
CASVB2	0.783	0.323	0.059	−0.109
CASVB3	0.272	0.883	−0.038	−0.316
CASVB4	0.752	0.370	0.058	−0.131

covalent bonds. The main differences from this classic picture include the small deformations of the orbitals in the bond-forming directions and the substantial overlaps between the different sp^x-like hybrids (0.53 – see Table II). It should be emphasized that the spin-coupled calculation converges to a situation with four symmetry-related pairs of orbitals without the incorporation of any constraints. The orbitals forming one of the four equivalent C—H bonds are shown in Figure 1. The overlap between these two orbitals is 0.69 (see Table II) and the corresponding electron spins are essentially singlet coupled. As such, the perfect pairing function strongly dominates the total spin function, with a weight of 90.0% in the Kotani basis (see Table III).

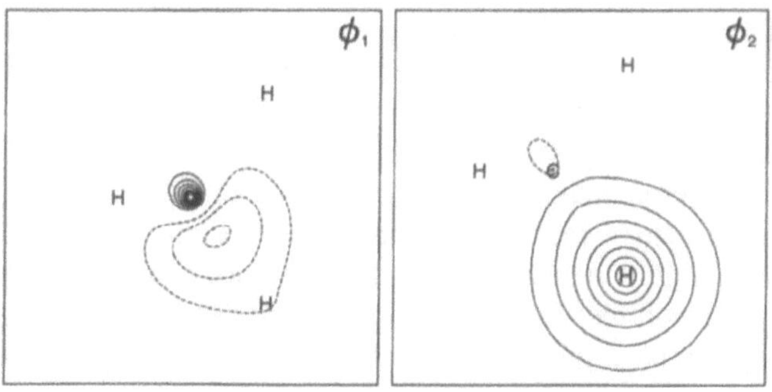

Figure 1. Symmetry unique spin-coupled orbitals for methane.

Table III. Weights of the perfect pairing spin function for the various calculations on methane.

Calculation	w_{PP}
SC	90.00%
CASVB1	57.82%
CASVB2	96.79%
CASVB3	64.85%
CASVB4	96.99%

The spin-coupled wavefunction retrieves 79.1% of the correlation energy appropriate to the CASSCF solution (see Table I). This difference can be attributed to two opposing factors: the exclusion of 'ionic' structure (i.e. those with doubly occupied orbitals) and the variation in the active orbital spaces. The former is the major contributor to the energy difference, as can be seen, for example, by augmenting the spin-coupled wavefunction with all possible ionic structures (SC+CI in Table I). This recovers 13 of the 17 millihartree separating the two wavefunctions.

The spin-coupled orbitals were expanded in the full space of the CASSCF active and virtual orbitals, and so the difference in orbital spaces can be gauged by examining the various components of each orbital. 0.35% and 0.56% of orbitals ϕ_1 and ϕ_2, respectively, is expanded in the CASSCF virtual space. These numbers are significantly larger than for other systems we have studied [8-11]. The remaining 4 millihartree separating SC+CI and CASSCF can be attributed to this difference. Alternatively, one may perform a restricted spin-coupled calculation, expanding the orbitals in only the active CASSCF space (SC^{CAS}). The difference between analogous wavefunctions based on different orbital spaces (SC^{CAS} and SC) of 3 millihartree is significantly greater than that found for other systems.

The different CASVB criteria for generating valence bond representations were applied to the CASSCF wavefunction describe above. The various symmetry-unique CASVB orbitals are plotted in Figure 2. CASVB2 and CASVB4 clearly produce orbital descriptions that are in harmony with the fully-variational spin-coupled orbitals. The CASVB carbon hybrids are rather more delocalized and this is reflected in an increased overlap between bond-forming orbitals (see Table II). This may be rationalized by the higher degree of singlet coupling of the spins associated with the participating electrons, as evidenced by the significantly increased weight of the perfect pairing spin function (see Table III).

The description generated by the CASVB1 and CASVB3 criteria is quite different. Both of these give highly localized carbon hybrids with only relatively small overlaps with the 1s-type orbitals on hydrogen (0.20 and 0.27, respectively – see Table II. As one might expect, the perfect pairing weights also decrease; indeed the spin functions correspond, to a good approximation, to quintet coupling of the spins associated with the carbon orbitals.

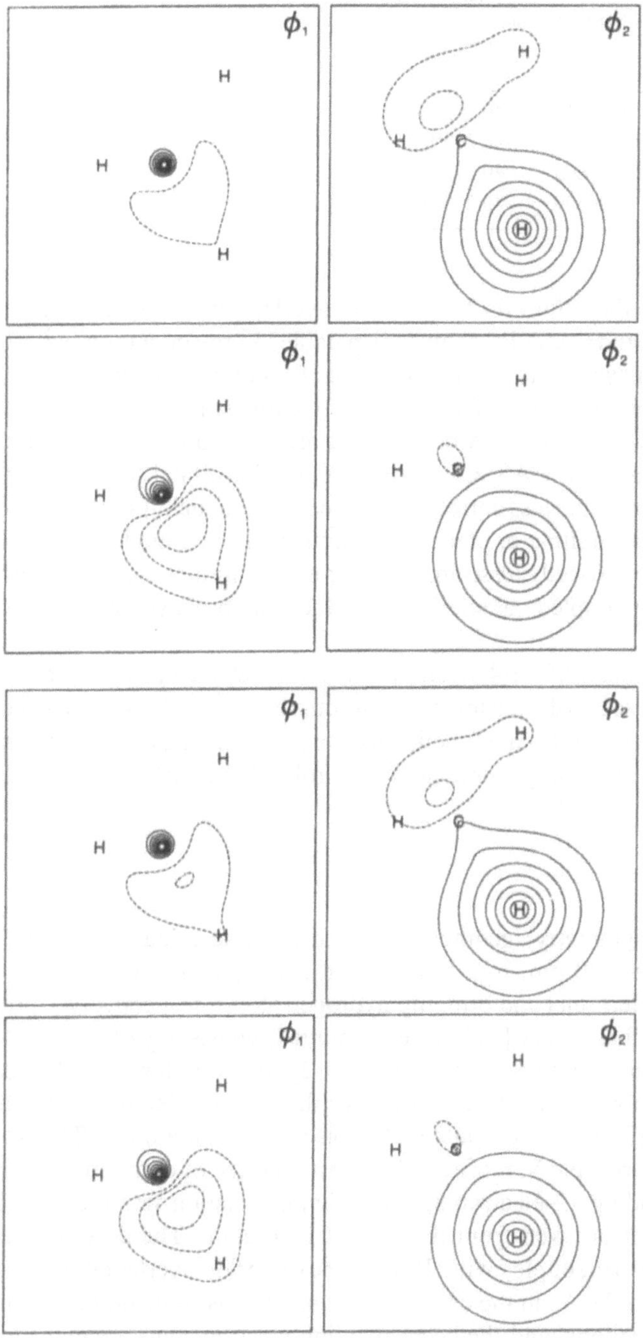

Figure 2. Symmetry-unique CASVB orbitals for methane. In order, the four rows correspond to CASVB 1-CASVB4.

Table IV. Values of S_{cov} and E_{cov} for the various calculations on methane.

Calculation	S_{cov}	E_{cov}
CASVB1	0.99613	−40.275304
CASVB2	0.99517	−40.272922
CASVB3	0.99605	−40.275607
CASVB4	0.99509	−40.273219

It is clear from the values for the covalent overlaps and energies listed in Table IV that CASVB 1 and CASVB3 show indisputable improvements on both fronts. It is therefore difficult simply to dismiss this type of solution as an 'accidental' feature of the optimization procedure. The most likely explanation for this phenomenon is that significantly contracted carbon orbitals are made possible by the increased electron correlation in the CASSCF wavefunction. This orbital set, however, seems to be inappropriate for the less correlated spin-coupled model. As such, no evidence for analogous local minima or stationary points could be found for the *full* spin-coupled treatment. Nonetheless, analogous rogue solutions have been obtained previously in spin-coupled calculations in which the basis set size was severely restricted.

The CASVB2 and CASVB4 solutions are more realistic in nature than CASVB 1 and CASVB3. It seems that basing the spin coupling on the CASSCF CI vector has the effect of restricting the wavefunction to more realistic regions of the parameter space, mainly by increasing significantly the role of the perfect pairing mode of spin coupling.

4.2. DIBORANE

The B_2H_6 (D_{2h} symmetry) geometry adopted in these calculations was: r_{BB} = 176.65 pm, r_{BH^t} = 120 pm, r_{BB^t} = 132.6 pm and $\angle$(B—B—H^t) = 121°, where H^t and H^b denote terminal and bridging hydrogen atoms, respectively. Correlation consistent pVDZ basis sets [20] were employed, consisting of $(9s4p1d/4s1p)$ Cartesian Gaussian functions contracted to $[3s2p1d/2s1p]$ for B/H. A '12 in 12' CASSCF calculation was then carried out for the valence electrons, keeping four electrons in optimized core orbitals ($\approx 1s^2$ on both boron atoms). This solution, labelled 'CAS A' in Table V, retrieves 90.32 millihartree of correlation energy.

The corresponding 12-electron spin-coupled calculation was then performed, with the core orbitals taken from the CASSCF solution. The description of the B—H^t bonds is very simple, with B(sp^x-type) hybrids overlapping H($1s$-type) orbitals (see Figure 3). As in the case of methane there is some deformation in the bond-forming directions, and the corresponding electron spins are close to being singlet coupled. It is clear that there are three important participating components for each of the bridging 3-centre 2-electron bonds. These are sp^x-type hybrids on

Table V. Energies for diborane. Details of the various calculations are given in the text.

Calculation	E/hartree	$(E - E_{CAS})$/millihartree
SCF	−52.815220	90.32
SC	−52.893038	12.51
CAS B	−52.898012	7.53
CAS A	−52.905543	(0)

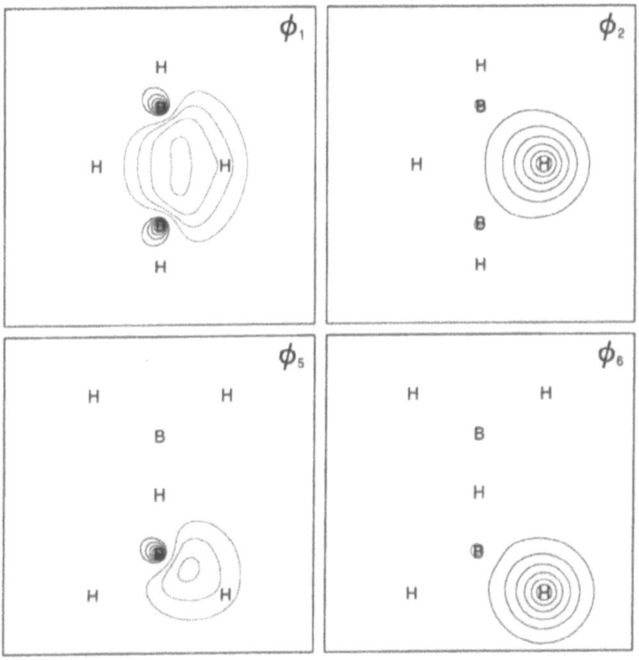

Figure 3. Symmetry-unique spin-coupled orbitals for diborane.

each boron atom and the $1s$ orbital on a bridging hydrogen. To form the two spin-coupled orbitals, two B(sp^x-type) hybrids 'fuse', leaving the $1s$ function on H^b relatively unchanged (see Figure 3). Again we find predominantly singlet coupling and the overall weight of the perfect pairing spin function is 95.6% in the Kotani basis. These findings are consistent with earlier results, including SOPP-GVB [24] and simpler spin-coupled treatments [25].

The bridging orbitals generated by CASVB interpretations of the 'CAS A' solution are shown in Figure 4. This description, although radically different, can be viewed as being formed from the same components as before. In this case, however, the orbital picture is closer to

$$\phi_1 \approx B_1(sp^x) + H^b(1s) \quad \text{and} \quad \phi_2 \approx B_2(sp^x). \tag{15}$$

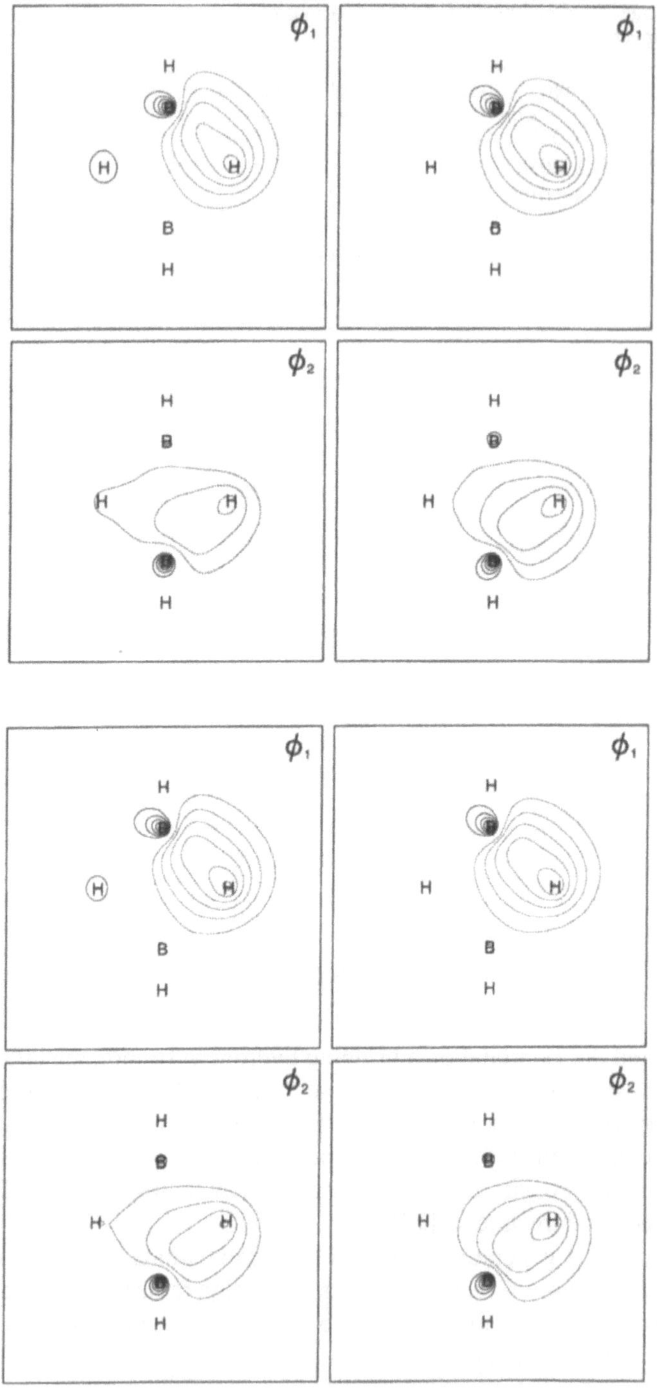

Figure 4. Symmetry-unique CASVB bridging orbitals for diborane based on 'CAS A'. In order, the four columns correspond to CASVB 1-CASVB4.

Of course, one may envisage any intermediate between these two extremes. The symmetry breaking inherent in this picture is best viewed as a consequence of forcing a single-configuration description on the bridging region. As such, symmetry adaptation of the wavefunction should be carried out by employing a multi-configuration wavefunction, for example by use of the projection operator

$$\hat{P}^{(i)} = (d_i/g) \sum_{R} \chi^{(i)}(R)^* \hat{R}. \quad (16)$$

This form is appropriate for a d_i-dimensional irreducible representation, $\Gamma^{(i)}$ in a group of order g, with elements $\hat{R}$, and characters $\chi^{(i)}(R)$. Further details regarding the symmetry-adaptation of this and other systems may be found elsewhere [11].

These CASVB interpretations could be useful to describe the process $2BH_3 \rightarrow B_2H_6$ [26]. They could also be combined with the more conventional spin-coupled solution to form an improved *multiconfigurational* description of the bridging region.

The discrepancy between the two pictures seems to be dictated by symmetry considerations. For 'CAS A', and its various CASVB interpretations, the bridging orbitals transform as $A_g + B_{2g} + B_{1u} + B_{3u}$ whereas those from the spin-coupled calculation transform as $2A_g + 2B_{3u}$. The optimal CASSCF solution is therefore incompatible with the traditional description of the 3-centre 2-electron bond. It proves straightforward to construct a secondary CASSCF solution with 'correct' symmetry properties. This 'CAS B' solution lies 7.5 millihartree higher than 'CAS A' (see Table V). Its CASVB representations, shown in Figure 5, are indeed in very good agreement with the spin-coupled results. Furthermore, it is also clear that the values for S_{VB} and E_{VB} (see Table VI) are significantly improved in this case. Clearly, 'CAS B' is better suited to interpretations in terms of a single covalent configuration than is 'CAS A'. Analogous observations apply to '4 in 4' treatments of the bridging region, although the CASSCF solutions are somewhat closer in energy (−52.840084 hartree and −52.838508 hartree, for A and B, respectively).

5. Discussion

We have outlined a new approach, CASVB, for variational optimization of general types of valence bond wavefunction. In terms of efficiency, the procedure competes favourably with the most efficient of the traditional strategies that involve the direct evaluation of nonorthogonal matrix elements. This is particularly true when molecular point group symmetry can be utilized. Furthermore, since the valence bond wavefunction in the CASVB approach can be viewed as a constrained form of CASSCF wavefunction, a multitude of techniques immediately becomes available for valence bond methods. Examples of such are: energy gradients, geometry optimization, inclusion of dynamical correlation, and 'direct' on-the-fly evaluation of integrals.

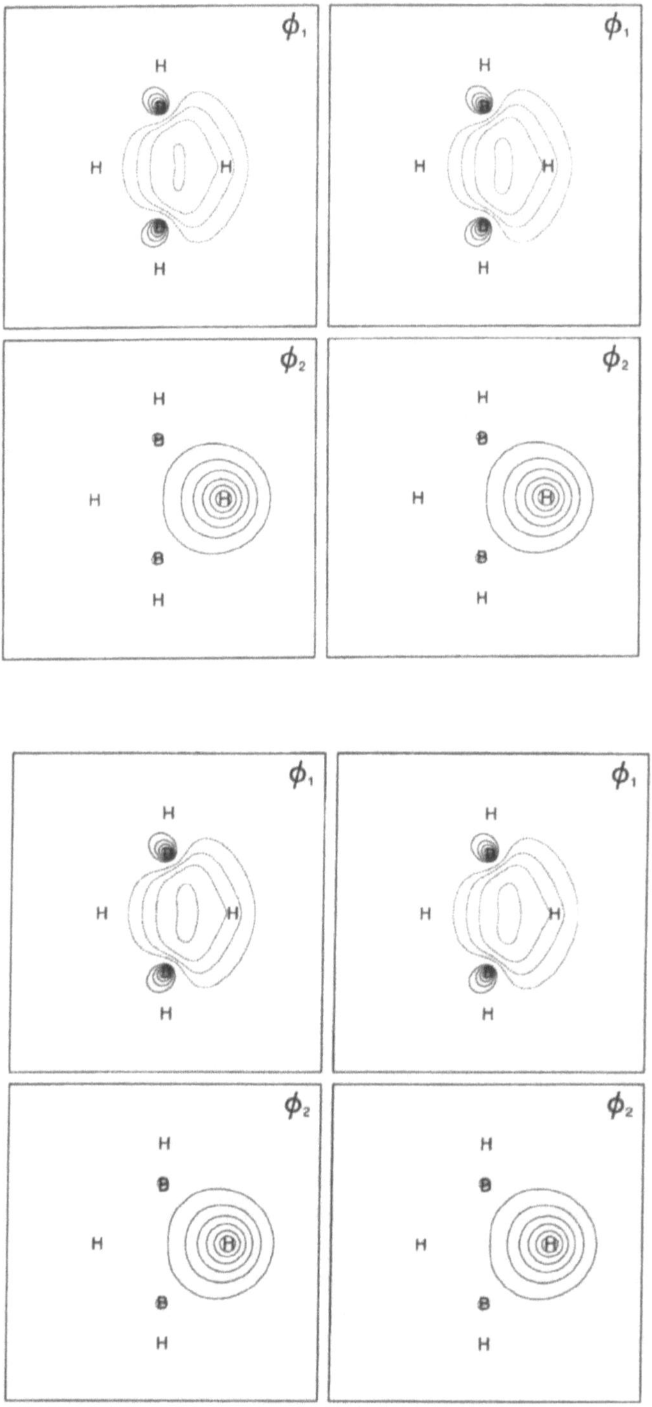

Figure 5. Symmetry-unique CASVB bridging orbitals for diborane based on 'CAS B'. In order, the four columns correspond to CASVB 1-CASVB4.

Table VI. Values of S_{VB} and E_{VB} for diborane. The various orbital sets are based on CASSCF solutions 'A' and 'B'.

Calculation	S_{VB}	E_{VB}
CASVB1 (A)	0.99571	−52.889322
CASVB2 (A)	0.99552	−52.889201
CASVB3 (A)	0.99557	−52.889782
CASVB4 (A)	0.99540	−52.889496
CASVB1 (B)	0.99862	−52.892407
CASVB2 (B)	0.99861	−52.892413
CASVB3 (B)	0.99861	−52.892450
CASVB4 (B)	0.99860	−52.892435

We have emphasized in this work an alternative use of the CASVB method, namely the interpretation of CASSCF wavefunctions in terms of valence bond concepts, exploiting the invariance of full CI expansions under linear transformations of the active orbitals. We have investigated overlap- and energy-based criteria for generating representations of CASSCF solutions in which the wavefunction is dominated by a very restricted number of valence bond structures. For the particular case where the spin-coupled wavefunction defines a representation of the 'N in N' CASSCF structure space, we find these criteria to be highly successful, retrieving on the order of 99% of the total wavefunction. A further finding is the very good qualitative agreement between analogous overlap-based and energy-based criteria. The former are, of course, computationally much cheaper and easier to implement.

The applications presented here were chosen so as to highlight some of the possible differences that may occur between CASSCF representations and fully-variational calculations. In general, these can be associated with:

1. the inclusion of further (ionic) structures, and
2. variations in the defining orbital spaces.

The former effect is the most important for the examples presented here. In order to quantify this, it is useful to consider the number of variational free parameters necessary for the definition of the active part of the wavefunction in the two cases. Since the active-active orbital rotations are redundant for the CASSCF wavefunction, the number of parameters is simply $N_{CAS} - 1$, where N_{CAS} is the length of the full CI expansion. This is given by the Weyl formula [27] as:

$$N_{CAS}(N, m, S) = \frac{2S+1}{m+1}\binom{m+1}{\frac{1}{2}N - S}\binom{m+1}{\frac{1}{2}N + S + 1}, \tag{17}$$

where m is the number of active orbitals. There are generally no redundant orbital parameters for the spin-coupled wavefunction and the number of structure para-

Table VII. Numbers of variational free parameters (excluding orbital rotations involving core or virtual orbitals) for 'N in N' CASSCF and spin-coupled wavefunctions.

N	S	CAS	SC
2	0	2	2
2	1	0	0
3	1/2	7	7
4	0	19	13
4	1	14	14
5	1/2	74	34
6	0	174	34
6	1	188	38
7	1/2	783	55
8	0	1763	69
8	1	2352	83
9	1/2	8819	113
10	0	19403	131
10	1	29699	179
11	1/2	104543	241
12	0	226511	263
12	1	382238	428
13	1/2	1288286	584
14	0	2760614	610
14	1	5010004	1182

meters is simply $f_S^N - 1$, where f_S^N signifies the number of linearly independent N-electron spin eigenfunctions for total spin S. We have

$$f_S^N = \binom{N}{\frac{1}{2}N - S} - \binom{N}{\frac{1}{2}N - S - 1}, \tag{18}$$

and the number of active spin-coupled parameters then becomes

$$N_{\text{parm}}^{\text{SC}}(N, S) = \min[N(N-1) + f_S^N, N_{\text{CAS}}(N, N, S)] - 1. \tag{19}$$

As an illustration, we compare in Table VIII the number of active parameters for analogous spin-coupled and CASSCF wavefunctions. The numbers coincide for very few electrons: the two methods are equivalent in these cases. The disparity between the two methods increases dramatically with increasing N and one should not expect a spin-coupled wavefunction to perform so well in this limit. The fact that the 12-electron spin-coupled calculation on diborane recovers about 80% of the CASSCF correlation energy, and the SC representation 99% of the CASSCF wavefunction, must therefore be said to be more than satisfactory.

The variation of orbital spaces is of course brought on by this difference in the active part of the wavefunction and so it is perhaps not surprising that this is only a minor effect in energy terms. However, this small variation in the orbital space controls whether or not the modern valence bond representations resemble the fully-variational results. It is useful to distinguish variations in the symmetries of the MOs from other variations in the orbital spaces. In the case of diborane, for example, it seems that the symmetry of natural orbitals with small occupation numbers determines the possibility of reproducing the fully-variational result. A related situation was found in a π-electron treatment of ozone [8], for which two rival CASSCF solutions correspond to the well-known hypervalent and diradical valence bond descriptions of the bonding. For methane, however, the symmetry of the CASSCF active MOs does *not* rule out a valence bond representation similar to the spin-coupled picture. It seems that more subtle correlation effects become important in this case, perhaps related to an exaggerated contraction of the orbitals associated with the carbon atom. In order to extract a spin-coupled-like picture it is necessary to extract the structure coefficients from the transformed CI vector. For all of these reasons, it therefore remains very important, to consider also fully-variational descriptions.

Appendix: A Simple Example of the Full CI Structure Transformation

We consider here a specific example of the transformation of full CI spaces [10, 28], taking the very simple case of a '2 in 2' expansion with $S = 0$. This could be appropriate, for example, for a description of H_2, and as such it is related to the early foundations of valence bond theory as set out by Heitler and London [29], and by Coulson and Fischer [30]. We can write down four distinct Slater determinants in this case:

$$\Phi_1 = |\overset{+}{\phi}_1\overset{-}{\phi}_1|, \quad \Phi_2 = |\overset{+}{\phi}_2\overset{-}{\phi}_1|, \quad \Phi_3 = |\overset{+}{\phi}_1\overset{-}{\phi}_2|, \quad \Phi_4 = |\overset{+}{\phi}_2\overset{-}{\phi}_2|, \tag{20}$$

where we have used the shorthand notation $\overset{+}{\phi}$ and $\overset{-}{\phi}$ for orbitals associated with α and β spin, respectively. We now consider the transformation to the orbital set $\{\phi'\}$, defined by

$$\phi'_\mu = \sum_\nu o_{\nu\mu}\phi_\nu \tag{21}$$

(cf Equation (3)). The corresponding structures could in principle be found by inserting this identity into the definition of the structures, for example:

$$\begin{aligned} \Phi'_3 &= |\overset{+}{\phi'_1}\overset{-}{\phi'_2}| = \left|\left(\sum_\nu o_{\nu 1}\overset{+}{\phi_\nu}\right)\left(\sum_\nu o_{\nu 2}\overset{-}{\phi_\nu}\right)\right| \\ &= o_{11}o_{12}\Phi_1 + o_{21}o_{12}\Phi_2 + o_{11}o_{22}\Phi_3 + o_{21}o_{22}\Phi_4. \end{aligned} \tag{22}$$

It would, however, be prohibitively expensive to generate $\mathbf{T}(\mathbf{O})$ in this manner for larger CI spaces. Rather than transforming all the orbitals in the structures simultaneously, it proves more efficient to employ *consecutive* updates of each orbital in turn. We therefore transform to an intermediate set of structures, defined in terms of the orbitals $\{\phi_1', \bar{\phi}_2\}$ the complete transformation then being achieved by the two steps $\{\Phi\} \rightarrow \{\Phi^{\mathrm{int}}\} \rightarrow \{\Phi'\}$. The first step can be achieved as before, for example:

$$\Phi_3^{\mathrm{int}} = |\overset{+}{\phi_1'}\overset{-}{\phi_2}| = \left|\left(\sum_\nu o_{\nu 1}\overset{+}{\phi_\nu}\right)\overset{-}{\phi_2}\right| \tag{23}$$

or alternatively in two steps using the update $\mathbf{O}_{21}(o_{21})$ followed by $\mathbf{O}_{11}(o_{11})$, as defined in Equation (4). The latter is more convenient computationally and, in our previously used notation, Equation (23) then becomes

$$\Phi_3^{\mathrm{int}} = (\hat{I} + (o_{11} - 1)\hat{E}_{11}^{\alpha})(\hat{I} + o_{21}\hat{E}_{21}^{\alpha})|\overset{+}{\phi_1}\overset{-}{\phi_2}| \tag{24}$$

(cf. Equation (6)). The update of ϕ_2 is carried out in a similar fashion, but it is clearly now inappropriate simply to substitute the original definition,

$$\phi_2' = \sum_\nu o_{\nu 2}\phi_\nu. \tag{25}$$

The intermediate structures are defined in terms of ϕ_1', so that we must use instead

$$\phi_2' = o_{12}'\phi_1' + o_{22}'\phi_2. \tag{26}$$

It is straightforward to solve Equation (26) for the os using the original definitions of ϕ_1' and ϕ_2'. This can be achieved, for example, by an LU decomposition of the orbital transformation matrix $\mathbf{O}$ [9, 10]. Using these coefficients we can then obtain the total transformation of the structure space, for example:

$$\Phi_3' = (\hat{I} + (o_{22}' - 1)\hat{E}_{22}^{\beta})(\hat{I} + o_{12}'\hat{E}_{12}^{\beta})(\hat{I} + (o_{11} - 1)\hat{E}_{11}^{\alpha})(\hat{I} + o_{21}\hat{E}_{21}^{\alpha})\Phi_3. \tag{27}$$

In general, therefore, we can generate the total transformation of the structure space by m^2 applications of very simple excitation operators.

References

1. D. L. Cooper, J. Gerratt, M. Raimondi, M. Sironi, and T. Thorsteinsson: *Theor. Chim. Acta.* **85**, 261 (1993).
2. B. Friis-Jensen and S. Rettrup: *Int. J. Quantum Chem.* **60**, 91 (1996).
3. F. W. Bobrowicz and W.A. Goddard: in *Modern Theoretical Chemistry*, Vol. 3, H. F. Schaefer (ed.), Plenum, p. 79 (1977).
4. J. M. Norbeck and R. McWeeny: *Chem. Phys. Lett.* **34**, 205 (1975).
5. P. W. Payne: *J. Chem. Phys.* **77**, 5630 (1982).

6. J. J. W. McDouall: *Theor. Chim. Acta.* **83**, 339 (1992).
   J. J. W. McDouall: *Theor. Chim. Acta.* **85**, 395 (1993).
7. C. Kadolkar, C. R. Sarma, and S. Rettrup: *Int. J. Quantum Chem.* **53**, 183 (1995).
8. T. Thorsteinsson, D. L. Cooper, J. Gerratt, P. B. Karadakov, and M. Raimondi: *Theor. Chim. Acta* **93**, 343 (1996).
9. T. Thorsteinsson: 'Development of Methods in Spin-Coupled Theory', Ph.D. thesis, University of Liverpool, U.K. (1995).
10. T. Thorsteinsson and D.L. Cooper: *Theor. Chim. Acta* **94**, 233 (1996).
11. T. Thorsteinsson, D.L. Cooper, J. Gerratt, and M. Raimondi: *Theor. Chim. Acta* **95**, 131 (1997).
12. B.O. Roos: *Adv. Chem. Phys.* **69**, 399 (1987).
13. H.-F. Werner and P. J. Knowles: *J. Chem. Phys.* **82**, 5053 (1985).
14. P. J. Knowles and H.-J. Werner: *Chem. Phys. Lett.* **115**, 259 (1985).
15. For a review of applications see, for example: D. L. Cooper, J. Gerratt, and M. Raimondi: *Chem. Rev.* **91**, 929 (1991).
16. R. McWeeny: *Proc. R. Soc. Lond. Ser. A* **223**, 306 (1954).
17. S. M. Goldfeld, R. E. Quant, and H. F. Trotter: *Econometrica* **34**, 541 (1966).
18. P. B. Karadakov, J. Gerratt, M. Raimondi, and D. L. Cooper: *J. Chem. Phys.* **97**, 7635 (1992).
19. MOLPRO is a package of *ab initio* programs written by H.-J. Werner and P. J. Knowles, with contributions from J. Almlöf, R. D. Amos, M. J. O. Deegan, S. T. Elbert, C. Hampel, W. Meyer, K. Peterson, R. Pitzer, A. J. Stone, P. R. Taylor, R. Lindh, M. E. Mura, T. Thorsteinsson and D. L. Cooper.
20. T. H. Dunning, Jr.: *J. Chem. Phys.* **90**, 1007 (1989).
21. F. Penotti, D. L. Cooper, J. Gerratt, and M. Raimondi: *J. Mol. Struct. (Theochem)* **169**, 421 (1988).
22. P. C. Hiberty and D. L. Cooper: *J. Mol. Struct. (Theochem)* **169**, 437 (1988).
23. L. Pauling: *Proc. Natl. Acad. Sci. U.S.A.* **14**, 359 (1928).
    L. Pauling: *J. Am. Chem. Soc.* **53**, 1367 (1931).
24. S. Wilson and J. Gerratt: *J. Mol. Phys.* **30**, 765 (1975).
25. D. L. Cooper, J. Gerratt, and M. Raimondi: *J. Mol. Struct. (Theochem)* **229**, 155 (1991).
    M. Sironi, M. Raimondi, D. L. Cooper, and J. Gerratt: *J. Phys. Chem.* **95**, 10617 (1991).
26. W. N. Lipscomb: *Science* **196**, 1047 (1977).
27. J. Paldus: in *Theoretical Chemistry: Advances and Perspectives*, Vol. 2, H. Eyring and D. G. Henderson (eds.), Academic Press, NY (1976).
28. P. Å. Malmqvist: *Int. J. Quantum Chem.* **30**, 479 (1986).
29. W. Heitler and F. London: *Z. Phys.* **44**, 455 (1927).
30. C. A. Coulson and I. Fischer: *Phil. Mag.* **40**, 386 (1949).

Molecular Engineering **7**: 87–108, 1997.

Studies of Chemical Structure, Spectroscopy and Electron Scattering Using Generalized Multistructural Wavefunctions

MARCO ANTONIO CHAER NASCIMENTO
Instituto de Quimica, Departamento de Fisico-Quimica, Universidade Federal do Rio de Janeiro, Cidade Universitaria, CT Bloco A sala 412, Rio de Janeiro, RJ 21949-900, Brazil.

(Received: 16 July 1997; in final form: 16 September 1997)

Abstract. Applications of the generalized multistructural wavefunction (GMS) to the calculation of molecular structure photoexcitation and photoelectron spectra of molecules, acidity of aliphatic alcohols and carboxylic acids, and electron scattering processes are presented. This wavefunction combines the advantages of the self-consistent field molecular orbital (SCF-MO) and valence bond (VB) models, preserving the classical chemical structures but optimizing the orbitals in a self-consistent way. It is particularly suitable to treat situations where the description of the molecular state requires localized wavefunctions. It also provides a very efficient and convenient way of treating the electron correlation problem, avoiding large configuration interaction (CI) expansions The final wave functions are much more compact and easier to interpret than those obtained by the conventional methods, using orthogonal orbitals.

Key words: Generalized multistructural wavefunction, molecular structure, molecular spectra, electron scattering processes, localized wavefunctions.

1. Introduction

The quantum theory of molecular structure has been developed using two distinct models, namely the valence-bond (VB) [1–8] and the molecular orbital (MO) [9–12] models. Although in principle equivalent, these models furnish quite different pictures of the molecules. In the VB model the identity of the constituent atoms of the molecule is to a greater extent preserved and the molecule is regarded as a set of atoms held together by *chemical bonds*, with each particular set of bonds joining the atoms defining a different *structure*. This is a very appealing model inasmuch as it represents the quantun mechanical translation of the classical ideas to which chemists have a deep-rooted attachment [13–21].

In the MO model the atomic nuclei are viewed simply as centers of charge furnishing a potential field for the motion of all the electrons of the molecule. The properties of the molecule are now described in terms of molecular orbitals which are, in general, delocalized over the entire molecule. The states of the molecule are fully specified once the occupation of the molecular orbitals is defined. In contrast to the VB model, the *chemical bond* is no longer viewed as a result of adjacent atoms pairing their electrons, although molecular orbitals could, in principle, be more or less localized in regions which would correspond to the classical chemical bonds. Nevertheless parameters such as bond orders [22–24], overlap populations and charge densities [25] were introduced in order to analyze the MO results in terms of quantities peculiar to the VB picture.

From the computational point of view, the MO model is much more tractable than the VB because the theory can be developed in terms of orthogonal orbitals, while the VB method presents the disadvantage of having to deal with nonorthogonal orbitals. That was, most certainly, the main reason why the MO model has been used by chemists in preference to its predecessor, the VB model.

Despite its tremendous success, the MO theory, in its simpler levels, is faced with serious problems in certain situations where the correct description of the state of the molecule requires localized wavefunctions. One of the best known examples is the case of core-ionized states of homonuclear diatomic molecules. In the case of the O_2 molecule, for instance, solving for the Hartree–Fock (HF) wavefunction with a $1s$ core hole localized on one oxygen atom yields an energy more than 10 eV lower than if the wavefunction is forced to retain the full molecular symmetry, with the singly-occupied core orbital delocalized onto both oxygens [26]. Many other examples of this phenomenon, generally known as *symmetry breaking*, can be found in the literature [27]. From the mathematical point of view this effect can be related to the more general problem of instabilities of the Hartree–Fock equations [28, 29]. In some cases this situation can be remedied with multiconfigurational wavefunctions [30], however using the GMS approach we can deal with this sort of problem even at the HF level of calculation. In the example of the O_2 molecule, although the broken symmetry solution gives a much better result than the symmetric one, it does not possess the full symmetry of the nuclear geometry, as it should for the correct wave function.

On the other hand, the VB model can very easily deal with these situations because symmetry breaking can be viewed as a tendency for the wave function to achieve the chemically intuitive localized structure. Nevertheless a one structure VB wavefunction for the ionized molecule with a localized hole presents the same symmetry problems as the MO function.

From either point of view, the correct way of dealing with such a situation is to construct a wavefunction with the proper symmetry, by taking a linear combination of the wavefunctions representing localized holes on the left and on the right oxygen atoms. From the VB point of view, combining the two localized structures poses no new problem to the model since the superposition of chemical structures is inherent to the theory. However, the process of combining two localized MO functions brings the nonorthogonality problem into the framework of the MO theory.

There might be situations where the hole state is not completely localized, in which case one could add the delocalized wavefunction to the linear combination of localized structures, and associate its coefficient to the degree of hole localization.

There are many other situations where the linear combination of MO wave functions can be useful. It is well known that many molecules possess electronic excited states of mixed (valence-Rydberg) character and in some cases this character is strongly dependent on one or more of the molecule's geometrical parameters. In the usual approach to this problem, diffuse functions are added to the valence basis set and the amount of diffuse character of the state is measured in terms of the spatial extension (second moments) of the excited orbital. However, since in these cases differential correlation effects can seldom be neglected, some sort of treatment, most often a configuration interaction (CI) calculation, is performed to take those effects into account. Again, second moments and also the weight of the different configurations in the CI expansion are used as a measure of the degree of mixed character. However, both criteria can be very deceptive and usually tend to exaggerate the diffuse nature of the excited state. This is so because the valence and the Rydberg-type orbitals are forced to mix characters due to the imposed orthogonality condition. This same problem could be approached in a totally different way, by solving separately for valence (nondiffuse functions) and Rydberg-type *structures* and taking linear combinations of the two wavefunctions. The coefficient of the linear combinations, determined variationally, could then be associated to the degree of valence and Rydberg-type character of the state.

Another interesting situation occurs when one of the atoms of a molecule presents near degenerated electronic configurations, as in the case of some transition metal atoms. For instance, the energy differences between the s^2d^{2n} and s^1d^{n+1} electronic configurations are $\Delta E \sim 0.03$, 0.10 and 0.83 eV for nickel, vanadium and iron) respectively. If we choose an HF-type calculation as the first step in our theoretical treatment, only one configuration can be used. The other configuration can be brought about at the CI level but at a considerable computational effort because the shapes and sizes of the optimized orbitals are already biased by the HF treatment. On the other hand, one could start from a multiconfigurational self-consistent field (MCSCF) calculation considering both electronic structures. However the orbitals obtained at that level are not the best for either one of the structures and any observed mixing of configurations could be a consequence of the process of averaging out the orbitals. In this case, as an alternative, one could combine the wavefunctions obtained from separate calculations, with the orbitals optimized for the proper electronic configurations.

Finally, an accurate description of the electronic structure and properties of molecules requires the proper treatment of the electronic correlation effects. The rigorous description of these effects is a difficult problem from the computational point of view, even for simple systems and, although formally known, the exact solution to the problem is seldom achieved because of the very complex nature of the calculations, regardless of the approach adopted. In the usual treatment an HF or MCSCF wavefunction is used as a zero-order function for a perturbation calculation (MBPT), or as a reference wavefunction either for a CI or else for a coupled-cluster (CC) calculation. In the VB approach, configuration interaction is introduced by the superposition of chemical *structures* representing different ways of pairing the electrons of the different atoms of the molecule. Consequently the correlation effects are introduced without destroying the picture provided by the classical chemical structures. Unfortunately the VB method turns out to be computationally even more complex than the methods based on the MO model because of the larger number of structures needed in the expansion and the nonorthogonality of the atomic orbitals. Another serious problem with the original VB model is that the atomic orbitals could not readjust their shapes to take account of the formation of chemical bonds. Orbital optimization within the VB approach has been considered by several authors [31–40] and it is now recognized that it can drastically reduce the number of structures needed in the VB expansion.

As a general method for taking account of all the different problems discussed above, we have recently introduced a generalized multistructural wave function (GMS) which combines the advantages of both models (SCF-MO and VB), preserving the classical chemical structures but optimizing their orbitals in a self-consistent way [41–49]. Each *structure* can be represented by an HF, MCSCF or CI wavefunction which contains the most important correlation effects, clearly described. The final wavefunction is obtained variationally and consists of a superposition of the wavefunctions representing the different *structures* considered. It is important to realize the very general meaning of the word *structure* in the GMS context. For instance, in the case of states with a mixed character, the excited state represented on a valence basis set would correspond to a structure, while its representation on a diffuse (Rydberg) basis corresponds to another structure.The same idea would apply to the case of molecules containing transition metal atoms with near degenerate states. As far as the treatment of the electronic correlation effects is concerned, because each structure can be represented by a correlated wavefunction containing the dominant (and clearly identifiable) effects, a very small number of structures will be needed in the final wavefunction, in order to describe the less important effects. The recovery of the residual, but not negligible effects, by the more usual approaches, usually requires an extremely large number of configurations in a CI or CC calculation or a high-order MPBT treatment. This is particularly true whenever some degree of bond or charge (polar structures) localization is needed in the MO description of a given state, in which case a large number of configurations in the conventional CI expansion is generally necessary to mimic those effects. On the other hand, the use of more than one bonding structure drastically reduces the number of configurations needed to achieve results comparable to the ones obtained by the conventional CI, MBPT or CC calculations. With the GMS wavefunction the correlation effects are therefore introduced without losing the simple picture provided by the chemical structures and at a moderate computational effort.

2. The Generalized Multistructural Wavefunction

Only a brief description of the GMS wavefunction will be presented here. A full discussion of the theory, with a detailed description of all the computational aspects involved in obtaining the solution of this kind of wavefunction, can be found elsewhere [42].

A GMS wavefunction can be defined as

$$\Psi_{\text{GMS}} = \sum_{I=1}^{N_{\text{struct}}} \sum_{I=1}^{N_{\text{sef}}} c_i^I \varphi_i^I, \tag{1}$$

where φ_i^I represents the ith spin-eigenfunction (N_{sef}) of the Ith bonding structure (N_{struct}) and the c_i^I its weight in expansion (1). In principle, there are no restrictions whatsoever on the form of the wavefunction φ_i^I. Each of the φ_i^Is can be individually optimized at the HF or multiconfigurational (generalized valence-bond method, GVB, complete-active-space self-consistent-field method, CASSCF, etc.) level, followed or not by a configuration interaction (CI) treatment, or else can be taken simply as a VB-type function. In this last case the Ψ_{GMS} is obviously identical to the VB function of the molecule.

Each one of the φ_i^I is represented in a basis of orthogonal orbitals $\{\phi_g^I\}$, properly optimized for the Ith structure. Although the orbitals of a given structure are taken to be orthogonal to each other,

$$\langle \phi_\alpha^I \mid \phi_\beta^I \rangle = \delta_{\alpha,\beta}, \tag{2}$$

no such restriction exists for the orbitals belonging to different structures I and II,

$$\langle \phi_\alpha^{\text{I}} \mid \phi_\beta^{\text{II}} \rangle = \mathbf{S}_{\alpha,\beta}^{\text{I,II}}. \tag{3}$$

However, when using GVB wave functions, the GVB orbitals describing a given structure are not necessarily orthogonal.

The coefficients c_i^I are obtained variationally by solving the equations,

$$\langle \delta\Psi_{\text{GMS}} | H - E | \Psi_{\text{GMS}} \rangle = 0, \tag{4}$$

$$(H - SE)C = 0, \tag{5}$$

where H and S are interaction supermatrices containing the diagonal (same structure) and the interstructural matrix elements. The matrix elements involving orbitals belonging to different structures are computed using a biorthogonalization procedure. In principle the $\{\varphi_i^I\}$ can either be optimized separately or in the presence of each other. Optimizing them separately is computationally simpler and in applications dealing with localized structures it seems that, as long as they are well localized (i.e., the overlap between their wavefunctions is small), there is no need to optimize them in the presence of each other.

3. Illustrative Examples

In order to illustrate the usage of the GMS wave functions, results of calculations for a variety of properties, of several different systems, will be presented.

3.1. THE DESCRIPTION OF MOLECULAR STRUCTURE

A good example of the usefulness of the GMS wave function to the description of chemical structures can be provided by the calculations performed on the so-called 1,3 dipolar molecules (e.g., O_3 and NO_2). The classical VB description of the ground state of the ozone molecule corresponds to a resonance hybrid of the zwitterionic structures (I) and (II) [50], while (III) corresponds to a diradical structure first proposed by Goddard *et al.* [51]:

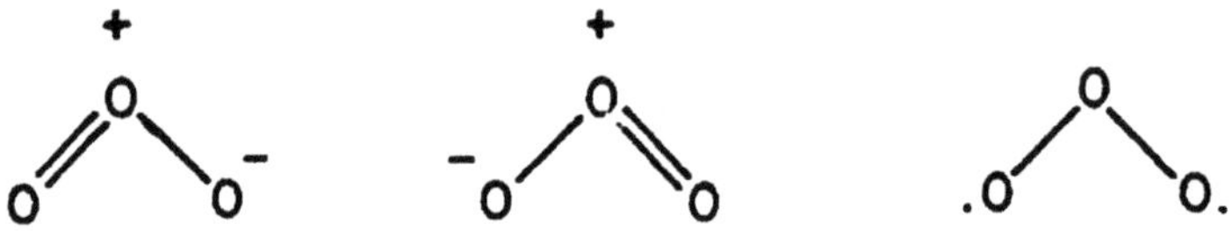

(I) (II) (III)

Analogous structures can be written for the NO_2 molecule. The classical VB and the molecular orbital descriptions of both molecules differ considerably. The HF description of the ozone molecule is known to be completely inadequate, leading incorrectly to a triplet ground state, with the first singlet state approximately 2 eV higher in energy. Goddard *et al.* [51], using the GVB method, were the first to give a proper description of the ozone singlet ground state. Subsequently, CI calculations based on the GVB wavefunction [52] not only confirmed the diradical character of the ground state but also showed that the ionic structures could be only associated with highly excited electronic states. In order to take account of the diradical character in the VB description, Harcourt and Rosso [53] introduced structure (III) (long bond between non-adjacent atoms) in their VB expansion. Their VB calculations showed that structure (III) was indeed dominant, but that the contribution of the zwitterionic structures was quite significant. This same description of the 0_3 molecule was obtained by projecting HF wavefunctions onto a basis of VB functions [54, 55].

For the NO_2 molecule, the projected wave functions show an even larger contribution of the zwitterionic structures and, as in the case of O_3, their contribution becomes more important as the quality of the basis set is improved. On the other hand, similarly to the ozone case, the MO description of the NO_2 molecule does not show any ionic character in the ground state wavefunction.

In an attempt to understand the differences in the VB and MO descriptions, GMS calculations were carried out for both molecules, using the Dunning [56] double-zeta (DZ) contraction of the Huzinaga's primitive Gaussian basis sets [57], augmented with d polarization functions ($\zeta_d(O) = 0.85$; $\zeta_d(N) = 0.80$), and at

Table I. Structural weights for the ground state wavefunction of the O_3 and NO_2 molecules.

Calculation	O^-—O^+=O	O=O^+—O^-	O•—O—O•
Ref. [54] (STO-3G)	0.184	0.185	0.593
Ref. [55] (4-31G)	0.231	0.231	0.476
Ref. [53]	0.308	0.308	0.793
Ref. [38c]	0.0	0.0	1.000
GMS (DZP)	1.4 (−4)	1.4 (−4)	1.000
	O^-—N^+=O	O=N^+—O^-	O•—N—O•
Ref. [54]	0.205	0.205	0.527
Ref. [55]	0.266	0.266	0.354
Ref. [38c]	0.0	0.0	1.000
GMS (DZP)	1.7 (−4)	1.7 (−4)	1.000

the respective experimental equilibrium geometries (O_3: r_e = 1.278 Å and θ_e = 116.8° [58]; NO_2: r_e = 1.196 Å and θ_e = 134.2° [59]). The GMS wavefunction for the ozone molecule [47] was obtained at the GVB(3/PP) level of calculation, correlating the three bond pairs for structures (I) and (II), and the two bond pairs plus the unpaired electrons for structure (III):

$$\Psi_{\mathrm{GMS}}(^1A_1) = c_1\varphi_{\mathrm{III}} + c_2(\varphi_{\mathrm{I}} + \varphi_{\mathrm{II}}). \tag{6}$$

An analogous GMS wave function was constructed for the $X\ ^2A_1$ ground state of the NO_2 molecule. Table I shows the structural weights for the ground state of both molecules compared to other calculations and Table II the energies of the various wavefunctions. From Table I it is clear that, after orbital optimization is considered, the contribution of the zwitterionic structures becomes negligible. Thus, a pure VB-type calculation for these molecules where the orbitals are allowed to readjust their shapes should not show any appreciable contribution of structures (I) and (II), in agreement with the MO description. This is exactly what was recently found by Malcolm and McDouall [38c]. On the other hand, the results of spin-coupled VB (SCVB) calculations [60] suggest a hypervalent description for both molecules, with two double bonds between the central and each of the terminal atoms. However, as discussed previously [61], that description results from an incorrect electronic distribution in the π state caused by the fact that the σ core was kept frozen during the calculations. More recent results seem to confirm this observation [39].

Therefore, the GMS results reconcile the VB and the MO descriptions and provide a clear explanation for the differences observed in previous calculations. Once orbital optimization is taken into account there is no need to consider the ionic structures. Besides the practical aspect of having to deal with a smaller number of structures, this result also suggests that for more complex molecules, some

Table II. Total energies for the O_3 and NO_2 molecules.

Structure	Level of calculation	Energy (Hartree)	
		$O_3(X\ ^1A_1)$	$NO_2(X\ ^2A_1)$
	HF	−224.305 781	−204.068 559
I or II	HF	−224 280 050	−204.053 582
I or II	GVB(3/2)	−224.341 382	−204.089 281
III	GVB(3/2)	−224.414 765	−204.131 997
I + II + III	GMS	−224.415 672	−204.132 488

highly unphysical structures, generally present in the VB expansions, can be also eliminated.

3.2. COMPUTING CORRELATION ENERGY

As an illustration of the efficiency of the GMS wave function at recovering correlation energy, let us examine the water molecule. This molecule was chosen as a test case because results of VB and highly sophisticated MO–CI calculations are available for comparison.

It was mentioned before that the usual treatment of electronic correlation effects, based on the MO model, starts from an HF or MCSCF wavefunction as the zero-order function for either a CI, MBPT or a coupled-cluster (CC) calculation. In the standard VB approach, orbital shape readjustments and correlation effects are introduced by superposing chemical structures representing different couplings and distributions of electrons among the various atoms of the molecule. In principle, for a given basis set, if all the electronic configurations, compatible with the spin and spatial symmetries of the state being considered, were included in a MO-type calculation and all possible Lewis-type structures were considered in the VB expansion, the final results should be equivalent. However, in practice, because of the very complex nature of the calculations, both expansions are truncated at a certain point, in order to satisfy our computational limitations.

From the MO point of view it is known that, for small molecules like water, a large fraction of the correlation energy, relative to an HF wavefunction and a given basis set, can be recovered by performing a CI calculation including all single and double excitations (CISD) from the reference configuration. Using a DZ quality basis, Saxon *et al.* [62] have found that a CISD calculation, involving just 361 configurations, accounts for 94.7% of the total correlation energy. However the recovery of the residual correlation energy, relative to the DZ basis, required a calculation involving 256 473 configurations, making it quite difficult to interpret the final wavefunction. On the other hand, the only all-valence VB calculation for the water molecule reported in the literature is the one by Peterson and Pfeiffer [63]. The importance of the ionic structures in their VB expansion, containing 10 covalent and 39 ionic structures, was found to be 62%, quite a large contribution.

However, the results of that calculation, when compared to the one by Saxon *et al.*, clearly indicate that many more structures would be needed in the VB expansion in order to achieve results comparable to the MO–CI ones. But increasing the basis of VB structures requires a great deal of computational work.

The idea behind the GMS method is to construct high quality wavefunctions, which incorporate the dominant correlation effects, but which are sufficiently compact to be easily interpreted. Thus, starting from a small number of VB structures, for which the orbitals have been optimized and the main correlation effects included, one should be able to construct very compact and yet high quality GMS wavefunctions.

From our previous discussion of the ozone molecule, it is known that once orbital optimization is considered, the importance of the ionic structures becomes negligible. Thus, for the water molecule, the importance of the ionic structures can be established at the HF level of calculation. The GMS (HF) wavefunction for the water molecule was constructed as the superposition of structures (IV)–(VII),

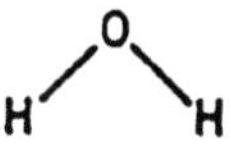

(IV)

(V)

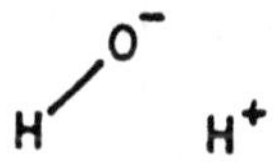

(VI)

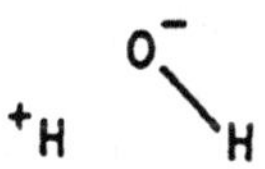

(VII)

$$\Psi_{\text{GMS}}(^1A_1) = c_1\varphi_{\text{IV}} + c_2\varphi_{\text{V}} + c_3(\varphi_{\text{VI}} + \varphi_{\text{VII}}). \tag{7}$$

Because oxygen is much more electronegative than hydrogen, ionic structures with the positively charged oxygen atom have not been considered. A procedure analogous to the one employed by Bagus and Schaefer [64] was used to solve for the charge-localized structures. As expected, after solving self-consistently for the wavefunction (7), the results indicated a negligible contribution ($c < 10^{-3}$) of the ionic structures. Therefore, only the covalent structures IV and V were considered at the higher levels of calculation. Subsequently a series of GMS calculations was performed, representing the covalent structures by more elaborate wavefunctions. First GVB(2/PP) wavefunctions were constructed, correlating the two bond pairs for structure V and the H—H plus the oxygen electron pair of structure VI. In another level of calculation the perfect-pairing scheme was relaxed and GVB(RCI) wavefunctions were constructed for each structure. Finally inter-pair correlations were introduced and each structure was represented by a CI wavefunction, constructed from the respective GVB(2/PP) wavefunctions and allowing: (a) up to quadruple excitations (full CI) in the GVB space (GVB-CI); and (b) full CI in the

Table III. Total energies for the H_2O molecule.

Calculation		Number of configurations	Energy (Hartree)
GMS[a]			
GVB(2PP)		8	−76.136 741
GVB-RCI		18	−76.164 835
GVB-CI		36	−76.226 770
GVB-POL-CI		350	−76.251 586
MO-CI			
Full-CI	Dz[b]	256 473	−76.157 861
	DZP[c]	6 740 280	−76.256 624
MR(CI)			
	DZP[c]	31 096	−76.254 108
VB			
	DZP	49[d]	−76.020 2

[a]Only structures (IV) and (V) are being considered.
[b]Ref. [62].
[c]Ref. [66]. The oxygen $1s$ electrons were not correlated.
[d]Number of structures in the VB expansion. Ref. [63].

GVB space plus single excitations to the virtual space, GMS(GVB-POLCI). Table III shows the GMS results compared to other selected calculations. It is important to emphasize that many other results are available in the literature [65]. However, the ones selected are the most appropriate to our discussions. From the results of Table III is clear that the GMS wavefunction can be extremely efficient at recovering correlation energy. At the GVB-RCI level of calculation the GMS wavefunction furnishes a better description than the DZ full-Cl calculation [62] and it only contains 18 configurations. In fact this is not a fair comparison inasmuch as the GMS wavefunctions are represented in a DZP basis set. However, the results from the GMS(GVB-POLCI) calculations are comparable to the ones obtained from full-Cl calculations using DZP quality basis set [66], despite of the tremendous difference in the number of configurations.

3.3. MOLECULAR SPECTROSCOPY

The GMS wavefunction can also be extremely useful in the field of molecular spectroscopy, particularly to describe molecules whose excited states exhibit symmetry-breaking solutions. Localized solutions can be obtained at any level of calculation and recombined to recover the full symmetry of the molecular wavefunction. In order to illustrate the applicability of the GMS methodology to this type of problem, we will discuss the photoelectron spectra of the *trans*-glyoxal molecule and the core excitation process of some diatomic molecules.

For the *trans*-glyoxal molecule [44, 67–69] the calculations were carried out with the Dunning DZP basis set [56] and at the HF optimized ground state geometry. Using the optimized geometry, self-consistent solutions for the ^2A_g, ^2B_u, ^2B_g and ^2A_u ion states of the molecule were obtained, at the HF level of calculation. From the respective HF wavefunctions, CI(SD) were performed, keeping the first seven occupied orbitals frozen. These calculations served not only to identify the important correlations effects but also as a basis for comparison with the GMS calculations. In order to investigate the symmetry-broken solutions, revealed by the HF calculations using C_s symmetry, GMS wavefunctions representing the ion state with localized (n^{-1} or π^{-1}) holes were constructed, at the HF, GVB and CI levels of calculation. This can be achieved by recombining the delocalized π and n orbitals as

$$\pi_{1,r} = (1/\sqrt{2})(1a_u \pm 1b_g) \tag{8}$$

and

$$n_{1,r} = (1/\sqrt{2})(7a_g \pm 6b_u), \tag{9}$$

and solving self-consistently for the different ion states [64]. In Equations (8) and (9), π_1 and n_r represent the π and n orbitals localized on the left and right C = O groups of the glyoxal molecule. Using Equation (1), the wavefunction for the ^2A_u state $[(1a_u)^1(1b_g)^2]$, for instance, can be written in terms of localized orbitals, as:

$$\psi(^2A_u) = \mathbb{A}[(\text{core})(\pi_1 - \pi_r)(\pi_1 - \pi_r) \times (\pi_1 + \pi_r) \ldots \alpha\beta\alpha] \tag{10}$$

or

$$\psi(^2A_u) = \mathbb{A}[(\text{core})(\pi_1\pi_1\pi_r + \pi_r\pi_r\pi_1) \ldots \alpha\beta\alpha], \tag{10'}$$

which can be translated to a localized two-structures GMS wave function:

$\psi(^2A_u) = C$ 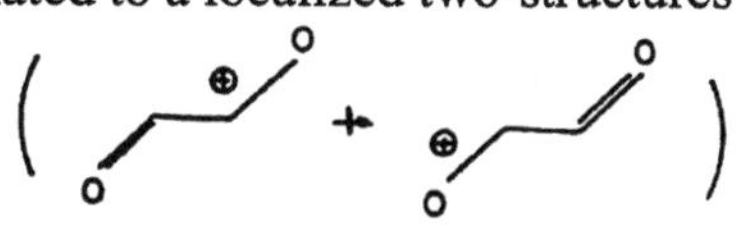 ,

where C is a normalization constant. In order to add some flexibility to the GMS wavefunction, the delocalized structures, with C_{2h} symmetry, can also be included in the expansion.

The localized structures can also be represented at the GVB(3/PP) level, correlating the two π bonds and the oxygen lone pair, for the n^{-1} states, or the two lone pairs and the π bond in the case of the π^{-1} states. These GVB wavefunctions can in turn be superimposed to generate the GMS(GVB) wavefunctions for each of the ion states. Ionization potentials at this level of calculation were obtained relative to a ground state represented at the GVB(4/PP) level of calculation. At a higher level of calculation, inter-pair correlation effects were taken into account

Table IV. Ionization potentials (eV) for the *trans*-glyoxal molecule compared to experiments.

State	Calculations					Experiment[a]
	HF(C_{2h})	HF(C_s)	GMS(HF)	CI(SD)	GMS(CI)	
^2A_g	11.38	10.38	9.90	10.79	10.25	10.6
			(3)[b]	(2529)	(38)	
^2B_u	13.40	(10.38)	11.12	12.06	12.40	12.19
			(3)	(2529)	(38)	
^2B_g	13.50	13.02	12.44	13.51	13.34	14.0
			(3)	(2491)	(34)	
^2A_u	15.74	(13.02)	14.58	15.17	15.33	15.4
			(3)	(2491)	(34)	

[a]Ref. [70].
[b]Number of configurations in the GMS or CI wavefunction.

using the GVB(3/PP) wavefunctions of each one of the ion states, the GVB(4/PP) representation of the ground state, and allowing a full-CI in the respective GVB spaces. These GVB(CI) wavefunctions were then used to construct the respective GMS(CI) wavefunctions.

Table IV shows the results obtained at the various levels of calculation compared to the experiments. From Table IV one sees that the GMS(CI) results are quite remarkable considering the compactness of the wavefunctions relative to the CI(SD) ones. It is clear from the results that the photoelectron spectra of this molecule can be well understood in terms of localized ionization processes and that the consideration of localized structures dramatically reduces the number of configurations needed relative to the CI calculations using symmetry orbitals.

The study of inner-shell excitation in molecules, by photons and electrons, has been the subject of increasing interest in the last few years, because of its many practical applications. From the theoretical point of view these processes are extremely interesting because once a core hole is created, several structural changes may take place and the correct description of the process will need a well balanced treatment of orbital relaxation, hole localization and electronic correlation effects. The effect of orbital relaxation on the electronic transition energy can, in principle, be easy established, but in order to take proper account of its effect on the transition intensity (optical (OOS) and generalized (GOS) oscillator strengths), nonorthogonal target wavefunctions must be used. Previous attempts at giving an indirect description of the effect of orbital relaxation on OOS and GOS values, by a CI calculation using the same set of orthogonal orbitals to represent both the ground and the excited states, have shown that this procedure may work well for valence transitions but not for inner-shell excitations [71]. The problem of correctly describing the core-hole produced in these processes is even more interesting because, as observed by several authors [64, 72, 73], in order to obtain transition energies in agreement with the experimental spectra the calculations must be performed by treating the core-hole as localized. However, this localization can only be achieved by artificially reducing the symmetry of the molecule (for instance, $D_{\infty h}$ to $C_{\infty V}$ for linear molecules). Nevertheless the Hamiltonian of the molecule still commutes with the symmetry operators of the full symmetry point group, implying that the core-hole states should transform like one of the irreducible representations of the *full* symmetry group of the molecule. Thus, in spite of the fact that the core-hole can actually be localized (and not merely as a consequence of a symmetry-breaking solution), there still remains the problem of describing it with wavefunctions which have the full molecular symmetry. A related question is how to define the degree of localization of the core-hole. Finally the effect of electronic correlation on both the transition energy and intensity has to be considered.

From the previous examples it is clear that the GMS wavefunction is very suitable for dealing with this type of problem. The effect of orbital relaxation on transition energies and intensities can easily be determined since orbital nonorthogonality is inherent to the GMS wavefunction. Also, the problem of describing localized hole-states while keeping the full symmetry of the molecular point group can easily be achieved by properly combining structures with holes localized in symmetric equivalent atoms of the molecule. The GMS wavefunction also provides an unambiguous way of defining the degree of localization of the hole-state. By adding the delocalized structure to the GMS expansion and variationally solving for the coefficients of the different structures, one can associate the degree of localization of the core-hole state to the coefficient of the localized structures. For instance, the simplest GMS wavefunction to describe the hole states of the N_2 molecule can be written as

$$\psi_{\mathrm{GMS}} = c_1\varphi\,\underset{\mathrm{I}}{(\mathrm{N}\overset{\circ}{—}\mathrm{N})} + c_2\{\varphi\,\underset{\mathrm{II}}{(\mathrm{N—N^{\circ}})} \pm \varphi\,\underset{\mathrm{III}}{(\mathrm{N^{\circ}—N})}\} \tag{11}$$

where structures I, II and III represent delocalized (I) and localized (II and III) hole states. The coefficient c_2 can therefore be used to define the degree of localization. Finally, since the structures can be represented at the HF, MCSCF, CI or any plausible combination of these levels of calculation, the influence of electronic correlation effects on the transition energies and intensities can also be studied.

Table V shows the results of GMS-CI inner-shell transition energies and intensities calculations for N_2, C_2H_2, CO_2 and O_2 molecules compared to the available experimental data. The transitions studied and the basis sets used are also specified in the table. For the CO_2 molecule the transition energies refer to excitations from the oxygen core electrons while for the O_2 molecule they correspond to ionization from the $1s$ orbital. Calculations at several different levels have been considered in order to establish the effect of relaxation, hole-localization and electronic correlation on the transition energies and intensities. A detailed description of the calculations and the basis set construction for the case of the N_2 molecule can be found in Ref. [43]. The same strategy was used for the other molecules. The degree

Table V. Transition energies (ΔE), optical oscillator strengths (f_{on}) and degrees of localization (d_l and τ_h^*/τ_h) for N_2, C_2H_2, CO_2 and O_2 molecules.

Transition	ΔE (eV) Theory	Experiment	f_{on} Theory	Experiment	d_l	τ_h^*/τ_h
$N_2(12s7p1d)/[9s, 5p, 1d]$						
$1\sigma_g \rightarrow 1\pi_g$	402.62	401.1[a]			0.78	0.75[b]
$1\sigma_u \rightarrow 1\pi_g$	402.55		0.200	0.200,[c] 0.18[d]	0.78	
$C_2H_2(12s7p1d; 5s2p)/[10s5p1d; 3s1p]$						
$1\sigma_g \rightarrow 1\pi_g$	287.49	285.81[a]			0.67	0.60[e]
$1\sigma_u \rightarrow 1\pi_g$	287.47		0.175		0.67	
$CO_2(12s6p1d)/[10s, 4p, 1d]$						
$1\sigma_g \rightarrow 2\pi_u$	535.99	534.4,[f] 535.3[g]	0.093	0.12–0.13[h]	0.86	
$1\sigma_u \rightarrow 2\pi_g$	535.99				0.86	
$O_2(12s7p1d)/[9s, 5p, 1d]$						
$1\sigma_g \rightarrow {}^2\Sigma_g^-$	544.07	544.2[i]			0.82	
$1\sigma_g \rightarrow {}^4\Sigma_g^-$	543.31	543.1[i]			0.82	

[a]Ref. [76]; [b]Ref. [75]; [c]Ref. [77]; [d]Ref. [78]; [e]Ref. [79]; [f]Ref. [80]; [g]Ref. [81]; [h]Ref. [82]; [i]Ref. [83].

of localization, d_l, given by the square of the coefficient c_2 plus the overlap between the wavefunctions φ_{II} and φ_{III} [74] is also shown in Table V. The last column of the table deserves further comment. Although localized holes are generally assumed by most experimentalists, it appears that the only evidence for this localization comes from theoretical calculations of ionization potentials, transition energies and line-shapes although, as previously discussed, this localization is imposed in order to achieve agreement with the experimental data. King *et al.* [75] have suggested that some information about the degree of localization could be obtained from the ratio τ_{h^*}/τ_h, where $\tau_h(-\hbar/\Delta E)$ is the characteristic *hopping time* between the orbitals $1\sigma_g(1s)$ and $1\sigma_u(1s)$ of homonuclear diatomic molecules for instance, differing by ΔE in energy, and τ_{h^*} is the lifetime of the hole state. However, because ΔE is in general very small and difficult to obtain experimentally, theoretical values of ΔE must be used in order to compute τ_h. Therefore, in order to use the King *et al.* criterion one has to rely on a combination of theoretical and experimental data which are most certainly obtained with different degrees of accuracy.

From the results in Table V one sees that, as far as transition energies and intensities are concerned, the GMS calculations give a very good description of the inner-shell excitation processes studied without resorting to reducing the symmetry of the molecules. Generally speaking, the transition energies are largely affected by hole localization, while relaxation effects dominate for the OOS orbital. As far as the nature of the hole-state is concerned, the GMS results indicate that, for the molecules studied, the holes exhibit a great deal of localization, in agreement with the assumption usually made by most experimentalists and with the estimates of King *et al.* where available. Also, the GMS results are consistent with the fact that, as the nuclear atomic charges increase, one should expect the inner electrons to be more tightly bound to the nuclei and therefore more localized.

3.4. ELECTRON SCATTERING PROCESSES

Several modern spectroscopic techniques make use of electron beams as the source of excitation. These techniques present some advantages over the standard ones which use photons to promote the excitation. For instance, using the same excitation source, i.e. an electron gun, spectra can be obtained in a wide range of energy, from the X-ray region to the near infrared. Also, because of differences in the selection rules, some transitions which do not pertain to the optical spectra can be observed in the electron-impact spectra, for instance spin and parity-forbidden transitions. In electron-impact spectroscopy (EIS) the quantities which characterize a transition are the energy, the intensity as measured by the generalized oscillator strength and the dependence of the GOS on the electron scattering angle. In fact, it is sometimes the behavior of the GOS with the scattering angle which permits the precise characterization of the transition. Therefore this kind of information is very important for the full interpretation of the EIS sprectra. Thus, any theoretical investigation of an EIS spectrum should also address the problem of calculating the angular dependence of the GOS.

This section presents the application of the GMS wavefunction to the study of inner-shell excitations of the N_2 molecule. Similar calculations have been performed for the CO_2 [48] and C_2H_2 [49] molecules. The calculations were performed at both HF and GMS levels, at the experimental equilibrium geometry. The generalized oscillators strengths were computed within the first Born approximation [84] taking relaxation effects directly into account by using wavefunctions optimized for each of the states involved in the transition. The basis set was the same used to compute the optical spectra (Table V) and the GMS wavefunctions have been constructed as described in the previous section. A detailed description of the calculations performed for the N_2 molecule can be found in Ref. [43].

Figure 1 shows the behavior of the sum of the GOS for the $(1\sigma_g \rightarrow 1\pi_g)$ and $(1\sigma_u - 1\pi_g)$ transitions with the scattering angle, compared to the experimental and other theoretical results. Because the two transitions are unresolved experimentally, the GOSs for both transitions were added in order to compare the theoretical and

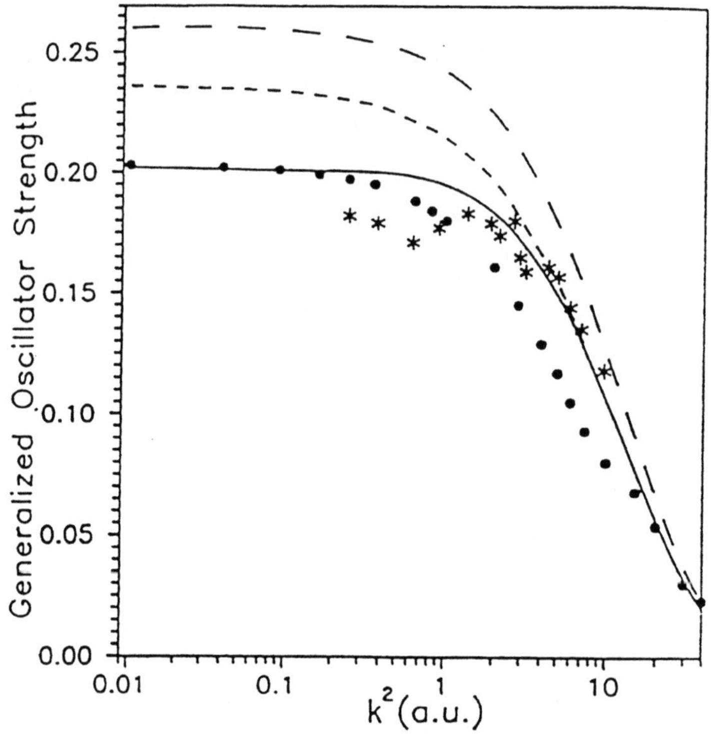

Figure 1. Total generalized oscillator strength for the preionization peak compared to experiments and other theoretical results: ——— GMS(CI); – – – (HFFC); - - - - Rescigno and Orel [85]; ● exp. of Camilloni *et al.* [77]; * exp. of Barbieri and Bonham [78].

experimental results. The curve designated as HFFC corresponds to a calculation at the HF level using the frozen-core approximation. From Figure 1 one observes that the GOS calculated by Rescigno and Orel [85] falls between our HFFC and GMS(CI) results. This is an expected result as they tried to mimic relaxation and hole-localization effects using a standard orthogonal CI calculation. Very good agreement is observed between the GMS(CI) GOS values and the experiments of Camilloni *et al.* [77] for small K^2 values and with the results of Barbieri and Bonham [78] for larger values of K^2. For larger scattering angles, which correspond to larger values of K^2, there is some disagreement between the two sets of experimental data. The GMS(CI) results support Barbieri and Bonham's results in this K^2 region but it must be considered that the FBA could make incorrect predictions for larger scattering angles, as noticed previously for some valence excitation processes. This same sort of agreement between the GMS results and the experimental data for N_2 was also observed for other molecules such as CO_2 and C_2H_2, which allow us to conclude that the GMS wavefunction can be a useful tool for examining electron scattering processes.

3.5. GAS-PHASE ACIDITIES OF ALIPHATIC ALCOHOLS AND CARBOXYLIC ACIDS

Carboxylic acids are relatively stronger acids than alcohols in the gas phase and in solution. The difference in the gas-phase acidities is usually attributed to a low energy content of the carboxylate anions, which would be stabilized by the resonance effect. This generally accepted interpretation has recently been challenged by Siggel and Thomas [86–89], who attributed the larger acidity of carboxylic acids to an unusually high energy content of the undissociated molecule. Those authors claim that the decisive factor for the higher acidity of the acids would be the inductive effect due to the carbonyl group. Dewar *et al.* [90], comparing the theoretical acidities of the vinylogues of formic acid, concluded that resonance, rather than inductive effects, was responsible for the enhanced acidities of the acids relative to the corresponding alcohols. More recently, Bordwell and Satish [91], comparing acidities in DMSO, for benzoic acid, tropolone and the enol of 1,3-cyclohexadione, also concluded that resonance in the carboxylate-like ions was the major effect making carboxylic acids more acidic than alcohols. Very recently Hiberty and Byrman [92] also examined this problem and concluded that resonance and inductive effects contribute with equal importance to the acidity of carboxylic acids.

Despite the many calculations of Brønstead acidities and proton affinities reported in the literature, for a large number of different systems, from the theoretical point of view there are still uncertainties concerning the relative importance of inductive and resonance effects in the acidity of these compounds. This section presents results of GMS calculations for methanol, ethanol, formic and acetic acids.

The Brønstead acidity of a compound can be defined as the variation of the Gibbs free energy (ΔG_T^0) for the reaction:

$$\mathrm{RO{-}H} \rightarrow \mathrm{RO^-} + \mathrm{H^+}.$$

However, because the entropic term (mainly due to the H^+) cancels out in a comparison of relative acidities, one may take ΔH_T^0 as a measure of the compound's acidity. This is also a convenient definition for the purpose of comparison with experimental enthalpies. From the quantum mechanical calculations one generally obtains ΔE_T^0, i.e., the variation in internal energy, at zero temperature and at the equilibrium positions of ROH and RO^-. In order to convert the internal energy to values comparable to experiments, the following equation can be used:

$$\Delta H_T^0 = \Delta E_{\mathrm{eq}}^0 + \Delta(PV) + \Delta(ZPE) + \Delta E_T^v + \Delta E_T^r + \Delta E_T^t \qquad (12)$$

where $\Delta(PV)$ is the contribution which converts internal energy to enthalpy and $\Delta(ZPE)$ is the variation in zero-point vibration energies. ΔE_T^v, ΔE_T^r and ΔE_T^t are terms which correct for the changes in vibrational, rotational and translational energies, respectively, when the system is brought from zero degrees to temperature

T. For pressures normally used in the measurements of gas-phase acidities, it is reasonable to consider reagents and products as ideal gases, and therefore,

$$\Delta(PV) = RT\,\Delta n = RT \tag{13}$$

for all reactions, since $\Delta n = 1$. The corrections due to translation and rotation can be treated classically, using the equipartition theorem. Thus, considering that in the final state one always has two species (RO^- and H^+) present and one in the initial state, and also that the proton has only translational degrees of freedom, one can write:

$$\Delta E_T^t = (3/2)RT \quad \text{and} \quad \Delta E_T^v = 0. \tag{14}$$

In units of RT, room temperature corresponds to an energy of 201.63 cm^{-1}. None of the species studied present normal vibrational modes of such low energy that could be excited when bringing the system from zero degrees to room temperature. Therefore, the population of the vibrational states should not change to a significant extent and one can take $\Delta E_T^v = 0$. Finally, the $\Delta(ZPE)$ correction was computed in the harmonic approximation as:

$$\Delta(ZPE) = \Delta(\Sigma(1/2)h\nu_i). \tag{15}$$

The vibrational frequencies were calculated using numerical derivatives (second derivatives) of the analytic gradient of the energy and used, unscaled, to compute the ZPE corrections. Although it is usual to scale vibrational frequencies obtained from HF wavefunctions by a factor $\sim$0.9 , the GMS wavefunctions were constructed using GVB wavefunctions for which there are insufficient calculations of vibrational frequencies to justify the use of that factor or any other.

The geometry of all the species studied were optimized at the GVB-PP level of calculation [93], correlating all the bonding electron pairs. In the case of the anions, the electron pair resulting from the heterolytic dissociation of the neutral molecule was also correlated. For the neutral species the calculations were performed with the 6-31G** basis set while for the anions the 6-31G**+ basis was used. The influence of diffuse functions in the description of the neutral species was tested and found to be negligible (<4 mhartree). Similarly, calculations for the methoxide ion showed an even smaller (0.4 mhartree) contribution of the diffuse functions on the hydrogen atoms. Besides the energy criterion, the influence of the latter diffuse functions on the geometry and atomic charges was tested and also found to be negligible.

In order to establish the relative importance of inductive and resonance effects for the acidity of the carboxylic acids, GMS(GVB-CI) wavefunctions of the type,

$$\psi_{\mathrm{GMS}} = N\{\varphi_{\mathrm{A}} + \varphi_{\mathrm{B}}\}$$

Table VI. Gas-phase acidities at the GVB + CI and GMS + CI levels of calculation.

	CH_3OH	HCOOH	C_2H_5OH	CH_3COOH
$\Delta E^0_{eq.}$	0.6206	0.5922	0.6149	0.6005
Corrections (eV)				
ΔE^t_T	0.038	0.038	0.038	0.038
ΔE^r_T	0.0	0.0	0.0	0.0
ΔE^v_T	0.0	0.0	0.0	0.0
$\Delta(PV)$	0.025	0.025	0.025	0.025
$\Delta(ZPE)$	−0.423	−0.381	−0.419	−0.374
Resonance:	0.0	−0.779	0.0	−0.753
ΔH^0_{28} (This work)				
(kcal/mol)	381.11	346.26	377.62	352.13
(eV)	16.52	15.02	16.37	15.27
(Hartree)	0.6073	0.5520	0.6018	0.5612
Ref. [88]				
(kcal/mol)	383.26	343.14	380.03	343.60
(eV)	16.62	14.88	16.48	14.90
(Hartree)	0.6108	0.5468	0.6056	0.5476
Experiment [94] (kcal/mol)	379.14	345.19	376.13	348.46
(eV)	16.44	14.97	16.30	15.11
(Hartree)	0.6042	0.5501	0.5994	0.5553

were constructed from the respective GVB wave functions for the carboxylate anions, where φ_A and φ_B represent the two resonant structures,

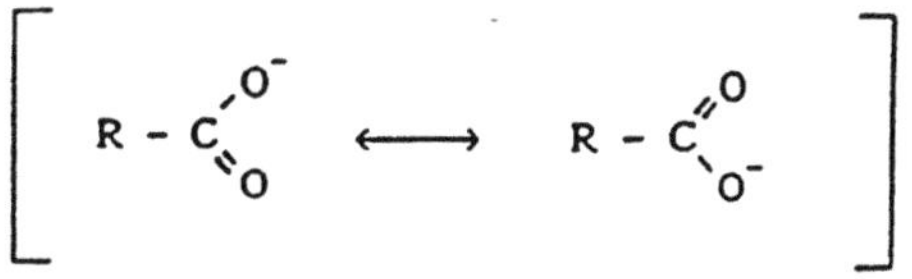

(A) (B)

For the neutral molecules and the alkoxide ions only one structure has been considered since resonance effects are negligible for those species. The relative importance of resonance and inductive effects for the acidity of the carboxylic acids can be established by assuming that, in the absence of resonance, the inductive effect of the carbonyl group would be mainly responsible for the stronger acidity of the acids relative to the corresponding alcohols. This can be accomplished by calculating the acidities of the carboxylic acids using only one of the localized structures (A or B) to represent the carboxylate anions (i.e., in the absence of resonance) and comparing them with the ones obtained using both structures. Equivalently one could compute the resonance energy associated to the pair of structures, A and B, and use it to correct the acidity calculated with only one of the resonant structures.

Table VI shows the acidity values computed at the GVB + CI and GMS(GVB + CI) levels of calculation, including all the corrections to ΔE^0_{eq}, compared with the experimental values and other theoretical calculations. From Table VI it is clear that once resonance effects are taken into account we obtain acidities, for the carboxylic acids, comparable to the experimental values, with the same degree of accuracy as for the alcohols. On the other hand, in the absence of resonance acetic acid is only slightly more acidic than ethanol and the difference in acidities between formic acid and methanol is only half of the difference between the experimental values. In conclusion we can say that resonance is not only important but the dominant effect responsible for the greater acidity, in the gas-phase, of the carboxylic acids relative to the corresponding alcohols.

Acknowledgements

It is pleasure to acknowledge Dr. Carlos Eduardo Bielschowsky, who adapted the GMS wave function to his FBA code, and to my former students Eduardo Hollauer, who wrote the GMS code and was involved in many of the earlier applications of the GMS wave function, Wely Brasil Floriano, Solange Regina Blaszkowski, Joaquim Delphino da Motta Neto and Marcelo Pessoa Miranda. The author also acknowledges CNPq and Finep for financial support.

References

1. W. Heitler and F. London: *Z. Phys.* **44**, 455 (1927).
2. G. Rumer: *Göttingen Nach.* 337 (1932).
3. L. Pauling; *Proc. Natl. Acad. Sci. USA* **18**, 293 (1932).
4. L. Pauling and G. W. Wheland: *J. Chem. Phys.* **1**, 362 (1933).
5. L. Pauling and J. Sherman: *J. Chem. Phys.* **1**, 606 (1933).
6. H. Eyring: *J. Am. Chem. Soc.* **54**, 3191 (1932).
7. J. C. Slater: *Phys. Rev.* **37**, 481 (1931); **38**, 1109 (1931).
8. L. Pauling: *J. Am. Chem. Soc.* **53**, 1367 (1931): **53**, 3225 (1931).
9. F. Hund: *Z. Phys.* **51**, 759 (1928).
10. F. Hund: *Angew. Chem. Intl. Ed. Engl.* **16**, 87 (1977).
11. R. S. Mulliken: *Phys. Rev.* **32**, 761 (1928).
12. E. Hückell: *Z. Phys.* **60**, 204 (1931): **73**, 310 (1932): **76**, 628 (1932).
13. C. A. Russell: *The History of Valence*, University, Leicester (1971).
14. J. Weyer: *Angew. Chem. Int. Ed. Engl.***13**, 591 (1974).
15. K. Hafner: *ibid.* **18**, 641 (1979).
16. L. Pauling: *J. Mol. Sci.* **2**, 1 (1981).
17. E. Frankland: *Phil. Trans.* **142**, 440 (1850).
18. A. Kekule: *Liebigs Ann. Chem.* **106**, 129 (1858).
19. A. S. Couper: *CR Acad. Sci.* **46**, 1157 (1858): *Phil. Mag.* **16**, 104 (1858).
20. A. M. Butlerov: *Z. Chem.* **4**, 549 (1861).
21. G. N. Lewis: *J. Am. Chem. Soc.* **38**, 762 (1916).
22. C.A. Coulson: *Proc. R. Soc. London, Ser. A* **169**, 413 (1939): *Proc. Cambridge Phil. Soc.* **36**, 201 (1940): *J. Chem. Soc.* 3111 (1954).
23. C. A. Coulson and G. S. Rushbrooke: *Proc. Cambridge Phil. Soc.* **36**, 193 (1940).

24. C. A. Coulson and H. C. Longuet-Higgins: *Proc. R. Soc. London, Ser. A* **191**, 39 (1947).
25. R. S. Mulliken: *J. Chem. Phys.* **23**, 1833 (1955); 1841 (1955).
26. P. S. Bagus and H. F. Schaefer III: *J. Chem. Phys.* **56**, 224 (1972).
27. See, for example, (a) J. McKelvey and W. J. Hehre: *Mol. Phys.* **25**, 983 (1973). (b) C. F. Jackels and E. R. Davidson: *J. Chem. Phys.* **64**, 2908 (1976). (c) D. Guenzburger:: *Chem. Phys. Lett.* **86**, 316 (1982). (d) S. Canuto, O. Goscinsky and M. Zerner: *ibid.* **68**, 232 (1979). (e) L. E. Nitzche and E. R. Davidson: *ibid.* **58**, 171 (1978). (f) R. L. Martin: *J. Chem. Phys.* **74**, 1852 (1981).
28. J. Paldus and A. Veillard: *Mol. Phys.* **25**, 983 (1973).
29. J. Cizek and J. Paldus: *J. Chem. Phys.* **53**, 821 (1970).
30. L. Engelbrecht and B. Liu: in *Energy Storage and Redistribution in Molecules*, Plenum, New York, p. 219 (1983).
31. A. F. Voter: *Ph.D. thesis*, California Institute of Technology (1983). A. F. Voter and W. A. Goddard III: *J. Chem. Phys.* **75**, 3638 (1981): *Chem. Phys.* **57**, 253 (1981).
32. J. Gerratt: *Adv. At. Mol. Phys.* **7**, 141 (1971). N. C. Pyper and J. Gerratt: *Proc. R. Soc. London, Ser. A* **355**, 407 (1977). D. L. Cooper, J. Gerratt, and M. Raimondi: *Adv. Chem. Phys.* **69**, 319 (1987). J. Gerratt and M. Raimondi: *Proc. R. Soc. London, Ser. A* **371**, 525 (1980). P. B. Karadakov, J. Gerratt, D. L. Cooper, and M. Raimondi: *J. Chem. Phys.* **97**, 7637 (1992).
33. R. McWeeny: *Int. J. Quantum Chem.* **34**, 25 (1988).
34. G. A. Gallup, R. L. Vance, J. R. Collins, and J. M. Norbeck: *Adv. Quantum Chem.* **16**, 229 (1982).
35. D. J. Klein and N. Trinajistic: *Valence Bond Theory and Chemical Structure*, Elsevier, Netherlands (1990).
36. J. H. van Lenthe and G. G. Balint-Kurti: *Chem. Phys. Lett.* **76**, 138 (1980). *J. Chem. Phys.* **78**, 5699 (1983).
37. J. Verbeek: *Ph.D. thesis*, Utrecht University (1990).
38. J. J. McDouall: *Theor. Chim. Acta* **83**, 339 (1992): **85**, 395 (1993). N. O. J. Malcolm and J. J. W. McDouall: *J. Comp. Chem.* **15**, 1365 (1994).
39. T. Thorsteinsson, D. L. Cooper, J. Gerratt, P. Karadakov, and M. Raimondi: *Theor. Chim. Acta* **93**, 343 (1966).
40. C. P. Byrman: Ph.D. thesis, Utrecht University (1995).
41. E. Hollauer and M. A. C. Nascimento: *Chem. Phys. Lett.* **181**, 463 (1991).
42. E. Hollauer and M. A. C. Nascimento: *J. Chem. Phys.* **99**, 1207 (1993).
43. C. E. Bielschowsky, E. Hollauer, and M. A. C. Nascimento: *Phys. Rev.* **A45**, 7942 (1992).
44. E. Hollauer and M. A. C. Nascimento: *Chem. Phys.* **177**, 79 (1993).
45. M. A. C. Nascimento: in E. Silveira (ed.), *Collision Processes of Ion, Positron, Electron and Photon Beams with Matter*, World Scientific, New Jersey, p. 247 (1992).
46. M. P. de Miranda, C. E. Bielschowsky, and M. A. C. Nascimento: *J. Phys. B* **28**, L15 (1995).
47. W. B. Floriano, S. R. Blaszkowski, and M. A. C. Nascimento: *J. Mol. Struct. (Theochem)* **335**, 51 (1995).
48. M. P. de Miranda and C. E. Bielschowsky: *J. Mol. Struct. (Theochem)* **282**, 71 (1993).
49. M. P. de Miranda, C. E. Bielschowsky, H. M. Roberty, and G. G. B. de Souza: *Phys. Rev.* **A49**, 2399 (1994).
50. G. W. Wheland: *Resonance in Organic Chemistry*, Wiley, New York (1955).
51. W. A. Goddard III, T. H. Dunning Jr., W. J. Hunt, and P. J. Hay: *Acc. Chem. Res.* **6**, 368 (1973).
52. P. J. Hay, T. H. Dunning, Jr., and W. A. Goddard, III: *J. Chem. Phys.* **62**, 3912 (1975).
53. R. H. Harcourt and W. Rosso: *Can. J. Chem.* **56**, 1093 (1978).
54. P. C. Hiberty and C. Leforestier: *J. Am. Chem. Soc.* **100**, 2012 (1978).
55. P. C. Hiberty and G. Ohanessian: *J. Am. Chem. Soc.* **104**, 66 (1982).
56. T. H. Dunning and P. J. Hay: in H. F. Schaefer (ed.), *Modern Theoretical Chemistry*, Plenum Press, New York, p. 1 (1977).
57. S. Huzinaga: *J. Chem. Phys.* **42**, 1293 (1965).
58. T. Tanaka and Y. Morino: *J. Mol. Struct.* **33**, 538 (1970).
59. C. R. Stevens and R. N. Zare: *J. Mol. Spectrosc.* **56**, 167 (1975).
60. D. L. Cooper, J. Gerrat, and M. Raimondi: *J. Chem. Soc., Perkin Trans. II* 1187 (1987).
61. W. B. Floriano: M.Sc. thesis. Instituto de Quimica, Universidade Federal do Rio de Janeiro (1992).

62. P. Saxe, H. F. Schaefer III, and N. C. Handy: *Chem. Phys. Lett.* **79**, 202 (1981).
63. C. Peterson and G. V. Pfeiffer: *Theor. Chim. Acta* **26**, 321 (1972)
64. P. S. Bagus and H. F. Schaefer, III: *J. Chem. Phys.* **56**, 224 (1972).
65. See, for instance, I. R. Levine: *Quantum Chemistry*, Prentice-Hall, Englewood Cliffs, New Jersey, pp. 474 and 510 (1991).
66. C. W. Bauschlicher, Jr. and P. R. Taylor: *J. Chem. Phys.* **85**, 2779 (1986).
67. W. von Niessen: *J. Am. Chem. Soc.* **99**, 7151 (1977).
68. Y. Osamura and H. F. Schaefer, III: *J. Chem. Phys.* **74**, 4576 (1981).
69. L. E. Nitzche and E. R. Davidson: *Chem. Phys. Lett.* **58**, 171 (1978). J. F. Graw and H. F. Schaefer, III: *J. Chem. Phys.* **83**, 1741 (1985).
70. D. W. Turner, C. Baker, A. D. Baker, and C. R. Brundle: *Molecular Photoelectron Spectroscopy*, Wiley-Interscience, New York, p. 252 (1970).
71. C. E. Bielschowsky, M. A. C. Nascimento, and E. Hollauer: *Phys. Rev.* **A42**, 5223 (1990).
72. T. Rescigno and A. E. Orel: *J. Chem. Phys.* **70**, 3390 (1979).
73. J. Muller, H. Agren, and O. Goscinski: *Chem. Phys.* **38**, 349 (1979).
74. B. H. Chirgwin and C. A. Coulson: *Proc. R. Soc., Ser. A* **201**, 196 (1958).
75. G. C. King, F. H. Read, and M. Tronc: *Chem. Phys. Lett.* **52**, 50 (1977).
76. R. N. S. Sody and C. E. Brion: *J. Electron Spectrosc.* **34**, 363 (1984).
77. R. Camilloni, E. Fainelli, G. Petrocelli, and G. Stefani: *J. Phys. B* **20**, 1893 (1987).
78. R. S. Barbieri and R. Bonham: *Phys. Rev.* **A45**, 7929 (1992).
79. M. Tronc, G. C. King, and F. H. Read: *J. Phys. B* **12**, 137 (1979).
80. G. R. Wright and C. E. Brion: *J. Electron Spectrosc.* **3**, 191 (1974).
81. D. M. Barrus, R. L. Blake, A. J. Burek, K. C. Chambers, and A. L. Pregenzer: *Phys. Rev.* **A20**, 1045 (1979).
82. R. McLaren, S. A. C. Clark, I. Ishii, and A. P. Hitchcock: *Phys. Rev.* **A36**, 1683 (1987).
83. K. Siegbahn, C. Nordling, G. Johanson, J. Hedman, P. F. Heder, K. Hamrin, U. Gelius, T. Bergmark, L. O. Wulm, and R. C. Brundle: *Molecular Photoelectron Spectroscopy*, Interscience, New York (1970).
84. S. Chung and C. C. Lin: *Appl. Opt.* **10**, 1790 (1971).
85. T. N. Rescigno and A. E. Orel: in F. A. Gianturco and G. Stefani (eds.), *Wavefunctions and Mechanisms for Electron Scattering Processes*, Springer-Verlag, Berlin, p. 215 (1984).
86. M. R. F. Siggel and D. T. Thomas: *J. Am. Chem. Soc.* **108**, 4360 (1986).
87. M. R. F. Siggel, D. T. Thomas, and A. Streitwieser: *J. Mol. Struct. (Theochem)* **165**, 309 (1988).
88. M. R. F. Siggel, D. T. Thomas, and L. J. Seathre: *J. Am. Chem. Soc.* **110**, 91 (1988).
89. M. R. F. Siggel, A. Streitwieser, and D. T. Thomas: *J. Am. Chem. Soc.* **110**, 8022 (1988).
90. M. J. S. Dewar and K. L. Krull: *J. Chem. Soc., Chem. Commun* 344 (1990).
91. F. G. Bordwell and A. V. Satish: *J. Am. Chem. Soc.* **116**, 8885 (1994).
92. P. C. Hiberty and C. P. Byrman: *J. Am. Chem. Soc.* **117**, 9875 (1995).
93. F. B. Brobowicz and W. A. Goddard, III: in H. F. Schaefer III (ed.), *Modern Theoretical Chemistry*, Plenum Press, New York, Vol. 3, p. 79 (1977).
94. J. E. Bartmess and R. T. McGiven: in M. J. Bowers (ed.), *Gas Phas Ion Chemistry*, Academic Press, New York, Vol. 2 (1979).

Molecular Engineering **7**: 109–147, 1997.

The Characteristic (Crossover) Temperature in the Theory of Thermally Activated Tunneling Processes

S. G. CHRISTOV
Institute of Physical Chemistry, Bulgarian Academy of Sciences, Sofia 1113, Bulgaria.

(Received: 12 September 1996; accepted 17 June 1997)

Abstract. We reconsider the concept of a characteristic (crossover) temperature and its applications to thermally activated tunneling processes. Alternative definitions of this concept are discussed. The analysis reveals the various aspects of the characteristic temperature and its derivatives in electron emission theory and reaction rate theory, including the proper dynamic and the stochastic approaches to the latter theory.

Key words: Tunnel effect, electron emission, reaction rates

1. Introduction

The quantum-mechanical tunnel effect is often related to the thermal activation of atomic systems in various physical, chemical and biological processes, such as electron emission from metals and semiconductors [1, 2], polaron motion in crystals [3, 4], gas-phase reactions involving hydrogen atoms [5, 6], proton transfer reactions in solution [7, 8], elementary acts in photosynthesis and vision [6, 9, 10], etc.

In thermally activated processes, the role of the tunnel effect depends essentially on temperature. Thus, a characteristic (crossover) temperature T_k can be defined at which the thermally averaged probabilities for overbarrier and tunneling transitions are equal. This has been done by Christov [11], who first calculated this characteristic temperature for the discontinuous rectangular barrier as a simple model for tunneling processes.

Somewhat latter, Goldanskii [12] treated the same problem for the continuous parabolic and Eckart barriers. He defined the characteristic temperature by the condition that the maximum of the thermally weighted tunneling probability lies at the barrier peak. However, he used the ordinary WBK approximation which is inaccurate in that energy range. Making use of a correct expression for the transition probability, Christov [13] calculated the characteristic temperature as defined in [11] for both parabolic and Eckart barriers. In this way, it was found that the correct results are twice as great than those found by Goldanskii [12]. These calculations were later extended to more general continuous one-dimensional barriers [14] and to many-dimensional potential energy surfaces which possess a saddle point [15, 16]. It is important to note that both alternative definitions of the characteristic temperature, proposed by Christov [11] and by Goldanskii [12], give nearly the same results, provided the same correct expression for the transition proability is used. These results coincide exactly only in the special case of infinitely high parabolic barriers [6, 15].

The characteristic temperature T_k introduced by Christov [11, 13] has been used in different areas of physics, chemistry aud biology [2, 4, 6–8, 10, 17–25] usually refered to as the 'Christov characteristic temperature' [7, 17–25]. The related quantity $T_k' = T_k/2$ introduced by Goldaiiskii [12] has another meaning which will be clarified below. The goal of this paper is to reconsider the concept of the characteristic (crossover) temperature and its applications in electron emission theory and reaction rate theory, including some new aspects of the recent development of the latter [24–30].

2. The Quantum-Mechanical Transition Probability

First we consider a one-dimensional potential energy barrier represented by a potential function $V(x)$, which has a maximum $E_c = V(x_m)$ at the point $x = x_m$. Denoting by $j_i(E_x)$ the incident paiticle current on the left and by $j_t(E_x)$ the transmitted current on the right of the barrier, one defines the transition probability $W(E_x)$ for a given particle energy E_x by the ratio j_t/j_i. The function $W(E_x)$ comprises both transitions over the barrier $(E_x > E_c)$ and through the barrier $(E_x < E_c)$, being a continuous function at all E_x. Exact expressions for $W(E_x)$ have been derived for a few barrier types such as the discontinuous rectangular and triangular barriers as well as for the continuous parabolic and Eckart barriers [7].

The general expression for the transition probability can be written in the quasiclassical form [6, 31, 32]

$$W(E_x) = \left[1 + A(E_x)\exp\left(\frac{2i}{\hbar}\int_{x_1}^{x_2} p_x(E_x)\,\mathrm{d}x\right)\right]^{-1},$$

$$p_x = \sqrt{2m(E_x - V(x)} \tag{1}$$

where p_x is the momentum of the particle, m is its mass, while x_1 and x_2 are the classical turning points at which $p_x = 0$ $(E_x = V(x))$ $(\hbar = h/2\pi$ where h is the Plank constant). The factor $A(E_x)$ depends on the form of the potential $V(x)$. It equals unity $(A = 1)$ if there are no other (real or complex) roots of $p_x = 0$ near x_1 and x_2 [31]. Such is the case of a parabolic function

$$V(x) = E_c - \frac{f_x^*}{2}(x - x_m)^2, \qquad f_x^* = |\mathrm{d}^2V/\mathrm{d}x^2| \tag{2}$$

which can be used to approximate any continuous barrier in its top region $(x \approx x_m)$. In the general case $A(E_x)$ tends to ∞ when $E_x \to 0$.

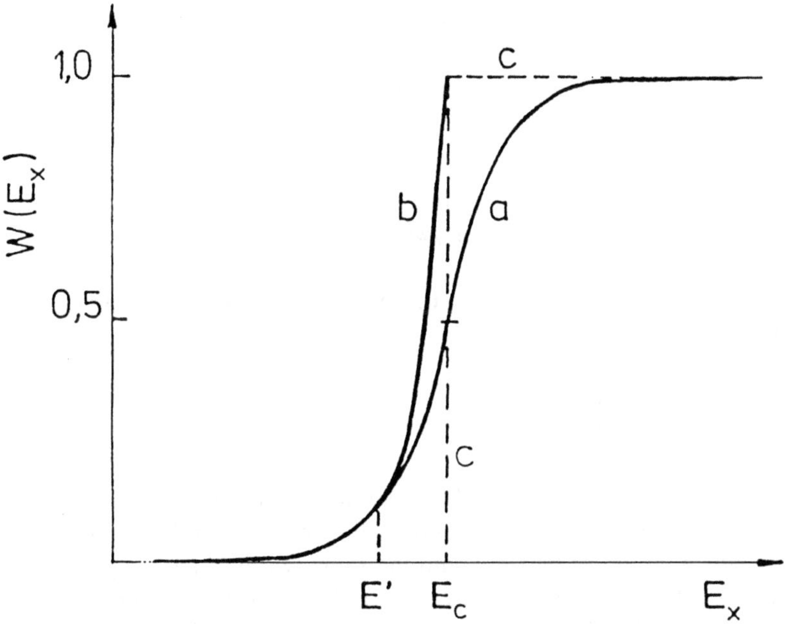

Figure 1. Transition probabilities $W(E_x)$ for a potential barrier $V(x)$ as a function of energy E_x. Curve (a) according to Equation (1) with $A(E_x) = 1$. Curve (b) according to Equation (3). (c) The step function is the classical transition probability ($W(E_x) = 0$ for $E_x < E_c$, $W(E_x) = 1$ for $E_x = E_c$). E_c is the barrier height, $E' = E_c - kT_k$ with T_k defined by Equation (23).

Expression (1) is valid for any value of the energy ($E_x \rightleftharpoons E_c$), hence for both tunneling transitions ($E_x < E_c$) at which x_1 and x_2 are real and p_x is imaginary, and overbarrier transitions ($E_x > E_c$) at which p_x is real while x_1 and x_2 are imaginary. For $E_x = E_c(x_1 = x_2, p_x = 0)$, one obtains $W(E_c) = 1/(1 + A(E_c))$ and for continuous barriers $W(E_c) \cong 1/2$. For $E_x \ll E_c$ with $A(E_x) = 1$ expression (1) turns into the usual quasi-classical Wentzel–Brillouin–Kramers (WBK) approximation

$$W(E_x) = \exp\left(-\frac{2i}{\hbar}\int_{x_1}^{x_2} p_x(E_x)\,\mathrm{d}x\right) \tag{3}$$

which for $E_x = E_c$ ($x_1 = x_2, p_x = 0$) gives $W(E_c) = 1$ instead of the correct result $W(E_c) = 1/2$. For $E_x = 0$ it has a finite value. However, the exact result is $W(E_x) = 0$ if the factor $A(E_x)$ is not omitted ($A^{-1}(0) = 0$) (see Figure 1).

The phase integral in (1) and (3) can be evaluated for a generalized Eckart barrier [33]

$$V(x) = B'\left[\frac{e^{2ax/l}}{(1+e^{2ax/l})^2} - C\right] + Q'\left[\frac{e^{2ax/l}}{1+e^{2ax/l}} - D\right] \quad (-l \le x \le l) \tag{4}$$

assuming $V(x) = 0$ for $x \leq l$ and $V(x) = Q$ for $x \geq l$. The constants

$$B' = B\left(\frac{e^{2a}+1}{e^{2a}-1}\right)^2, \quad Q' = Q\frac{(e^a + e^{-\alpha})^2}{e_x - e^{-2\alpha}},$$

$$C = \frac{e^{2n}}{(1+e^{2a})^2}, \quad D = \frac{1}{1+e^{2a}}$$

are functions of the parameter $a \geq 0$. Equation (4) describes a barrier of width $2l$ and height

$$E_c = E_c^0 + \frac{Q}{2}\left(1 + \frac{Q}{8E_c^0}\right) \tag{5}$$

where $E_c^0 = B/4$ and $B > Q$. For a symmetric barrier $(Q = 0) E_c = E_c^0$. At fixed values of the constants l, Q and B, the variation of the parameter a from 0 to ∞ yields a family of barriers with the same width $2l$ and height E_c located at $x = x_m = Ql/4E_c^0$. For $a = \pi$, $B' \cong B$, $Q' \cong Q$, $C \cong 0$ and $D \cong 0$, so that Equation (4) practically coincides with the expression of the Eckart barrier [33]. In the limit $a \to 0$ the function (4) turns into the equation [14]

$$V(x) = E_c^0\left(1 - \frac{x^2}{l^2}\right) + \frac{Q}{2}\left(1 + \frac{x}{l}\right) \tag{6}$$

which describes a truncated parabolic barrier in the range $-1 \leq x < l$.

The quasiclassical approximation $(A = 1)$ to expression (1) is always good for the upper part of the barrier potential (4) which can be well approximated by the parabolic function (2). Evaluating the phase integral in (1), yields [6, 14]

$$W(E_x) = [1 + \exp(\sigma' - \alpha' - \beta')]^{-1} \tag{7}$$

where

$$\alpha' = \frac{l'}{h}\sqrt{2mE_x'}, \quad \beta' = \frac{l'}{h}\sqrt{2m(E_x' - Q')}, \quad \sigma' = \sqrt{\frac{8mB'l'^2}{h^2} - 1} \tag{8}$$

with

$$l' = \frac{\pi l}{a}, \quad E_x' = E_x + BC + Q'D.$$

The approximation (7) is valid for any energy value $(E_x \rightleftharpoons E_c)$ if $\alpha' \gg 1$, $\beta' \gg 1$, $\sigma' \gg 1$. For the Eckart barrier $(a = \pi) l' = l$, $E_x' = E_x$ $(C = 0, D = 0)$, and $Q' = Q$, so that (7) represents a good approximation to the exact expression [33]

$$W(E_x) = \frac{\mathrm{ch}\,2\pi(\alpha+\beta) - \mathrm{ch}\,2\pi(\alpha-\beta)}{\mathrm{ch}\,2\pi(\alpha+\beta) + \mathrm{ch}\,2\pi\sigma} \tag{9}$$

where $\beta = \beta'$, $\alpha = \alpha'$ and $\sigma = \sigma'$ for $a = \pi$. For a parabolic barrier ($a = 0$) expression (7) turns into [14]

$$W(E_x) = [1 + \exp(2\pi(E_c - E_x)/h\nu^*)]^{-1} \tag{10}$$

where $\nu^* = (f_x^*/m)^{1/2}/2\pi$, $f_x^* = 2E_c^0/I^2$ is the absolute value of the imaginary barrier frequency ν^*. The same result is obtained [34] by direct use of the parabolic potential (6) to evaluate the phase integral in (1). Note that for a nontruncated (infinitely high) parabolic barrier (2) formula (10) represents an exact expression for the transition probability [7, 31] for any energy value ($E_x \rightleftharpoons E_c$).

The curvature $f_x^* = |\mathrm{d}^2V/\mathrm{d}x^2|(x = x_m)$ in the top region of the barrier potential (4) varies from $2E_c^0/l^2$ (for $a = 0$) to ∞ (for $a \to \infty$). For finite values of parameter a the barrier (4) can be approximated in that region ($x \approx x_m$) by the parabolic function (2) making use of the expression [14]

$$f_x^* = \frac{4Ba^2}{l^2}\left[\frac{2-(\delta^{1/2}-\delta^{-1/2})^2}{(\delta^{1/2}+\delta^{-1/2})^2}\right] + \frac{4Q\mu a^2}{l^2}\left[\frac{\delta^{1/2}-\delta^{-1/2}}{(\delta^{1/2}+\delta^{-1/2})^3}\right] \tag{11}$$

$$\delta = \frac{B+\mu Q}{B-\mu Q}, \quad \mu = \frac{(e^a+e^{-a})^2}{e^{2a}-e^{-2a}}.$$

For a symmetric barrier ($Q = 0$), the second term vanishes. If $a = 0$, $f_x^* = 2E_c^0/l^2$ (note that $B = 4E_c^0/(1-4C)$). This simple expression for f_x^* holds for any finite value of a with corresponding values of E_c^0 and l. Therefore, the quasiclassical fomula (10) for $W(E_x)$ can be used for finite values of the barrier curvature f_x^* in the transition region from the classical ($E_x > E_c$) to the tunneling regime ($E_x < E_c$).

This is valid, in particular, for the Schottky potential

$$V(x) = V_a - e_0^2/4Dx - e_0Fx \tag{12}$$

which represents the potential energy of an electron as a function of the distance x from the metal (semiconductor) surface in the presence of an electric field with strength F, e_0 being the electron charge and D the dielectric constant of an insulator in the metal/insulator system. This potential is used in the theory of electron emission from metals in vacuum ($D = 1$) or dielectrics [1, 2]. It represents a barrier with maximum

$$E_c = V_a - (e_0^3F/D)^{1/2} \tag{13}$$

at $x_m = \frac{1}{2}(e_0 DF)^{1/2}$ and a top curvature

$$f^* = (e_0 D)^{1/2} F^{3/2}. \tag{14}$$

The parabolic approximation (2) in the region $x \approx x_m$ is very good except at low field strengths [2]. Thus, in the top energy range formula (10) can be used to calculate the transition probability in electron emission phenomena [2].

We now consider a many-dimensional potential energy surface $V(\tilde{x})$ ($\tilde{x} \equiv \{x_1, x_2, \ldots, x_i, \ldots, x_z\}$) which has a saddle point at $\tilde{x} = \tilde{x}^*$. Using normal mode coordinates, we assume $\mathrm{d}^2V/\mathrm{d}x_n^2 > 0$ along x_n, so that the potential

$$V(x_n) = E_c - \frac{f_n^*}{2}(x_n - x_n^*)^2 \tag{15}$$

represents a parabolic barrier with curvature f_x^* in the saddle-point region. In particular, the surface $V(\tilde{x})$ can be represented by the equation

$$V(\tilde{x}) = \frac{1}{2}(V_1 + V_2) - \frac{1}{2}[(V_1 - V_2)^2 + 4V_{12}^2]^{1/2} \tag{16}$$

where

$$V_1(\tilde{x}) = \sum_{i=1}^{z} \frac{f_i}{2} x_i^2, \qquad V_2(\tilde{x}) = \sum_{i=1}^{z} \frac{f_i}{2}(x_i - x_i^0)^2 + Q \tag{16'}$$

are two interesting z-dimensional paraboloids and $V_{12} > 0$ is a coupling constant. The minima of $V_1(\tilde{x})$ and $V_2(\tilde{x})$ are located at $\tilde{x} = 0$ and $\tilde{x} = \tilde{x}^0$, respectively, and $Q = V_2(\tilde{x}^0) - V(0) > 0$. There is a line x of minimum energy normal to x (the classical reaction coordinate) that connects the points $\tilde{x} = 0$ and $\tilde{x} = \tilde{x}^0$ passing through the saddle-point $\tilde{x} = \tilde{x}^*$. In the region $\tilde{x} \approx \tilde{x}^*$, the reaction coordinate coincides with the normal mode coordinate x_n. The barrier curvature is found to be [16]

$$f_n^* = \frac{\sum_{i=1}^{z} f_i E_r^{(i)}}{2V_{12}} \left(1 - \frac{Q^2}{E_r^2}\right)^{3/2} - \frac{\sum_{i=1}^{z} f_i^2 E_r^{(i)}}{\sum_{i=1}^{z} f_i E_r^{(i)}} \tag{17}$$

where

$$E_r^{(i)} = \frac{f_i}{2}(x_i^0)^2, \quad E_r = \sum_{i=1}^{z} E_r^{(i)}.$$

This expression is valid if V_{12} is not very small, so that f_n^* has a finite value. The transition probability from point $\tilde{x} = 0$ to point $\tilde{x} = \tilde{x}^0$ through the saddle point

$\tilde{x} = \tilde{x}^m$ at energies $E_x \approx E_c$ can be calculated by expression (10) using (16) for $V(x) = V(x_n)$ and (17) for $f^* = f_n^*$. This possibility is very useful in the reaction rate theory in which the potential surface (16) is adequate to many phenomena in dense media [6, 16, 29, 38, 56, 57].

3. The Characteristic (Crossover) Temperature

In thermally activated events, one needs the weighted transition probability

$$F(E_x) = W(E_x) f(E_x, T) \tag{18}$$

where

$$f(E_x, T) = \frac{1}{kT} \exp(-E_x/kT) \tag{19}$$

represents a Boltzmann energy distribution among the particle current j_i incident on the left of a one-dimensional barrier with height $E_{c'}$. We assume that on the right of the barrier its height is $E_c - Q$ $(Q \neq 0)$. The total (thermally averaged) probability is

$$P = \int_0^\infty F(E_x)\,\mathrm{d}E_x = \int_0^\infty W(E_x) \exp(-E_x/kT)\,\mathrm{d}E_x/kT. \tag{20}$$

In the classical approximation

$$W(E_x) = 0 \quad \text{for} \quad E_x < E_c, \qquad W(E_x) = 1 \quad \text{for} \quad E_x > E_c$$

so that expression (20) yields

$$P^{\mathrm{cl}} = \exp(-E_c/kT). \tag{21}$$

Using formula (10) for $W(E_x)$ from (20) and (21), one obtains for a parabolic barrier (6) $(Q < 0)$

$$\kappa_t \equiv \frac{P}{P^{\mathrm{cl}}} = \int_{-E_c}^{\infty} \frac{e^{-\epsilon_x/kT}}{1 + e^{-2\epsilon_x/kT_k}} \mathrm{d}\epsilon_x/kT \tag{22}$$

where $\epsilon_x = E_x - E_c$ and the quantity

$$T_k = h\nu^*/\pi k \tag{23}$$

is introduced at this point for convenience. The ratio $\kappa_t = P/P^{\mathrm{cl}}$ is a definition of the tunneling correction to the classical rate of transitions. If $Q > 0$, E_c is replaced by $E_c - Q$.

The characteristic temperature can be calculated from the condition [11, 13]

$$P'/P'' = \kappa'/\kappa'' = 1 \tag{24}$$

where $\kappa' = P'/P^{\text{cl}}$ and $\kappa'' = P''/P^{\text{cl}}$ so that

$$\kappa' = \int_{-E_c}^{0} F(\epsilon_x)\,\mathrm{d}\epsilon_x,$$

$$\kappa'' = \int_{0}^{\infty} F(\epsilon_x)\,\mathrm{d}\epsilon_x \tag{25}$$

correspond to the total (average) probabilities P' and P'' for tunneling ($\epsilon_x < 0$) and overbarrier transitions ($\epsilon_x > 0$). The integrand $F(\epsilon_x)$ is given by (18) and (19) by replacing E_x by $\epsilon_x = E_x - E_c$.

Thus, for the parabolic barrier (6)

$$F(\epsilon_x) = \frac{1}{kT} e^{-\epsilon_x/kT}/(1 + e^{-2\epsilon_x/kT_k}) \tag{26}$$

represents the integrand in (22). Setting $y = -\epsilon_x/E_c$, $\alpha = E_c/kT$ and $\beta = 2E_c/kT_k$ from (25), we find

$$\kappa' = \alpha \int_0^1 \frac{e^{\alpha y}\,\mathrm{d}y}{1 + e^{\beta y}}, \quad \kappa'' = \alpha \int_{-\infty}^{0} \frac{e^{\alpha y}\,\mathrm{d}y}{1 + e^{\beta y}}. \tag{27}$$

For $\beta = 2\alpha (T = T_k)$ evaluation of the integrals gives [13]

$$\kappa' = \arctan e^{\alpha} - \frac{\pi}{4}, \quad \kappa'' = \frac{\pi}{4}. \tag{28}$$

If $\alpha \gg 1 (E_c \gg kT)$, $\arctan e^{\alpha} = \pi/2$, therefore, $\kappa' = \kappa''$. In this way we prove that for a very high parabolic barrier the characteristic temperature T_k, defined by (24), is given exactly by expression (23). This result also holds for $Q > 0$ in which $\alpha = (E_c - Q)/kT$, so that $E_c - Q \gg kT$ is the condition of its validity [13].

The same result can be derived using the definition of Goldanskii [12] by determination of the maximum of $F(\epsilon_x)$. Using (26) from $\mathrm{d}F/\mathrm{d}\epsilon_x = 0$, one obtains the equation [6, 15a]

$$\epsilon_x^{(m)} = E_x^{(m)} - E_c = -\frac{kT_k}{2} \ln \frac{\frac{1}{2}(T_k/T)}{1 - \frac{1}{2}(T_k/T)} \tag{29}$$

which is valid for $T > T_k/2$. For $T = T_k$, $\epsilon_x^{(m)} = 0 (E_x^{(m)} = E_c)$, so that the maximum of the thermally weighted transition probability $F(\epsilon_x)$ lies exactly at the barrier peak ($E_x^{(m)} = E_c$) Goldanski [12] has used the incorrect expression

(3) for the transition probability $W(E_x)$ which is valid only for $E_x \ll E_c$. For the parabolic barrier this expression gives

$$F(\epsilon_x) = \frac{1}{kT} \exp\left\{-\epsilon_x \left(\frac{1}{kT} - \frac{2}{kT_k}\right)\right\} \tag{30}$$

which directly follows from (26) for $\epsilon_x = E_x - E_c \gg kT/2(E_x < E_c)$. The equation $dF/d\epsilon_x = 0$ has three solutions: $\epsilon_x = \infty(E_x \gg E_c)$, $\epsilon_x = 0(E_x = E_c)$ and $T = T_k/2$. The first two are inconsistent with the condition of validity of expression (30) $\epsilon_x \ll 0(E_x \ll E_c)$. None of these three solutions defines the position of a real maximum of $F(\epsilon_x)$ like Equation (29). Formally, the solution $T = T_k/2$ coincides with the result of Goldanskii [12]

$$T'_k = T_k/2 = h\nu^*/2\pi k. \tag{31}$$

However, its derivation in this way is obviously incorrect.

Another kind of 'derivation' of (31) is based on Equations [26, 28]

$$\exp(-E_c/kT) = \exp(-2E_c/kT_k)$$

where the Arrhenius factor is set equal to the tunneling factor for a parabolic barrier foi $E_x = 0$ in the quasiclassical approximation. This equation has the meaning of an extrapolation of the classical (overbarrier) transition probability (21) $(E_x > E_c)$ in the low temperature range $(T \to 0)$ in which the tunneling probability is independent of energy $(E_x \to 0)$. This extrapolation yields the intersection point of two straight lines which evidently has no direct physical sense. Moreover, the nonzero value of the tunneling probability for $E_x = 0$ is an artefact of the quasiclassical approximation, as disussed in the derivation of expression (3).[1] An interpretation of the quantity T'_k, defined by (31) in the framework of the ordinary tunneling theory of thermally activated processes, will be given below, where the actual relatious between T_k and $T'_k = T_k/2$ will be discussed.

For this purpose we first calculate the integral in (22) to obtain the exact expression [6]

$$\kappa_t = \frac{\pi\delta/2}{\sin(\pi\delta/2)} - \frac{\delta/2}{1-(\delta/2)} e^{b(\delta-2)} \sum_{n=1}^{\infty} (-1)^n \frac{1-(\delta/2)}{(n+1)-(\delta/2)} e^{-2nb} \tag{32}$$

---

[1] A more accurate expression based on the above extrapolation approach is

$$T'_k = \frac{T_k}{2} \frac{E_c}{E_c - \epsilon_0}$$

where $\epsilon_0 = h\nu/2$ is the zero-point energy of vibration in a potential well on the left of the barrier. The tunneling probability for $E_x = \epsilon_0$, calculated by the WBK approximation (3), differs by a factor of about 2 from the exact results [Christov, S.G., *Phil. Mag. B* **49**, 325 (1984)].

where $\delta = T_k/T$ and $b = E_c/kT_k$. This expression was first derived by Bell [35] for a symmetric parabolic barrier ($Q = 0$) and by Christov, S.G. [34] for an asymmetric barrier ($Q > 0$) for which $b = (E_c - Q)/kT_k$. We observe that if $\delta/2 = n + 1$ ($n = 0, 1, \ldots, \infty$) both the first term and the corresponding term in the series are infinite. However, their sum is finite and equals $(-1)^{n+1}(E_c/kT)$ [7]. Therefore, expression (32) is valid for any value of $\delta = T_k/T$. In particular, if $\delta = 2(T = T_k/2)$, a direct evaluation of the integral in (22) yields [35, 34]

$$\kappa_t = \ln(1 + e^{2b}), \quad b = E_c/kT_k. \tag{33}$$

The same result follows from expression (32) [7]. If $\delta < 2(T > T_k/2)$ and $b \gg 1(E_c \gg kT_k)$, the second term in (32) can be neglected, so that

$$\kappa_t = \frac{\pi\delta/2}{\sin(\pi\delta/2)}, \quad \delta = T_k/T. \tag{34}$$

For $\delta = \frac{1}{2}(T = 2T_k)$, this formula gives $\kappa_t = 1.12$, for $\delta = 1(T = T_k)$, $\kappa_t = \pi/2$ and for $\delta = 2(T = T_k/2)$, $\kappa_t = \infty$, instead of the correct result from (33), $\kappa_t = 2b$ for $b \gg 1(E_c/kT_k \gg 1)$.

An approximation to expression (32) is [34, 36]

$$\kappa_t = \frac{2}{2-\delta}\left(1 - \frac{\delta}{2}e^{b(\delta-2)}\right) \tag{35}$$

which is valid for all values of $\delta = T_k/T$. For $\delta = 1$, it gives

$$\kappa_t = 2\left(1 - \frac{1}{2}e^{-2b}\right) \tag{36}$$

so that for $2b \gg 1(E_c \gg kT_k/2)\kappa = 2$ instead of the more correct result of expression (34) $\kappa = \pi/2 \cong 1.57$. For $\delta = 2(T = T_k/2)$ from (35), we get

$$\kappa_t = 1 + 2b, \quad b = E_c/kT_k \tag{37}$$

in agreement with the result of expression (33) for $2b \gg 1$. Therefore, for $\delta > 2(T < T_k/2)$ formula (35) is a very good approximation to the more complicated formula (32) [34]. When $b = E_c/kT_k \gg 1$, the tunneling correction is very large in the temperature range $T \leq T_k/2$.

The above analysis shows that the characteristic temperature $T_k' = T_k/2$, introduced by Goldanskii [12] by Equation (31) is related to the divergence of expression (34) at $\delta = 2(T = T_k/2)$, which results from the approximation involved in this expression by neglecting the second term of the exact expression (32) for κ. From a physical point of view T_k' has no precise meaning in contrast to T_k, which is defined exactly by condition (24). Nevertheless, the quantity $T_k' = T_k/2$ can be used conventionally for a definition of the temperature range $T < T_k/2$ of large tunneling [6, 7, 13–15], in which the tunneling correction, according to (33) or (37), can be very large ($\kappa \geq 2E_c/kT_k \gg 1$).

Formula (23) is an exact expression for the characteristic temperature T_k for a very high parabolic barrier ($E_c \gg kT$), however, it can be used as an approximation for any barrier which can be described in the top region by the parabolic function (2). This is the case for potential function (4) for any finite value of parameter $a > 0$. It has been shown [14, 15] that the parabolic approximation is very good in the energy range $E_x > E_c - kT_K$, so that the quantity $E_c - E_x = kT_k$ can be considered as a characteristic energy of this barrier type. The same is valid for the Schottky potential (12) at sufficiently high values of F[2]. The exact expression for T_k for such barriers has the form [13, 14, 37]

$$T_k = \alpha h \nu_x^* / \pi k \tag{38}$$

where $\alpha \approx 1$. Thus, for instance, for the Eckart barrier ($a = \pi$ in (4)) $\alpha = 7/8$, so that setting $\alpha = 1$ is a good approximation.

The exact derivation of the characteristic temperature T_k for a many-dimensional potential energy surface, which possesses a col, has shown [6, 15] that expression (38) with $\alpha = 1$ is sufficiently accurate if the curvature of the classical reaction coordinate x in the col region is small, so that the most probable tunneling path at $T \approx T_K$ is close to x. Then the parabolic approximation (15) is a basis for a determination of the barrier frequency $\nu^* = (1/2\pi)(f_n^*/m)^{1/2}$ for finite values of the barrier curvature f_n^*. In particular, this is the case of the potential surface (16) if V_{12} is not too small.

The characteristic temperature T_k, defined by (23), can be used on the basis of Equations (32–37) for an approximate delimitation of four temperature ranges [6, 7, 15]: (1) classical range $T > 2T_k$ ($\kappa \approx 1$); (2) range of small tunneling $2T_k > T > T_k$ ($\kappa = 1$ to $\pi/2$); (3) range of moderate tunneling $T_k > T > T_k/2$ ($\kappa = \pi/2$ to $2E_c/kT_k$); (4) range of large tunneling $T < T_k/2 (\kappa > 2E_c/kT_k)$. This conventional division is appropriate for any barrier which can be approximated near the top by a parabola, although for $T < T_k/2$ the tunneling correction κ_t is usually smaller than that for the parabolic barrier (6). The transition range between the classical regime ($T > 2T_k$) to the large tunneling regime ($T < T_k/2$) can be quite wide. Thus, for $T_k = 300$ K it lies between 150 K and 600 K. Within this transition range the actual crossover is $T = T_k$ at which, according to condition (24), the probability for tunneling (P') equals that for overbarrier transfer (P'').

4. Calculation of the Average Transition Probability

We consider a one-dimensional barrier $V(x)$ which can be approximated by a parabola in the energy range $E_x > E_c - kT_k$ as is the case for potential functions (4) and (12). In this range formula (10) for $W(E_x)$ can be used. It has been shown [37] that for $E_x < E_c - kT_k$ the usual WBK approximation (3) is very good.

Therefore, expression (20) for the the thermally averaged transition probability can be written as a sum

$$P = P_1 + P_2 = \int_0^{E_c - kT_k} e^{-S(E_x,T)/\hbar}\, \mathrm{d}E_x/kT + \\ + \int_{E_c-kT_k}^{\infty} \frac{e^{-E_x/kT}}{1 + e^{2(E_c - E_x)/kT_k}} \frac{\mathrm{d}E_x}{kT} \tag{39}$$

where the full action is

$$S(E_x, T) = 2i \int_{x_1}^{x_2} p_x(E_x)\, \mathrm{d}x + \frac{\hbar E_x}{kT} \tag{40}$$

and $p_x = [2m(E_x - V(x))]^{1/2}$.

The function $S(E_x, T)$ usually has a minimum. Then the first term in (39) can be calculated by an improved method of steepest descent [37, 38] which yields

$$P_1 = \left(\frac{2\pi\hbar}{S''(E_m)}\right)^{1/2} \frac{\Phi(y_m^{(1)}) + \Phi(y_m^{(2)})}{kT} e^{-S(E_m)/\hbar} \tag{41}$$

where $S''(E_m) = (\mathrm{d}^2 S/\mathrm{d}E_x)_{E_x = E_m}$, E_m is the value of E_x at which $S(E_x, T)$ has a minimum ($S' = \mathrm{d}S/\mathrm{d}E_x = 0$), defined by the equation

$$S'(E_m)kT/\hbar = -1. \tag{42}$$

The error integrals

$$\Phi(y) = \frac{2}{\sqrt{2\pi}} \int_0^y e^{-t^2/2}\, \mathrm{d}t \tag{43}$$

have the arguments

$$y_m^{(1)} = E_m[S'(E_m)/\hbar]^{1/2}, \quad y_m^{(2)} = (E_c - kT_k) - E_m.$$

Evaluation of the second term in (39) gives

$$P_2 = \left(\frac{\pi\delta/2}{\sin(\pi\delta/2)} - \frac{\delta/2}{1 - \delta/2}\right) e^{-E_c/kT}, \quad (T > T_k/2) \tag{44a}$$

$$P_2' = 2\left(1 + \frac{1}{2}(\delta - 2) - \cdots\right) e^{-E_c/kT}, \quad (T \cong T_k/2) \tag{44b}$$

$$P_2'' = \frac{\delta/2}{(\delta/2) - 1} e^{\delta - 2} e^{-E_c/kT}, \quad (T < T_k/2) \tag{44c}$$

where $\delta = T_k/T$.

Equations (39–44) allow P to be calculated at all temperatures. The first term in (39) dominates for $T < T_k/2$ $(P_1 > P_2)$ and the second one for $T > T_k/2$ $(P_2 > P_1)$, the crossover being $T \cong T_k/2$ $(P_1 \approx P_2)$. Indeed, Equation (29) shows that the maximum of $F(E_x)$ (the integrand in (20)) lies in the range $E_x > E_c - kT_k$ when $T > T_k/2$ $(E_m^{(x)} = E_c - kT_x$ for $T_k/T = 2/(1+e^{-2}) \cong 2)$. In the classical temperature range $T > 2T_k$ the maximum of $F(E_x)$ lies in the energy range $E_x > E_c + kT_k/2$, so that tunneling can be neglected. Then, Equation (39) yields the Arrhenius equation (21) $(P = P^{\mathrm{cl}})$. In the temperature range of moderate tunneling $(2T_k > T > T_k/2)$ the maximum of $F(E_x)$ is between $E_x = E_c + kT_k/2$ and $E_x = E_c - kT_k$. In this range the expression

$$P = \kappa_t e^{-E_c/kT}, \quad \kappa_t = \frac{(\pi/2)(T_k/T)}{\sin(\pi/2)T_k/T} \tag{45}$$

is a good approximation to Equation (39) if $(T_k/T > 3/2)$. The tunneling correction κ_t coincides with formula (34). In the temperature range of large tunneling $(T < T_k/2)$, the second term in (39) is small compared to the first one when the maximum of $F(E_x)$ is in the energy range $E_x < E_c - kT_k$. In this case the approximation to (39)

$$P = \left[\frac{2\pi\hbar}{S''(E_m)}\right]^{1/2} \frac{1 + \Phi(y_m^{(1)})}{kT} \exp(-S(E_m, T)/\hbar) \tag{46}$$

is valid.

This approach has been applied to evaluate the average transition probabilities of the potential barriers, described by Equations (4) [37] and (12) [2]. The same is done for potential (16) in the particular case in which

$$V_1(x) = \frac{f_x}{2}x^2, \quad V_2(x) = \frac{f_x}{2}(x - x_0)^2 + Q. \tag{47}$$

The barrier frequency is given by [16]

$$\nu_x^* = \nu_x \left[\frac{E_r^{(x)2}}{2|V_{12}|}\left(1 - \frac{Q^2}{E_r^{(x)}}\right)^{3/2} - 1\right]^{1/2} \tag{48}$$

where $\nu_x = (f_x/m)^{1/2}/2\pi$ is the vibration frequency in the potential well $V_1(x)$ and $E_r^{(x)} = \frac{1}{2}f_x x_0^2$ is the 'reorganization energy' of the oscillator, x_0 being the distance between the equilibrium positions $x = 0$ and $x = x_0$ in the wells $V_1(x)$ and $V_2(x)$. This formula implies that $|V_{12}|$ is sufficiently large in order to use the parabolic approximation (2) in the barrier region (the crossing point of $V_1(x)$ and $V_2(x)$). Using (48), the characteristic temperature $T_k = h\nu_x^*/\pi k$ can be calculated for the harmonic oscillator model considered. The full action (40) is found under the condition $|V_{12}| \ll E_c$ [4, 39]. In the particular case of a symmetric barrier ($Q = 0$), it reads

$$S(E_m, T) = \frac{2E_c}{\pi\nu_x} \tanh \frac{h\nu_x}{4kT}. \tag{49}$$

The second derivative is

$$S''(E_m, T) = [\pi\nu_x E_c \operatorname{sech}(h\nu_x/4kT) \tanh(h\nu_x/4kT)]^{-1} \tag{50}$$

where

$$E_m = E_c \operatorname{sech}^2(h\nu_x/4kT). \tag{51}$$

Using Equations (48–51), one can calculate the average transition probability P on the basis of Equations (39–45).

5. Electron Emission Theory

We begin with the application of the characteristic temperature in the theory of electron emission from metals in insulators (vacuum or dielectric) which is governed by the Schottky potential (12). The characteristic temperature can be found by taking into account that in the energy $E_x > \mu + 2kT$, where μ is the Fermi energy, the Fermi–Dirac energy distribution turns into the classical Boltzmann distribution, E_x being the electron energy for motion along the x-coordinate normal to the metal surface. Therefore, formula (23) for T_k is valid with

$$\nu^* = (f^*/m_c)^{1/2}/2\pi$$

where the barrier curvature is given by (14). This yields [2]

$$T_k = \frac{h(e_0 D)^{1/4}}{\pi^2 k m_c^{1/2}} F^{3/4} \tag{52}$$

D and m_c being the dielectric constant and the effective electron mass on the lower edge of the conduction band of the insulator (for vacuum $D = 1$ and $m_c = m_0$ where m_0 is the free electron mass).

The density of the emitted electron current is given by the expression

$$j = \frac{4\pi m_c e_0 kT}{h^3} \int_0^\infty W(E_x) \ln\{1 + \exp[-(E_x - \mu)/kT]\} \, \mathrm{d}E_x$$

where the transition probability $W(E_x)$ is obtained from (1) with $A(E_x) = 1$ and (12). The integral can be evaluated in the whole temperature range [2a] by taking into account that for $(E_x - \mu)/kT > 2$

$$\ln\{1 + \exp[-(E_x - \mu)/kT]\} \cong \exp[-(E_x - \mu)/kT].$$

In this energy range an approach similar to that discussed in Section 4 can be used, while in the lowest energy range $(E_x - \mu)/kT < 2$ another approach, based on the exact energy distribution, must be applied [2] in order to calculate the corresponding part of the current density. Thus, the total current is represented as a sum of three terms, corresponding to the energy ranges $E_x > E_c - kT_k$, $\mu + 2kT < E_x < E_c - kT_x$ and $E_x < \mu + 2kT$. In this way [2], a very good accuracy of the current calculation is achieved for the entire ranges of field and temperature, the maximal error being less than 10% [2].

In the temperature range of moderate tunneling $T > T_k/2$, the tunneling correction (34) is applicable, therefore, the current density can be written as [2]

$$j = \frac{(\pi/2)(T_k/T)}{\sin[(\pi/2)(T_k/T)]} j_{\mathrm{RS}} \tag{53}$$

where

$$j_{\mathrm{RS}} = \frac{4\pi m_c e_0 (kT)^2}{h^3} e^{-(\chi - e_0^{3/2} F^{1/2})/kT} \tag{53'}$$

is the Richardson–Schottky equation for thermionic emission ($\chi = W_a - E_f$ is the metal–insulator work function). Equation (53) coincides with an expression first derived by Murphy and Good [1]. In this way we find that the condition $T > T_k/2$ defines the 'extended thermionic emission region', while the condition $T > 2T_k$ is the definition of the proper 'thermionic emission region', in which $j = j_{\mathrm{RS}}$.

The characteristic temperature T_k for electron emission from metals in vacuum has been tabulated for different values of the electric field strength using the formula (52) by Bermond *et al.* [18]. Similar tabulated data for electron emission from metals in dielectrics and semiconductors have been presented by Vodenicharov [42]. For instance, one finds [2a] that for the emission from metals in vacuum at a field strength $F = 2.10^7$ V/cm the value of T_k is 3100 K, therefore, expression (53) is valid for the temperature range $T > T_k/2$ = 1550 K. For emission from metals in an insulator with $D = 8.9$ and $m_c = 1.5m_0$ at $F = 1.104$ eV one obtains T_k = 466 K [42], hence one can use Equation (53) in the range $T > T_k/2 = 233$ K.

Exact calculations of the normal energy distribution of the electrons emitted from a metal in a vacuum [2a] have shown that at $T = T_k$ the maximum of the distribution lies precisely at the barrier peak ($E_x = E_c$) and is symmetric with respect to it (Figure 2). These results correspond to both definitions of the characteristic temperature T_k introduced by Goldanskii [12] and Christov [11, 13], as discussed in Section 3. The coincidence of these definitions is due to the fact that the Schottky potential (12) is well approximated by the parabolic function (2) in the energy range $E_x > E_c - kT_k$, when the field strength is sufficiently high [2a]. At $T = T_k/2$ the maximum of the energy distribution lies somewhat below the energy value $E_x = E_c - kT_k$, while for the ideal parabolic barrier, according to

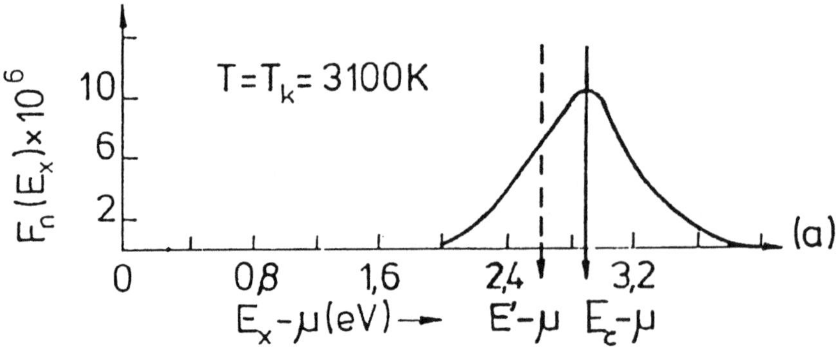

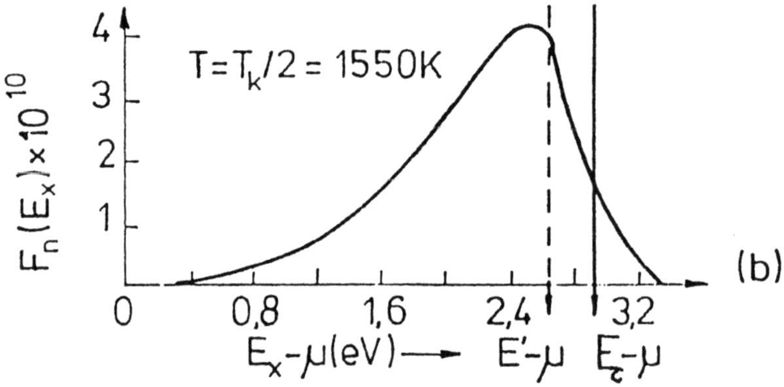

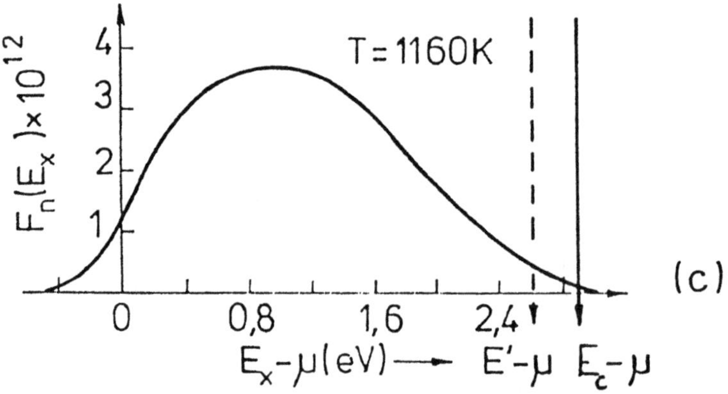

Figure 2. Exact normal energy distribution $F_n(E_x)$ of the electrons emitted from a tungsten cathode in vacuum [2a,b] for three temperatures (field strength 2×10^9 V m^{-1}, work function $\chi = 4.5$ eV); $E'_x = E_c - kT_k$ with T_k defined by Equation (23).

(29), it lies somewhat above this value. We conclude that for $T < T_k/2$ the electron emission occurs in the so called 'temperature-field $(T - F)$ emission region'.

We recall that expression (34) for the tunneling correction yields the values $\kappa_t = \pi/2$ for $T = T_k$ and $\kappa_t \cong 2E_c/kT_k$ for $T = T_k/2$, so that in the latter case we find for electron emission in vacuum that κ_t ranges between 50 and 100 for E_c = 5–10 eV and $T_k \cong 3000$ K ($F = 10^7$ V/cm).

All these results of the study of electron emission phenomena are in complete accord with the conclusions in Section 3 that the characteristic temperature T_k is the actual crossover between the classical and the tunneling regime, while the derivative temperature $T_k' = T_k/2$ can be regarded as crossover between the temperature ranges of moderate and large tunneling.

It is remarkable that the tunneling correction $\kappa_t = j/j_{RS}$ to the Richardson–Schottky equation can be represented as function of the ratio T_k/T not only in the temperature range $T > T_k/2$ (extended thermionic emission region) but also below $T = T_k/2$ ($T - F$ emission region), although an analytical expression like (34) for κ_t covering the entire temperature range $T \leq T_k/2$ for the Schottky barrier has not been derived (Figure 3). Thus, the characteristic temperature T_k proves to be an important parameter in the unified theory of electron emission which has been confirmed experimentally [18, 19, 40, 41, 43] in wide field and temperature ranges, including the 'field emission region' in which the emitted electron current becomes independent of temperature.

6. Reaction Rate Theory: Dynamic Approach

In the general case of a many-dimensional potential energy surface $V(\tilde{x})$, which has a col the reaction rate (or the rate constant) at temperature T is given by the equation [6]

$$v = \frac{kT}{hZ} \sum_n \sum_{n_x} W_n(E_{n_x}) g_n e^{-(E_n + E_{n_x})/kT} \Delta E_{n_x}/kT \tag{54}$$

where Z is the full partition function of the initial state. E_{n_x} and E_n are the energies for motion along the reaction coordinate x and normal to it, respectively, in the reactions region of configuration space $\tilde{x} \equiv \{x_1, x_2, \ldots, x_i, \ldots,\}$, $\mathbf{n}_x$ and $\mathbf{n}$ are the corresponding quantum numbers, $W_n = \sum_{n'} W_{n,n'}$ is the total transition probability from initial state $\mathbf{n}$ to any final state $\mathbf{n}'$ and $\Delta E_{n_x} = E_{n_x+1} - E_{n_x}$ is the spacing between the energy levels E_{n_x} (g_n is a statistical weight factor). If $\Delta E_{n_x}/kT \ll 1$, the sum over $\mathbf{n}_x$ can be replaced by an integral over $E_x(\Delta E_{n_x} \to \mathrm{d}E_x)$.

The derivation of the rate Equation (54) is based on the assumption that the thermal equilibrium among the reactants exists only in the initial state of the system. The most useful formulation of the reaction rate can be derived by making use of the law of energy conservation

$$E = E_{n_x} + E_n = E_x^+ + \epsilon_n^+ = E_c + \epsilon_x^+ + \epsilon_n^+ \tag{55}$$

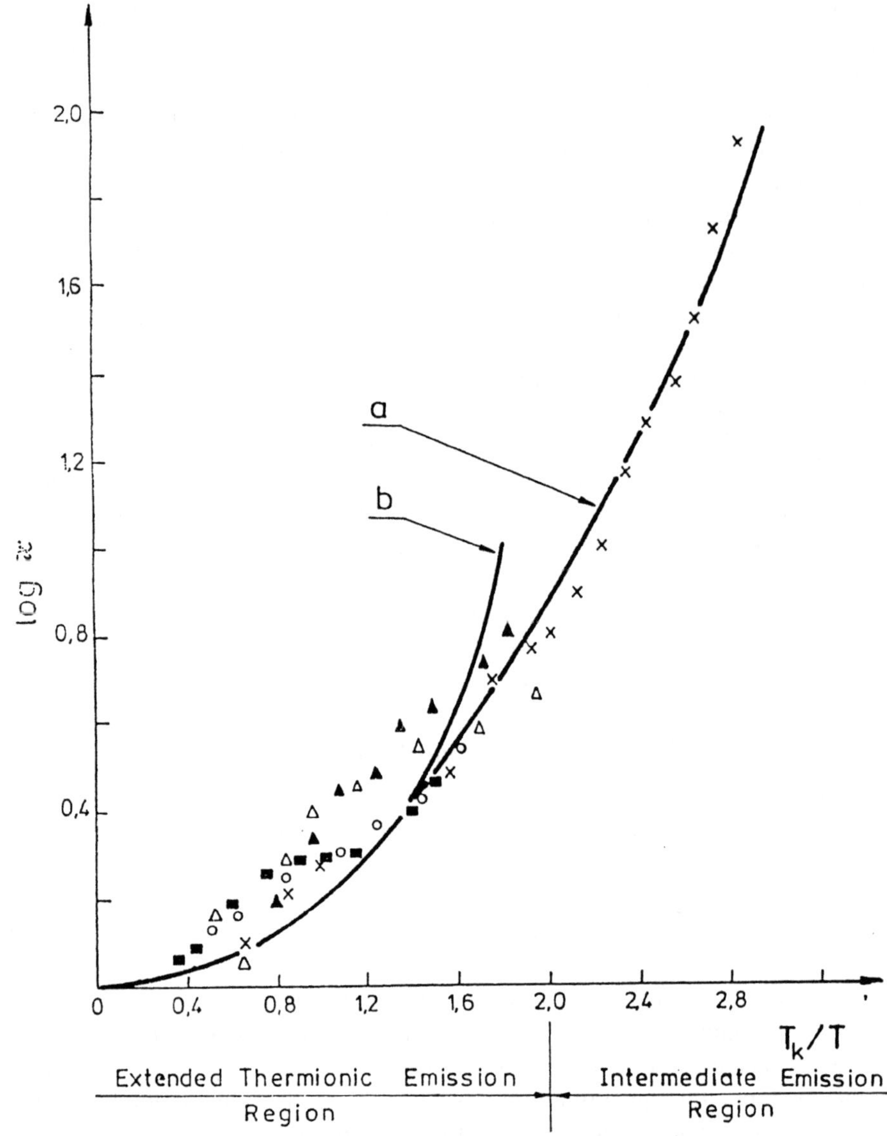

Figure 3. Tunneling correction $\kappa = j/j_{\mathrm{RS}}$ to the Richardson–Schottky equation for electron emission from an aluminium cathode in an insulator (Al_2O_3) as universal function of T_k/T. (a) Exact calculations. (b) According to Equation (53). Symbols represent experimental data for different values of temperature, field strength, work function and dielectric constant [2b].

where E is the total energy, $\epsilon_x^+ \rightleftharpoons 0$ is the 'kinetic' energy of translational x-motion at the saddle point ($x = x^+$) and ϵ_n^+ is the vibration (rotation) energy normal to x at $x = x^+$. This relation implies a virtual adiabatic motion along x but it is not restricted by any assumption of a real thermal equilibrium in the transition state (saddle point region). Using (55), Equation (54) can be transformed into expression [6, 15b]

$$v = \kappa_{ac} \frac{kT}{h} \frac{Z^{\neq}_{ac}}{Z} e^{-E_c/kT} \tag{56}$$

where

$$\kappa_{ac} = \int_{-\infty}^{\infty} \bar{W}(\epsilon_x^+) e^{-\epsilon_x^+/kT} \, \mathrm{d}\epsilon_x^+/kT, \quad \bar{W}(\epsilon_x^+) = \sum_n W_n(\epsilon_x^+) f(\epsilon_n^+, T) \tag{57}$$

and

$$f(\epsilon_n^+, T) = g_n e^{-\epsilon_n^+/kT} / Z^{\neq}_{ac}, \quad Z^{\neq}_{ac} = \sum_n g_n e^{-\epsilon_n^+/kT} \tag{58}$$

$Z^{\neq}_{ac}$ being the partition function of the transition state (activated-complex) formally introduced in the above equations.

The tunneling correction can be defined by the ratio

$$\kappa_t = \kappa_{ac} / \kappa_{ac}^{\mathrm{scl}} \tag{59}$$

where the semiclassical factor

$$\kappa_{ac}^{\mathrm{sct}} = \int_{-\infty}^{\infty} \bar{W}^{\mathrm{scl}} e^{-\epsilon_x^+/kT} \, \mathrm{d}\epsilon_x^+/kT, \quad \bar{W}^{\mathrm{scl}} = \sum_n W_n^{\mathrm{sct}}(\epsilon_x^+) f(\epsilon_n^+, T) \tag{60}$$

involves the transition probabilities W_n^{sct} as calculated by classical trajectory methods [6]. Thus, the factor κ_{ac} in (56) can be written as

$$\kappa_{ac} = \kappa_t \kappa_{ac}^{\mathrm{scl}}. \tag{61}$$

It involves the nonseparability of the reaction coordinate x and the nonequilibrium effects in the intermediate stages of the reaction.

Equation (56) represents a formal generalization of the semiclassical transition state (activated-complex) theory in which $\kappa_{ac} \doteq 1$. This theory implies: (1) a classical motion along the reaction coordinate ($\kappa_t = 1$); (2) its separability (no recrossing of the saddle-point region); and (3) real thermal equilibrium in that region (transition state). A frequently used approximation is to set $\kappa_{ac}^{\mathrm{scl}} = 1$ and $\kappa_t > 1$ by assuming the parabolic potential (15) to be adequate in the transition state in which $x \equiv x_n$. Then, in range of moderate tunneling ($T > T_k/2$) formula (34) for $\kappa_{ac} = \kappa_t$ can be used. According to this $\kappa_t \cong 1$ if $T > 2T_k$, which defines the condition of validity of the familiar Eyring activated complex theory (provided $\kappa_{ac}^{\mathrm{sct}} = 1$).

Mortensen [44] has first calculated both κ_{ac} and $\kappa_{ac}^{\mathrm{scl}}$ for the bimolecular gas phase reaction $H_2 = H \rightarrow H + H_2$ and related isotopic reactions using a corrected semiempirical potential energy surface constructed by Weston [45]. Similar calculations have been done by Christov and Parlapanski [6], who also computed the tunneling correction $\kappa_t = \kappa_{ac}/\kappa_{ac}^{\mathrm{sct}}$. The characteristic temperature was found to be $T_k = 1240$ K for $H_2 + H$ and $T_k = 896$ K for $D_2 + D$ reactions [6]. For $T_k = 300$ K ($T < T_k/2$) one obtains $\kappa_t = 8.65$ and $\kappa_t = 3.64$, respectively. For $T \cong T_k$, the values of κ_t agree well with formula (34). The tunnel effect also plays a considerable role in proton transfer reactions in solution [6–8] where an estimation of the characteristic temperature is also possible on the basis of simple models [8].

Another useful formulation of reaction rate theory, first introduced by Hirschfelder and Wigner [46], is based on the adiabatic potential

$$V_n(x) = V(x) + \epsilon_n(x) \tag{62}$$

where $V(x)$ is the electronic energy along the reaction coordinate x and $\epsilon_n(x)$ is the vibration-rotation energy in quantum state $\mathbf{n}$ as a function of x. The potential $V_n(x)$ in the general case, has a maximum $E_v^{(n)}$ at a point $x = x^+$, which does not coincide with the position $x = x^+$ of the maximum of $V(x)$. Therefore, the transition state depends on the quantum number $\mathbf{n}$. The total energy can be written as [6]

$$E = E_{n_x} + E_n = E_c^{(n)} + \epsilon_{x,n}^+ \tag{63}$$

where $\epsilon_{x,n}^+ \rightleftharpoons 0$ is the 'kinetic' translation energy along x at the point $x = x_n^+$. Using these relations, the basic rate Equation (54) can be represented in two different equivalent ways. The first one is simply [6, 47, 48]

$$v = \frac{kT}{hZ} \sum_n \int_0^\infty W_n(E) e^{-E/kT} \, \mathrm{d}E/kT \tag{64}$$

and the other one [6]

$$v = \frac{kT}{hZ} \sum_n \kappa_n g_n e^{-E_c^{(n)}/kT} \tag{65}$$

where

$$\kappa_n = \int_{-\infty}^{\infty} W_n(\epsilon_{x,n}^+) e^{-\epsilon_{x,n}^+/kT} \, \mathrm{d}\epsilon_{x,n}^+/kT. \tag{66}$$

Equation (65) with (66) can be transformed into [6, 50]

$$v = \kappa_{ac}^+ \frac{kT}{h} \frac{Z_{ac}^+}{Z} e^{-E_c^0/kT} \tag{67}$$

with

$$\kappa_{ac}^{+} = \sum_{n} \kappa_n f(\epsilon_c^{(n)}, T), \quad f(\epsilon_c^{(n)}, T) = g_n e^{-\epsilon_c^{(n)}/kT} / Z_{ac}^{+} \tag{68}$$

and

$$Z_{ac}^{+} = \sum_{n} g_n e^{-\epsilon_c^{(n)}/kT}, \quad \epsilon_c^{(n)} = E_c^{(n)} - E_c^{0}. \tag{69}$$

In this formulation the factor κ_{ac}^{+} is the thermal average of the factors κ_n defined by (66) over the quantum states of the activated complex, $f(\epsilon_c^{(n)})$ being the occupation probability of quantum state **n** and Z_{ac}^{+} is the partition function of the activated complex in this representation.

Equation (67) is a formal generalization of the semiclassical transition state (activated-complex) theory in which $\kappa_{ac}^{+} = 1$. This theory presumes a very slow classical motion along the reaction coordinate x (no tunneling) in which the quantum states of the vibration-rotation motions, normal to x, remain constant, so that the adiabatic potentials (62) really govern the reaction course without recrossing of the transition regions and assure thermal equilibrium in those regions.

The tunneling correction for any quantum state **n** can be defined by the ratio

$$\kappa_t^{(n)} = \kappa_n / \kappa_n^{\mathrm{scl}} \tag{70}$$

where κ_n is given by (66), while

$$\kappa_n^{\mathrm{scl}} = \int_0^{\infty} W_n^{\mathrm{scl}}(\epsilon_{x,n}^{+}) e^{-\epsilon_{x,n}^{+}/kT} \, \mathrm{d}\epsilon_{x,n}^{+}/kT \tag{71}$$

is the semiclassical factor in which $\epsilon_{x,n}^{+} \geq 0$ (no tunneling through the barrier potential (62)), hence the transition probability. W_n^{scl} takes into account only the coupling of the reaction coordinate with the nonreactive vibration-rotation modes. Neglecting the coupling in the transition region ($x \approx x_n^{+}$) means that we use the harmonic approximation for both reactive and nonreactive motions in that region. Then $\kappa_t^{(n)} = \kappa_n$ can be calculated by means of (66) making use of expression

$$W_n(\epsilon_{x,n}^{+}) = [1 + \exp(-2\pi\epsilon_{x,n}^{+}/h\nu_n^{*})]^{-1} \tag{72}$$

for the transition probability of a parabolic barrier $V_n(x)$, where ν_n^{*} is the barrier frequency. Combining (66) and (72), we can calculate $\kappa_t^{(n)} = \kappa_n$ in the same way as is done in Section 3 for the parabolic barrier $V(x)$. The characteristic temperature $T_k^{(n)}$ can be defined in a similar manner, thus obtaining the expression

$$T_k^{(n)} = h\nu_n^{*}/\pi k. \tag{73}$$

If $T > T_k^{(n)}/2$, the tunneling correction is

$$\kappa_t^{(n)} = \frac{(\pi/2)(T_k^{(n)}/T)}{\sin[(\pi/2)(T_k^{(n)}/T)]}. \tag{74}$$

In the general case in which the reaction coordinate x is nonseparable, the average tunneling correction can be defined as

$$\bar{\kappa}_t = \kappa_{ac}^{+}/(\kappa_{ac}^{+})^{\text{scl}} \tag{75}$$

where κ_{ac}^{+} is given by (68) and (69), while

$$(\kappa_{ac}^{+})^{\text{scl}} = \sum_n \kappa_n^{\text{scl}} f(\epsilon_c^{(n)}) \tag{76}$$

is defined in the same way by replacing κ_n by κ_n^{scl}. Thus the factor κ_{ac}^{+} in the rate Equation (67) can be written as

$$\kappa_{ac}^{+} = \bar{\kappa}_t(\kappa_{ac}^{+})^{\text{scl}} \tag{77}$$

where $(\kappa_{ac}^{+})^{\text{scl}}$ takes into account the nonseparability of the reaction coordinate and the departure from thermal equilibrium in the transition state.

A resonance state approach to transition state theory, recently proposed by Truhlar and Garrett [30], shows a way of an approximate treatment of the mode coupling in the transition state region. The rate constant is represented by Equation (64) and the transition probability is written as

$$W_n(E) = \{1 + \exp[2\pi(E_n - E)/\Gamma_n]\}^{-1} \tag{78}$$

where E_n is considered as the resonance energy and $\Gamma_n = h\nu_n^*$ as the width of the half-maximum of the resonance feature. Both E_n and Γ_n enter the expression

$$E_{\text{res}}^{(n)} = E_n - i\Gamma_n/2 \tag{79}$$

of the complex energy of the resonant state **n**.

Expression (78) is identical with formula (10) for the parabolic barrier since $E_n = E_c + \epsilon_n^{+}$ and $E = E_x + \epsilon_n^{+}$. It is exact in the harmonic approximation in which

$$E_n = \sum_{n_i} E_{n_i}, \quad E_{n_i} = \left(n_i + \frac{1}{2}\right) h\nu_i$$

is the energy of a system of harmonic oscillators corresponding to the nonreactive modes in the transition state. Introducing anharmonic coupling with the reaction coordinate in the simplest case of a single harmonic vibration means using the effective potential

$$V(x,y) = E_c - \frac{f_x^*}{2}(x-x^+)^2 + \frac{f_y}{2}(y-y^+) + V_1(x,y) \tag{80}$$

where $x = x^+$ and $y = y^+$ are the coordinates of the saddle-point and $V_1(x,y)$ is the coupling term between x-motion and y-vibration. The coupling leads to changes of both the resonance energy $E_n = (n_y + \frac{1}{2})h\nu_y$ and the resonance width $\Gamma_n = h\nu_n^*$, which can be calculated using either perturbation theory or adiabatic transition state theory [30]. This has been demonstrated in the cases of cubic anharmonicity

$$V_1(x,y) = \tfrac{1}{2} f_{xy}(x-x^+)(y-y^+)^2$$

and biquadratic anharmonicity

$$V_1(x,y) = \tfrac{1}{2} f_{xy}(x-x^+)^2(y-y^+)^2.$$

The results of both approaches agree in the two cases in the limit where $\nu_y^2 \gg (\nu^*)^2$, which is the condition of a slow motion along the reaction coordinate that assures the vibrational adiabaticity.

Assuming the validity of expression (78) in the presence of mode coupling, allows an extension of the definition for $T_k^{(n)}$ related to the adiabatic formulation of transition state theory. Thus, using the new resonance state approach [30], we can derive an expression for $T_k^{(n)}$ by determination of the maximum of the integrand. $F_n(E) = W_n(E)e^{-E/kT}$ in any of the integrals of the rate Equation (64) making use of the formula (78) for $W_n(E)$. This yields

$$E = E_m = E_c^{(n)} - \frac{kT_k^{(n)}}{2} \ln \frac{\frac{1}{2}(T_k^{(n)}/T)}{1 - \frac{1}{2}(T_k^{(n)}/T)} \tag{81}$$

where

$$T_k^{(n)} = \Gamma_n/\pi k. \tag{82}$$

Thus, we find that for $T = T_k^{(n)}$ the position $E = E_m^{(n)}$ of the maximum of the thermally weighted transition probability $W_n(E)$ lies at the barrier peak ($E_m^{(n)} = E_c^{(n)}$) of the adiabatic potential (62). This corresponds to the definition of the characteristic temperature introduced by Goldanskii [12]. One obtains the same by using the Christov definition [11, 13] $P_n' = P_n''$ where

$$P_n' = \int_0^{E_c^{(n)}} W_n(E)e^{-E/kT}\,\mathrm{d}E/kT, \quad P_n'' = \int_{E_c^{(n)}}^{\infty} W_n(E)e^{-E/kT}\,\mathrm{d}E/kT.$$

Using expression (78) again for $W_n(E)$, we find that the integrand in both integrals is symmetric with respect to $E_c^{(n)}$, when $E_m^{(n)} = E_c^{(n)}$, therefore $P_n' = P_n''$.

In the framework of the resonant state approach the mode coupling is introduced in expression (78) through E_n and Γ_n, therefore, the definition (82) of the characteristic temperature $T_k^{(n)}$ is valid in the presence of coupling. However, in this way one neglects the coupling in the semiclassical limit by setting $\kappa_n^{\rm scl} = 1$, so that from (70) $\kappa_t^{(n)} = \kappa_n$. This approximation gives an upper limit of κ_t for bimolecular reactions such as $H_2 + H \rightarrow H + H_2$, since $\kappa_{ac}^{\rm scl} > 1$ [6, 44, 46]. In the temperature range of moderate tunneling $T > T_k/2$, formula (74) for $\kappa_t^{(n)}$ can be used, which allows an approximate evaluation of the average tunneling correction $\bar{\kappa}_t = \kappa_{ac}^{+}$ defined by (68) and (69).

A very useful method for studying many-dimensional tunneling is based on Feynman's path integral formulation of quantum mechanics [59]. In the one-dimensional case the quasiclassical expression (3) for the tunneling probability is used and the full (Euclidean) action is defined by (40), where $\beta = 1/kT$ plays the role of the imaginary time $\tau = it$ in the propagator $\exp(-i\hat{H}t)$, where $\hat{H} = \hat{p}_x^2/2m + V(x)$ is the Hamiltonian for motion along the reaction coordinate x. Tunneling is described in real time t as a periodic classical motion in the inverted potential $V_{\rm in} = -V(x)$, so that the equation of motion

$$m\,\mathrm{d}^2x/\mathrm{d}t^2 = \mathrm{d}V(x)/\mathrm{d}t$$

is solved with periodic boundary conditions $x(t+\theta) = x(t)$, where the period is $\theta = \hbar\beta = \hbar/kT$. The solutions $x(t)$ are stationary points of the full action (40). If the potential $V(x)$ has a minimum at $x = 0$ and a maximum at $x = x_m^0$, both points represent trivial solutions, so that only the periodic trajectory $x_b(t)$, called 'bounce' or 'instanton' solution, is related to tunneling through the barrier potential $V(x)$. With increasing temperature, the amplitude of this periodic motion decreases in such a way that at a definite temperature T_0 it reduces to the point $x(t) = x_m^0$ [28, 66] (Figure 4).

It is interesting that the 'cross-over temperature' T_0 exactly coincides with the temperature $T_k' = T_k/2$, defined by (31), which was first introduced by Goldanskii [12]. Therefore, it can be defined as the temperature at which the 'instanton' disappears. However, these findings do not at all mean that at $T = T_0$ tunneling 'switches off', as stated by Benderskii *et al.* [68]. This happens rather at $T = T_k = 2T_0$ than at $T = T_0 = T_k/2$. Indeed, the first (tunneling) term of the full action (40) vanishes at $E_x = E_c$ and not at $E_x \cong E_x - kT_k$ which corresponds, according to (29), to the maximum of the thermally weighted transition probability at $T = T_k$ and $T \cong T_k/2$, respectively. Therefore, at $T = T_k$, $S(E_x, T) = \hbar E_c/kT$. Actually, tunneling does not disappear even at $T > T_k = 2T_0$, the tunneling correction being $\kappa_t = \pi/2$ at $T = T_k$ and $\kappa_t = 1.12$ at $T = 2T_k (= 4T_0)$ as follows from (34). Therefore, there is no definite temperature at which tunneling 'switches off', but there exists a wide transition region between $T_0 \equiv T_k' = T_k/2$ and $T_k'' = 2T_k$, in

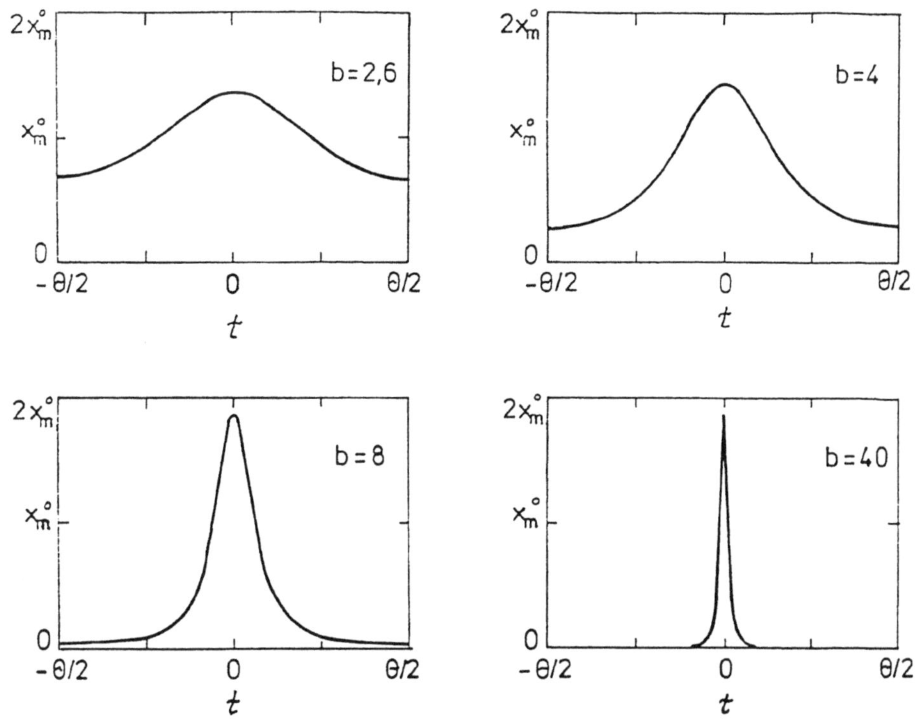

Figure 4. Bounce (instanton) trajectories $x_b(t)$ in a cubic metastable potential $V(x) = 2\pi^2\nu_x^2x^2 - \frac{1}{3}u_x x^3$ [28]; $x_m^0 \equiv (2\pi\nu_x)^2/u_x$ is the constant (saddle point) solution; $b = \theta\nu_x = h\nu_x/2\pi kT$. With increasing temperature T (decreasing b) $x_b(t)$ tends to x_m^0 and coalesces with it at $T = T_0$ ($\theta = \theta_0 = \hbar/2\pi kT_0$) where $T_0 = T_k/2$.

which the tunneling correction ranges between $\kappa_t \cong 2E_c/kT_k$ ($T \cong T_k/2$) and $\kappa_t \cong 1$ ($T \cong 2T_k$) [see Equations (33) and (34)]. This corresponds to the gradual increase of the transition probability (1) from 0 to 1 with increasing energy and the related gradual decrease of the ratio $P'/P'' = \kappa'/\kappa''$ of the thermally averaged transition probabilities for subbarrier and overbarrier transitions from ∞ to 0 with increasing temperature [see Equations (18–25)].

The ratio P'/P'' is the natural measure of the extent of tunneling. Therefore, there is no other, more reasonable, universal definition of the 'crossover' temperature than that based on the equality $P'/P'' = 1$ as given by (24) and (25). The related temperatures $T_0 \equiv T_k' = T_k/2$ and $T_k'' = 2T_k$ have, in general, a less transparent physical meaning. The particular role of T_0 in the instanton method is associated with the quasiclassical approximation (3) of the tunneling probability, which gives incorrect values for $E_x > E_c - kT_k$ ($W(E_x) = 1$ for $E_x = E_c$) like the incorrect values for $E_x \to 0$ (non-zero value for $E_x = 0$). Therefore, the definition of T_0, as the temperature at which the instanton disappears, is closely related to an artefact of the quasiclassical approximation as well as definition (31) results from another artefact of the same approximation as discussed in Section 3.

7. Reaction Rate Theory: Stochastic Approach

The stochastic rate approach originates from Kramers [51] who introduced the simple model of a particle moving in a double-well potential $V_0(x)$ subject to the Brownian motion of the medium (solvent). A review by Hanggi *et al.* [28] presents a detailed account of the development of this model. (See also the more recent review of Benderskii *et al.* [68]).

The classical motion of the particle along the x-coordinate is described by the generalized Langerin equation (GLE)

$$\frac{\mathrm{d}p_x}{\mathrm{d}t} = -\frac{\partial V}{\partial x} - \int_0^t p_x(t')\eta(t-t')\,\mathrm{d}t' + F(t) \tag{83}$$

where $p_x(t)$ is the momentum at time t, $F(t)$ is a zero-centred random force, $\langle F(t)\rangle = 0$, and

$$\eta(t) = \langle F(0)F(T)\rangle/kT \tag{84}$$

is the time-dependent friction coefficient.

On the basis of Equation (83) Grote and Hynes [52] have derived the rate equation

$$v = \nu_x \frac{\nu'}{\nu_x^*} e^{-E_c/kT} \tag{85}$$

where ν_x is the frequency of the particle vibration in the initial state (well 1), ν_x^* is the barrier frequency of the potential $V_0(x)$, E_c is the barrier height and ν' is the positive root of equation

$$\frac{\nu'}{\nu_x^*} = \frac{\nu_x^*}{\nu' + \hat{\eta}(\nu')} \tag{86}$$

in which $\hat{\eta}(\nu')$ is the Laplace transform of $\eta(t)$, as defined by (84).

The GLE (83) can be derived from a microscopic model in which the particle is coupled bilinearly to the medium represented by a system of $z-1$ harmonic oscillators, so that the potential energy of the system (particle + medium) is given by the function [53]

$$V(x,\tilde{y}) = V_0(x) + \frac{1}{2}\sum_{i=1}^{z-1}\left(2\pi\nu_i y_i - \frac{C_i}{2\pi\nu_i}x\right)^2 \tag{87}$$

where $\tilde{y} = \{y_i\}$ is a set of (mass-weighted) medium coordinates, ν_i are the corresponding vibration frequencies and C_i are the particle–medium coupling constants. The particle potential $V_0(x)$ can be obtained from Equation (16) with

$$V_1(x) = \tfrac{1}{2}(2\pi\nu_x)^2 x^2, \quad V_2(x) = \tfrac{1}{2}(2\pi\nu_x)^2(x - x_0)^2 + Q \tag{88}$$

being two intersecting parabolic curves. The barrier frequency ν_x^* depends on the coupling constant V_{12} through the relation [16]

$$\nu_x^* = \nu_x \left(\frac{E_r^{(x)}}{2C_x} - 1\right)^{1/2}, \quad C_x = |V_{12}| \left[1 - \left(\frac{Q}{E_r^{(x)}}\right)^2\right]^{3/2} \tag{89}$$

where $E_r^{(x)} = \frac{1}{2}(2\pi\nu_x)^2 x_0^2$ is the reorganization energy of the particle oscillator.

The potential surface (87) has a line of minimum energy defined by the equation

$$y_1 = (C_i/2\pi\nu_i)^2 x, \quad (i = 1, 2, \ldots, \bar{z} - 1)$$

which follows from the condition $\partial V/\partial y_i = 0$ (for all i). This is a straight line q, which is related to the particle coordinate x by Equation (54) (Figure 5)

$$q = x\left[1 + \sum_{i=1}^{z-1} C_i^2/(2\pi\nu_x)^4\right]^{1/2}. \tag{90}$$

In the absence of coupling (all $C_i = 0$) $q \equiv x$.

The potential $V(q)$ along q results from Equation (16) with

$$V_1(q) = \tfrac{1}{2}(2\pi\nu_q)^2 q^2, \quad V_2(q) = \tfrac{1}{2}(2\pi\nu_q)^2(q - q_0)^2 + Q \tag{91}$$

begin two interesting parabolic curves corresponding to (88). The barrier frequency ν_q^* is given by the relation

$$\nu_q^* = \nu_q \left(\frac{E_r^{(q)}}{2C_q} - 1\right)^{1/2}, \quad C_q = |V_{12}| \left[1 - \left(\frac{Q}{E_r^{(q)}}\right)^2\right]^{3/2} \tag{92}$$

where $E_r^{(q)} = \frac{1}{2}(2\pi\nu_q)^2 q_0^2$ is the reorganization energy along q. It can be shown that the particle-medium coupling in expression (87) does not modify the particle potential $V_0(x)$, so that $E_r^{(q)} = E_r^{(x)}$ and $C_q = C_x$ [54]. The barrier height E_c is also independent of the coupling. Therefore, from (89) and (92) the important relations follow

$$\nu_q^*/\nu_x^* = \nu_q/\nu_x = \left[1 + \sum_{i=1}^{z-1} C_i^2/(2\pi\nu_i)^2\right]^{-1/2} \tag{93}$$

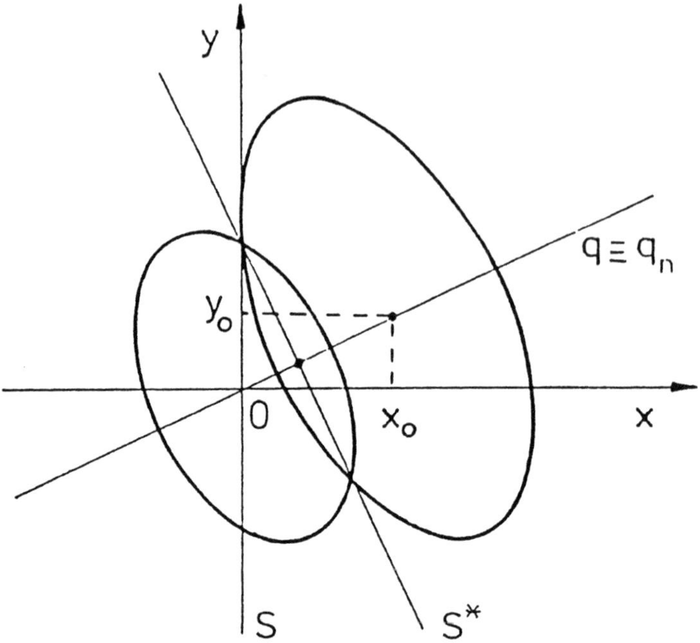

Figure 5. Horizontal cross-section of the potential surface $V(x, y)$ defined by Equations (87) and (88) in two dimensions (x, y). The intersection plane S^* of the paraboloids $V_1(x, y)$ and $V_2(x, y)$ (transition state of the medium) involves a saddle point situated between the minima positions (0, 0) and (x_0, y_0). The line $q \equiv q_n$ defines the minimum energy path of the medium. In an absence of coupling it coincides with the particle coordinate x (Equation 90) but the x-coordinates of three extrema remain unchanged (then S^* turns into S).

hence $\nu_q < \nu_x$ and $\nu_q^* < \nu_x^*$. The second equation results from $E_r^{(q)} = E_R^{(x)}$ [54].

The characteristic temperature in absence of coupling (all $C_i = 0$) is

$$T_k^{(x)} = h\nu_x^*/\pi k \tag{94}$$

and in the presence of coupling (all $C_i \neq 0$)

$$T_k^{(q)} = h\nu_q^*/\pi k. \tag{95}$$

Evidently $T_k^{(q)} < T_k^{(x)}$, therefore the particle–medium coupling leads to a decrease of the role of nuclear tunneling for the potential surface (87).

The rate constant can be calculated by application of the usual dynamic approach in two equivalent ways depending on the definition of the reaction path. The most natural way is based on the findings that the potential surface (87) has a separable rectilinear reaction coordinate q defined by relation (90), therefore, the transition probability $W(E_{n_q})$ is independent of the quantum state $\mathbf{n}$ of the nonreactive modes. Thus, we can write $Z = Z_q Z_n$, where $Z_n = \sum_n \exp(-E_n/kT)$, so that, replacing the x-coordinate by q from the basic rate Equation (54), one obtains

$$v = \frac{kT}{hZ_q} \sum_{n_q} W(E_{n_q}) e^{-E_{n_q}/kT} \Delta E_{n_q}/kT. \tag{96}$$

Assuming a quasicontinuous energy spectrum in the initial state (well 1) means that $\Delta E_{n_q} = h\nu_q \ll E_c$, where E_c is the barrier height. Then, replacing the sum over n_q by an integral yields

$$v = \frac{2kT}{h} \sinh \frac{h\nu_q}{2kT} \int_0^\infty W(E_q) e^{-E_q/kT} \, \mathrm{d}E_q/kT \tag{97}$$

where the partition function $Z_q = [2\sinh(h\nu_q/2kT)]^{-1}$ is introduced. The integral represents the thermally averaged transition probability (20) calculated in Section 4 for all temperature.

In particular, in the temperature range of moderate tunneling $T > T_k^{(q)}/2$, we have

$$v = 2\kappa_t \sinh(h\nu_q/2kT) e^{-E_c/kT} \tag{98}$$

where

$$\kappa_t = \frac{(\pi/2)(T_k^{(q)}/T)}{\sin[(\pi/2)(T_k^{(q)}/T)]} = \frac{h\nu_q^*/2kT}{\sin(h\nu_q^*/2kT)}. \tag{99}$$

In the high temperature range $T > 2T_k^{(q)}$ $(h\nu_q^*/2kT \ll 1)$ $\kappa_t = 1$ and $h\nu_q/2kT \ll 1$; then Equation (98) turns into

$$v^{\mathrm{cl}} = \nu_q e^{-E_c/kT} \tag{100}$$

or using relations (93)

$$v^{\mathrm{cl}} = \nu_x \frac{\nu_q^*}{\nu_x^*} e^{-E_c/kT}. \tag{101}$$

This equation is identical to (85) if we set $\nu' = \nu_q^*$. It has been first derived by Pollak [55a], making use of a version of transition state theory to be discussed below. He also showed that relation (86) is valid with $\nu' = \nu_q^*$, which proves the identity of Equations (85) and (101).

Our derivation of the rate Equation (101), based on a proper dynamic approach, shows that transition state theory is not necessary for this derivation; however, its stochastic interpretation is possible only by means of relation (86), derived by Pollak [55a] with $\nu' = \nu_q^*$. In this way, it becomes clear that the frictionless motion of the system (particle + medium) along the q-coordinate is related to the particle motion along the x-coordinate subject to friction [57].

Equation (101) can be related to transition state theory in a formal and somewhat artificial way if we assume that the reaction coordinate in the initial state coincides with the particle coordinate x, while in the transition state it is identical with the line of minimum energy q of the whole system (particle + medium). Thus, two different reaction coordinates **x** and **q**, which are related by Equation (90), must be used in the absence and presence of coupling, respectively. At this condition, from the exact expression (56), taking into account relations (57 to 61), one derives for the oscillator model considered the rate equation

$$v = \frac{2kT}{h}\kappa_t k_{ac}^{\rm scl} \sinh\frac{h\nu_x}{2kT}\prod_{i=1}^{Z-1}\frac{\sinh(h\nu_1/2kT)}{\sinh(h\nu_i^+/2kT)}e^{-E_c/kT} \tag{102}$$

where ν_i and ν_i^+ are the medium vibration frequencies in the initial state (no coupling) and the transition state (coupling involved), respectively.

The rate Equations (97) and (102) are equivalent if $\kappa_{ac}^{\rm scl} = 1$. Indeed, we can write

$$E = E_q + E_n = E_x + E_y$$

so that the total partition function in the initial state Z can be expressed in two equivalent ways,

$$Z = Z_q Z_n = Z_x Z_y$$

where Z_x and Z_y and the partition functions of the particle and the medium in the absence of coupling. Then, Equation (97) can be written as

$$v = \frac{kT}{h}\frac{Z_n}{Z_x Z_y}\int_0^\infty W(E_q)e^{-E_q/kT}\,\mathrm{d}E_q/kT. \tag{103}$$

Here, $Z_n = Z_{ac}^{\neq}$ because the medium vibration frequencies in the initial and transition states in the presence of coupling are equal (complete separability of the reaction coordinate q). Therefore, this equation can be considered as quantum generalization of transition state (activated complex) theory. Introducing the corresponding harmonic oscillator partition functions, one arrives at the expression (102) with $\kappa_{ac}^{\rm scl} = 1$, where

$$\kappa_t = P_q/P_q^{\rm cl} = \int_0^\infty W(\epsilon_q)e^{-\epsilon_q/kT}\,\mathrm{d}\epsilon_q/kT \tag{104}$$

is the exact tunneling correction, P_q and $P_q^{\rm cl} = \exp(-E_c/kT)$ being the thermally averaged quantum and classical transition probabilities, respectively, and $\epsilon_q = E_q -$

E_c. It should be stressed that this expression can be calculated at all temperatures by use of the results in Section 4 for $P \equiv P_q$ by simply replacing E_x by ϵ_q.

Equations (101) and (102) ($\kappa_{ac}^{\rm scl} = 1$) can be combined, so that using an earlier procedure [29], we obtain

$$v = \kappa_q \nu_x \frac{\nu_q^*}{\nu_x^*} e^{-E_c/kT} \tag{105}$$

where

$$\kappa_q = \frac{2kT}{h\nu_q^*} \kappa_t \sinh\left(\frac{h\nu_x}{2kT}\right) \prod_{i=1}^{Z-1} \frac{\sinh(h\nu_i/2kT)}{\sinh(h\nu_i^+/2kT)} \tag{106}$$

is the full quantum correction to the classical rate Equation (101) ($\kappa_q \geq 1$). Note again that the tunneling correction κ_t can be computed by (104) at any temperature. In particular, for $T > T_k^{(q)}/2$ it reduces to expression (99). Then, Equation (106) coincides with the result of Pollak [55b] derived in a different way.[2] He has also shown that in the temperature range $T > T_k^{(q)}/2$ ($h\nu_q^*/2kT < 1$) Equation (105) allows a stochastic interpretation, which results in expression [29]

$$v = \nu_x \frac{\nu_q^*}{\nu_x^*} \left[\prod_{m=1}^{\infty} \frac{\nu_x^2 + \nu_m^2 + \nu_m \hat{\eta}(\nu_m)}{-(\nu_x^*)^2 + \nu_m^2 + \nu_m \hat{\eta}(\nu_m)} \right] e^{-E_c/kT}. \tag{107}$$

In the temperture range $T < T_k/2$, the rate Equation (105) is valid especially if the parabolic approximation for the potential $V(q)$ is used, which means using the exact expression (32) for $\kappa_t^{(q)}$ ($T_k = T_k^{(q)}$). The only reason to abandon this approximation is that it may not work well at low temperatures, where tunneling occurs predominantly in the energy range $E_q < E_c - kT_k$, as discussed in Section 4. Therefore, the condition $T > T_k^{(q)}/2$ ($h\nu_q^*/2kT < 1$) that restricts the validity of expression (98) for $\kappa_t^{(q)}$ concerns only a mathematical problem, which can be resolved in the rate theory making use of the approach presented in Setion 4.

The usual stochastic approach to tunneling in the low temperature range $T < T_k^{(q)}/2$, based on the potential (87), makes use of Feynman's path integral formulation of quantum mechanics [58]. One considers the tunneling path for which the effective (Euclidean) action functional [59, 60]

---

[2] In Ref. [29] the potential function

$$V(x, y) = V_0(x) + \frac{1}{2} \sum_{i=1}^{Z-1} [(2\pi\nu_i)^2 y_i^2 - C_i y_i x]$$

is used instead of (87). The results of both models agree only in the weak coupling limit $\sum[C_i^2/(2\pi\nu_i)^2] \ll (2\pi\nu_x)^2$, as shown in Ref. [57].

$$S(x,\theta) = \int_{-\theta/2}^{\theta/2} \left(\frac{1}{2}p_x^2 + V_0(x)\right) \mathrm{d}t + \frac{1}{2}\int_{-\theta/2}^{\theta/2} x(t)\,\mathrm{d}t \int_{-\theta/2}^{\theta/2} \times \times\, \eta(t-t')x(t')\,\mathrm{d}t' \tag{108}$$

has an extremal value. The first term describes the frictionless particle motion along the x-coordinate and the second one is related to the energy loss due to the particle-medium coupling, $\eta(t)$ being the time-dependent friction coefficient as a periodic function with period $\theta = \hbar/kT$. The extremal action path is a solution of the equation [28, 59]

$$\frac{\mathrm{d}^2 x}{\mathrm{d}t^2} = \frac{\mathrm{d}V_0}{\mathrm{d}x} + \int_{-\theta/2}^{\theta/2} x(t')\eta(t-t')\,\mathrm{d}t' \tag{109}$$

where the first term describes, in real time, the particle tunneling as frictionless classical motion governed by the inverted potential $V_0(x)$ and the second term considers the influence of the time-dependent friction on this motion. The non-trivial bounce (instanton) solution $x_b(t)$ represents a periodic motion with period $\theta = \hbar/kT$ [28, 63, 67]. With increasing the temperature it approaches the trivial constant solution $x(t) = x_m^0$, where x_m^0 is the position of the maximum of the particle potential $V_0(x)$ [28] (see Figure 4). Both solutions, $x_b(t)$ and x_m^0, coalesce at a temperature

$$T_0 = h\nu'/2\pi k \tag{110}$$

where ν' is the positive root of (86). Since for the potential surface (87) $\nu' \equiv \nu_q^*$, it is clear that T_0 is identical to the temperature $T = T_k^{(q)}/2$, where $T_k^{(q)}$ is defined by (95). Thus, T_0 is the temperature at which the 'instanton' disappears as in the case of frictionless x-motion, discussed in Section 6.

Therefore, the characteristic temperature $T_k^{(q)}$ has the same physical meaning in the usual dynamic and the stochastic rate theories. The same is valid for the related temperature $T_0 = T_k^{(q)}/2$, although it allows a more specific interpretation in the framework of the path integral approach to the stochastic theory. Hauggi *et al.* [27, 28] have investigated the fluctuation modes $x_n(t)$ around the trivial solution $x(t) = x_m^0$ at $T > T_0$ and have found that the mode $x_t(t)$ corresponds to the frequency $\nu_1 = \nu_{-1}$ that becomes zero at $T = T_0$. The eigenvalue equation for $x_n(t)$ for $n = \pm 1$ reduces to

$$\nu_0^2 + \nu_0\hat{\eta}(\nu_0) = (\nu_x^*)^2$$

where $\nu_0 = 2\pi kT_0$. Therefore, T_0 can be defined as

$$T_0 = h\nu_0/2\pi k \tag{110'}$$

It has been shown [28] that $\nu_0 = \nu'$, so that the definitions (110) and (110′) are equivalent. Both definitions are related to some specific features of the imaginary-time functional-integral methodology. However, they have no direct relation to tunneling.

In fact, the full extremal action

$$S(E_\theta) = \varphi(E_\theta) + \theta E_\theta \tag{111}$$

where $\varphi(E_\theta)$ is the 'small' action, coincides with the action functional (108), when it is evaluated with the bounce solution $x_b(t)$ that has a period $\theta = \hbar/kT$, E_θ being the corresponding total energy of the whole system (particle + medium). Expression (111) implies the validity of the usual WBK approximation (3) for the tunneling probability, hence the condition $E \ll E_c$ $(E < E_c - kT_k)$. As shown in Section 3, the incorrect use of this approximation leads to expression (31) for $T'_k \equiv T_0 = T_k/2$. Therefore, an application of Equations (108) and (109) in the temperature range $T > T_0$ is an extrapolation which can also lead to incorrect results (artifacts). The most general physical meaning of $T_0 = T_k/2$ is that at $T = T_0$ the maximum of the thermally weighted tunneling probability along the system reaction coordinate q (the maximum of the integrand in (97)) lies near the characteristic energy $E_q = E_c - kT_k$.[3]

The rate equation in the low temperature range $T < T_0 = T_k^{(q)}/2$ has the form [28, 63–65]

$$v = Be^{-S(E_\theta)/\hbar} \tag{112}$$

where $S(E_\theta)$ is defined by (111), while the prefactor B depends on the form of the particle potential $V_0(x)$.

An expression of this type can be derived by making direct use of the full separability of the reaction coordinate q for the potential surface (87) on the basis of (97) and the WBK approximation (3) for the tunneling probability $W(E_q)$, which yields

$$v = \frac{2kT}{h} \sinh \frac{h\nu_q}{2kT} \int_0^\infty e^{-S(E_q,T)/\hbar}\, \mathrm{d}E_q/kT \tag{113}$$

with the full action

$$S(E_q, T) = 2i \int_{q_1}^{q_2} p_q(E_q)\, \mathrm{d}q + (\hbar E_q/kT) \tag{114}$$

corresponding to expression (40) for the potential $V(x)$. Evaluation of the integral in the same way yields the rate equation

$$v = \frac{2kT}{h} \sinh(h\nu_q/2kT) \frac{(1 + \Phi(y_m^{(1)}))}{2kT[S''(E_q^{(m)})]^{1/2}} e^{-S(E_q^{(m)})/\hbar} \tag{115}$$

[3] As discussed in Section 3 for the parabolic barrier, this maximum lies somewhat above $E_x = E_c - kT_k$. However, for the Schottky potential (12) it lies somewhat below $E_x = E_c - kT_k$ [2a].

corresponding to (46), where $E_q^{(m)} = E_m$ is defined by (42) and $\Phi(y_m^{(1)})$ by (43).

An equivalent equation can be derived on the basis of the adiabatic formulation of rate theory, Equation (64), by retaining only the first term of the sum ($n = 0$), which dominates at low temperatures. This means using the adiabatic potential (62) for $n = 0$ with $\epsilon_0 = h\nu/2$. The momentum $p_q(E_q)$ can be written as

$$p_q(E_q) = [2m(E_q - V(q))]^{1/2} \equiv p(E) \equiv [2m(E - V_0(q))]^{1/2}$$

where $E = E_q + \epsilon_0$ and $V_0(q) = V(q) + \epsilon_0$, therefore, the full action reads

$$S(E,T) = \varphi(E) + (\hbar E/kT), \quad \varphi(E) = 2i\int_{q_1}^{q_2} p(E)\,dq. \tag{116}$$

Then, the rate constant can be expressed by

$$v = \frac{kT}{h}\frac{Z_{ac}^{\neq}}{Z}\int_0^{\infty} e^{-S(E,T)/\hbar}\,\mathrm{d}E/kT \tag{117}$$

where $Z = Z_q Z_n$ and

$$Z_n = Z_{ac}^{\neq} = \prod_{i=1}^{Z-1}[2\sinh(h\nu_i^{+}/2kT)]^{-1} \tag{118}$$

is the partition function of the transition state. Note again that $\nu_i^{+} = \nu_i$ because of the full separability of the reaction coordinate q. Using the relations

$$S'(E) = \varphi'(E) + (\hbar/kT), \quad S''(E) = \tau'(E) \tag{119}$$

where $\tau(E)$ is the imaginary vibration period, after the steepest descent evaluation of the integral in (117), we derive the equation

$$v = \frac{1 + \Phi(y_m^{(1)})}{2Z|h\tau'(E_m)|^{1/2}}e^{-S(E_m)/h}\prod_{i=1}^{Z-1}[2\sinh(h\nu_i^{+}/2kT)]^{-1} \tag{120}$$

where $\tau'(E_m) = (\mathrm{d}\tau/\mathrm{d}E)_{E=E_m}$. This equation coincides with previous results [28, 66] if we set $\Phi(y_m^{(1)}) = 1$,[4] by taking into account the complete separability of the tunneling coordinate q.

An alternative expression is obtained, on the basis of Equation (102) with $\kappa_{ac}^{\mathrm{scl}} = 1$ and $\kappa_t^{(q)}$ given by (104), if the integral is calculated in the same way to yield

$$\kappa_t = \frac{1 + \Phi(y_m^{(1)})}{kT|h\tau'(\epsilon_m)|^{1/2}}e^{-S(\epsilon_q^{(m)})/h}e^{E_c/kT} \tag{121}$$

[4] The appearance of the error integrals in Equations (41), (46) and (120) is a result of an improvement of the method of steepest descent [37, 38], which allows a considerable improvement of its accuracy.

where $\epsilon_q^{(m)} = E_q^{(m)} - E_c$. The resulting rate equation can be considered as a quantum generalization of transition state theory in the temperature range of large tunneling ($T < T_k/2$), as discussed above in the derivation of Equation (102). We recall that this approach implies that the initial and transition states are defined in the absence and presence of particle-medium coupling, respectively. A similar result is obtained from (117) if we write $Z = Z_x Z_y$ as in (103) and introduce the harmonic oscillator partition functions for the particle (Z_x) and the medium (Z_y) in the absence of coupling. Then, the tunneling correction in (102) is expressed by

$$\kappa_t = \frac{1 + \Phi(y_m^{(1)})}{kT|h\tau'(E_m)|^{1/2}} e^{-S(E_m)/\hbar} e^{E_c/kT} \tag{122}$$

which is equivalent to (121).

The transition temperature range $T \approx T_0 = T_k^{(q)}/2$ has been investigated by the imaginary free energy method [28, 63, 66]. A unified treatment covering all temperatures is proposed by Hanggi and Hontcha [28, 65] on the basis of the periodic orbit approach [67] and Miller's quasiclassical transition state theory for nonseparable systems [61]. However, the potential function (87) describes a separable system at any value of the coupling constants ($C_i \geq 0, i = 1, 2, \ldots, z-1$). Therefore, we again can use the results of Section 4, thus obtaining the equation

$$v = 2\sinh\left(\frac{h\nu_q}{2kT}\right) \frac{\Phi(y_m^{(1)}) + \Phi(y_m^{(2)})}{2[hS''(E_q^{(m)})]^{1/2}} e^{-S(E_q^{(m)})/\hbar} + \frac{\nu_q^*}{\pi} e^{-E_c/kT} \tag{123}$$

which is valid for $T \approx T_k^{(q)}/2 (h\nu_q^* \approx 2\pi kT)$. For $T < T_k^{(q)}/2$ we regain expression (115) by neglecting the small second term and setting $\Phi(y_m^{(2)}) = 1$. We recall that the analytical results for E_m, $S(E_m)$, $S''(E_m)$ and $y_m^{(1)}$, given by Equations (49–51) for potential (16) with (47) ($Q = 0$), directly apply to expression (123) by replacing ν_x by ν_q. Then, the relations (86) ($\nu' = \nu_q^*$) and (93) giving

$$\nu_q = \nu_x \frac{\nu_q^*}{\nu_x^*} = \nu_x \frac{\nu_x^*}{\nu_q^* + \hat{\eta}(\nu_q^*)} \tag{124}$$

can be used for a direct stochastic interpretation of Equations (115) and (123).

Equation (123) can be recast in the form (102) with $\kappa_{ac}^{\text{scl}} = 1$ and

$$\kappa_t = \left[\frac{h}{S''(E_m)}\right] \frac{\Phi(y_m^{(1)}) + \Phi(y_m^{(2)})}{kT} e^{-S(E_q^{(m)})/\hbar} e^{E_c/kT} + \frac{h\nu_q^*}{\pi kT}. \tag{125}$$

With this expression for the tunneling correction the rate Equation (102) represents the quantum transition state theory formulation for the intermediate range $T \approx$

$T_k^{(q)}/2$ between the temperature ranges of moderate tunneling ($T > T_k/2$) and large tunneling ($T < T_k/2$).

8. Conclusions

This paper has presented a reconsideration of the various aspects of the concept of a characteristic (crossover) temperature and its applications to thermally activated processes. The most general and clear definition of this concept, proposed by Christov [11, 13], consists in that at this temperature $T = T_k$ the thermally averaged probabilities for subbarrier (tunneling) and overbarrier (classical) transitions are equal. The alternative definition of Goldanskii [12], according to which at $T = T_k$ the maximum of the thermally weighted tunneling probability lies at the barrier peak $E_x = E_c$, gives the same result, Equation (23), for a parabolic barrier if the exact expression (10) for the transition probability $W(E_x)$ is used. The reason that Goldanskii has obtained another expression, Equation (31), is due to the incorrect use of the ordinary WBK approximation (3) for $W(E_x)$, which is valid only for $E_x \lll E_c$ ($E_x < E_c - kT_k$). The characteristic temperature of Goldanskii, $T_k' = T_k/2$, has no precise and clear physical sense, being the temperature at which the maximum of the thermally weighted transition probability lies near the energy value $E_x = E_c - kT_k$ [6, 15a]. In any case, this is not 'the temperature below which the tunnel effect plays an appreciable role', as stated in the original paper by Goldanskii [12], since tunneling is actually appreciable at $T_k' < T < 2T_k''$.

Nevertheless, the derivatives of T_k, $T_k' = T_k/2$ and $T_k'' = 2T_k$ can be used for a conventional delimitation of four temperature ranges: $T > 2T_k$ (negligible tunneling), $2T_k > T > T_k$ (small tunneling), $T_k > T > T_k/2$ (moderate tunneling) and $T < T_k/2$ (large tunneling) [6, 7, 15a].

The usefulness of this division appears first of all in the theory of electron emission, as discussed in Section 5, where it is shown that the characteristic temperature T_k defined by (52), is a basic parameter of that theory. The same holds for reaction rate theory based either on a proper dynamic (Section 6) or a stochastic approach (Section 7). In the first case the characteristic temperature T_k is estimated directly from the properties of the potential energy surface in the saddle point region. Such estimations show that for gas phase reactions involving hydrogen, the tunneling corrections to transition state theory at $T \approx 300$ K can be considerable [6, 44]. The same is valid for proton transfer reactions in solution [7].

In the framework of the adiabatic transition state theory, the characteristic temperature $T_k^{(n)}$ is defined for any vibration state $\mathbf{n}$ by formula (73). An extension of this definition is based on a recent resonance state approach to transition state theory [30] which yields expression (82).

In the stochastic approach to reaction rate theory (Section 8), the characteristic temperature T_k, defined by (23), has exactly the same physical meaning, account has been taken for the fact that the GLE (83) can be derived from a Hamiltonian including the potential surface (87). Expressions (94) and (95) represent the definitions of T_k in the absence and presence of particle–medium coupling, respectively. The delimitation of four temperature ranges of different extent of tunneling clearly appears in the rate equations derived from a particular formulation of transition state theory, based on the potential surface (87). The existence of a completely separable reaction coordinate q in the configuration space of that surface, defined by (90), has been overlooked except in the recent papers of the author [29, 57, 69]. Use of this reaction coordinate greatly simplifies the calculation of the tunneling correction at all temperatures (Equations 99, 122, 125). This possibility shows that for the potential surface (87) it is not necessary to use the more complicated 'bounce' (instanton) methodology [28, 61, 68] nor the periodic orbit approach [61, 65, 68], which are appropriate for nonseparable systems.

The specific meaning of the Goldanskii 'crossover' temperature $T_k' = T_k/2$ in the framework of the instanton approach to low temperature processes consists in that at this temperature the instanton disappears. However, this result does not at all mean that at $T = T_k' = T_k/2$ the tunnel effect 'switches off' [68]. Indeed, there is a wide transition temperature range between $T_k' = T_k/2$ and $T_k'' = 2T_k$, in which tunneling plays a considerable role. Within this range, the actual 'crossover' between the tunneling and classical regime is the temperature T_k, defined by (24) and (25). The particular role of $T_k' = T_k/2$ in the instanton theory is related to the restrictions of the quasiclassical approximation (3) used in this theory, which is invalid for $T > T_k/2$. In any case, beyond instanton theory, there is no correct derivation of Goldanskii characteristic temperature $T_k' = T_k/2$, although it can be considered in a conventional sense as crossover between the temperature ranges of moderate and large tunneling. In contrast to this, the characteristic temperature T_k, based on the general definition (24), has an immediate and clear physical meaning as crossover from the proper classical to the tunneling temperature range.

References

1. A. Sommerfeld and H. Bethe: in *Handbuch der Physik*, 2. Aufl, Band 24, Springer, Berlin (1933). E. Guth and C. J. Mullin: *Phys. Rev.* **61**, 339 (1942). E. L. Murphy and R. H. Good: *Phys. Rev.* **102**, 1464 (1956).
2. S. G. Christov: (a) *Phys. Stat. Sol.* (a) **17**, 11 (1966). *ibid.* **21**, 159 (1967). *ibid.* **32**, 509 (1969). *ibid.* **7**, 371 (1971). (b) *Contemp. Phys.* **13**, 199 (1972). (c) *Surf. Sci.* **70**, 32 (1978).
3. T. Holstein: *Ann. Phys.* **8**, 325, 343 (1959). *Phil. Mag. B* **37**, 49, 499 (1978).
4. S. G. Christov: *Phys. Rev. B* **26**, 6918 (1982). M. Georgiev, A. D. Gochev, S. G. Christov, and A. Kyuldjiev: *Phys. Rev. B* **26**, 6936 (1982).
5. H. S. Johnston: *Gas Phase Reaction Rate Theory*, Ronald Press, New York (1966). E. M. Mortenson: *J. Chem. Phys.* **48**, 4029 (1968). D. G. Truhlar and A. Kuppermann: *J. Chem. Phys.* **52**, 3841 (1970). *ibid.* **56**, 2232 (1972). *ibid.* **59**, 395 (1973).
6. S. G. Christov: *Collision Theory and Statistical Theory of Chemical Reactions*, Springer, Berlin (1980).
7. R. P. Bell: *The Tunnel Effect in Chemistry*, Chapman and Hall, London (1980).
8. S. G. Christov: *Chem. Phys.* **168**, 327 (1992).
9. J. J. Hopfield: *Proc. Natl. Acad. Sci.* **31**, 695 (1978). J. Jortner: *J. Chem. Phys.* **64**, 4860 (1976). A. D. Gochev: *C.R. Acad. Bulg. Sci.* **31**, 695 (1978).

10. S. G. Christov: in J. Maruani (ed.), *Molecules in Physics, Chemistry and Biology*, Vol. 3, Kluwer Academic Press, Dordrecht (1989).
11. S. G. Christov: *Z. Electrochem.* **62**, 567 (1958).
12. V. I. Goldanskii: *Dokl. Akad. Nauk SSSR* **124**, 1261 (1959).
13. S. G. Christov: *Dokl. Akad. Nauk SSSR* **136**, 663 (1960).
14. S. G. Christov: *Ann. Phys. (Germany)* **12**, 20 (1963).
15. S. G. Christov: (a) *Ber. Bunsenges. Phys. Chem.* **76**, 507 (1972). (b) *ibid.* **78**, 537 (1974).
16. S. G. Christov: *Phil. Mag. B* **52**, 71, 91 (1985).
17. F. Gutmann: *Nature* **219**, 1350 (1968). *J. Res. Inst. Catalysis*, Hokkaido Univ. **17**, 96 (1969). *Jap. J. Appl. Phys.* **8**, 1417 (1969).
18. J. M. Bermond, M. Lenoir J. M. Prulhier and M. Drechsler: *Surf. Sci.* **42**, 306 (1974).
19. M. Vodenicharova: *Phys. Stat. Sol.* (a) **28**, 263 (1975).
20. C. M. Vodenicharov: *Phys. Stat. Sol.* (a) **42**, 785 (1977).
21. C. Grimbert and A. Laforgue: *Colloq. Int. Chem. Quant. Express*, Latina, Salamanca (1977).
22. J. Wysocki and C. Vodenicharov: *Phys. Stat. Sol.* (a) **50**, 411 (1978).
23. F. Gutmann, H. Keyzer, and L. E. Lions: *Organic Semiconductors*, Part B, R. E. Krieger (1983).
24. R. Daudel, G. Leroy, D. Peeters, and M. Sana: *Quantum Chemistry*, Wiley (1983).
25. G. Leroy, M. Sana, and A. Tinant: *Canad. J. Chem.* **63**, 1447 (1985).
26. E. Bunks and J. Jortner: *J. Phys. Chem.* **84**, 337 (1980).
27. P. Hanggi, H. Grabert, G. L. Ingold, and U. Weiss: *Phys. Rev. Lett.* **55**, 761 (1985).
28. P. Hanggi, P. Talkner, and M. Borkovec: *Rev. Mod. Phys.* **62**, 251 (1990).
29. S. G. Christov: *Mol. Phys.* **76**, 1242 (1992).
30. D. G. Truhlar and B. C. Garett: *J. Phys. Chem.* **96**, 6515 (1992).
31. E. Kemble: *The Fundamental Prinepiles of Quantum Mechanics*, 2nd. ed., McGraw Hill, New York (1958) (1st ed. 1937).
32. S. G. Christov: *Phys. Stat. Sol.* (a) **42**, 583 (1970).
33. C. Eckart: *Phys. Rev.* **35**, 1303 (1930).
34. S. G. Christov: *Z. Electrochem.* **64**, 840 (1960).
35. R. P. Bell: *Trans. Faraday Soc.* **55**, 1 (1959).
36. R. P. Bell: *Proc. Roy. Soc.* (*London*) *Ser. A* **148**, 241 (1935).
37. S. G. Christov: *Ann. Phys.* (Germany) **15**, 87 (1965).
38. S. G. Christov: *Phys. Rev. B* **26**, 6918 (1982).
39. S. G. Christov: *Ber. Bunsenges. Phys. Chem.* **79**, 357 (1975).
40. C. M. Vodenicharov and S. G. Christov: *Phys. Stat. Sol.* (b) **25**, 387 (1974).
41. C. M. Vodenicharov, A. A. Stanchev, and M. Lozev: *Phys. Stat. Sol.* (a) **39**, K1 (1977).
42. C. M. Vodenicharov: *Phys. Stat. Sol.* (a) **29**, 223 (1975).
43. R. Vanselow: *Appl. Phys.* **2**, 229 (1973).
44. E. M. Mortensen: *J. Chem. Phys.* **48**, 4029 (1968). **49**, 3526 (1968).
45. R. Weston: *J. Chem. Phys.* **31**, 892 (1952).
46. J. O. Hirschfelder and E. Wigner: *J. Chem. Phys.* **7**, 619 (1939). M. A. Eliason and J. O. Hirschfelder: *J. Chem. Phys.* **30**, 1426 (1959).
47. L. Hofacker: *Z. Naturforsch.* **18A**, 607 (1963).
48. R. A. Marcus: *J. Chem. Phys.* **43**, 2138 (1966). *ibid.* **46**, 959 (1967).
49. D. G. Truhlar, A. D. Isaacson, and B. C. Garett: in M. Baer (ed.), *Theory of Chemical Reaction Dynamics*, Vol. 4, CRC Press, FL, (1985).
50. A. Kuppermann: *J. Phys. Chem.* **83**, 171 (1979).
51. H. A. Kramers: *Physica* **7**, 284 (1940).
52. R. F. Grote and J. T. Hynes: *J. Chem. Phys.* **73**, 2715 (1980).
53. R. Zwanzig: *J. Stat. Phys.* **9**, 215 (1973).
54. S. G. Christov: *Int. J. Quant. Chem.* **52**, 7219 (1994).
55. E. Pollak: (a) *Chem. Phys. Letters* **127** 178 (1986). (b) *J. Chem. Phys.* **85**, 865 (1986).
56. S. G. Christov: *Int. J. Quant. Chem.* **36**, 391 (1989).
57. S. G. Chistov: *Chem. Phys. Letters* **188**, 532 (1992).
58. R. P. Feynmann and A. Hibbs: *Quantum Mechanics and Path Integrals*, McGraw-Hill (1965).
59. R. P. Feynmann: *Statistical Mechanics*, Benjamin (1972).

60. A. O. Caldeira and A. J. Legett: *Ann. Phys. N.Y. Acad. Sci.* **149**, 374 (1983).
61. W. H. Miller: *J. Chem. Phys.* **62**, 1899 (1975).
62. A. O. Caldeira and A. J. Legett: *Ann. Phys. N.Y. Acad. Sci.* **149**, 374 (1983).
63. A. L. Larkin and A. A. Ovehinnikov: *Zh. Exp. Teoret. Phys.* **82**, 2708 (1985).
64. Y. I. Dakhnovskii, A. A. Ovehinnikov, and Z. Smedarchina: *Phys. Lett. A* **149**, 43 (1990).
65. P. Hangii and W. Hontcha: *Ber. Bunsenges. Phys. Chem.* **95**, 378 (1991).
66. P. S. Riseborough, P. Hanggi, and E. Freidkin: *Phys. Rev. A* **32**, 489 (1985).
67. M. C. Gutzwiller: *J. Math. Phys.* **12**, 343 (1971).
68. V. A. Benderskii, V. I. Goldanskii, and D. E. Makarov: *Physics Reports* **233**, 195 (1993).
69. S. G. Christov: *Int. J. Quant. Chem.* **64**, 393 (1997).

Molecular Engineering **7**: 149–160, 1997.

An Improved Method of Relaxation Rate Calculation in Double-Well Potential Systems

L. MIHAILOV, M. KIRTCHEVA and A. MANOV
Institute of Solid State Physics, Bulgarian Academy of Sciences, 72 Tzarigradsko Chaussee, 1784 Sofia, Bulgaria.

(Received: 12 September 1996; accepted 17 June 1997)

Abstract. The relaxation behavior of double-well potential systems is considered on the basis of numerical calculation of eigenvalue spectrum of the Schrödinger equation. Depending on the temperature, the relaxation behavior of these systems is either quantum-mechanical or classical. The quantum-mechanical behavior is quite similar to that of two-level systems for which the tunneling model has been developed. On the other hand, the reaction rate method has been adapted for calculation of the relaxation rates of non-radiative transitions. We propose an improvement to this method, which makes it applicable to any double-well potential system. A 'tunneling degree' parameter is introduced in order to describe the specific relaxation behavior of these systems in the framework of both the conventional tunneling model and the classical Debye relaxation. A comparative analysis is conducted for α-quartz (E' defect renters, H^- ion at E'_4 centers) and high-temperature superconducting cuprates (apical O(A) atoms). The calculations presented allow one to give a theoretical description of the relaxation response in these materials under the action of an external field.

Key words: Schrödinger equation, double-well potential, tunneling system, reaction rate, Debye relaxation.

1. Introduction

Finding the correlation between the relaxation response under the action (elastic or electromagnetic) of an external field and the relaxation properties of double-well potential systems (DWPS) is not a new problem. Theoretical investigations during the last few years have made it clear that the existence of the standard tunneling systems (TS) in amorphous solids [1, 2] is not sufficient to explain the acoustic and dielectric loss features in the low temperature range. In accordance with the soft atomic potential (SAP) method [3], in amorphous solids there also exist relaxation system (RS), in addition to the well known TS. This method considers RS as DWPS which contribute to the observed properties at higher temperatures, compared to TS, where the relaxation corresponds to thermally activated overbarrier transitions, described by the Arrhenius law. It is assumed [4–6] that in high-temperature superconducting cuprates (HTSC) and α-quartz there exist DWPS with, depending on temperature, either quantum-mechanical or classical relaxation behavior. Such behavior is similar to the conventional TS at low temperature and to the classical RS at higher temperature. In addition, these systems may also relax via thermally activated tunneling. TS and RS in such non-amorphous materials (and maybe in amorphous solids, too) have a common physical nature and allow one to consider these systems as tunneling–relaxation ones (TRS). E'_1, E'_2–E'_4 centers and H^- ion at a E'_4 center in α-quartz [5, 6], centers in some polycrystalline superconductors (SC) and normal metals [7], as well as O(A) atoms in HTSC [4] are examples of TRS in non-amorphous solids.

The adiabatic electron transfer is a characteristic of the TRS considered. This is the reason why the configurational transitions dominate over the electron ones. As a result of the wave–TRS interaction, both resonant and relaxation absorption of the incident waves takes place. However, in acoustic experiments it is difficult to observe resonant absorption even at superlow temperatures because the typical intensity of the acoustic waves is high enough to equalize the population of the two levels in TS and in this way to 'shut off' the resonant interaction [1]. The relaxation absorption cannot be saturated by such a mechanism. It is known that the possible relaxation rates of processes contributing to the wave propagation, related to the relaxation via thermal phonons field, are: one-phonon (direct) processes, first- and second-order Raman processes (with negligibly small contribution), and thermally activated Arrhenius-type processes. The adapted reaction rate method [8] includes three main techniques for the relaxation rate calculation – related to ground-state tunneling, to quasiclassical, and to classical transitions. The present work introduces a modified scheme with phonon-assisted tunneling rate.

The aim of this paper is to propose an approach to calculating the eigenvalue spectrum and corresponding wave functions of DWPS and to obtain the relaxation rate and response of these systems under the action of an external field. The calculation method employed is applied to α-quartz and $YBa_2Cu_3O_{7-\delta}$ (YBCO) where E′ centers and apical oxygen atoms O(A), respectively, are considered as TRS.

2. Numerical Calculation of One-Dimensional Schrödinger Equation

The one-dimensional single-particle Schrödinger Equation (SE) is:

$$\frac{\partial^2 \chi(E,Q)}{\partial E^2} + \frac{2M}{\hbar^2}[E - U(Q)]\chi(E,Q) = 0, \tag{1}$$

where M is particle mass, Q is configuration coordinate, $\chi(E,Q)$ is wave function and E is energy. Equation (1) has to be solved numerically in order to obtain solutions for DWPS. The numerical calculation method is applicable for any type of potential $U(Q)$. In this paper we have used anharmonic DWP in accordance with the pseudo-Jahn–Teller effect (PJTE) description of broken-symmetry centers such as E′ series centers in α-quartz [5, 6] and O(A) in HTSC [4]:

$$U(Q) = \frac{1}{2}\left(KQ^2 - \frac{1}{2}\sqrt{(2GQ+D)^2 + E_{gu}^2}\right) - E_{\min}, \tag{2}$$

where $K = M\omega_0^2$ is spring constant, ω_0 is bare oscillator frequency, G is electron–phonon coupling constant, E_{gu} is energy splitting of the mixed electronic states, D is an extrinsic parameter introduced to account for any additional perturbations to the particular PJTE, and $E_{\min}$ is a constant fixing the initial energy level.

Setting $R_e = 2/\hbar\omega_0$ and $R_c = \sqrt{M\omega_0/\hbar}$, as well as dimensionless coordinates and energies ($q = R_c Q, \varepsilon = R_e E$), one obtains the dimensionless SE:

$$\frac{\partial^2\chi(\varepsilon,q)}{\partial\varepsilon^2} + [\varepsilon - q^2 + \delta(q)]\chi(\varepsilon,q) = 0. \tag{3}$$

In the case of $\delta q = 0$ Equation (3) has an analytical solution (linear harmonic oscillator (LHO) problem). However, a non-zero value of the anharmonic term δq may cause drastic changes in the energy levels. For the considered potential (2), δq is:

$$\delta(q) = \frac{1}{2}\sqrt{(4q\sqrt{\varepsilon_{\rm JT}} + d)^2 + \varepsilon_{gu}^2} - \varepsilon_{\min}, \tag{4}$$

where $\varepsilon_{\rm JT} = R_e E_{\rm JT}$, $E_{\rm JT} = G^2/2K$, $d = R_e D$, $\varepsilon_{gu} = R_e E_{gu}$, $\varepsilon_{\min} = R_e E_{\min}$.

For $\delta q = 0$ the numerical solution coincides with the theoretical levels ($2E/\hbar\omega_0 = \varepsilon = 2n+1$, $n = 0,1,2,\ldots$) and the wave functions of LHO. When $\delta q \gg 0$ the potential transforms into two deep separate wells (two weakly interacting LHO). Then, one should expect LHO-like results (double degenerated LHO levels), but with a modified condition for the energy quantisation:

$$\frac{2}{\hbar\omega}E = 2n+1, \tag{5}$$

where ω is the renormalised phonon frequency determined from the curvature of the potential energy minimum. At $D \to 0$ the renormalised frequency is:

$$\omega = \omega_0\sqrt{1 - E_{gu}^2/16E_{\rm JT}^2}. \tag{6}$$

The numerical calculation of Equation (3) confirms condition (5) (for example, E_1' and E_2'–E_4' broken-symmetry centers in α-quartz). In some investigated centers, even the lowest energy levels are substantially different from the 'harmonic' ones (5) (for example, H^- ion at E_4' center in α-quartz).

The initial conditions in the numerical calculation of Equation (3) are $\chi_{q=-\infty} = (\partial\chi/\partial\varepsilon)_{q=-\infty} = 0$. The limits $q_{\pm\infty}$ of the numerical integration of Equation (3) are determined for each level according to the symmetry requirements: $\chi_{q=+\infty} \sim \chi_{q=-\infty} = 0$. The accuracy in the calculated energies is about 1 cm^{-1}, which is less than the resolution of the modern spectrometers. In addition, the experimental accuracy of ω_0, which is an input parameter in the calculations, is also ~1 cm^{-1}.

Thus, one may obtain the eigenvalues E_n and eigenfunctions $\chi_n(Q)$. All energy levels common to both wells are split (tunnel split for the subbarrier energy levels),

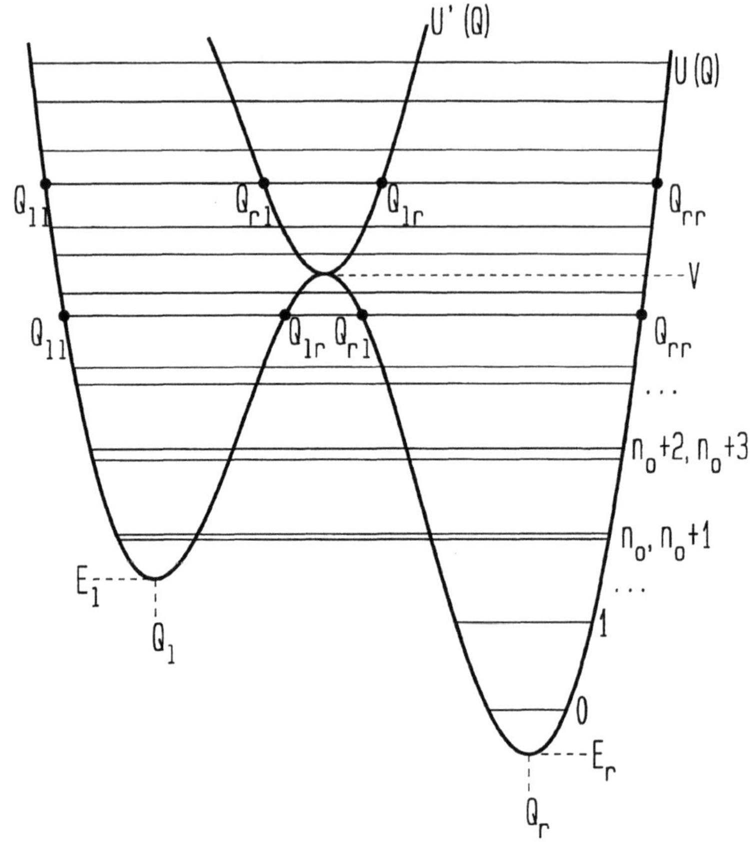

Figure 1. Asymmetric DWP $U(Q)$ with energy levels E_n, turning points Q_{lr}, barrier height V, coordinates of the well bottoms $(Q_{l/r}, E_{l/r})$. $U'(Q)$ is the reversed potential energy.

as is shown in Figure 1. The lowest pair of tunnel split levels have quantum numbers n_0 and $n_0 + 1$. DWP (2) is applied to interpret cluster-calculated energy profiles for E'_1, $E'_{2,4}$ centers and H^- ion at E'_4 center [9]. Using numerical calculation of SE, a thorough characterization is performed of E′ series centers, H^- ion at E'_4 center, as well as O(A) centers. Figure 2 shows the energy potentials and levels corresponding to E'_1, $E'_{2,4}$ centers, H^- ion at E'_4 center with PJT parameters taken from [10], and O(A) centers with PJT parameters from [4].

3. Relaxation Behavior of Double-Well Potential Systems

DWPS are strongly relaxing systems. There are two modes of inter-well transition along the adiabatic energy profile (2): (i) from the lower to the upper surface by a vertical Frank–Condon transition and *vice versa*, and (ii) non-radiatively, via tunneling or classical jumps across the interwell barrier along the lower branch. The

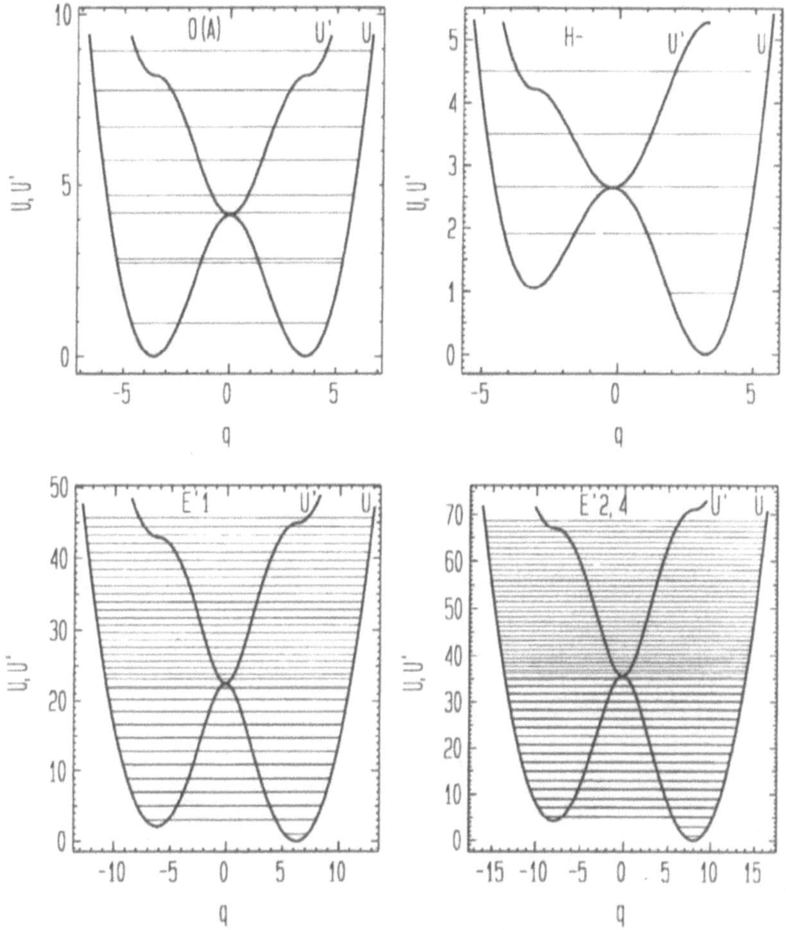

Figure 2. Dimensionless double well potential energy $U(q)$, $U = 2/(\hbar\omega)U$; $q = (\sqrt{M\omega_0/(\hbar)}Q$ with numerically calculated vibronic levels of E'_1, $E'_{2,4}$ centers, H^- ion at E'_4 center in α-quartz, and O(A) in YBCO. The factor $2/\hbar\omega$ (not $2/\hbar\omega_0$) turns U dimensionless, which clarifies the treatment of the levels (if the harmonic approximation is valid, then the levels must be positioned at odd ordinate numbers). $U'(q)$ is the corresponding reversed potential energy.

result of both transition ways is the charge transfer from one atom to another, which is accompanied by a large configuration relaxation. These switch-over transitions (for example E'_1–E'_1 and E'_2–E'_4 in α-quartz) lead to acoustic or dielectric losses.

3.1. Relaxation rate calculation

The interwell transfer rate can be calculated on the basis of numerically computed energy levels using the adapted for non-radiative transitions reaction rate method

(RRM) [11]. According to this method, TRS may relax via tunneling in the ground state, thermally activated tunneling in remaining sub-barrier states, and classical transitions in overbarrier states [11, 12]. Three temperature intervals are defined correspondingly: 0–T_k, T_k–$4T_k$, $4T_k$–∞, where $2T_k$ is Christov's characteristic temperature. In the present calculations RRM is improved in order to calculate the temperature dependence of the relaxation rate of the lowest pair of tunnel split energy levels. This relaxation rate is obtained in accordance with the standard tunneling model (TM). The contribution from the remaining (sub- and overbarrier) energy levels is computed numerically using a quasi-classical approximation. The relaxation rates obtained are close to those calculated by RRM.

In general, the relaxation rate τ^{-1} is determined by the total transition probability W per unit time along the potential energy profile U. This transition probability is a sum over all energy levels with weight coefficients A_n (thermal occupation factors):

$$\tau^{-1} = W = \sum_{n=n_0}^{\infty} W_n, \qquad W_n = A_n T_n \nu_n, \tag{7}$$

where T_n is transition probability, and

$$\nu_n = \begin{cases} \dfrac{E_{n-1} - E_n}{2h}, & n > 0 \\ \dfrac{E_{n+1} + E_n}{2h}, & n = 0. \end{cases} \tag{8}$$

It is well known that the occupation probability of the level n is:

$$A_n = \frac{1}{S} \exp\left(-\frac{E_n - E_0}{kT}\right), \qquad S = \sum_{n=0}^{\infty} \exp\left(-\frac{E_n - E_0}{kT}\right), \tag{9}$$

where S is the statistical sum of a large canonical ensemble. $A_n\nu_n$ is interpreted as a probability for the closeness to the barrier at a given energy in the interval $h\nu_n$. The definition given for $h\nu_n$ (8) makes the interval $h\nu_n$ a smooth function on n. This will not be the same and does not have a physical meaning, if $h\nu_n = E_{n+1} - E_n$.

As has been mentioned above, the lowest pair of tunnel split energy levels (see Figure 1) n_0 and $n_0 + 1$ 'form' the classical TS. n_0 is the number of those energy levels which lie between E_l and E_r. This number depends on the asymmetry energy $|E_l - E_r|$, where $E_l = U(Q_l)$ and $E_r = U(Q_r)$. The normalised occupation probabilities of the levels n_0 and $n_0 + 1$ are $\mu_{n_0} = 2A_{n_0}$, $\mu_{n_0+1} = 2A_{n_0+1}$, respectively. The tunneling transition probability (between levels n_0 and $n_0 + 1$) W^{TS} taking into account the temperature occupations ($W_n = T_n A_n \nu_n$) is:

$$W^{\mathrm{TS}} = W_{n_0} + W_{n_0+1}, \qquad \begin{matrix} W_{n_0} = \mu_{n_0} \dfrac{\Delta_0^2 \Delta_E \gamma^2}{2\pi\hbar^4 \rho v^5}\eta \\ W_{n_0+1} = \mu_{n_0+1} \dfrac{\Delta_0^2 \Delta_E \gamma^2}{2\pi\hbar^4 \rho v^5}(1 + \eta) \end{matrix},$$

$$\eta = \frac{1}{\exp\left(\dfrac{\Delta_E}{kT}\right) - 1}, \quad \Delta_E = E_{n_0+1} - E_{n_0}, \tag{10}$$

where Δ_0 is tunnel splitting of a 'symmetric' PJT potential (D = 0), ρ is mass density, v is elastic wave velocity, $\gamma = \mathcal{D}\Delta_E/2\Delta$, $\mathcal{D}$ is deformation potential/dipole moment in the case of interaction with elastic/electromagnetic field, respectively, $\Delta = \sqrt{\Delta_E^2 - \Delta_0^2}$. In the special case of pure TS $\mu_{n_0} = \mu_{n_0+1} \approx 1$ the transition probability (including both transition directions) is:

$$\tau_d^{-1} = W^{\mathrm{TS}} = W_{n_0} + W_{n_0+1} = \frac{\Delta_0^2 \Delta_E \gamma^2}{2\pi\hbar^4 \rho v^5} \coth\left(\frac{\Delta_E}{2kT}\right), \tag{11}$$

which is the well known formula for the 'direct processes' in the tunneling model [1].

In order to obtain the transition probability W^{RS} for the remaining energy levels $(n > n_0 + 1)$, it is necessary to know the wave functions for each well separately. Numerical integration of SE gives some general wave function, which is not factorised. The linear combination of Wentzel–Kramers–Brillouin (WKB) functions is an imitation of a numerically integrated solution (for example, the number of roots of the oscillating solution is the same). For this reason the WKB method (quasi-classical approximation) is used for determining the transition probabilities:

$$T_n = \frac{1}{1 + x_n},$$

$$x_n = \begin{cases} \dfrac{V^2}{4(V - E_n)} \sinh^2\left(\dfrac{\sqrt{2K}}{\omega_0} \displaystyle\int_{Q_{lr}(n)}^{Q_{rl}(n)} \sqrt{V - U(Q)}\, \mathrm{d}Q\right) & E_n < V \\ \dfrac{-V^2}{4(V - E_n)} \sin^2\left(\dfrac{\sqrt{2K}}{\omega_0} \displaystyle\int_{Q_{lr}(n)}^{Q_{rl}(n)} \sqrt{U'(Q) - V}\, \mathrm{d}Q\right) & E_n > V \end{cases}, \tag{12}$$

where $n \geq n_0 + 2$, $Q_{lr}(n)$ and $Q_{rl}(n)$ are turning points, V is energy of the barrier top, $U'(Q)$ is 'reversed' potential (reflected image of $U(Q)$ relative to V) (see Figure 1):

$$U'(Q) = \begin{cases} 2V - U(Q) & Q_l < Q < Q_r \\ 2(V - E_l) + U(Q) & Q < Q_l \\ 2(V - E_r) + U(Q) & Q > Q_r \end{cases} \tag{13}$$

The calculated temperature dependencies of W (7) for E'_1, E'_2, E'_4 centers, H^- ion at E'_4 center in α-quartz, and O(A) atoms in HTSC, are shown in Figure 3. The values of the input parameters and the most important derivative parameters are listed in Table I. The frequencies of the incident field ω_i are taken from experiments (see Refs. [4–6, 10]).

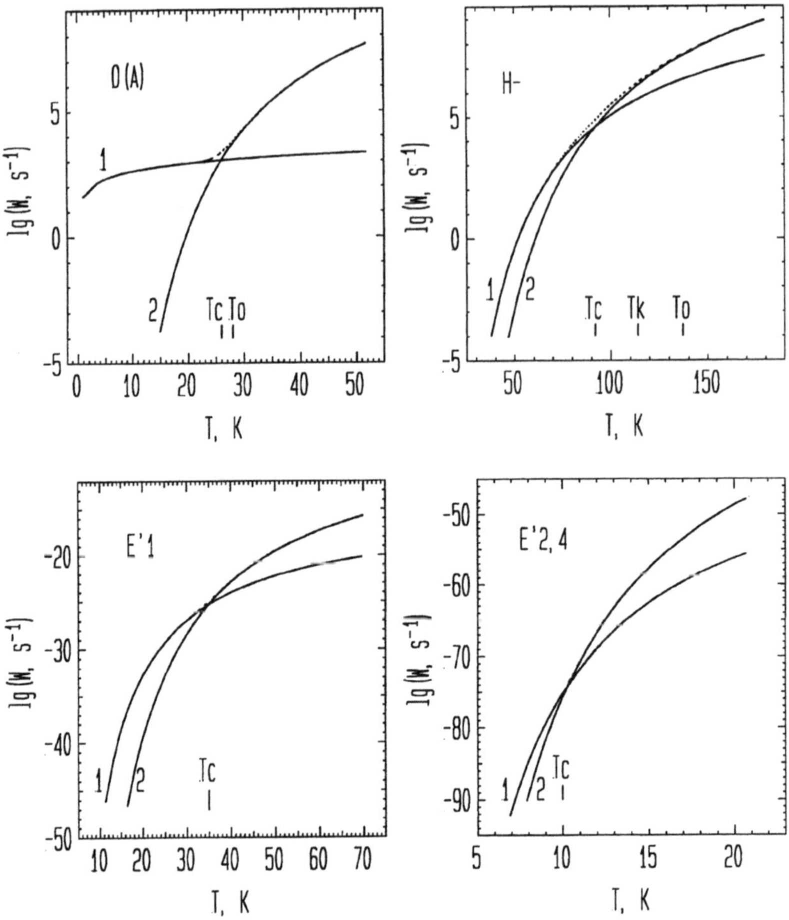

Figure 3. Temperature dependencies of the relaxation rate for E_1', $E_{2,4}'$ centers, H^- ion at E_4' center in α-quartz and O(A) in YBCO (dotted lines). The components W^{TS} (curves 1) and W^{RS} (curves 2) are drawn with solid line. Their crossing point defines T_c.

3.2. Tunneling degree

It is clear (Figure 3) that DWPS with higher barrier heights relax much more slowly than those with lower barriers, which are similar to TS. It is seen that the TS contribution dominates at lower temperature, while the RS contribution dominates at higher temperature. These two contributions are equal at some characteristic temperature T_c. This temperature can be compared with Christov's characteristic temperature

$$2T_k = \left(\frac{4E_{\rm JT}}{E_{gu}} - 1\right)\frac{\hbar\omega_0}{k\pi}$$

Table I. Input parameters and most important derivative parameters.

Quantity	O(A)	H^- at E'_4	E'_1	$E'_{2,4}$
Input parameters				
E_{gu}, eV	1.5	7.4	1.8	1.84
G, eV/Å	3.27	11.57	9.67	4.67
D, eV	1.8×10^{-5}	0.105	0.072	0.08
ω_0, eV	0.066	0.2	0.069	0.038
K, eV/Å^2	8.34	27.6	31.9	8.0
$\mathcal{D}$, eV	0.36	1.0	1.0	1.0
ρ, g/cm^3	2.3	2.65	2.65	2.65
v, m/s	6400	3300	3300	3300
$\omega_i/2\pi$, Hz	1050	$5 \times 10^{+6}$	$5 \times 10^{+6}$	$5 \times 10^{+6}$
N, Å^{-3}	1×10^{-5}	7.5×10^{-7}	7.5×10^{-7}	15×10^{-7}
Derivative parameters				
Δ_0, eV	1.76×10^{-6}	7.38×10^{-5}	2.45×10^{-17}	3.03×10^{-28}
E_{JT}, Ev	0.64	2.42	1.47	1.36
ω, eV	0.0535	0.129	0.0657	0.0358
E_{min}, eV	−0.86	−3.87	−1.64	−1.56
T_k, K	173	228	575	276
T_c, K	26	92	35	10
T_0, K	28	137	671	608
$\lambda(T_0)$	0.224	0.0957	2.43×10^{-23}	2.04×10^{-45}

and the position of the relaxation absorption peak T_0. The calculated temperature dependencies of the relaxation rates in the investigated systems (E$'$ defect centers, H^- ion at E'_4 centers and O(A) atom in YBCO) demonstrate that these systems are mixed tunneling–relaxation systems in the vicinity of T_c. For this reason the parameter 'tunneling degree' is introduced in order to calculate the relaxation losses in accordance with both TM and Debye relaxation ratio. One can rewrite (7) as:

$$W = \sum_{n=n_0}^{\infty} W_n = W^{TS} + W^{RS}, \quad W^{TS} = \sum_{n=n_0}^{n=n_0+1} W_n, \quad W^{RS} = \sum_{n=n_0+2}^{\infty} W_n, \tag{14}$$

where W^{TS} is the pure tunneling system contribution and W^{RS} is the pure relaxation system contribution. The tunneling degree λ is defined as:

$$\lambda = \frac{W^{TS}}{W^{TS} + W^{RS}} = \begin{cases} 1, & \text{TS} \\ 0, & \text{RS.} \end{cases} \tag{15}$$

In the special case of a pure tunneling system, only the lowest tunnel split level (for $n = n_0, n_0 + 1$) is occupied, i.e. $W_{n_0} = W_{n_0+1} \approx 0.5$ ($W^{TS} = 1, W^{RS} = 0$)

and $\lambda = 1$. In the case of a pure relaxation system, all energy levels have nearly the same occupation, i.e. $W^{\mathrm{RS}} \gg W_{n_0} + W_{n_0+1} = W^{\mathrm{TS}}$ and $\lambda \to 0$. As is seen from (14) and (15), the tunneling degree λ depends on temperature. At the characteristic temperature T_c, $W^{\mathrm{TS}} = W^{\mathrm{RS}}$ ($\lambda = 0.5$). As one may expect, for all investigated centers the value of T_c is smaller than T_k (half of Christov's characteristic temperature).

As is known, for classical RS the relaxation strength is proportional to the factor $\mathcal{D}^2$. In the case of TS this factor is

$$\mathcal{D}^2 \frac{\Delta^2}{\Delta_E^2} \times \operatorname{sech}^2\left(\frac{\Delta_E}{2kT}\right).$$

Because of the treatment of DWPS as mixed TRS this factor is constructed as a sum of two terms with weight coefficients λ and $\lambda - 1$:

$$\mathcal{D}^2 \left(\lambda(T) \frac{\Delta^2}{\Delta_E^2} \operatorname{sech}^2\left(\frac{\Delta_E}{2kT}\right) + [1 - \lambda(T)]\right), \tag{16}$$

The first term in (16) accounts for the tunneling system contribution to the wave energy losses, while the second term accounts for the classical relaxation system contribution to them. Equation (16) can be proved more rigorously. The simple treatment given corresponds to the DWPS consideration as TRS. One may use the first term in (16) when the energies of both tunneling levels are separated from the remaining energy levels. For this reason an additional weight-factor may be introduced in (16), which accounts for the level distribution (something like μ_{n_0+1}). The question of the introduction of an additional weight factor is open for discussion.

The investigated TRS O(A), H^- at E'_4, E'_1 and $E'_{2,4}$ are ordered by the decreasing of the tunneling degree (see Table I). It is useful to analyze the temperature dependence of the tunneling degree parameter λ (Figure 4). The difference between the E′ series centers and O(A)/H^- ions is clearly seen. The E′ center is a typical classical relaxation system at temperature near the position of the absorption peak T_0 (~600 K). The tunneling system absorption part in (16) is negligibly small, because at very low temperature (at $T < T_c = 35$ K for E'_1 and $T < T_c = 10$ K for E'_4 center) the system is a pure tunneling one. The reason is the very large value of the barrier height (containing ~10 split levels) and asymmetry energy of the order of 1–2 vibronic quanta. The O(A) atom is a typical tunneling system because of the lower barrier (containing two split levels) and the smaller asymmetry energy. The H^- ion at E'_4 acts as a mixed tunneling–relaxation system because the transition between the two limiting cases ($\lambda \sim 0.5, T \sim T_c$) occurs in a broader temperature range and is very near to the absorption peak temperature T_0.

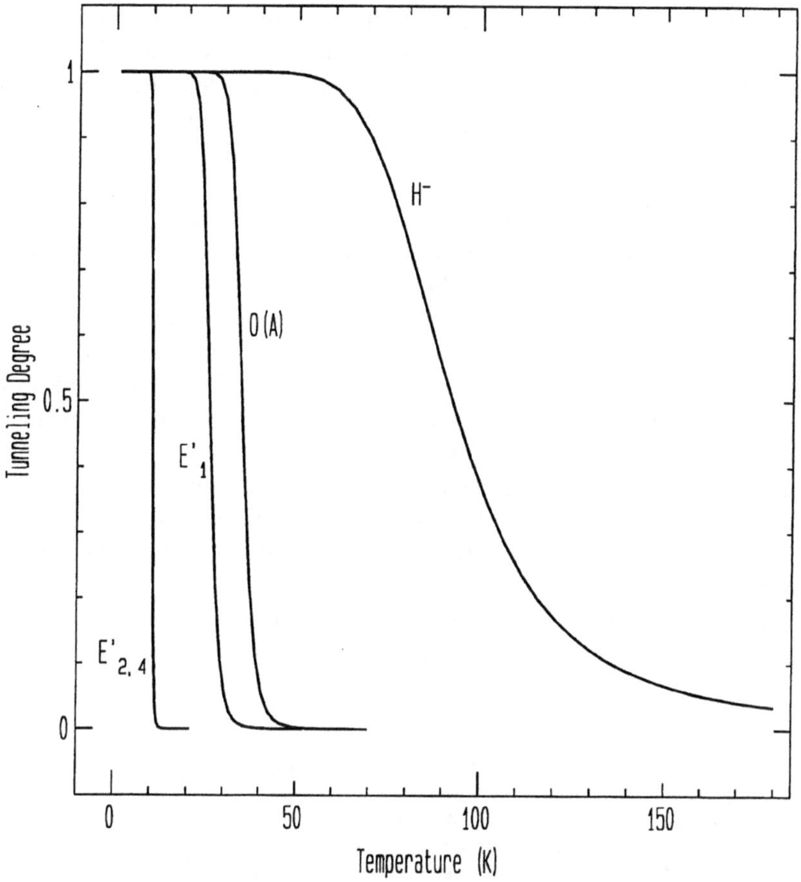

Figure 4. Temperature dependencies of the tunneling degree λ of E'_1, $E'_{2,4}$ centers, H^- ion at E'_4 center in α-quartz, and O(A) in HTSC.

4. Conclusion

This paper has proposed an improvement of the relaxation rate method for calculation of the relaxation rates. A 'tunneling degree' parameter has been introduced in order to connect the specific relaxation behavior of these systems with both the conventional tunneling model and the classical Debye relaxation. This parameter defines the system (at a given temperature) as a tunneling or a relaxation one.

The proposed approach for calculating the eigenvalue spectrum and corresponding wave functions of double-well potential systems allows one to obtain the relaxation times and the relaxation response of these systems under the action of an external field.

Interwell transfer rates and relaxation times can be calculated using the standard reaction rate method. In the presented calculations the reaction rate method has been modified in order to calculate the temperature dependence of the relaxation times of the lowest couple of tunnel split energy levels. This relaxation time is obtained in accordance with the standard tunneling model. The remaining energy level contribution is determined by the Wentzel–Kramers–Brillouin quasi-classical approximation. The relaxation times obtained are close to those calculated by the standard reaction rate method.

A comparative analysis of the relaxation behavior in α-quartz (E' defect centers, H^- ion at E'_4 centers) and high-temperature superconducting cuprates (apical O(A) atom in YBCO) has been reported. From the analysis of the temperature dependencies of the tunneling degree parameter one can clearly see the difference between the E' series centers and O(A) and H^- ion. The E' center is a typical classical relaxation system at temperatures near the position of the relaxation absorption peak. The tunneling system absorption part is negligibly small, because the system is a pure tunneling one at very low temperature. O(A) atom is a typical tunneling system because of the lower barrier and the smaller asymmetry energy. H^- ion at E'_4 acts as a mixed tunneling–relaxation system because the transition between the two limiting cases occurs in a broader temperature range and is very near to the relaxation absorption peak temperature.

The present calculations are useful for careful theoretical description of the relaxation response in these materials under the action of an external field.

References

1. W. Phillips: *Amorphous Solids – Low Temperature Properties. Topics in Current Physics* **24**, Springer, Berlin-Heidelberg-New York (1981).
2. S. Hunklinger: *Acta Phys. Slov.* **40**, 14 (1990).
3. D. Parshin: *Fizika twerdogo tela (Russ.)* **7**, 1809 (1994).
4. L. Mihailov, A. Manov, and A. Vavrek: *J. Alloys and Compounds* **211/212**, 292 (1994).
5. A. Manov: *J. Alloys and Compounds* **211/212**, 349 (1994).
6. A. Manov and V. Georgiev: *Acoustic Letters* **14**, 160 (1991).
7. P. Esquinazi, R. König, D. Valentin, and F. Pobell: *J. Alloys and Compounds* **211/212**, 27 (1994): R. König, P. Esquinazi, and F. Pobell: *Physica B* **194-195**, 417 (1994).
8. M. Georgiev and A. Manov: *Czech. J. Phys. B* **38**, 83 (1988).
9. J. Isoya, J. A. Weil, and L. E. Halliburton: *J. Chem. Phys.* **74**, 5436 (1981). J. K. Rudra, W. B. Fowler, and F. J. Feigl: *Phys. Rev. Lett.* **55**, 2614 (1985). L. E. Halliburton: *Cryst. Latt. Defect Amorph. Mater.* **12**, 163 (1985).
10. A. Manov: Ph.D. Dissertation, Institute of Solid State Physics (1989).
11. S. G. Christov: *Collision Theory and Statistical Theory of Chemical Reactions, Lecture Notes in Chemistry, Vol. 18*, Springer, Berlin-Heidelberg-New York (1980).
12. L. Mihailov and A. Manov: *Z. Phys. B* **98**, 27 (1995).

Molecular Engineering **7**: 161–167, 1997.

Correlation Effects in the Double Proton Transfer of the Formic Acid Dimer

HENRYK CHOJNACKI
Wyb. Wyspiańskiego 27, 50-370 Wrocław, Institute of Physical and Theoretical Chemistry, Wrocław Technical University, Poland.

(Received: 12 September 1996; accepted 17 June 1997)

Abstract. The role of correlation effects in the potential energy hypersurface for the double proton transfer in the hydrogen bond of the formic acid dimer has been studied at the non-empirical level. The calculations were performed in different large as well as for the first time in complete basis set at the correlation level. The possible reasons of incorrect results of quantum chemical calculations are considered.

Key words: Proton transfer, correlation effects, formic acid dimer.

1. Introduction

The formic acid dimer, well characterized in the gas phase, has attracted great interest from the theoretical point of view as a model hydrogen bonded system. It is an example where double proton transfer may occur. This kind of reaction seems to play an important role in the quantum chemical interpretation of some biological processes like mutations, aging and cancerogenic action [1]. From this point of view the formic acid dimer was the object of quantum chemical studies for testing some theoretical approaches as well as for interpretation of double proton transfer mechanism [2–6]. Unfortunately, because of incomplete basis sets and correlation effects, diversified results may be obtained in quantum chemical calculations. In the case of the barrier height its value ranges in different calculations from 5 to 60 kcal/mole [7].

On the other hand, in some crystalline carboxylic acids NMR measurements have demonstrated the coexistence of two interconverting molecular conformations, taking place by double proton transfer processes (Figure 1). In this process the reactant C_{2h} symmetry is transformed to its C_{2h} product via a transition state D_{2h}. For some carboxylic acids, the solid state NMR studies showed that the activation energy for this reaction is of the order of 1.2 kcal/mole [8, 9]. It means that it is much lower than the respective values obtained both from semi-empirical and *ab initio* quantum chemical calculations.

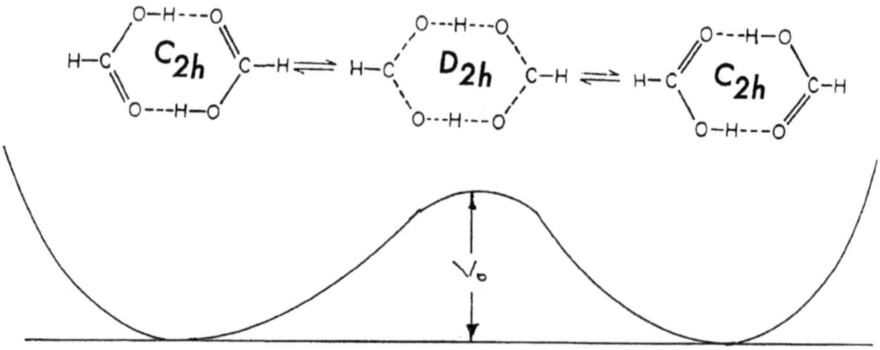

Figure 1. Initial equilibrium, transition and final states of the formic acid dimer. The potential barrier height is denoted by V_o.

2. Method

The electronic structure of the formic acid dimer has been studied at the non-empirical level with full geometry optimization both for equilibrium (C_{2h} symmetry) and transition state (D_{2h} symmetry). The potential hypersurfaces for the proton transfer process were studied by using the GAUSSIAN-94 package [10] for different basis sets at the correlation MP2 level including complete basis set CBS5* (6D, 10F) [11]. It consists of 666 primitives and 472 contracted Gaussians for the formic acid dimer. The configuration interaction calculations were performed for both singly and doubly excited configurations within optimized basis set [12]. The interelectronic correlation energy for the formic acid dimer has been evaluated by the Sinanoglu and Pamuk effective pair correlation energy method [13].

3. Results and Conclusions

Both basis set and electron correlation effects were found to be important in the description of the hydrogen bond of the formic acid dimer. In the CBS5* basis set [11] the barrier height of 17.100 kcal/mole for the proton transfer at the Hartree–Fock level was found. The role of the correlation effects can be clearly seen when improving the basis set (Figure 2). The extrapolated barrier lowers asymptotically to 8.0 kcal/mole i.e. to the same value as that for the complete basis set CBS5* (the last point to the right).

The Hartree–Fock energy for the equilibrium geometry (R(O· · ·O) = 2.795 Å) is −377.7418610 a.u. whereas for the transition state geometry (R(O· · ·O) = 2.378 Å it is −377.7146002 a.u. So far as we know these energies are the lowest in the literature. The correlation energy at the MP2 level amounts to −1.3189828 a.u. whereas that of Sinanoglu and Pamuk [13] method is −1.797825 a.u. In this situation within the complete basis set we take into account ca. 73% of the total correlation energy. At the correlation level the respective hydrogen bond lengths are

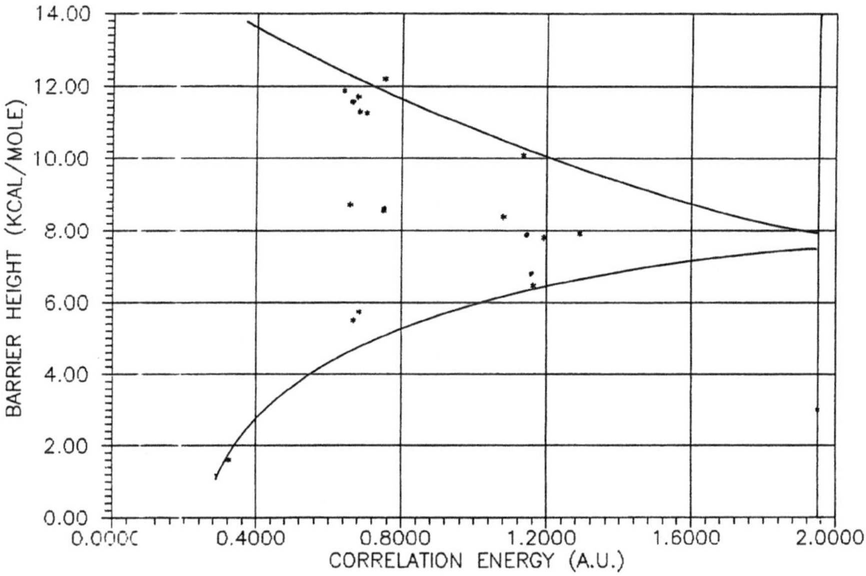

Figure 2. Barrier height dependence on the correlation energy in different basis sets. The vertical line denotes the correlation energy estimated by Sinanoglu and Pamuk method [13].

2.703 Å and 2.415 Å, respectively. As can be seen, there is an important shortening of the O···H—O bond in the transition state. The experimental hydrogen bond length for the equilibrium geometry resulting from neutron diffraction studies [15] is 2.703 Å. In general, the results seem to be reliable when at least 60% of the correlation energy is taken into account in the calculations.

The freezing of some core orbitals may lead to different, unreasonable results both in the case of Moeller–Plesset level (Figure 3) as well as in the configuration interaction calculations (Figure 4) (for different basis sets). Thus, many accidental results could be generated when no full configuration space is considered in the calculations.

It should be noted that the DFT method with the BLYP functional [9] within the same complete basis set leads to the barrier height of 4.378 kcal/mole (without the zero-point energy correction). It seems that in this case the barrier height may be underestimated

4. Possible Proton Transfer Mechanism

In the case of moderate potential barrier height, both classical and quantum mechanical tunneling processes may be important in the proton transfer mechanism [14]. Thus, the total transfer probability P may be expressed as the sum of the two terms

$$P = P_{clas} + P_{tun}$$

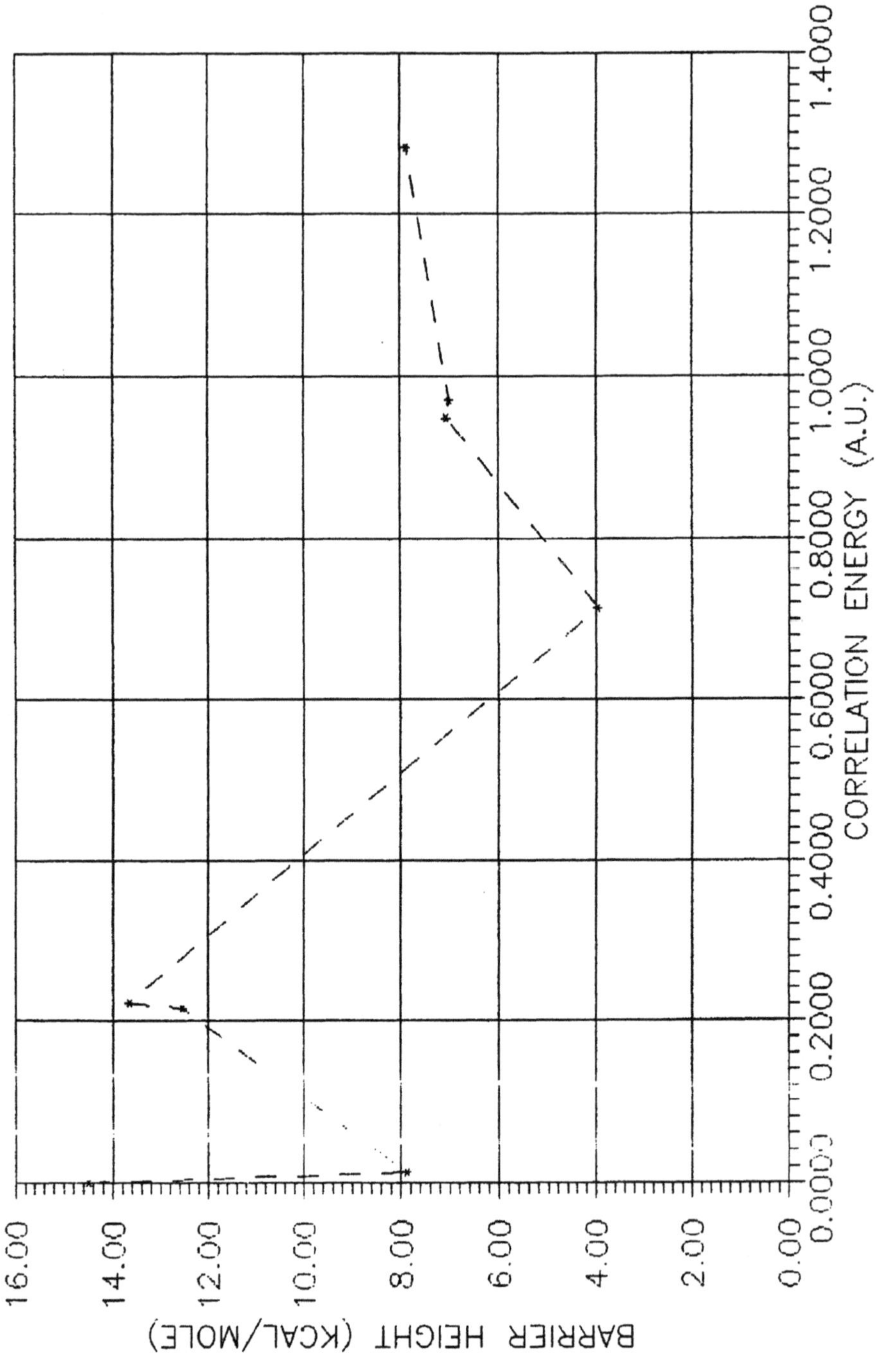

Figure 3. Dependence of the barrier height on the correlation energy evaluated by Moeller-Plesset method with frozen core orbitals within the $6s3p1d$ contracted basis set [12] for oxygen and carbon and $3s2p2d$ for hydrogen, respectively.

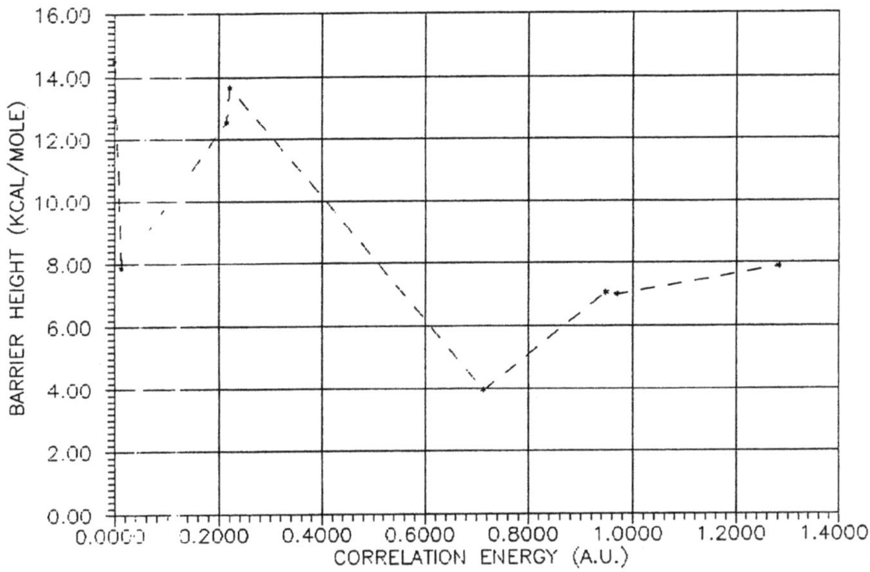

Figure 4. Dependence of the barrier height on the correlation energy evaluated by configuration interaction method (all single and double substitutions substitutions) with frozen core orbitals within the $6s3p1d$ contracted basis set [12] for oxygen and carbon and $3s2p2$ for hydrogen, respectively.

where the P_{clas} is the Boltzmann and P_{tun} denotes the quantum mechanical tunneling probability. In the case of rectangular potential barrier, the total probability may be expressed as

$$P = \exp(-V_o/kT) + (1 - P_{clas})\exp(-2/\hbar\sqrt{2m(V_o - E)}l)$$

with activation energy V_o and barrier width l. With the above assumptions the temperature dependence of the total transfer probability is different for low barriers and high ones. In the first case the classical mechanism dominates and a linear Arrhenius plot should be found, whereas in the second case (Figure 5) the deviation in this kind of plot may be observed.

It seems that the temperature dependence of the transfer probability enables the interpretation of the temperature dependence of the proton spin-lattice relaxation time T_1. At low temperatures the tunneling mechanism predominates whereas at temperatures higher then 125 K rather classical transfer is the most important process [9].

Taking into account the zero-point energy correction of 5.3 kcal/mole, the potential barrier for the proton transfer of the gas phase formic acid dimer is of the order of 3 kcal/mole. In this situation the linear Arrhenius plot for the temperature dependence of the proton transfer should be observed.

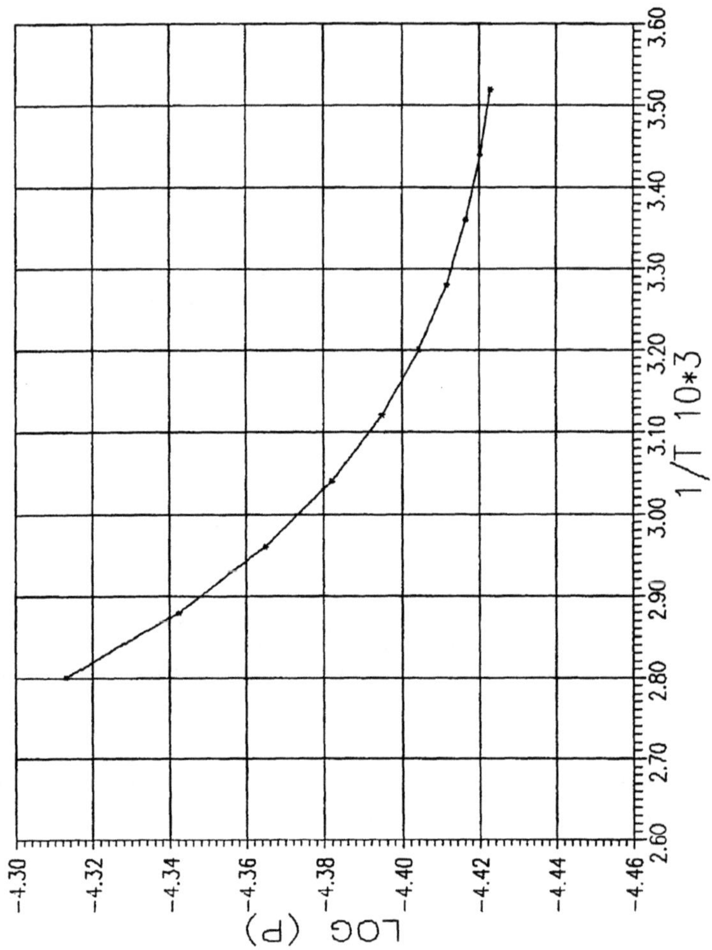

Figure 5. Temperature dependence of the proton transfer probability for the barrier height $V_o = 8.0$ kcal/mole. Classical transfer and quantum mechanical tunneling are taken into account.

Acknowledgements

This work has been partly sponsored by the D3 COST program and Wrocław Technical University.

References

1. P. O. Loewdin: *Adv. Quantum Chem.* **2**, 216 (1967).
2. S. Hayashi, J. Umemura, S. Kato, and K. Morokuma: *J. Phys. Chem.* **88**, 1330 (1984).
3. C. Mijoule, M. Allavena, J. M. Leclercq, and Y. Boutellier: *Chem. Phys.* **109**, 207 (1986).
4. H. Chojnacki, J. Andzelm, D. T. Nguyen, and W. A. Sokalski: *Comp. Chem.* **19**, 181 (1995).
5. L. Turi: *J. Phys. Chem.* **100**, 11285 (1996).

6. P. H. Smit, J. L. Derissen, and F.B. van Duijneveldt: *J. Chem. Phys.* **69**, 4241 (1978).
7. H. Chojnacki, J. Lipiński, and W. A. Sokalski: *Int. J. Quantum Chem.* **19**, 339 (1981).
8. B. H. Meier, F. Graf, and R. R. Ernst: *J. Chem. Phys.* **76**, 767 (1982).
9. S. Nagaoka, T. Terao, F. Imashiro, A. Saika, N. Hirota, and S. Hayashi: *J. Chem. Phys.* **79**, 4694 (1983).
10. M. J. Frisch, G. W. Trucks, H. B. Schiegel, P. M. W. Gill, B. G. Johnson, M. A. Robb, J. R. Cheeseman, T. Keith, G. A. Petersson, J. A. Montgomery, K. Raghavachari, M. A. Al-Laham, V. G. Zakrzewski, J. V. Ortiz, J. B. Foresman, C. Y. Peng, P. Y. Ayala, W. Chen, M. W. Wong, J. L. Andres, E. S. Replogle, R. Gomperts, R. L. Martin, D. J. Fox, J. S. Binkley, D. J. Defrees, J. Baker, J. P. Stewart, M. Head-Gordon, C. Gonzalez, and J. A. Pople: GAUSSIAN 94, Revision B.3. Gaussian, Inc., Pittsburgh PA (1995).
11. G. A. Petersson, A. Bennett, T. G. Tensfeldt, M. A. Al-Laham, W.A. Shirley, and J. Manzaris: *J. Chem. Phys.* **89**, 2193 (1988).
12. H. Chojnacki: unpublished results.
13. O. Sinanoglu and H. Pamuk: *Theor. Chim. Acta* **27**, 289 (1972).
14. H. Chojnacki: in *Quantum Science – Methods and Structure*, J. L. Calais, O. Goscinski, J. Linderberg, and Y. Ohrn (Eds.), Plenum Press, New York (1976).
15. A. Almenningen, O. Bastiansen, and T. Motzfeldt: *Acta Chem. Scand.* **23**, 2848 (1969).

Molecular Engineering **7**: 169–183, 1997.

Effect of the Vibrational/Rotational Energy Partitioning on the Energy Transfer in Atom–Triatomic Molecule Collisions

I. ROSENBLUM, E. I. DASHEVSKAYA, E. E. NIKITIN* and I. OREF
Department of Chemistry, Technion – Israel Institute of Technology, Haifa 32000, Israel.

(Received: 12 September 1996; accepted 17 June 1997)

Abstract. The effect of the initial partitioning of the molecular energy between vibrational and rotational modes of a triatomic molecule on the collisional energy transfer is studied for a model atom–triatomic molecule system. We considered the collisions of thermal bath Ar atoms with SO_2 molecules, and used the trajectory calculations for determining the energy transfer for three different samplings of initial conditions of the molecule. The first sampling method generated the microcanonical distribution over all states, entering into the vibrational and rotational manifolds, while two others produced distributions with relatively lower values of the rotational energies. It is shown that both the average energy transfer per collision and the mechanism of the energy exchange are significantly affected by the vibrational/rotational energy partitioning before the collisions. Relative decrease in the rotational energy results in the decrease of the averaged energy transfer and progressively emphasizes the role of active rotation as the gateway for translation-vibration energy exchange.

Key words: Unimolecular reactions, collisional energy transfer, sampling methods

1. Introduction

The distribution of the internal energy of the molecule over different degrees of freedom is known to have a considerable influence on the energy transfer in collisions [1–4]. In some cases a particular distribution of energy within the isolated vibrationally excited molecule can significantly enhance the efficiency of the energy transfer. The study of supercollisions [1, 2] reveals that, for a supercollision to occur, a large local fluctuation of energy must be present in the 'collision site'. A detailed investigation of the energy transfer and intramolecular relaxation mechanisms [3] showed that the efficiency of such excitation dramatically depends on the type of vibrational mode on which the fluctuation is located: while the fluctuation on one of the high-frequency modes has virtually no effect on the energy transfer, the lowest-frequency mode presents a sort of a gateway for this transfer: therefore, the excitation on this mode strongly increases the transferred energy compared to its statistical mean. Another example of such influence is the so-called active rotation of the molecule. It was found recently [5] that under some conditions this mode, weakly coupled to the vibrational reservoir of a molecule, can also

* Author for correspondence.

represent a kind of a doorway step in transferring the kinetic energy of colliding particles into the vibrational energy of the molecule. This phenomenon indicates that the energy transfer mechanism is affected by the intramolecular distribution of energy: whether the 'true vibrational' or the 'active rotational' route predominates is determined by the modes on which the excitation is concentrated and on the extent of its deviation from the mean value.

In view of these effects, the way the initial conditions are sampled for trajectory calculations of energy transfer becomes very important. A review of different methods of sampling is presented in Ref. [6]. These methods can be subdivided into two categories. In most methods it is assumed that the phase space of the internal motion of a molecule (vibrations and overall rotations) before a collision is populated uniformly under additional constraints imposed by the conservation of the total energy E and total angular momentum J [7] (equiprobable microcanonical distribution). In the second type of sampling techniques some definite quantities (such as the total angular momentum [8] or the energies of normal modes in the absence of rotations [9]) are fixed and the remainder of the energy is equidistributed over the remaining degrees of freedom. These latter methods, however, are not general and, in most cases, not precise: they allow one to approximately preset one certain energy [8] or to neglect some degrees of freedom [9]. In the present paper we suggest a new, more general, method allowing one to specify the energies of any desired degrees of freedom using the analytical probabilities predicted by a rigid rotor–harmonic oscillator (RRHO) model.

Depending on the restrictions and implications of various sampling methods, they can produce different distributions of energy over the molecule. Therefore, the mechanism and the efficiency of the collisional energy transfer are supposed to be noticeably sensitive to the sampling method used for the generation of initial conditions. We check this hypothesis by way of example, using collisions of an Ar atom with the nonlinear triatomic SO_2 molecule, prepared by different sampling methods. We demonstrate that the result of collision is considerably affected by the energy distribution, resulting from the way in which the initial conditions were sampled.

2. The Model

The SO_2 molecule possessing total molecular energy E (E = 10 or 100 kcal/mol) was prepared microcanonically using different sampling methods. A detailed description of these methods is presented in Section 3. As a result, three sets of initial conditions were obtained. The first one (I), generated with the use of the 'successive sampling' method, corresponds to the proper microcanonical distribution of the total energy over the molecular degrees of freedom. The second system (II), sampled with the Schranz–Nordholm–Freasier–Nyman (SNFN) method [7] without consideration of rotations, is supposed to produce the distorted distribution in which the weight of vibrational energy is overestimated. The third set

(III) was also obtained using the SNFN method but with an artificially small value of the weighting parameter; this set was intended to produce a distribution that dramatically underestimates the weight of adiabatic rotational energy E_j.

The model used for trajectory calculations is described in Ref. [5]. Briefly, the potential of the free SO_2 molecule was constructed as a superposition of atom–atom pair harmonic potentials. The interaction between SO_2 and Ar was taken as a sum of pairwise Lennard–Jones potentials between Ar and the atoms of the molecule (the so-called SLJ model, [12]). The relative velocity of the collider was sampled from the three-dimensional Maxwell flux distribution [13] at the bath gas temperature T = 300 K. The trajectories started from a mass centers separation of 5 Å and were calculated during approximately 25 ps.

The quantities recorded were the average values of the different components of energy transferred in collision. The rotational part of the total energy, E_r, which is essentially inseparable from the vibrational part, E_ν, was reasonably defined as the average time-dependent instantaneous rotational energy $\langle E_r(t)\rangle$ [14, 15]. For a symmetric top molecule, which is an acceptable approximation for SO_2 [5, 12], E_r can be subdivided into the 'adiabatic' part, E_j, and the 'active' part, E_k:

$$E_r = B_{\text{eff}} j^2 + (A_{\text{eff}} - B_{\text{eff}})k^2 \tag{1}$$

where A_{eff} and B_{eff} are the effective rotational constants, and k is the mean projection of j onto the line connecting the two oxygen atoms. The use of the mean projection k in Equation (1), rather than the instantaneous projection j_z of j onto this line, is related to the fact that for a slightly asymmetric top the projection j_z oscillates with a small amplitude about its mean value. These oscillations are related to the exchange of the energy between rotations about three principal axes of inertia. One can ignore this phenomenon provided that deviations of j_z from k are small. On a much longer time-scale, corresponding to thousands of vibrational periods, the dynamical quantity k suffers a change. This change is related to the intramolecular vibrational–rotational energy redistribution (IVRR) which is due to the coupling between vibrational and rotational motion. Another quantity used in calculations is the so-called effective vibrational energy, $E_{\nu,\text{eff}} = E - E_j$. This magnitude consists of the 'true' vibrational energy, E_ν, and the 'active' part E_k of rotational energy, coupled with vibrations. Therefore, the change of this value in the collisions may be due to two different mechanisms: the genuine 'vibrational' transfer or the transfer of active rotational energy which afterwards, in the inter-collision time, relaxes into the vibrational reservoir and thus represents the 'gateway' for the vibrational energy transfer. We will show that both of these mechanisms occur, and the initial distribution of internal energy determines which of them prevails.

3. Microcanonical Sampling Methods

3.1. SUCCESSIVE SAMPLING METHOD

In this section we present the 'successive sampling' approach which is suitable for the initial specification of the energies of both active and adiabatic rotations, E_k and E_j. This method is based on the known analytical probabilities of the energy distribution over different modes and therefore is exact for separable systems, when the notion of a mode is well defined. With nonzero coupling between modes, this method is approximate. However, if the coupling is weak – in the sense that it does not drastically affect the pattern of the spectrum of a separable system – the successive sampling method can still be used for trajectory calculations. We present a description of the successive sampling method below.

i. We begin with the distribution function $P(E_\nu;\, E)$ over the vibrational energy E_ν for a rotating system of oscillators (three vibrations and three rotations) for a given total energy E, calculated considering the limitation connected with the conservation of the total angular momentum. This distribution function is calculated by integrating the product of densities of energy levels for oscillators, active rotation and adiabatic rotation:

$$\begin{aligned} &P_E(E_\nu)\,\mathrm{d}E_\nu \\ &\quad = \frac{E_\nu^2\,\mathrm{d}E_\nu}{N}\int_{E_k=0}^{(E-E_\nu)\cdot((r-1/r))} \times \\ &\quad \times \int_{E_j=0}^{E-E_\nu-E_k} \delta(E-E_\nu-E_k-E_j)\,\frac{\mathrm{d}E_k}{\sqrt{E_k}}\,\mathrm{d}E_j \\ &\quad = \frac{105}{16}\cdot\frac{E_\nu^2}{E^2}\sqrt{1-\frac{E_\nu}{E}}\,\frac{\mathrm{d}E_\nu}{E} \end{aligned} \tag{2}$$

where r is the ratio $A_{\mathrm{eff}}/B_{\mathrm{eff}}$.

ii. In sampling the vibrational energies E_{ν_1}, E_{ν_2} and E_{ν_3} of three normal modes we follow Hase and Buckowski [9]. Since rotations were not considered in their work, we used the total vibrational energy E_ν instead of the full molecular energy E. In short, two first energies were sampled successively from the corresponding distributions

$$P_{E_\nu}(E_{\nu_1})\mathrm{d}E_{\nu_1} = 2(E_\nu - E_{\nu_1})\mathrm{d}E_{\nu_1}/E_\nu^2 \tag{3}$$

$$P_{E_\nu,E_{\nu_1}}(E_{\nu_2})\mathrm{d}E_{\nu_2} = \mathrm{d}E_{\nu_2}/(E_\nu - E_{\nu_1}). \tag{4}$$

When E_{ν_1} and E_{ν_2} are fixed, E_{ν_3} is determined as $E_{\nu_3} = E_\nu - E_{\nu_1} - E_{\nu_2}$.

iii. Now, the normal coordinates q_i should be sampled for individual oscillators, possessing force constants k_i, so as to satisfy their theoretical distributions for harmonic oscillators:

$$P_{E_{\nu_i}}(q_i)\mathrm{d}q_i = \frac{2\mathrm{d}q_i}{\pi\sqrt{(2E_{\nu_i}/k_i) - q_i^2}}. \tag{5}$$

The simplest way to do this, as well as to obtain the conjugate velocities $\dot{q}_i$ simultaneously, is [9] to choose them as

$$q_i = \sqrt{\frac{2E_{\nu_i}}{k_i}}\cos(2\pi R), \qquad \dot{q}_i = \sqrt{\frac{2E_{\nu_i}}{\mu_i}}\sin(2\pi R) \tag{6}$$

where μ_i is the effective mass of the normal mode and R is a random number uniformly distributed in the interval (0,1).

iv. The next step is the transformation of normal modes coordinates $\{\mathbf{q}\}$ and of conjugated velocities $\{\dot{\mathbf{q}}\}$ to Cartesian Jacobi (i.e. mass-center-separated – 'mcs') coordinates $\{\mathbf{x}^{\mathrm{mcs}}\}$ and to the corresponding vibrational velocities $\{\mathbf{v}^{\mathrm{vib\ mcs}}\}$. The procedure for the determination of the transformation matrix $\mathbf{C}$ is similar to that described elsewhere (see, for example, Ref. [16]); still, it contains some distinctions connected with the introduction of instantaneous, i.e. non-equilibrium, Eckart conditions. For this reason we would like to dwell on it.

Let $c_{i\beta\lambda}$ be the coefficients of the matrix $\mathbf{C}$ transforming the normal modes coordinates and momenta into $3n - 3$ Cartesian 'mcs' coordinates and velocities:

$$x_{i\beta}^{\mathrm{mcs}} = x_{i\beta}^{0\ \mathrm{mcs}} + \sum_{\lambda=1}^{3n-6} c_{i\beta\lambda}\cdot q_\lambda, \tag{7}$$

$$\nu_{i\beta}^{\mathrm{mcs}} = \sum_{\lambda=1}^{3n-6} c_{i\beta\lambda}\cdot \dot{q}_\lambda \tag{8}$$

where $x_{i\beta}^{0\ \mathrm{mcs}}$ are the equilibrium 'mcs' coordinates, q_λ and $\dot{q}_\lambda$ are the chosen coordinates of the normal modes and their derivatives, the subscript $i = 1, n-1$ corresponds to the number of Jacobian 'mcs' particles (see Ref. [16]), and the subscript $\beta = x, y, z$ corresponds to the number of Cartesian axes. The summation is accomplished over all $3n - 6$ normal modes λ.

These $3(n-1)(3n-6)$ coefficients $c_{i\beta\lambda}$ should be found from three types of conditions. First, since q_λ are the normal modes, the inverse transformation from $\{\mathbf{x}^{\mathrm{mcs}}\}$ to $\{\mathbf{q}\}$ must put the potential energy in the form $U = \frac{1}{2}\mu_0\sum_{\lambda=1}^{3n-6}\omega_\lambda^2 q_\lambda^2$. This requirement gives $(3n-6)(3n-7)/2$ equations:

$$\sum_{i\alpha k\beta}\left(\frac{\partial^2 U}{\partial x_{i\alpha}^{\mathrm{mcs}}\partial x_{k\beta}^{\mathrm{mcs}}}\right)_{q=0} c_{i\alpha\lambda}c_{k\beta\mu} = 0 \qquad \lambda,\mu = 1, \quad 3n-6, \quad \lambda \neq \mu. \tag{9}$$

Second, the condition that the vibrational kinetic energy, T_ν, must assume the form

$$T_\nu = \frac{1}{2}\mu_0 \sum_{\lambda=1}^{3n-6} (\dot{q}_\lambda)^2$$

leads to the requirement of orthogonalization of these coefficients:

$$\sum_{i\beta} \mu_i c_{i\beta\lambda} c_{i\beta\mu} = \mu_0 \delta_{\lambda\mu}, \qquad \lambda,\ \mu = 1, \quad 3n-6 \tag{10}$$

where μ_0 is the normalization constant of the dimension of mass, and the summation is accomplished over all $i = 1, n-1$ and $\beta = x, y, z$. The number of equations is equal to the number of all combinations of normal modes, i.e. to $(3n-6)(3n-5)/2$.

The third type of condition arises from the requirement to separate the coupled rotations and vibrations as much as possible, i.e. to minimize the vibrational angular momentum

$$\mathbf{L} = \sum_i \mathbf{r}_i^{\text{mcs}} \times \boldsymbol{\mu}_i \boldsymbol{\nu}_i^{\text{mcs}}.$$

Since the vibrational angular momentum depends on normal modes, it is impossible to make it vanish for all possible configurations; however, it is possible to do so for a certain configuration. Usually it is done for the equilibrium configuration [16]; the corresponding conditions we call the equilibrium Eckart conditions. However, in our case it is more convenient to make the vibrational angular momentum equal to zero for the current state for which the normal modes $\{\mathbf{q}\}$ were sampled. This leads to the so-called instantaneous Eckart conditions providing the general 'collinearity' of the velocity and position vectors:

$$\sum_i^{n-1} \mu_i \left[x_{i\beta}^{0\ \text{mcs}} c_{i\gamma\lambda} - x_{i\gamma}^{0\ \text{mcs}} c_{i\beta\lambda} + \sum_{\mu=1}^{3n-6} (c_{i\beta\mu} \cdot c_{i\gamma\lambda} - c_{i\gamma\mu} \cdot c_{i\beta\lambda}) \cdot q_\mu \right] = 0,$$

$$\beta,\ \gamma = x, y, z; \quad \lambda = 1, 3n-6. \tag{11}$$

These $3(3n-6)$ conditions complete the set of $3(n-1)(3n-6)$ equations needed for the determination of the coefficients $c_{i\beta\lambda}$.

Thus, the Equations (9)–(11) produce all the necessary coefficients $c_{i\beta\lambda}$, and therefore provide the transformation of normal modes and velocities to the Jacobi coordinates and velocities by the use of Equations (7) and (8). The further transformation to the Cartesian coordinates $\{\mathbf{x}_n\}$ and vibrational velocities $\{\mathbf{v}_n^{\text{vib}}\}$ of individual atoms is trivial, if the condition of the mass center being located in the origin is imposed. The arbitrariness which is still left is eliminated by three consecutive rotations of the whole molecule through three random Euler angles.

v. Afterwards, two rotational energies E_{jB} and E_{jC} about the corresponding molecular axes of inertia are sampled successively according to the probability densities

$$P_{E,E_\nu}(E_{jB})\mathrm{d}E_{jB} = \frac{\mathrm{d}E_{jB}}{2\sqrt{E_{jB}(E-E_\nu)}} \tag{12}$$

$$P_{E,E_\nu,E_{jB}}(E_{jC})\mathrm{d}E_{jC} = \frac{\mathrm{d}E_{jC}}{\sqrt{E_{jC}}\sqrt{E-E_\nu-E_{jC}-E_{jB}}}. \tag{13}$$

When E_{jB} and E_{jC} are fixed, the third component of rotational energy, E_{jA}, is determined from the equation $E_{jA} = E - E_\nu - E_{jB} - E_{jB}$. The adiabatic and the active parts of rotational energy are equal, correspondingly, to

$$E_j = B_{\text{eff}}J^2 = B_{\text{eff}} \cdot (J_A^2 + J_B^2 + J_C^2) = E_{jB} + E_{jC} + E_{jA}/r \tag{14}$$

$$\begin{aligned} E_k = E_R - E_j &= (E_{jA} + E_{jB} + E_{jC}) - (E_{jB} + E_{jC} + E_{jA}/r) \\ &= (r-1)/r \cdot E_{jA}. \end{aligned} \tag{15}$$

vi. The above set of rotational energies determine the components of the angular momentum about the principal axes of inertia:

$$J_m = \sqrt{2E_m I_m} \tag{16}$$

where $m = A, B, C$, and I_m are the instantaneous values of the principal momenta of inertia.

vii. With the above values of angular momentum components, the projections of the angular velocities onto the principal axes of inertia are calculated:

$$\omega_m = J_m/I_m. \tag{17}$$

The directions of these axes in the space-fixed frame are found by the diagonalization of the instantaneous tensor of inertia. Thus, the angular velocity vector $\boldsymbol{\omega}$ is completely determined.

The vector of angular velocity obtained in this way is used to calculate the rotational contribution to the linear velocity of an atom n:

$$\mathbf{v}_n^{\text{rot}} = \boldsymbol{\omega} \times \mathbf{r}_n \tag{18}$$

where $\mathbf{r}_n$ is the radius-vector of the nth atom from the origin (such choice renders the zero velocity of the mass center). The resulting set of rotational velocities $\{\mathbf{v}_n^{\text{rot}}\}$ makes the total rotational energy E_r equal to $E - E_\nu$. Besides, the instantaneous Eckart conditions ensure that the so-called vibrational velocities $\{\mathbf{v}_n^{\text{vib}}\}$ are collinear to the radius-vectors $\mathbf{r}$, and therefore orthogonal to the rotational velocities $\{\mathbf{v}_n^{\text{rot}}\}$ determined by equation (18). Consequently, the total velocity of each atom (defined as $\mathbf{v}_n = \mathbf{v}_n^{\text{vib}} + \mathbf{v}_n^{\text{rot}}$) yields the required value of the kinetic energy $T = E - U$, U being the molecular potential energy:

$$T = \sum \frac{m_n \nu_n^2}{2} = \sum \frac{m_n \nu_n^{\text{vib}^2}}{2} + \sum \frac{m_n \nu_n^{\text{rot}^2}}{2} = E_r + T_\nu = E - U. \quad (19)$$

A comparison of distributions obtained by the successive sampling method with theoretical predictions of RRHO model are shown in Figure 1. Since the agreement is very good,[1] we conclude that the molecule obtained can be considered prepared in a proper microcanonical state with respect to its energy E ('system 1').

3.2. MARKOV CHAIN METHOD OF SCHRANZ–NORDHOLM–FREASIER–NYMAN (SNFN)

The essence of the Markov chain approach developed by Schranz *et al.* [6, 7] is as follows. The microcanonical state of the molecule is fully specified by an appropriately selected set of spatial coordinates $\mathbf{q}$ and conjugated momenta $\mathbf{p}$. The required spatial coordinates are periodically selected from a Markov chain of spatial states, based on the acceptance–rejection procedure with a Metropolis-type microcanonical weighting function

$$W_g(\mathbf{q}) \propto [E - U(\mathbf{q})]^g \quad (20)$$

where $U(\mathbf{q})$ is a potential energy corresponding to the spatial coordinates $\mathbf{q}$, and g is a weighting parameter whose value will be discussed below. The momenta $\mathbf{p}$ are randomly selected from a normal distribution and then uniformly scaled so that the resulting $\{\mathbf{p}, \mathbf{q}\}$ state has the correct molecular energy E.

For a rotating nonlinear molecule with s vibrational degrees of freedom the appropriate value of the weighting parameter is [6]:

$$g = \frac{s+1}{2}. \quad (21)$$

With this weighting parameter the Markov chain method gives the proper microcanonical distribution coinciding with that obtained by the successive sampling method described above. However, in the present work we use a Markov procedure for generating two sets with distorted microcanonical distributions.

The first one, corresponding to the weighting parameter

$$g = \frac{s-2}{2}, \quad (22)$$

[1] A rather large deviation of the E_k distribution from its theoretical form arises from the impossibility to take into account the infinity of the distribution when considering a finite number of microcanonical states.

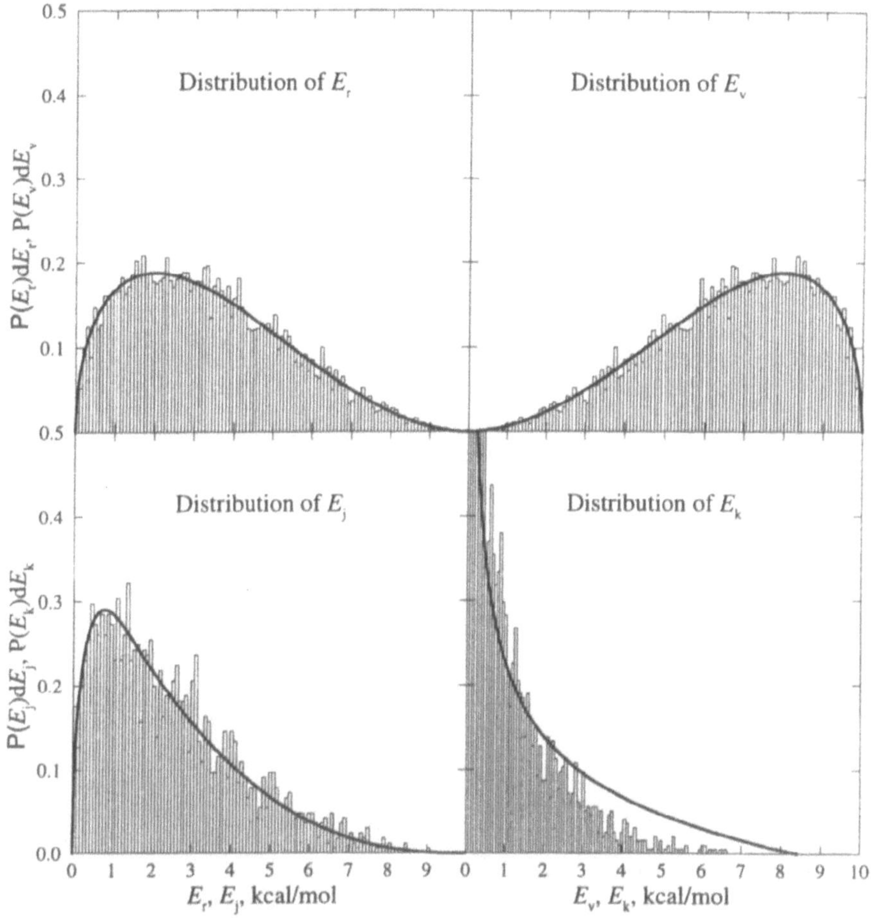

Figure 1. Distribution of different components of energy calculated by the successive sampling method (histograms) and by the theoretical RRHO model (full lines). Total molecular energy $E = 10$ kcal/mol.

is appropriate for a non-rotating system of s oscillators moving in the field of a potential $U(\mathbf{q})$ and possessing the energy E. In this case the rotation of the system is not taken into account explicitly in the form of the weighting function, but can be included implicitly by considering E as the total energy which includes the rotational energy. The corresponding distributions of energy compared to those obtained theoretically in the RRHO approximation are presented in Figure 2. The collisional system in which the SO_2 molecule is prepared with this weighting function is denominated as 'system 2'.

Another value of the weighting function was meant to produce even more distorted distribution, considerably underestimating the weight of adiabatic rotational energy E_j. For this purpose the parameter g was artificially decreased relative to

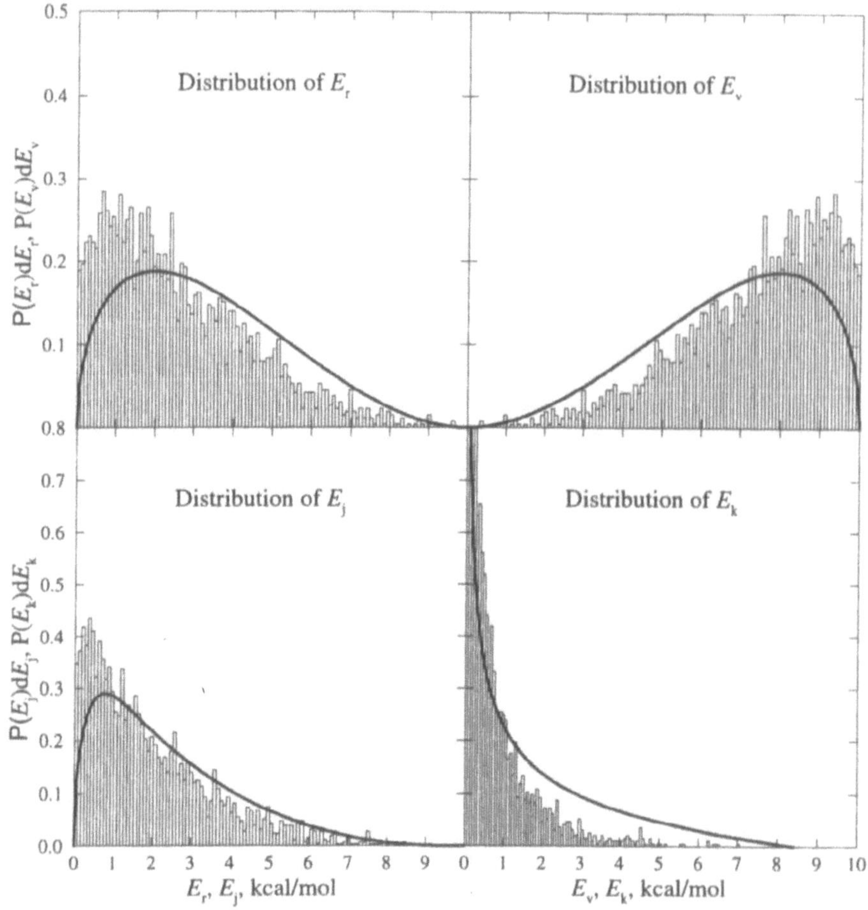

Figure 2. The same as in Figure 1, but for the Markov chain SNFN method with the weighting parameter $g = (s - 2)/2 = 1/2$.

that determined by (22): whereas its proper and distorted values (Equations (21) and (22)) for triatomic nonlinear molecule are 2 and 1/2, correspondingly, here we accepted the lower value $g = 0$, suitable for a linear chain of three atoms. The distributions of different energy components obtained with this value are shown in Figure 3; the according collisional system is denominated as 'system 3'.

4. Discussion and Conclusion: Dependence of the Energy Transfer in Collisions on the Microcanonical Sampling Method

For three systems corresponding to different initial distributions of energy over molecular degrees of freedom we calculated the average quantity of various components of energy transferred in collisions of SO_2 molecule with Ar. The results

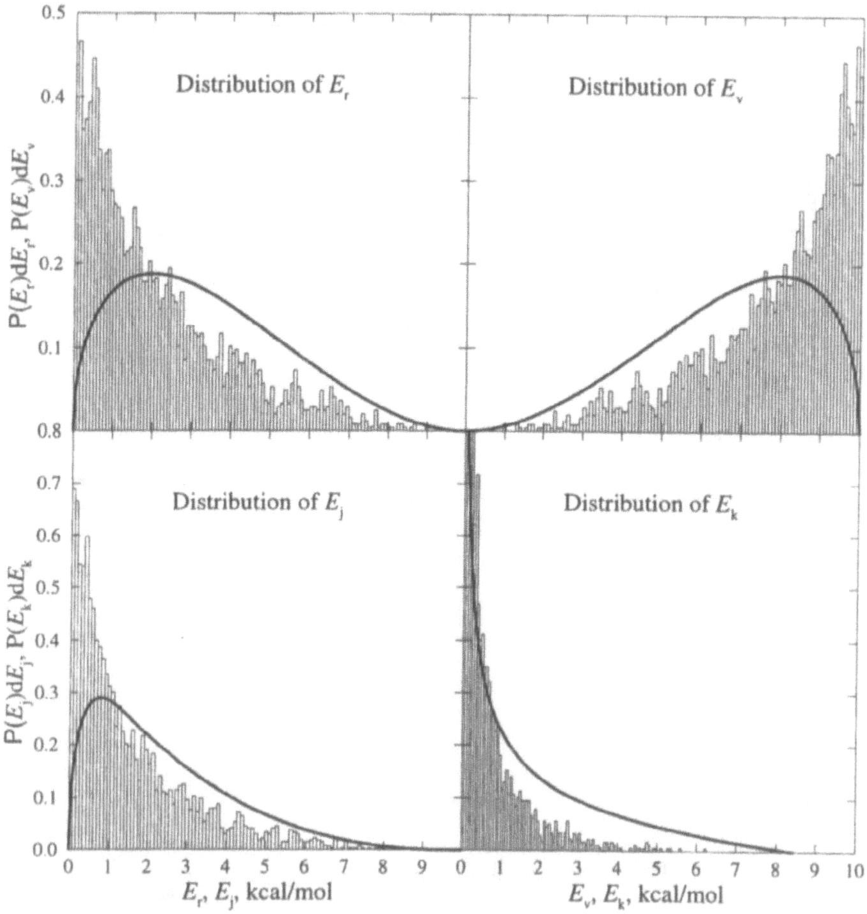

Figure 3. The same as in Figure 2, but with the artificial weighting parameter $g = 0$.

for the molecular energies $E = 10$ and 100 kcal/mol and the bath gas temperature $T = 300$ K are presented in Figure 4.

This figure shows that the increase of the relative weight of small rotational energies E_j, as we pass from system 1 to system 3, decreases the transition efficiency of all *rotational* components of energy. For the adiabatic rotational energy E_j it follows directly from the form of distribution (Figures 1–3): the smaller the initial average value of E_j in the microcanonical ensemble, the less effective is its collisional removal. Since the allowed value of E_k is limited by that of E_j, the decrease of $|\langle \Delta E_j \rangle|$ in systems 2 and 3 also causes the decrease of the E_k loss and, for high molecular energies (Figure 4b), the loss can even turn to the gain of active rotational energy.

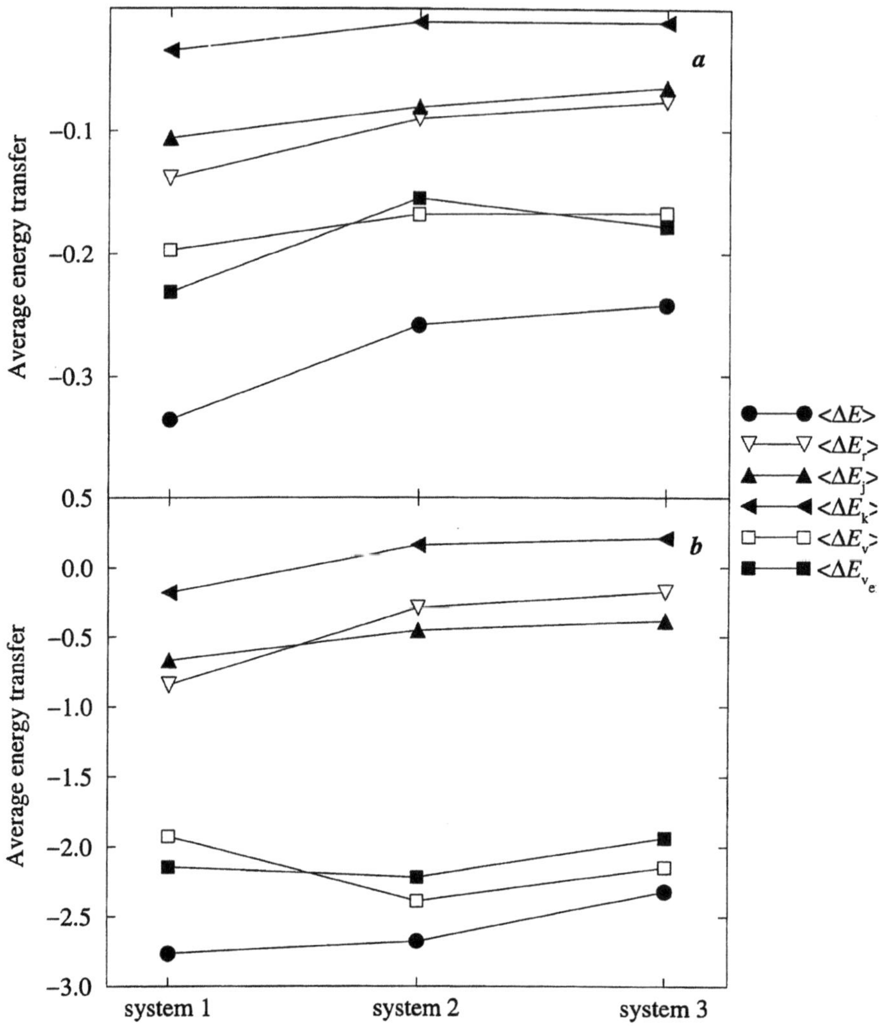

Figure 4. Dependence of the collisional energy transfer on the sampling method. Total molecular energy $E = 10$ kcal/mol (a) or $E = 100$ kcal/mol (b), heat bath temperature $T = 300$ K, harmonic intramolecular potential. Systems 1–3 are defined in the text.

As far as the transfer of *vibrational* energy E_ν is concerned, the dependence of the collision efficiency on the initial distribution becomes more complicated. From one side, the transition from system 1 to system 3 increases the deviation of the initial E_ν magnitude from its equilibrium microcanonical value; the average vibrational energy becomes larger, and its loss in collisions is supposed to grow. However, the simultaneous decrease of the average E_j may cause the opposite effect on the vibrational energy transfer. This phenomenon is the result of the Ehrenfest

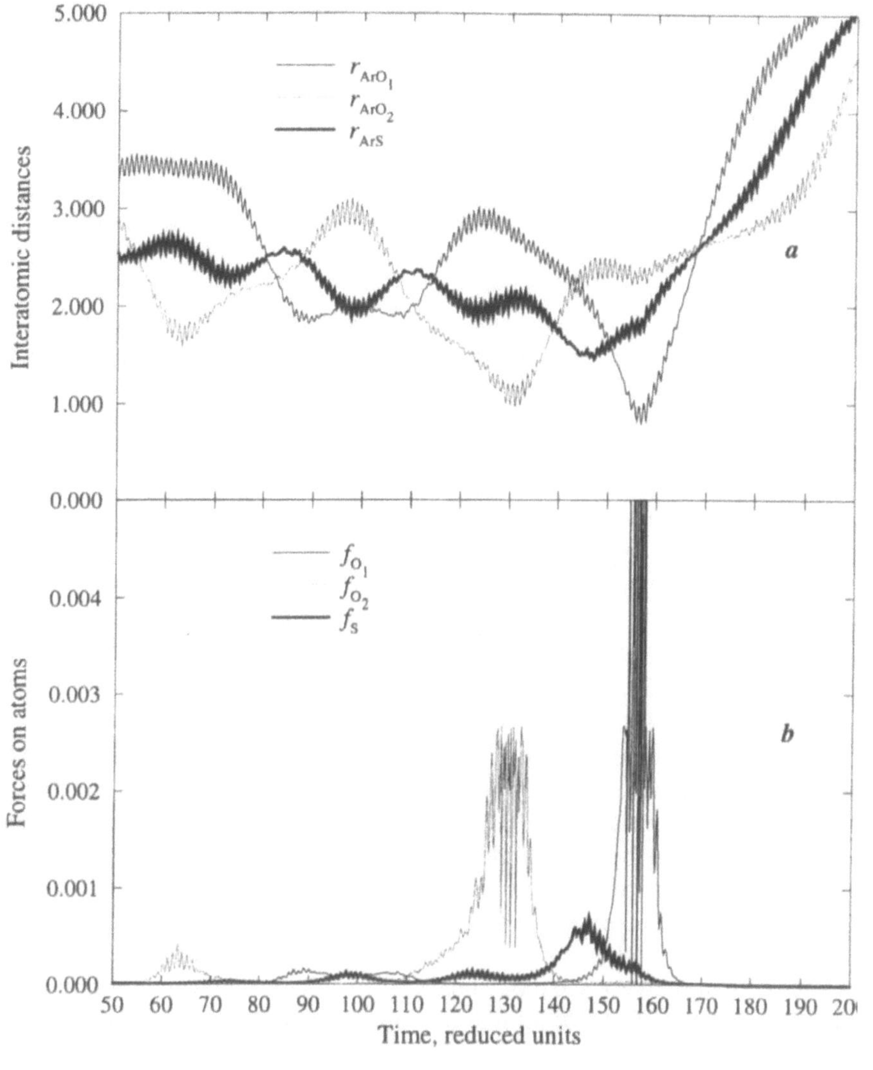

Figure 5. Change of different interatomic distances (a) and of the forces acting on atoms (b) along an arbitrary trajectory with the molecular angular momentum close to zero. Molecular energy $E = 100$ kcal/mol, heat bath temperature $T = 300$ K, harmonic intramolecular potential.

adiabatic principle of mechanics [17, 18], also known as the energy gap law. It states that under near-adiabatic conditions, when the collision duration τ is much larger than the inverse of vibrational frequency ω, the transition probability and, therefore, the average quantity of the transferred energy, decrease exponentially with the Massey parameter, $\omega\tau$. It can be shown [19, 21], that the smaller is the angular momentum J (or the adiabatic rotational energy E_j), the slower is the rotation of the molecule, the larger is the collisional duration τ (according to [19, 20], τ is approximately proportional to J^{-1}). For sufficiently small E_j we tend to nearly adiabatic conditions, when the transfer of vibrational energy is almost suppressed. This effect is illustrated in Figure 5 corresponding to the angular momentum close to zero. For so slowly rotating molecule the collider becomes trapped into a sort of a long-living complex, so that the interaction force oscillates with the vibrational frequency of the molecule; in such a case, the vibrational energy transfer, if any, is reversible and averages out during the collision time. As a result of these two factors, one enhancing and other hindering the loss of E_ν when the distortion of microcanonical distribution increases, the dependence of $|\langle\Delta E_\nu\rangle|$ on the sampling method becomes non-monotonous. On the whole, it can be concluded that the role of the adiabaticity factor is larger for low molecular energies, when the absolute value of the E_ν deviation is rather small; as the molecular energy grows, the equilibrium factor becomes more significant.

Comparison of $\langle\Delta E_\nu\rangle$ and $\langle E_{\nu_{\text{eff}}}\rangle$ indicates that, for most initial conditions, including the considerable deviations of the energy distributions from equilibrium, the transfer of the 'effective' vibrational energy is determined by the transfer of its 'true vibrational' part, E_ν, i.e. the 'vibrational' mechanism [3] of $E_{\nu_{\text{eff}}}$ transfer prevails. However, if the contribution of small angular momenta becomes very large (system 3) and the molecular energy is low, the role of the aforementioned adiabaticity factor in reducing the transfer of the 'true' vibrational energy E_ν may become considerable. In this case the participation of $\langle\Delta E_k\rangle$ and $\langle\Delta E_\nu\rangle$ in the transfer of $E_{\nu_{\text{eff}}}$ becomes comparable, i.e. the role of active rotations as a gateway for the vibrational energy transfer [5] grows. For a very large distortion and under the conditions when the transfer of the true vibrational energy is noticeably impeded, this mechanism may become dominant.

We conclude, therefore, that both the quantity of the energy, transferred in collisions, and the mechanism of this transfer are significantly affected by the initial distribution of energy over the molecular degrees of freedom. This confirms the necessity of the very accurate choice of the microcanonical sampling method, intended for the selection of initial conditions and being an essential prerequisite for all trajectory calculations.

Acknowledgments

We acknowledge very fruitful discussions with Prof. J. Troe and Dr. H.W. Schranz. This research was supported by the Fund for the Promotion of Research at the Technion.

References

1. I. Oref: *Chem. Phys.* **187**, 163 (1994).
2. I. Oref: *Adv. Chem. Kinetics and Dynamics* **2B**, 285 (1995).
3. I. Rosenblum, E. I. Dashevskaya, E. E. Nikitin, and I. Oref: *Chem. Phys.* **213**, 243 (1996).

4. H. W. Schranz and J. Troe: *J. Phys. Chem.* **90**, 6168 (1986).
5. I. Koifman, E. I. Dashevskaya, E. E. Nikitin, and J. Troe: *J. Phys. Chem.* **99**, 15348 (1995).
6. G. Nyman, S. Nordholm, and W. Schranz: *J. Chem. Phys.* **93**, 6767 (1990).
7. H. W. Schranz, S. Nordholm, and F. Freasier: *Chem. Phys.* **108**, 69 (1986).
8. H. W. Schranz, S. Nordholm, and G. Nyman: *J. Chem. Phys.* **94**, 1487 (1991).
9. L. Hase and D. G. Buckowski: *Chem. Phys. Lett.* **74**, 284 (1980).
10. T. Lenzer, K. Luther, J. Troe, R. G. Gilbert, and K. F. Lim: *J. Chem. Phys.* **103**, 626 (1995).
11. V. Bernshtein and I. Oref: *J. Chem. Phys.* **104**, 1958 (1996).
12. H. Hippler, H. W. Schranz, and J. Troe: *J. Phys. Chem.* **90**, 6158 (1986).
13. K. F. Lim and R. G. Gilbert: *J. Phys. Chem.* **94**, 72 (1990).
14. J. Jellinek and D. H. Li: *Phys. Rev. Lett.* **62**, 241 (1989).
15. J. Jellinek and D. H. Li: *Chem. Phys. Lett.* **169**, 380 (1990).
16. E. B. Wilson, J. C. Decius, and P. C. Cross: *Molecular Vibrations*, McGraw-Hill, New York–Toronto–London (1955).
17. B. Stevenson: *Collisional Activation in Gases*, Pergamon, New York (1967).
18. E. E. Nikitin: *Theory of Elementary Atomic and Molecular Processes in Gases*, Clarendon Press, Oxford (1974).
19. I. Rosenblum: *Chem. Phys. Lett.*, submitted.
20. R. G. Gilbert: *Austr. J. Chem.* **48**, 1787 (1995).
21. S. C. Farantos and N. Flytzanis: *J. Chem. Phys.* **87**, 6449 (1987).

Molecular Engineering **7**: 185–217, 1997.

Fragmentation of Ar_3^+: The Role of Rotational and Vibrational Predissociation Dynamics

E. BUONOMO and F. A. GIANTURCO
Department of Chemistry, The University of Rome, Cittá Universitaria, 00185 Rome, Italy.

(Received: 12 September 1996; accepted: 17 June 1997)

Abstract. The quantum study of the fragmentation dynamics of argon trimer ions has been carried out using two different *ab-initio* potential energy surfaces. Different computational methods have been employed and the effects of the total angular momentum of the system have been evaluated. It is found that the observed metastability can be explained with the lengthening of the lifetimes produced by the centrifugal barrier.

Key words: Cluster dynamics, rotational predissociation, vibrational predissociation, quantum structures, interaction forces.

1. Introduction

The study of the properties of rare gas ionized clusters is the subject of many recent theoretical and experimental studies [1]. One of the standard procedures for producing such clusters is to use a nozzle beam expansion for the production of neutral clusters of various sizes, then electron impact and photoionization to ionize the neutral clusters, which are further separated by mass spectrometry [2]. The excess energy associated with the ionization process usually leads to multiple fragmentations and, depending on the experimental conditions, associative reactions could also take place.

In this paper we will focus our attention on the metastability often observed in the experiments, in particular in ionized argon clusters, which have been measured many times by different groups in the last two decades, showing that the ionized cluster distribution obtained after the fragmentation cannot be related to what is collected at the detector [2]. In those systems, the metastability starts with the first terms of the clusters series, $n = 2$ [3–5] and $n = 3$ [6–8], and it has been observed up to $n = 60$ [9, 10], both measuring decay rates [3, 4, 6–8] and kinetic energy release distributions [5, 9, 10]. Furthermore, observations on the decay process for Ar_n^+ [11] showed a sequential decay with loss of one atom, with the only exception of Ar_4^+ where the decay proceeds with the evaporation of two neutral atoms, probably with a single-step process.

In order to try to understand the dynamics that presides over the observations made on such systems, one clearly needs to know several structural and dynamical aspects of the processes involved. One may reasonably assume that there is a fast dynamics following the ionization event, taking place on a time scale much shorter than that associated to the metastability (μs), where different electronic surfaces are accessed by the electron impact and fast neutral fragments may leave the cluster, carrying out part of the excess energy. The outcome of the fast processes, then, determines the actual pathways of energy storage and sets up the initial conditions to study processes on longer time scales, such as the metastability.

One of the most frequently studied systems is the ionized argon trimer, on which a large number of experimental and theoretical studies have been conducted: in particular, it has been found experimentally [12, 13], using scattering deflection techniques, that Ar_3^+ cannot be produced from direct ionization following electron impact. Recently, the electron impact on Ar_3^+ has also been studied theoretically [13] by a mixed quantum classical treatment in which nine diabatic electronic surfaces, calculated with the 'Diatomics in Molecules' method, were involved, confirming the experimental findings. The observations imply that the observed ionized clusters are formed from fragmentations of larger ions, with a strong reduction of the internal energy content; furthermore, from a model based on phase space theory [14], it has been shown that, as a consequence of multiple fragmentations, the final rotational temperature of small clusters like Ar_3^+ is very high, with distributions of total angular momenta peaking at very large values. Therefore, it is reasonable to assume that the metastability may be explained with dynamical processes taking place on the electronic ground state of the system, with all the available degrees of freedom possibly involved further taking into account the role of the total angular momentum of the system. Even when one assumes only one electronic surface, the dynamical study is only partially simplified as the determination of an accurate potential energy surface is a computationally demanding task and the heavy masses involved, together with the strong interaction expected for Ar_2^+ and Ar_3^+, will lead to a large density of internal states whose dynamical couplings with the dissociative coordinate determines the break-up event.

In this work, we describe the computational studies carried out on the ionized argon trimer, where the exchange of energy between the internal degrees of freedom and the dissociative coordinate is explicitly taken into account: in the next section the potential energy surface of the system is described, while Section 3 reports the theoretical model for the quantum dynamics of the fragmentation. Section 4 shows the results of our calculation and Section 5 finally summarizes our conclusions

2. The Ar_3^+ Potential Energy Surface

There have been many theoretical studies of the structural properties of ionized argon clusters, both with *ab-initio* (for Ar_3^+ [15–18]) and with semiempirical methods (for Ar_n^+ [19]). In particular, a great effort has been invested in the identification of the charged unit in clusters with a small number of atoms (($4 \leq n \leq 30$) for which there are many experimental results such as photodissociation in the visible

[20, 21] with measurements of photoabsorption cross-sections [20] and kinetic energy distributions of the fragments produced by photodissociation [21] and photoabsorption measurements in the UV [22]

The experimental evidence, in fact, suggests the presence of a charged core that acts as a cromophore, around which there is a distribution of neutral argon atoms; in the visible range, the photoabsorption spectra [20] of the different cluster sizes are similar to those measured for Ar_3^+ [23, 24], red-shifted with respect to the trimer, while the absorption in the UV for Ar_n^+ with $3 \leq n \leq 7$ [22] is slightly blue-shifted with respect to Ar_2^+. The former has been taken as the best evidence for the presence of the trimer core as a charged unit, in agreement with some theoretical work, while the latter supports the hypothesis of the dimer core as a charged unit.

In order to fully understand the experimental results, it seems that the calculations of the equilibrium configuration of the clusters in the ground electronic state and some of the electronic excited states are not sufficient. According to some theoretical studies [25, 26], in fact, the collision of the ionized clusters with a photon requires the inclusion of the nuclear motion to fully explain the experimental findings. The main reason for such behaviour resides in the flatness of the potential energy surface around the equilibrium configuration and there are many *ab-initio* theoretical studies [16–18] that have indeed found such a feature in the ionized trimer, with a PES weakly dependent on the asymmetric stretching coordinate in the region of the absolute minimum of the surface.

As a consequence, the equilibrium configuration that decides the type of cromophore is still a matter of debate; indeed, it is widely accepted that the trimer has an equilibrium linear structure, but it is not clear whether the configuration is symmetric or antisymmetric. Such disagreement between different calculations can be attributed to the efficacy of the methods employed in estimating the correlation energy. At any rate, the difference of energy between the symmetric and antisymmetric configuration are very small because, as mentioned before the potential energy surface along the asymmetric stretching coordinate is found to be very flat according to all the methods employed.

On the other hand, there are still few theoretical studies in which the energy surface have been calculated as a function of the complete set of nuclear degrees of freedom: in some cases, in fact, the potential energy surfaces of Ar_3^+ calculated with *ab initio* methods have been generated for some particular configurations only, and they are therefore not appropriate for use in dynamical calculations.

The first extensive study of the Ar_3^+ potential energy surface including the full dependence on the degrees of freedom of the system was carried out by Böhmer and Peyerimhoff[16], using the *Multi Reference Double Configuration Interaction* (MRDCI). The potential energy curves of the isolated dimer ion were calculated and the spectroscopic parameters for the ground state ($^2\Sigma_u^+$) and the first excited level were obtained giving, for the ground state, a well depth of 1.19 eV with the minimum located at $4.64a_0$ and a vibrational frequency of 289 cm^{-1}. It is worth noting that the binding energy is much larger than the neutral dimer (0.012 eV) with an equilibrium distance much lower: this can be taken as an indication of the strong structural changes that take place after the ionization process of the neutral cluster. The reported values are also in reasonable agreement with the experimental results, reporting a well depth of 1.3 eV and a frequency of 310 cm^{-1} [27]. Recent calculations [18] using an extensive CI method have shown very good agreement with the spectroscopic constants for Ar_2^+, pointing out the importance of the spin–orbit coupling in the evaluation of the well depth.

The MRDCI calculations for Ar_3^+, R have been carried out using Jacobian coordinates, commonly employed in scattering calculations. Here r describes the internuclear distance of Ar_2^+, R is the distance of the approaching Ar atom and γ is defined as $\gamma = \arccos(\hat{\mathbf{R}}\hat{\mathbf{r}})$. The points of the PES have been calculated for R varying between $4a_0$ and $20a_0$, for each of the four chosen internuclear distances ($r = 4.14, 4.64, 5.14, 6.14a_0$) and for each of the three orientations, $\gamma = 0, \gamma = 45°$ and $\gamma = 90°$. The absolute minimum of the surface was found in the collinear orientation with $r = 4.95a_0$ and a dissociation energy of 0.16 eV, separated by 0.01 eV only from the asymmetric minima; another minimum was found in the perpendicular orientation (T shape) with a well depth of 0.03 eV, at $R = 6.9a_0$ and $r = 4.64a_0$. In the linear configuration, therefore, there is a marked elongation of the Ar_2^+ distance in order to reach the equilibrium configuration of Ar_3^+: according to the original paper, there is a redistribution of charge over the third Ar atom with the formation of a chemical bond. In the perpendicular orientation, instead, Ar_2^+ is left unchanged by the interaction with the neutral atom, still retaining its ionic nature, and the attractive interaction is due to the polarisability of the argon atom in the presence of the the charged dimer. In the other orientation, $\gamma = 45°$, the interaction is weak and strongly dependent on the internuclear distance, with a repulsive behaviour for the smallest r and attractive for the largest value.

The available points for the equilibrium distance of Ar_2^+ were interpolated by us [28] to get a rigid-rotor surface, in order to carry out dynamical calculations in which the coupling between the dissociative coordinate and the bending motion of the cluster, described by γ, was explicitly taken into account. For each orientation γ, the available interaction energy points were fitted to a Morse function

$$V_\gamma = D_{e,\gamma}\{\exp[-2\alpha_\gamma(R - R_{e,\gamma}) - 2\exp[-\alpha_\gamma(R - R_{e,\gamma})\} \qquad (1)$$

where $D_{e,\gamma}, \alpha_\gamma$ and $R_{e,\gamma}$ are the usual Morse parameters defined for each γ and the potential energy surface was generated by inversion, i.e. by calculating the coefficients of the Legendre polynomials expansion with three terms ($\lambda = 0, 2$ and 4) in the following expression

$$V(R,\gamma) = \sum_{\lambda=0}^{4} V_\lambda(R)P_\lambda(\cos\gamma) \qquad (2)$$

for fixed values of R. The surface obtained by such a procedure is shown in Figure 1; the x axis coincides with the internuclear axis of Ar_2^+ and the contour level are

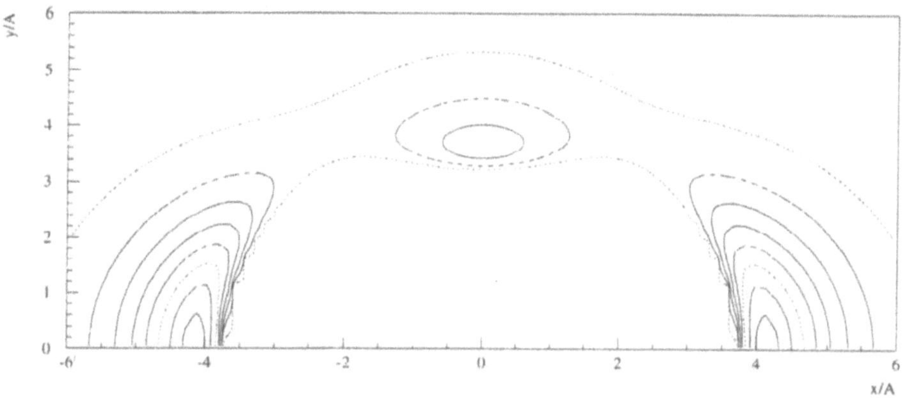

Figure 1. Rigid-rotor ground potential energy surface interpolated from MRDCI calculation; the distances are in Å units and the x-axis coincides with the Ar_2^+ internuclear axis, with the origin of the Cartesian coordinate in the center of mass of the charged dimer.

equally spaced by -0.01 eV, starting from the outermost level that corresponds to -0.01 eV. The reported figure shows the accuracy of the fitting: one can clearly see the two minima at the linear and perpendicular configurations, with the former about three times deeper than the latter, with a good agreement with the original surface (reported in Ref. [29]).

At any rate, the interpolated surface has a dependence on γ that is probably too weak, with only three terms in the expansion (2) over the Legendre polynomials: the same could be said for the dependence on the internuclear distance of Ar_2^+ of the PES points, with only four values over a very large grid. For these reasons, it was necessary to carry out the calculation of the PES over a more dense grid in order to describe all the features of the hypersurface. At the same time, Ar_3^+ is a complicated system to study with *ab-initio* methods, because of its 53 electrons and the need to include the correlation energy contributions.

Recently, methods based on the *Density Functional Theory* (DFT) have become popular and have been widely used to study a large number of systems. Such methods are usually computationally very efficient and can also give accurate results, provided that there is a careful choice of the functionals used in describing the exchange and correlation energy in the DFT framework: the detailed analysis of different functionals applied to ionized rare gas clusters has been the subject of two recent analyses [30] in which the applicability of such methods has been underlined. In order to generate the Ar_3^+ potential energy surface on a more dense grid, we have used a DFT technique with the exact Hartree–Fock exchange and the correlation energy by Lee, Yang and Parr [31], including the gradient correction to the correlation energy. The surface was generated [32] using the GAUSSIAN 92 [33] package of programs and the electronic wavefunction was expanded over the 6-31G** basis set of Gaussian functions that have been already employed for

Table I. Molecular parameters of Ar_2^+.

	r_e/a_0	D_e/(eV)	ω_e/cm^{-1}
exp.	–	1.33 ± 0.02 [27]	317
MRDCI [16]	4.64	1.19	291
POL-CI [15]	4.69	1.24	293
CI [18]	4.57	1.38	304
MP4 [17]	4.56	1.34	–
DFT [32]	4.69	1.24	289

Moeller Plesset [17] calculations at the fourth order (MP4) of the linear Ar_3^+. A first estimate of the quality of the DFT approach that has been used can be given from Table I, in which different spectroscopic parameters of the ionized argon dimer are reported: r_e the equilibrium internuclear distance of Ar_2^+, D_e is the dissociation energy and ω_e is the vibrational frequency that, in the DFT calculations, has been taken from the second derivative at the equilibrium point. The agreement with the more extensive configuration interaction methods and with large basis set MP4 calculations [17] is fairly good, pointing out the accuracy of DFT techniques when the interaction between partners is sufficiently strong.

The calculations [32] have been extended to Ar_3^+ for which the PES has been calculated over a grid composed of nine equidistant angles from 0° to 90° and seven values of the Ar_2^+ internuclear distance, from $4.2a_0$ to $5.3a_0$. For each pair (r, γ) about 100 points of the PES have been calculated, varying R from 2.5 Å to 15 Å and the correct asymptotic behaviour was imposed by fitting the curve to the following analytical form:

$$V_{\text{asym}} = -\frac{\alpha}{2R^4}\left(1 + \frac{\beta(r,\gamma)}{R^6}\right) \tag{3}$$

where the first term is the long range charge-induced dipole interaction while the second is a first order correction at intermediate internuclear distances to such behavior, where the Ar atom is interacting with a charge distributed over two atoms (Ar_2^+), instead of a point charge placed at the center of mass of Ar_2^+. The absolute minimum of the full PES is found in the asymmetric linear configuration, at $R = 7.9a_0$ and $r = 4.73a_0$, in good agreement with the reported asymmetric minimum of the MRDCI calculations[16], located at $R = 7.82a_0$ and $r = 4.64a_0$. Furthermore, the symmetric structure minimum is located at $r = 5.05a_0$, in good agreement with the MRDCI calculations and POL-CI calculations ($r = 5.04a_0$) [15]; the energy is 0.03 eV higher than the absolute asymmetric minimum. The DFT calculations are able to describe the very flat valley that has been found in the region of the absolute minimum by CI calculations, suggesting that little energy is involved in deforming the asymmetric structure where the charge is located preferentially over two atoms in the symmetric structure.

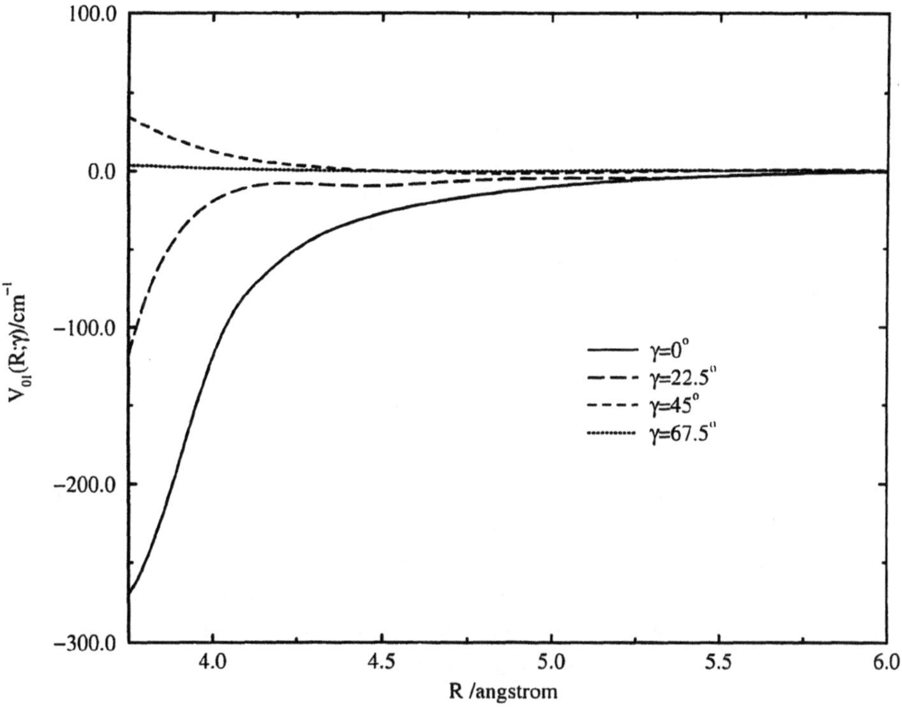

Figure 2. Computed vibrational coupling matrix elements between $v = 1$ and $v = 0$ vibrational states of the ionic core, reported for different orientations γ.

The dependence of the DFT potential on the Ar_2^+ internuclear distance for different orientations is in qualitative agreement with the MRDCI calculations, giving the deepest well in the linear region with a strong dependence on r and shallow well in the perpendicular region with a weak dependence on the Ar_2^+ internuclear distance. Those features are shown in Figure 2, where the off-diagonal vibrational matrix elements of the interaction potential have been reported at different orientations: they are obtained from the following quadrature over the internal coordinate r

$$V_{v,v'} = \int \chi_v(r)V(r;R,\gamma)\chi_{v'}(r)\,\mathrm{d}r \tag{4}$$

where χ_v are the vibrational functions of Ar_2^+ and $V(r,R,\gamma)$ is the interaction energy. We can see that the coupling changes by more than one order of magnitude from the linear to the perpendicular orientation: as we shall see in Section 4, such feature produces a dynamical effect related to the possibility of preparing the system to be within a limited range of orientations.

3. Theoretical Methods

3.1. CLOSE COUPLING AND HELICITY DECOUPLING METHODS

The methods discussed in this subsection are based on the Fano formalism [34] for treating the resonances. In particular, the *Close Coupling* (CC) method was originally developed to study the predissociation of Van der Waals complexes [35], in which the resonances are determined by the overlap between a discrete zero order bound state and the 'exact' wavefunction, both being calculated numerically. Assuming that only one electronic surface is involved in the process, the Hamiltionian is given by

$$H = -\frac{\hbar^2}{2m}\frac{\partial^2}{\partial R^2} - \frac{\hbar^2}{2\mu}\frac{\partial^2}{\partial r^2} + \frac{\hat{j}^2}{2mR^2} + \frac{\hat{j}^2}{2\mu r^2} + V(R, r, \gamma) \tag{5}$$

where $m = 2m_{\mathrm{Ar}}/3$ and $\mu = m_{\mathrm{Ar}}/2$, $\hat{l}$ and $\hat{j}$ are the angular momenta associated with $\hat{\mathbf{R}}$ and $\hat{\mathbf{r}}$, leading to the total angular momentum $\hat{J} = \hat{l} + \hat{j}$, and $V(R, r, \gamma)$ is the interaction potential, discussed in Section 2.

The starting point of the CC method is the calculation of the initial set of bound states, unperturbed wavefunctions Φ_q, with eigenvalue E_q via a variational method that employs a convenient stretching–bending basis set

$$\Phi_q^{JMp}(\mathbf{r}, \mathbf{R}) = \sum_{n,v,j,|\Omega|} a^q_{nvj|\Omega|}\phi_n(R)\chi_v(r)\Theta^{JMp}_{j|\Omega|}(\hat{\mathbf{r}}, \hat{\mathbf{R}}) \tag{6}$$

here M is the projection of $\mathbf{J}$ onto the space-fixed (SF) z-axis and $|\Omega|$ its projection onto the body-fixed (BF) rotating axis $\mathbf{R}$. The $\chi_v(r)$ are the numerically-obtained vibrational wavefunctions for the states supported by the non-rotating dimer core potential and the Θs form a free-rotor basis set in the BF frame, with $p = \pm 1$ being the parity under inversion [36]. The stretching functions required in latter expansion, $\phi_u(R)$, have been computed starting from the diagonal coupling potential (4) for $v = 0$ to yield an angular-dependent set of functions

$$\left[-\frac{\hbar^2}{2m}\frac{\partial^2}{\partial R^2} + V_{0,0}(R, \gamma)\right]\bar{\phi}_n(R|\gamma) = E_n(\gamma)\bar{\phi}_n(R|\gamma) \tag{7}$$

using different angular values. Since the global set so obtained is not orthogonal, the final stretching numerical basis is obtained after Graham–Schmidt orthogonalisation.

The next step involves the evaluation of the 'exact' wavefunctions $|\Psi_{fE}\rangle$ by writing them down an expansion over unknown R-dependent coefficients

$$|\Psi_{fE}\rangle \equiv \Psi^{JMp}_{vj|\Omega|E}(\mathbf{r}, \mathbf{R}) = \sum_{v',j',|\Omega|} \Psi^{vj|\Omega|}_{v',j'|\Omega'|E}(R)\chi_{v'}(r)\Theta^{JMp}_{j'|\Omega'|}(\hat{\mathbf{r}}, \hat{\mathbf{R}}). \tag{8}$$

The latter coefficients can therefore be obtained by solving the corresponding set of CC rotovibrational equations at energies close to the E_g eigenvalue of the expansion (6)

$$\left\{-\frac{\hbar^2}{2m}\frac{\partial^2}{\partial R^2}+E_v+B_{vj}-E\right\}\Psi^{vj|\Omega|}_{vj|\Omega|E}(R)$$

$$=-\sum_{v'j'|\Omega|'}\{\delta_{|\Omega||\Omega'|}\langle\Theta^{JMp}_{j|\Omega|}|V_{vv'}|\Theta^{JMp}_{j'|\Omega'|}\rangle\Psi^{vj|\Omega|}_{v'j'|\Omega'|E}(R)+$$

$$+\frac{\delta_{jj'}\delta_{vv'}}{2\mu R^2}\langle\Theta^{JMp}_{j|\Omega|}|\hat{l}^2|\Theta^{JMp}_{j'|\Omega'|}\rangle\Psi^{vj|\Omega|}_{v'j'|\Omega'|E}(R)\}. \tag{9}$$

The above radial eigenfunctions fulfill the standard boundary conditions for outgoing waves [35] and the other terms in the equations are defined as

$$B_v=\langle\chi_v|\frac{\hbar^2}{2\mu r^2}|\chi_v\rangle \tag{10}$$

$$V_{vv'}(R,\gamma)=\langle\chi_v|V(r,R,\gamma)|\chi_{v'}\rangle. \tag{11}$$

The latter coupling potential can be further expanded in Legendre polynomials

$$V_{vv'}(R,\gamma)=\sum_\lambda V^\lambda_{vv'}(R)P_\lambda(\cos\gamma). \tag{12}$$

The third step in the calculations now involves the evaluation of the overlap integrals between the discrete and 'exact' wavefunctions that will provide us with the coefficients and then with the linewidth of the resonance [34]. For each of the resonance positions E_r around the selected E_q we can therefore write the following cross section

$$\sigma(q,E)\sim\pi\sum_{v,j,|\Omega|}|\langle\Phi^{JMp}_q|\Psi^{JMp}_{vj|\Omega|E}\rangle|^2\sim\frac{\Gamma}{(E-E_r)^2+\Gamma^2} \tag{13}$$

and the numerical procedure involves the calculation of the overlaps in a small energy range around the E_q eigenvalues obtained from the diagonalisation of the Hamiltonian of Equation (5) followed by the fitting to Lorentzian functions of the overlaps calculated at different energy to determine the resonance energy and the linewidth, from which one can determine the associated lifetime

$$\tau=\hbar/2\Gamma. \tag{14}$$

The final distribution over final states is given by

$$P^{JMp}_{vj|\Omega|}=\frac{|\langle\Psi^{JMp}_{vj|\Omega|E}|\Phi^{JMp}_q\rangle|^2}{\sum_{v'j'|\Omega'|}|\langle\Psi^{JMp}_{v'j'|\Omega'|E}|\Phi^{JMpP}_q\rangle|^2}. \tag{15}$$

The computational effort in using the CC method become increasingly high as the total angular momentum J increases, because of the large number of $|\Omega|$ states that are involved in the discrete state expansion and in the CC continuum equations. In the Hamiltonian operator, expressed in the chosen basis set, only the angular operator $\hat{l}^2$ couples states with different tumbling quantum number giving a tridiagonal structure for the Hamiltonian matrix

$$h^{Jj}_{\Omega,\Omega}(R) = [J(J+1) - 2\Omega^2 + j(j+1)]/R^2 \tag{16}$$

$$h^{Jj}_{\Omega,\Omega\pm1}(R) = \{[J(J+1) - \Omega(\Omega\pm1)]^{1/2} \cdot [j(j+1) - \Omega(\Omega\pm1)]^{1/2}\}/R^2 \tag{17}$$

where one sees that the changing of the helicity index induced by Equation (17) corresponds to the Coriolis coupling terms in the full kinematics of the problem [37]. The matrix elements of the off-diagonal helicity coupling are the smallest contributions in the calculation of the discrete state for the present system, implying that their importance is neglegible in the interaction region, but they are still quite important in the long range radial region. The *Helicity Decoupling* approximation assumes that the effects of the $\hat{l}^2$ angular operator can be approximated by its diagonal elements

$$\langle j\Omega JM||\hat{l}|j'\Omega' JM\rangle \sim \delta_{jj'}\delta_{\Omega\Omega'}[J(J+1) + j(j+1) - 2\Omega^2] \tag{18}$$

both when searching for the discrete solutions, via diagonalisation of a simplified Hamiltonian in Equation (6), and when searching for the continuum solutions $\Psi^{vj|\Omega|}_{vj|\Omega|E}$ via Equation (9). In this way, the tumbling quantum number $|\Omega|$ is conserved. Such an approximation markedly reduces the number of coupled equations and considers that the larger diagonal coupling in $|\Omega|$ by the potential (2nd term on the r.h.s. of Equation (9)) gives a physically reliable correction via only the centrifugally diagonal term of Equation (16). Furthermore, the neglect of the non-diagonal elements of Equation (17) becomes more valid as $|\Omega|$ approaches the value of $\pm j$ or of $\pm J$.

The HD numerical procedure follows the same steps as the CC method, once each sum over the helicity quantum number is removed.

3.2. DIABATIC GOLDEN RULE APPROXIMATION (DGR)

If the motion of the diatomic fragment is fast as compared with the other mode, it is reasonable to assume that the zeroth order Hamiltonian [38, 35]

$$H = \frac{\hbar^2}{2m}\frac{\partial^2}{\partial R^2} + \frac{\hat{l}^2}{2mR^2} + B_v\hat{j}^2 + E_v + V_{vv}(R,\gamma) \tag{19}$$

correctly describes the dynamics. The discrete solutions can be written as

$$\Psi_{vi} = \chi_v(r) \sum_n \sum_{j'|\Omega'|} a^{vi}_{j'|\Omega'|} \phi_n(R) \Theta^{JM}_{j'|\Omega'|}(\hat{r}, \hat{R}) \tag{20}$$

therefore in the Hamiltonian matrix only the diagonal vibrational $V_{vv}(R, \gamma)$ is involved; the continuum equations are given by

$$\Psi^{j|\Omega|}_{vE} = \chi_v(r) \sum_{j'|\Omega'|} \phi^{vE,j|\Omega|}_{j'|\Omega'|}(R) \Theta_{j'|\Omega'|}(\hat{r}, \hat{R}) \tag{21}$$

and, after substituting into the Schrödinger, one obtains the close coupling equations in Equation (9), where the sum over the vibrational states in the right-hand side is restricted to just one v value.

In this method, a bound state calculated for the vibrational state v could be not bound with respect to a low lying vibrational state v' and the vibrational coupling between the discrete state and the continua of v' leads to the fragmentation. The linewidth associated to this process can be determined from the *Fermi Golden Rule* for which the resonance energy is set equal to the energy of the predissociative state and the linewidth is given by

$$\Gamma_{vi} = \pi \sum_{v' \leq v, j', |\Omega'|} |\langle \Psi^{j'|\Omega'|}_{v'E} | V(r, R, \gamma) | \Psi_{vi} \rangle|^2 \tag{22}$$

where the continuum wavefunctions are calculated on the energy shell, i.e. $E = E_{vi}$ and are normalized to the energy. The coupling in the latter equation is only determined by the potential, depending on the off-diagonal vibrational matrix elements $V_{vv'}(R, \gamma)$. The final distribution of the fragment can be obtained from

$$P^{v,i}_{v',j'} = \pi \sum_{,|\Omega'|} \frac{|\langle \Psi^{j'|\Omega'|}_{v'E} | V(r, R, \gamma) | \Psi_{vi} \rangle|^2}{\Gamma_{vi}}. \tag{23}$$

This approximation has been applied to study the vibrational predissociation of Van der Waals systems, for which there is usually a large separation between the vibrational mode of the diatomic fragment and the cluster modes. In that case, the DGR method is able to give accurate results.

3.3. ANGULAR ADIABATIC APPROXIMATION (AA)

The *Angular Adiabatic Approximation* has been used to study both the rotational predissociation of different Van der Waals systems [39], using rigid-rotor potential surfaces and the vibrational predissociation after the diabatic separation of the vibrational motion of the diatomic fragment. It is based on the different time scales of the stretching motion and the bending motion of the cluster, that is controlled by the rotational constants associated with $\hat{\mathbf{r}}$ and $\hat{\mathbf{R}}$, with the lifetimes calculated from the Fermi Golden Rule.

Assuming the rigid rotor Hamiltonian,

$$H = -\frac{\hbar^2}{2\mu}\frac{\partial^2}{\partial R^2} + \frac{\hbar^2 \hat{l}^2}{2\mu R^2} + B_e \hat{j}^2 + V(R,\gamma) \tag{24}$$

and that the stretching motion is faster than the bending one, we can separate adiabatically the two modes by writing the discrete wavefunction of the cluster as

$$\Psi^d_{nm}(\mathbf{R},\hat{\mathbf{r}}) = \phi_n(R,\gamma)F_{nm}(\hat{\mathbf{R}},\hat{\mathbf{r}}) \tag{25}$$

where n and m are the stretching and bending quantum number, respectively.

The adiabatic approximation then consists in neglecting the dependence on γ of the ϕ_n functions that hence become 'transparent' to the action of the angular momentum operators, i.e.

$$\hat{l}^2\Psi^d_{n,m} = \phi_n \hat{l}^2 F_{n,m}$$

$$\hat{j}^2\Psi^d_{n,m} = \phi_n \hat{j}^2 F_{n,m}. \tag{26}$$

Therefore, the functions ϕ_n can be defined by the terms of the Hamiltonian acting on the variable R, with eigenvalues that depend on the orientation γ

$$\left\{-\frac{\hbar^2}{2\mu}\frac{\partial^2}{\partial R^2} + V(R,\gamma)\right\}\phi_n(R;\gamma) = W_n(\gamma)\phi_n(R;\gamma). \tag{27}$$

After substituting the $\Psi^d_{n,m}$ in the Schrödinger equation

$$H\Psi^d_{n,m} = E_{n,m}\Psi^d_{n,m} \tag{28}$$

and using the approximation made on the action of the angular momentum operators, we obtain the bending equation

$$\left[-\frac{\hbar^2}{4\mu}\{A_n(\gamma),\hat{l}^2\} + B_e\hat{j}^2 + W_n(\gamma)\right]F_{n,m}(\hat{\mathbf{R}},\hat{\mathbf{r}}) = E_{n,m}F_{n,m}(\hat{\mathbf{R}},\hat{\mathbf{r}}) \tag{29}$$

where $\{\ ,\ \}$ is the anticommutator and $A_n(\gamma)$ is the mean value of $1/R^2$ in the ϕ_n for a given orientation γ:

$$A_n(\gamma) = \langle\phi_n(R;\gamma)|\frac{1}{R^2}|\phi_n(R;\gamma)\rangle. \tag{30}$$

The last integral is introduced in order to improve on the use of a simple average $\bar{R}$. Since A_n does not commute with the $\hat{l}^2$ operator, the hermiticity of the term is forced by using the anticommutator.

It is interesting to note that the angular dependent eigenvalues W_n, become the effective potentials in the bending motion. The equation Equation (29) can be solved expanding the $F_{n,m}$ functions on a 'body-fixed' (BF) basis set [36]:

$$F_{nm}(\hat{\mathbf{R}}, \hat{\mathbf{r}}) = \sum_{j,|\Omega|} C_{j,|\Omega|}^{mm} \langle \hat{R}, \hat{r} | JMj|\Omega|p \rangle \tag{31}$$

which have been already described in the CC method.

The calculation of the continuum functions is done using the 'Infinite Order Sudden Approximation' (IOSA) [37], where, as in the adiabatic approximation of discrete states, the wavefunction is expressed as a product of a dissociative function, parametrically dependent on the orientation, and a rotational function. Considering the rotational sudden approximation

$$\hat{l}^2 = \bar{l}(\bar{l}+1) \tag{32}$$

we can write the continuum wavefunction as

$$\Psi_{j|\Omega|,\epsilon\bar{l}}^c = \phi_{\epsilon,\bar{l}}(R;\theta) \langle \hat{\mathbf{R}}, \hat{\mathbf{r}} | JMj|\Omega|p \rangle \tag{33}$$

where the ϕ function is obtained as a solution of the single channel, one-dimensional differential equation

$$\left[-\frac{\hbar^2}{2\mu} \frac{d^2}{\mathrm{d}R^2} + V(R,\gamma) + \hbar^2 \frac{\bar{l}(\bar{l}+1)}{2\mu R^2} \right] \phi_{\epsilon\bar{l}}(R;\gamma) = \epsilon \phi_{\epsilon\bar{l}}(R;\gamma) \tag{34}$$

with the energy of the continuum state given by

$$E(j) = \epsilon + B_e j(j+1) \tag{35}$$

and using the real boundary conditions.

The linewidths are calculated from the Fermi Golden Rule

$$\Gamma_{n,m} = \pi \sum_{j,|\Omega|} |V_{nm \to j|\Omega|}^{dc}|^2 \tag{36}$$

where the couplings V^{dc} are given by

$$V_{nm \to j|\Omega|}^{dc} = \langle \Psi_{j|\Omega|,\epsilon}^c (\mathbf{R}, \hat{\mathbf{r}}) | H | \Psi_{nm}^d (\mathbf{R}, \hat{\mathbf{r}}) \rangle \tag{37}$$

and the different contributions to the coupling can be obtained substituting the discrete and continuum wavefunctions and the Hamiltonian operators, thus giving

$$\begin{aligned} V_{nm \to j|\Omega|}^{dc} &= B_e \langle \Psi_{j|\Omega|,\epsilon}^c | \hat{j}^2 | \Psi_{nm}^d \rangle + \\ &\quad + \frac{\hbar^2}{2\mu} \langle \Psi_{j|\Omega|,\epsilon}^c | \hat{l}^2 | \Psi_{nm}^d \rangle + V_{\text{pot}} \end{aligned} \tag{38}$$

where the last term, V_{pot}, is due to the adiabatic potential couplings

$$V_{\text{pot}} = \langle JMj|\Omega|p|\langle\phi_{\epsilon,\bar{l}}(R;\gamma)|\phi_n(R;\gamma)\rangle_R W_n(\gamma)|F_{nm}(\hat{\mathbf{R}},\hat{\mathbf{r}})\rangle. \tag{39}$$

Also in this case, the continuum wavefunctions are energy normalized and is is possible to obtain the final distribution of the fragment introducing the partial linewidth,

$$\Gamma_{nm\to j|\Omega|} = \pi|V^{dc}_{nm\to j|\Omega|}|^2 \tag{40}$$

and the ratio of the last quantity to the total linewidth gives the probability of the final fragment state $|j_f, |\Omega|\rangle$.

4. Results

We have carried out dynamical calculations to study the following fragmentation process:

$$\text{Ar}_3^+(v_i = 1, 0, j_i) = \text{Ar}_2^+(v_f = 0, j_f) + \text{Ar} \tag{41}$$

where the dimer can either be vibrationally excited [32, 40, 41] or be in its vibrational ground state inside the cluster [28, 40]: in the former case, the excess energy, stored into the dimer core, can be transferred into the dissociative motion either from the vibrational mode or from the rotational mode (which can be seen as a sort of hindered bending motion inside the cluster). In the latter case, with $v_i = 0$ only the rotational predissociation mechanism can take place. Given the strong anisotropy of the system, it is not surprising that the rotational quantum number j will be no longer a 'good' quantum number since the methods we have discussed in the previous section describe the metastable state as an expansion over such states; on the other hand, the predissociative complexes have a large energy content that is strongly dependent on the rotational energy of the dimer core: therefore, the quantum number j has been used in Equation (41) to describe the rotational energy of the dimer. The vibrational quantum number v, instead, is found to be almost conserved in the close coupling and helicity decoupling calculations, and it is possible to classify the metastable states with the vibrational level of the isolated dimer by projecting the corresponding wavefunctions over one particular vibrational state of the dimer and selecting those states with a projection larger than 0.8 on the latter. Such a procedure accumulates a very large fraction of the calculated predissociative states, given the large difference between the vibrational frequency of Ar_2^+ and the frequencies of the other modes.

The basis set used in the calculation includes three vibrational functions, from $v = 0$ to $v = 2$, 30 rotational states, from $j = 1$ to $j = 59$, and 14 stretching functions obtained from the Graham–Schmidt orthogonalization of six solutions of Equation (7) for $\gamma = 0$ and one for the remaining six orientations. The main purpose of the calculations has been the analysis of the effect of the temperature of the overall rotational motion on the lifetimes of the clusters: thus, very high values of the total angular momentum have been calculated, up to $J = 50$ in the helicity decoupling approximation and $J = 30$ in the angular adiabatic approximation. The term of the Hamiltonian operator that is changed by the total angular momentum does not involve the internuclear distance of Ar_2^+ and therefore there are neglegible effects on the stretching coordinate in the region of the minima with respect to the other terms. The basis set for the calculation of the metastable states has not been modified with increasing J in the description of the the stretching and of the dimer coordinate; as we have already discussed, the main effect on the number of functions in the basis set is given by the Ω multiplicity that increases dramatically with the total angular momentum J.

The dissociative coordinate has been described by 2048 points in the radial range 2.5 Å $\leq R \leq$ 21 Å and all the available nine orientations have been used in the calculations, expanding the vibrational couplings over the first nine Legendre polynomials of even symmetry.

The results we have obtained are reported according to the different dynamical effects that we have found in the low J range and in the high J range and will be discussed in the next two subsections.

4.1. THE ROLE OF THE HELICITY QUANTUM NUMBER

In this subsection the results obtained from the calculation performed for a limited range of the total angular momentum J are discussed [32, 40, 41]; in particular, we have carried out calculations for $J \leq 8$ using the vibrational diabatic [41] and the helicity decoupling [40] methods while the close coupling method has been utilized up to $J = 2$ [32, 41].

The lifetimes and the resonance positions of the metastable states with in the energy range between 0 and 200 cm^{-1} from the vibrational level $v = 0$ of Ar_2^+ and a projection larger that 0.8 over $v = 1$ have been calculated with the close coupling method. In the energy range examined, only the rotational states of the dimer belonging to the vibrational ground state of Ar_2^+ are open.

Since the actual energy distribution of the metastable states in not known, we assumed an equal probability of the predissociating states in the metastable ions: therefore, the average lifetimes have been calculated for each distribution of metastable states (obtained at different J values) and, consistent with the equiprobability assumption, the variance of the distribution has been taken as a measure of its spread.

For larger J values, the computational effort required in the CC calculations becomes demanding and the *Diabatic Golden Rule* approximation has been employed, for which the discrete and continuum equations are simplified assuming the vibrational motion as diabatic. Using this approach, calculations of lifetimes

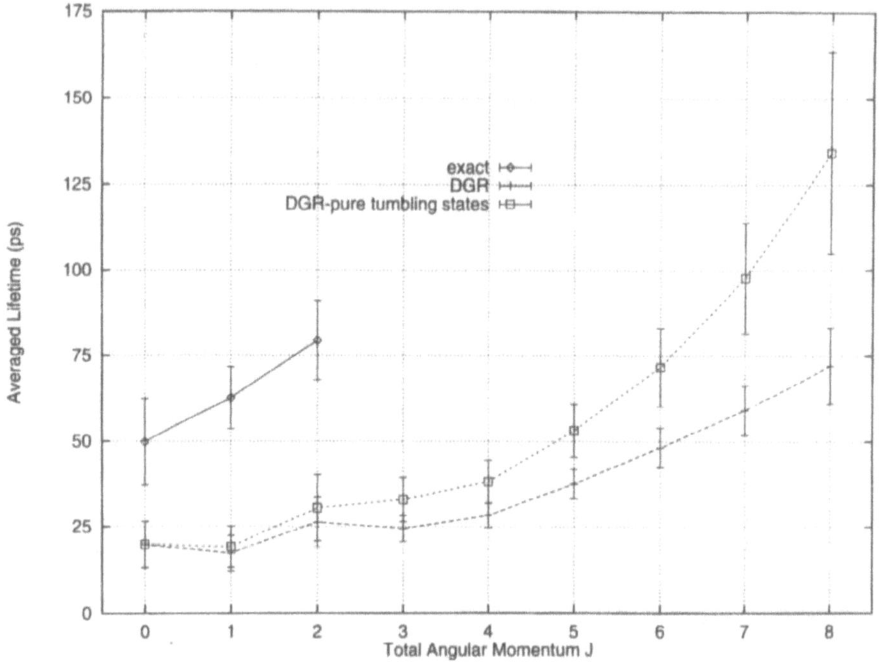

Figure 3. Computed average lifetime as a function of the total angular momentum of the system, using the CC method ('exact') and the DGR approximation. The results obtained from DGR calculations which include only states with large projection over a particular $|\Omega|$ value are also reported.

have been carried out up to $J = 8$ in the same energy range considered in the CC calculations.

The results obtained using the two methods outlined above are reported in Figure 3, where we see that the fragmentation process is very efficient at low J, with average lifetimes of the order of $0.1ps$, far from the metastable time windows involved in the experiments. One can also note that the DGR approximation gives lifetimes that are about 50% smaller than the average lifetimes obtained from CC calculations; as the energy shifts of the resonance in the CC calculations are small, the errors may probably be attributed to the role of the closed channels that are missed going from the CC to the DGR approach. At any rate, we can take the results obtained from the DGR calculations as being representative of the general trend of the average lifetime over the broad energy range chosen, as we will show below. One can also note from Figure 3 that both methods give an increase of the average lifetimes with J; such an increase is also much more marked when one takes into account only the predissociative states with a very large projection (>0.8) over one single $|\Omega|$ value.

We can get more insight into the process at hand from the results reported in Figure 4, where average lifetimes have been calculated for the distributions of metastable states with the same mean value of the tumbling quantum number (calculated as the closest integer to the actual value), showing a clear increase of r with $|\Omega|$ These features can be explained from the behaviour of the matrix elements of the $|JMj\Omega\rangle$ rotational states of the off-diagonal vibrational couplings, reported in Figure 5, that are explicitly involved in the calculation of the linewidth as it can be seen the Fermi Golden Rule expression, Equation (22). The figure shows marked differences between the curves calculated for different values of $|\Omega|$, with $|\Omega| = 4$ giving rise to matrix elements much smaller with respect to $|\Omega| = 0$. Thus, as the j value increases, such differences tend to become smaller. The γ dependence of the rotational wavefunction in the *body-fixed* frame is given by the associated Legendre polynomials $P_j^{|\Omega|}$ which have a behaviour such that they are close to zero in a range of low angles if $|\Omega| \neq 0$; the range becomes wider as $|\Omega|$ increases (for fixed values of j) while, when fixing $|\Omega|$, the range is shorter with the increase of j. If we further remember that, as reported in Figure 2, the off-diagonal vibrational couplings have a strong angular dependence with the largest coupling at very low angles, we can then conclude that the dependence on the tumbling quantum number of the average lifetimes in Figure 4 is caused by an 'orientational effect', because of the particular angular dependence of the vibrational couplings that is also reflected in the behaviour of the lifetimes with varying J, as shown in Figure 3. Furthermore, the preceeding discussion can also be applied to the CC results of Figure 3, considering that the average lifetimes have been calculated for $J \leq 2$ where the rotational barrier is too low to play any role.

At any rate, the DGR approximation is not appropriate for studying the effects of the rotational barrier, which is expected to be important for very high J values, as the number of basis set functions increases quickly with J; therefore, in the high J range, the *helicity decoupling* approximation is more appropriate as the basis set does not depend on the $|\Omega|$ degeneracy. Before going into the details of the HD calculations, it is interesting to get further insights on its applicability. The top part of Figure 6 therefore shows the histograms with the number of 'pure' pure states and the mean values states for each $|\Omega|$, calculated for $J = 8$: we note that the distributions of the mean value states is not symmetric and has a maximum for $|\Omega| = 3$; furthermore, the smallest differences between the mean value states and the 'pure' states are obtained for the maximum and minimum values of $|\Omega|$. The mixing of different tumbling states is caused by the $\hat{l}^2$ operator and the remaining terms of the Hamiltonian that act indirectly. Therefore, as the number of off-diagonal $|\Omega|$ couplings is reduced to one for $|\Omega| = 0$ and $|\Omega| = 8$, which are only coupled with $|\Omega| = 1$ and $|\Omega| = 7$, respectively, the states corresponding to those tumbling quantum numbers are the most pure of the distribution: moreover, the asymmetry of the mixed state distribution can be related to the number of basis functions with a fixed $|\Omega|$ value, since only the rotational states with $j \geq |\Omega|$ can contribute for a given value of the tumbling quantum number.

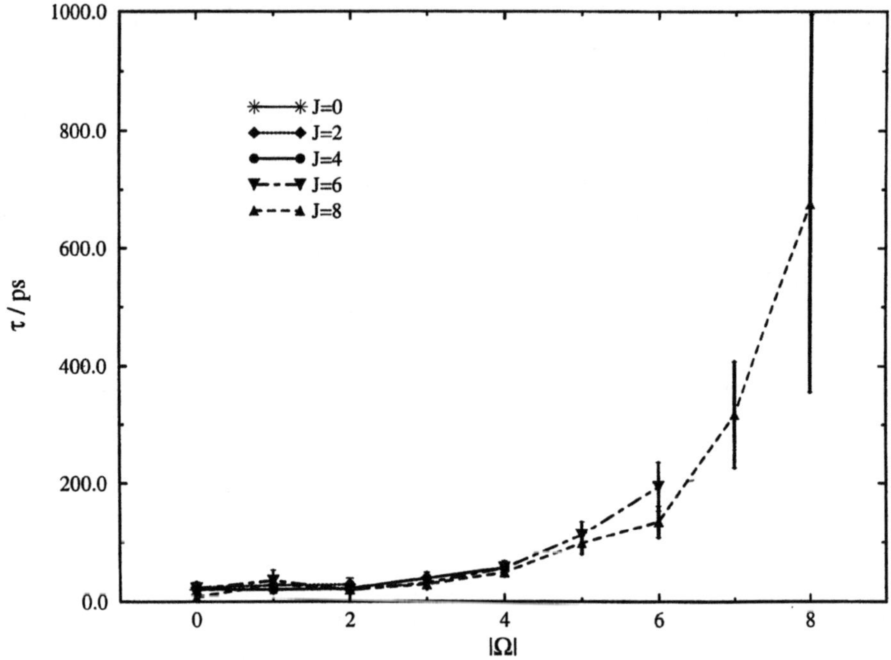

Figure 4. Computed DGR average lifetimes (in 10^{-12}s) as function of the mean value of the helicity operator, for $J = 8$.

It is also interesting to measure the weight of the 'pure' $|\Omega|$ states with respect to the total number of states for each J value: in the bottom part of Figure 6, the distributions of 'pure' tumbling states and the total number of states calculated with the DGR approximation are reported. In the chosen energy range, the total number of states increases almost linearly with J while the number of 'pure' states reaches its maximum at $J = 6$ and decreases for larger J. Thus, it is reasonable to apply the HD approximation in the range $J \leq 8$ where a consistent fraction of the predissociative states has a large projection over one particular $|\Omega|$. The distribution of tumbling quantum number of the final states, instead, is quite different, showing a weak tendency to a preferential decay in the channels where $|\Omega|$ is conserved, as we can see from the two examples reported in Figure 7. We recall that the final state distribution is directly dependent on the off-diagonal vibrational coupling. We can conclude that the main reason to apply the helicity decoupling approximation stands in its similarity to the exact CC method and we can expect similar results for the order of magnitude of the lifetimes, even considering that the approximation coincides with the CC method at $J = 0$.

Thus, we have carried out HD calculation up to $J = 8$, using the same basis set as for the CC calculations, in the same energy range and using predissociative states with a large projection over the vibrational state $v = 1$ of the dimer. The

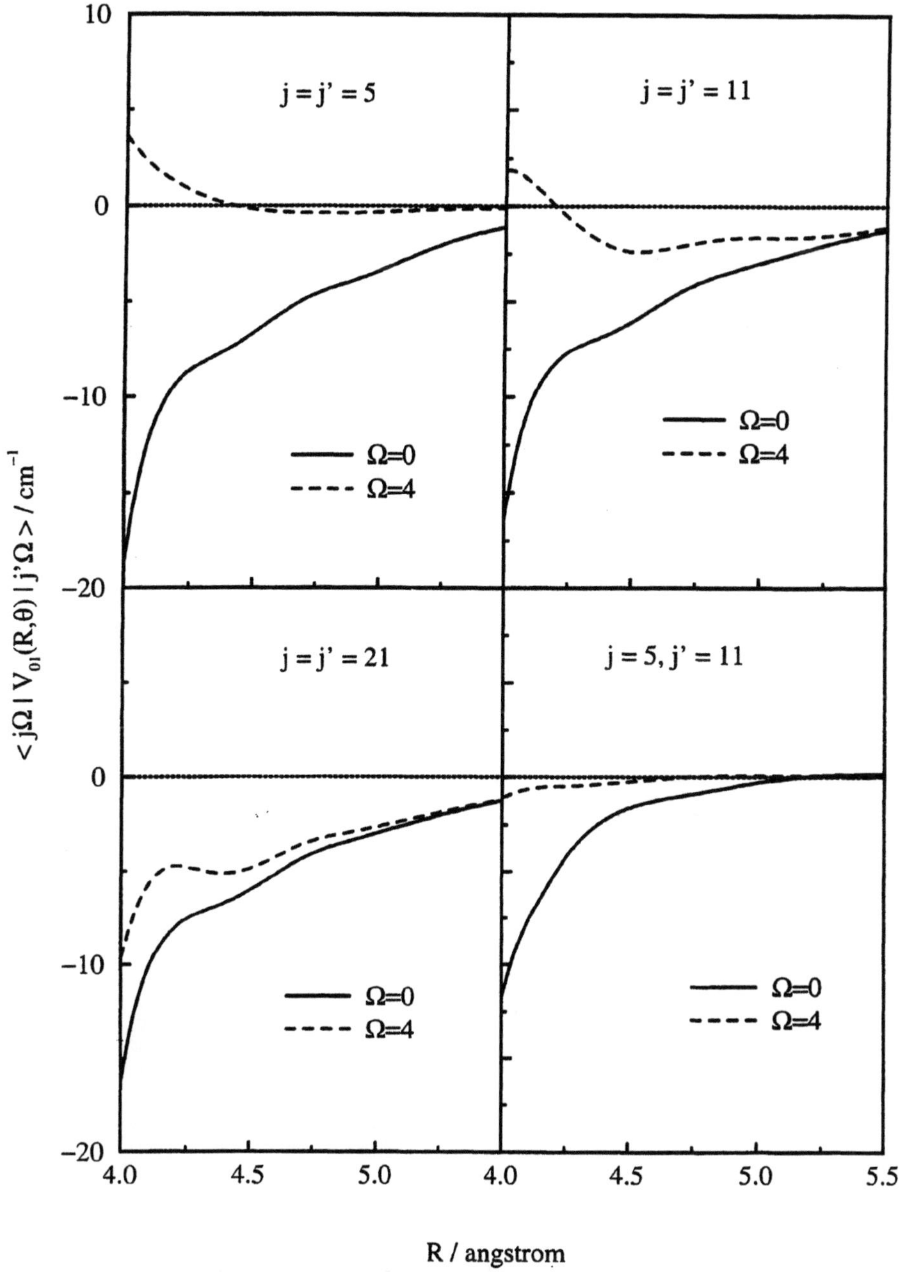

Figure 5. Matrix elements between $v = 1$ and $v = 0$ vibrational state of the dimer core, calculated for $J = 4$ and for different $|\Omega|$ values.

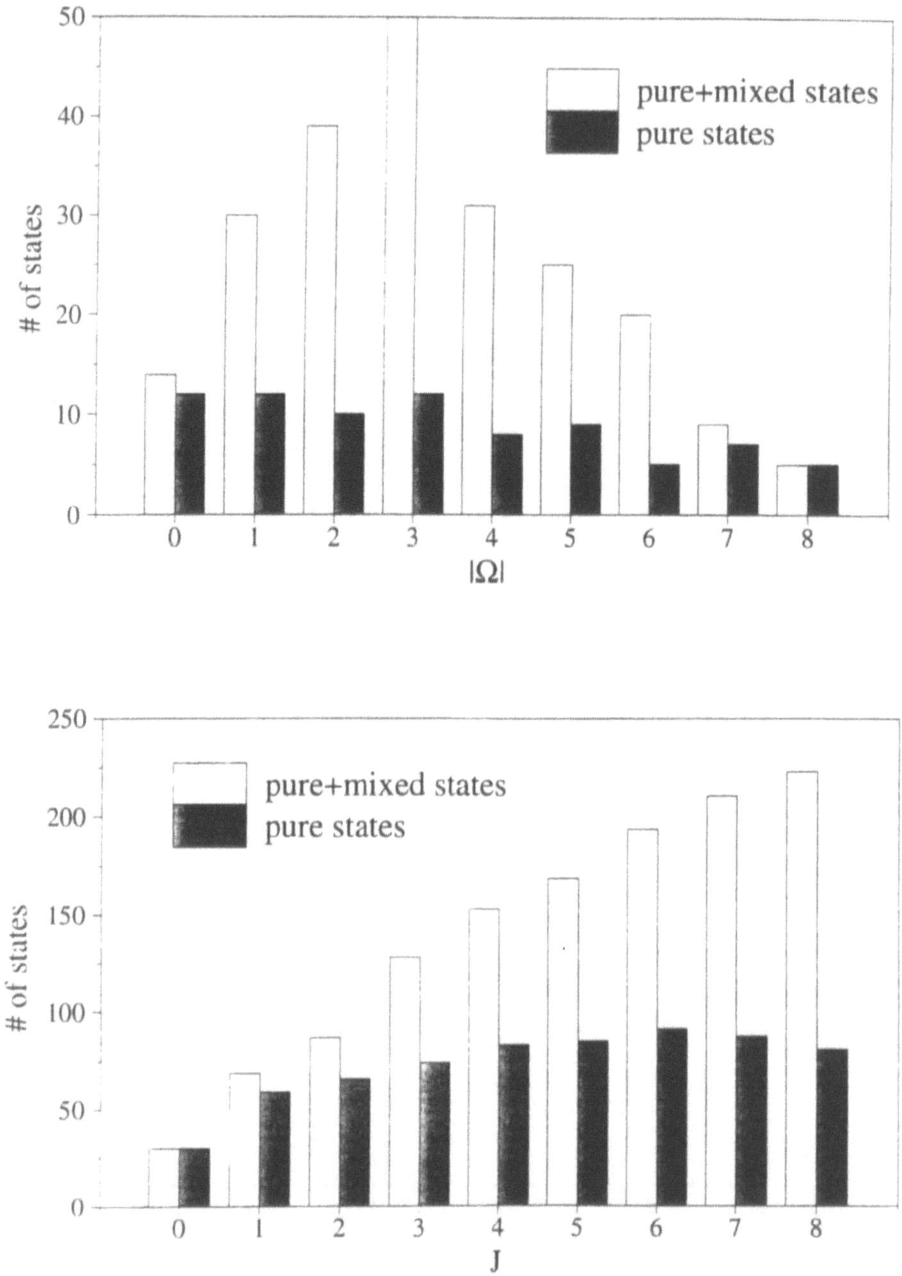

Figure 6. *Upper part.* Distributions of states with a given mean value of helicity and states with large projection on a given $|\Omega|$, calculated with the DGR approximation for $J = 8$ in the energy window reported in the text as a function of $|\Omega|$. *Lower part.* Distributions of states in the chosen energy window and of 'pure' helicity states calculated with the DGR approach, as a function of the total angular momentum J.

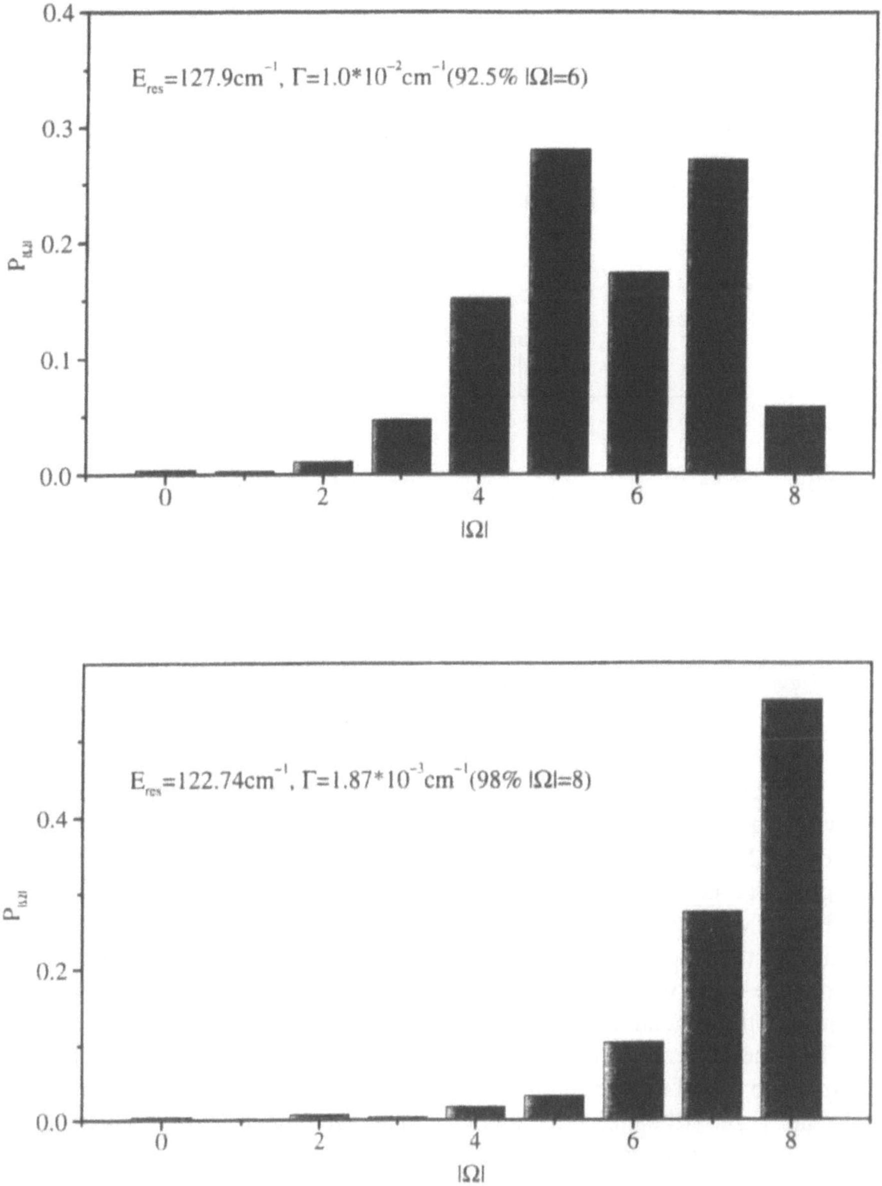

Figure 7. Calculated probability of fragmentation in the final dimer state $|\Omega|$ (summed over the final j) for two metastable state obtained for $J = 8$.

distributions of lifetimes which we have obtained at $J = 6$ and $J = 8$ are reported in Figure 8, for three different values of the tumbling quantum number: we can see that the lifetimes increase when going from low to high $|\Omega|$ in a rather monotonic

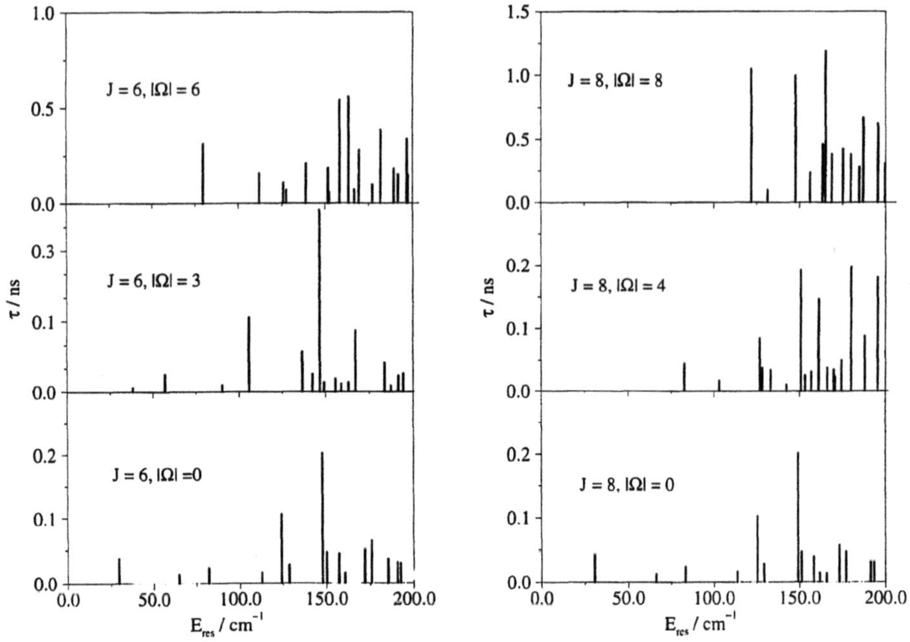

Figure 8. Calculated lifetimes (in ns) in the HD approximation as a function of the resonance energy, reported for $J = 6$ and $J = 8$ and for different values of $|\Omega|$.

way. Further, the resonances are distributed over the whole energy range for low $|\Omega|$ while, for instance, at $J = 8$ and $|\Omega| = 8$ there are resonances only beyond 100 cm^{-1}. Moreover, we can observe that the number of resonances in the present energy range is reduced for large $|\Omega|$ with respect to the smallest, at $J = 8$, confirming what we have shown in Figure 7 for the case of DGR calculations. The shifting of the resonances toward higher energy with $|\Omega|$ seems to be due to the fact that the states with $j < |\Omega|$, with the smallest mean energy, cannot be included because of the angular momentum restrictions.

The general trend of the average lifetimes for each J value are in Figure 9, in comparison with the close coupling calculations and the DGR calculations in which only the 'pure' $|\Omega|$ states were included. The agreement with the close coupling results is quite good, even matching the single resonances of the distribution, with the discrepancy at $J = 2$ that is basically due to just one resonance. We also note that the DGR results for 'pure' states have the same behaviour as the HD results, with the two trends of the average lifetimes running parallel after $J = 4$.

Thus, we can attribute the increase of the lifetimes to the orientational effect even in this case, confirming the DGR picture. Given the good comparison between different methods, we have to remark that the vibrational predissociation of Ar_3^+ is indeed a very fast process and that is only slightly altered by increasing the total angular momentum. If we take the chosen energy range as representative of the

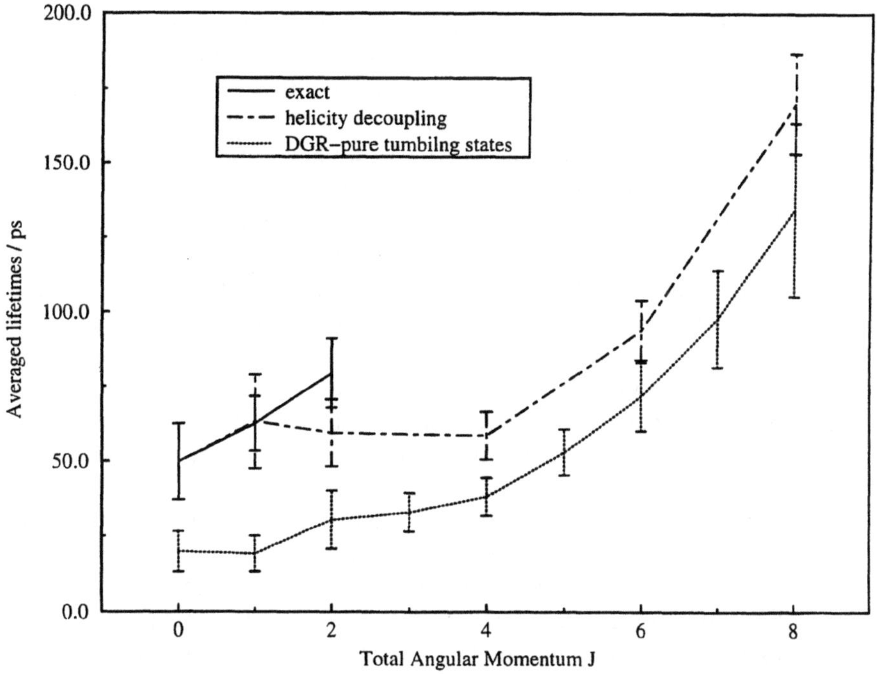

Figure 9. Average lifetime (in ps) calculated with CC (exact), HD and DGR approximations, as a function of the total angular momentum. The DGR results are obtained only from the 'pure' tumbling states.

metastable energy windows and further consider that there will be a limited range of $|\Omega|$ with metastable states in a given energy range, we can conclude that the orientational effect cannot be the main cause of the observed metastability because, as we have seen, the average lifetimes increase on a linear time scale, very far from the experimental time windows.

4.2. THE ROLE OF THE CENTRIFUGAL BARRIER

As we have already mentioned, in order to carry out dynamical calculations to test the efficiency of the rotational barrier in delaying the fragmentation process, it is necessary to consider very high J values. From the discussion in the previous subsection, the helicity decoupling method appears to be the most appropriate to study the dissociative dynamics in the high J range. In fact, the removal of the $|\Omega|$ degeneracy produces equations with complexity independent of J; in any case, as $|\Omega|$ is conserved, there will be a set of equations to solve for each value of $|\Omega|$. The results reported in Figure 8 show that the resonance energies are shifted to higher energy if the tumbling quantum number increases: as the effect of the barrier should be heightened for metastable states with energy close to zero, it is reasonable to restrict the range of tumbling quantum numbers to the lowest values. Furthermore, as has been already reported [41] in the DGR calculations, in which the $|\Omega|$ values are mixed, there is no remarkable difference between the lifetimes obtained from the 'pure' $|\Omega|$ states and from the states classified according to their mean value of $|\Omega|$, or for the lowest $|\Omega|$. As we have already observed, for $|\Omega| = 0$ the predissociative states contain a large fraction which is 'pure' with respect to the tumbling quantum number in the DGR results reported in Figure 7. For such reasons, we decided to carry out HD calculations at very high J values only for $|\Omega| = 0$ [40], where the coupling with the other quantum numbers is reduced and the lifetimes are, on the average, similar to those which can be obtained when including the mixing of different quantum numbers.

The distributions of lifetimes obtained from the HD calculations are reported in Figure 10: the top panel reports the results obtained for the rotational predissociation, where the metastable states with a large projection over the ground vibrational state of the dimer. The calculations have been carried out for $J = 30$, 40 and 50 in the energy range from 0 to 20 cm^{-1}: we can observe that, starting from $J = 30$, there are states with long lifetimes, with a fraction of them falling within the experimental metastable windows. It is also interesting to note the resonances with a very long lifetime (longer than 1 s) that correspond to metastable states that are completely trapped by the rotational barrier: the energy range of those states becomes broader with the increase of J. Thus, the rotational barrier is effective only in a limited range of energy: beyond that energy range, the lifetimes of the dissociative state are very short, similar to the values obtained for lower J. We can conclude that the rotational barrier is able to delay the rotational predissociation process. The same is true for the vibrational predissociation, as we can see from the bottom panel of Figure 10: here the energy range of the calculation is necessarily wider, as we are dealing with metastable states that belong to the $v = 1$ vibrational state of the dimer, which are locally bound with respect to the $v = 1$ asymptote and have the spacing of real bound states, associated with the frequencies of the stretching and bending motion of the cluster. The figure indicates that the distributions for $J = 30$, 40 and 50 show resonances with very short lifetimes, but comparison with the upper panel indicates that none of the metastable states fall in the energy range of the barrier. As it is easy to predict the way the position of the resonances shift with the energy, we were able to find some states (for $J = 46$, 48 and 49) which were both vibrationally excited and which had very long lifetimes, showing that the vibrational predissociation process can also be delayed by the rotational barrier. Such results are not surprising if we consider that the the off-diagonal vibrational couplings depend in the long-range part on R^{-6} while the barrier depends on R^{-2} and, therefore, more strongly influences the dissociation process at larger internuclear distances.

The rotational barrier also has an interesting effect on the final distributions of the dimer fragment, as we can see from Figure 11, where the final rotational distribution of the fragments are reported for resonances falling in the two different

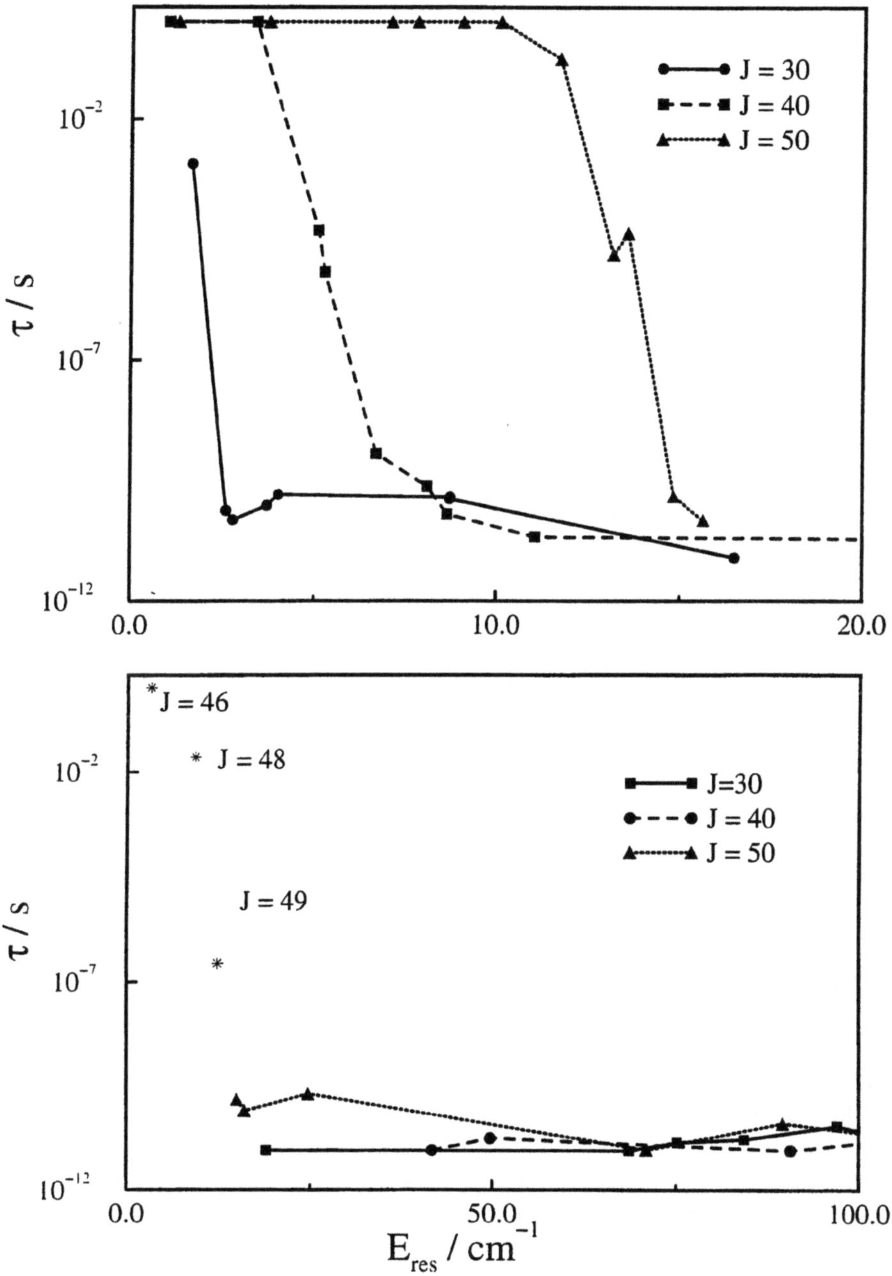

Figure 10. Computed metastable lifetimes using the HD approximation for $|\Omega| = 0$, for different J values. *Top.* Purely rotational predissociation. *Bottom.* Vibrational predissociation.

J regions that we have identified in the previous calculations. The top part shows the initial and final distribution of a particular metastable state, calculated for $J = 0$ and $J = 20$ and with the same stretching – bending quantum number: we can see that the initial distributions are exactly the same, even if the metastable states have different energy because of the shift of their positions with the increase of J. The final distributions also look similar and we can note that there is a finite probability of getting the dimer product in any of the open rotational states. The bottom part of Figure 11 shows, instead, the final rotational distribution obtained from calculations at very high total angular momenta: the figure indicates that there is a significant probability of obtaining the dimer in the lower lying rotational states and that it becomes exponentially larger as J increases. Such observation leads to a description of the dissociation, in presence of the rotational barrier, as a process where the two fragments tend to be separated with the maximum amount of kinetic energy associated to the relative motion, predicting for such observables an energy range of the order of 0.1 cm^{-1}.

A good description of the rotational predissociation process was already obtained by us using the *Angular Adiabatic* approximation [28], within the rigid rotor surface discussed in Section 2. Although the results cannot be directly compared with the three-dimensional calculations we have just discussed, such a simplified approach can give interesting details on the dynamics of the fragmentation process. We have already noted that the vibrational frequency of the dimer is much larger than the frequencies of the other mode: it is also interesting to report that the average frequency of the bending motion, described by the coordinate R, is about 40 cm^{-1} while the Ar_2^+ rotational constant is only 0.15 cm^{-1} and the rotational constant of the $\hat{\mathbf{R}}$ motion is 0.04 cm^{-1}, when taking $\bar{R} = 7.7a_0$. Thus it becomes reasonable to assume that the bending motion of the cluster, which is controlled by the two rotational constants, can be separated from the fast stretching motion and can be treated adiabatically. Therefore, the conditions under which one should apply the angular adiabatic approximation are possibly satisfied in this situation.

The adiabatic potentials obtained from the solution of Equation (27) are reported in Figure 12: the rigid-rotor surface supports seven bound stretching states that show two minima for $\gamma = 0°$ and $\gamma = 90°$ with a barrier around $\gamma = 60°$, reflecting the topology of the rigid rotor PES. The adiabatic potentials have an embedded structure and one can predict that there will be a strong mixing of bending levels of the different stretching states: therefore, the assumption of separability of motions is not working perfectly and needs some refinements. We have also applied a first order correction to this treatment, allowing the stretching–bending states to be mixed in a *Configuration Interaction* picture and the calculations we have done reproduced results which are qualitatively similar.

The solution of the bending equation has been obtained including 60 rotational states in the expansion (31) for $J = 0$ and the same rotational levels have also been employed for very larger J, leading to a basis set composed of thousands of functions. From the shape of the adiabatic potentials, the bending mode of the cluster can

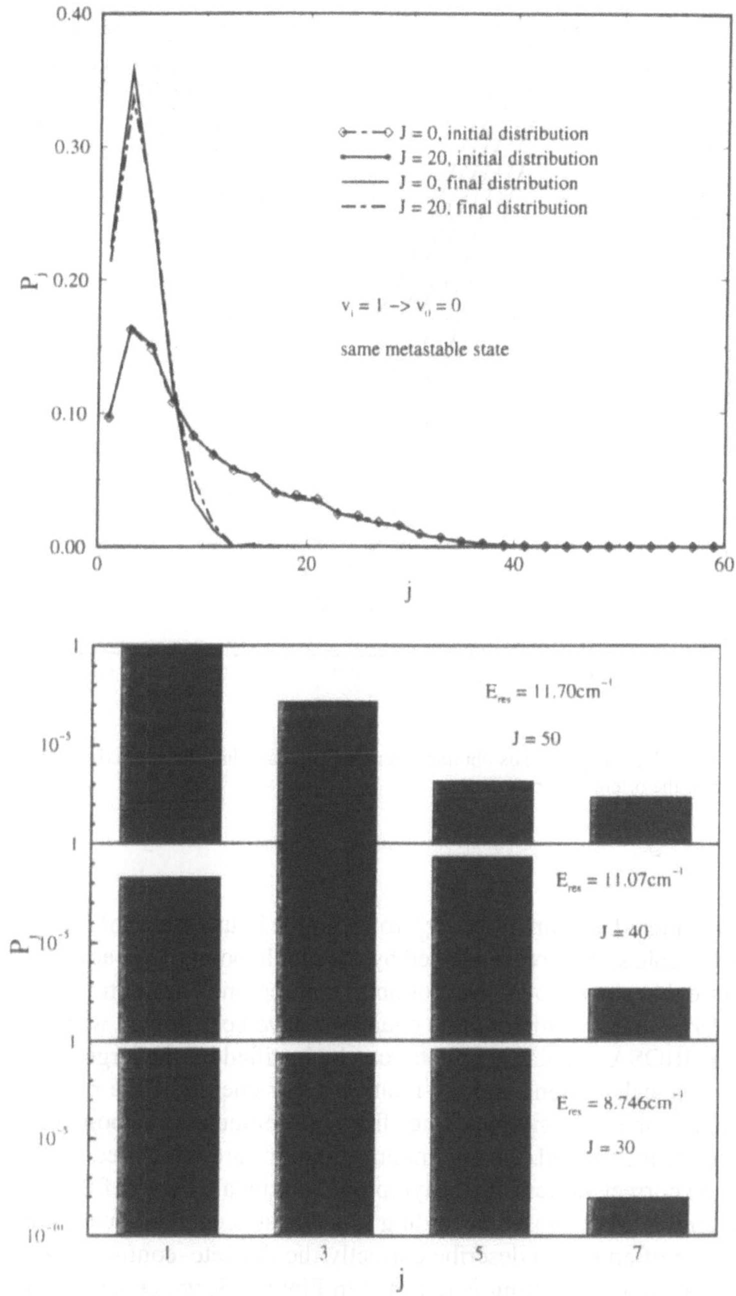

Figure 11. Upper part. Initial and final rotational state probabilities for two metastable states with the same index, calculated with the HD approximation for $J = 0$ and $J = 20$ at $|\Omega| = 0$. *Lower part.* Final rotational state probability as a function of the rotational state, for three different metastable states calculated for $J = 30$, 40 and 50.

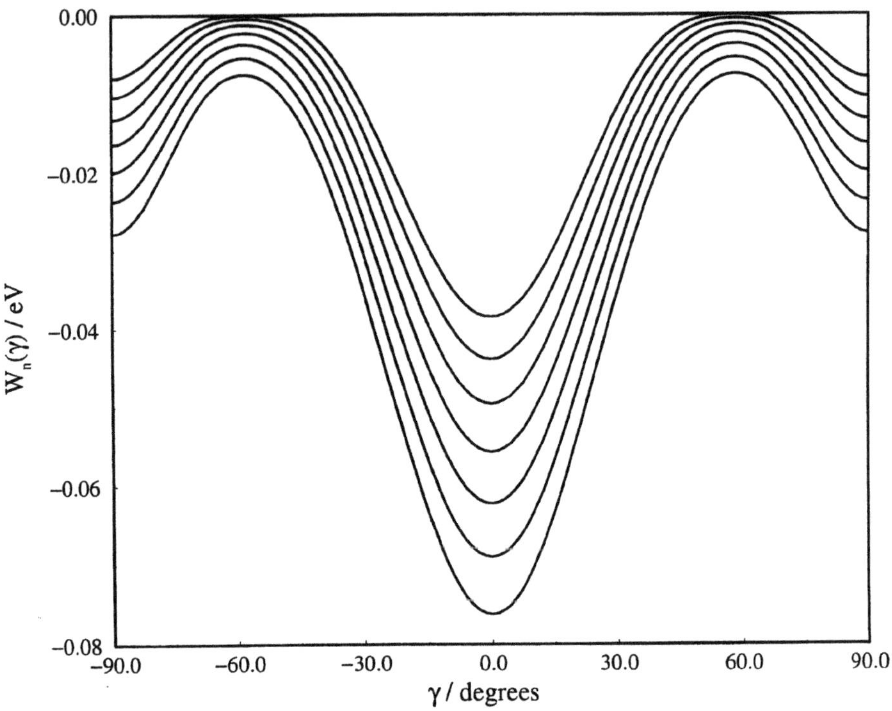

Figure 12. Adiabatic potentials obtained from the angular adiabatic approximation in eV, as a function of the orientation γ.

be seen as a hindered motion of the Ar_2^+ rotor for the bound states of the system but also the metastable states are influenced by the saddle point of the adiabatic potentials, a feature that slows down the motion in that region. The continuum *Rotation Infinite Order Sudden Approximation* equations have been solved with $\hat{l} = J$: the choice of the RIOSA angular parameter can be justified by the large total angular momenta of the calculations and the relatively low energy of the predissociative states, leading to the most simple choice from the computational point of view. As we have already mentioned, the continuum equations are integrated *on the energy shell*, with the correct values of the asymptotic rotational energy of the dimer. The parametric angles were chosen according to a Gauss–Legendre quadrature, using a large number of angles to describe correctly the discrete–continuum couplings. An example of such a coupling is reported in Figure 13, which shows the potential coupling for a metastable state obtained at $J = 20$. The coupling shows that exponentially larger overlap between the discrete and continuum wavefunctions is obtained at the saddle point region, where, as underlined above, the bending motion is slower. Therefore, the AA approximation calculations give a description of the

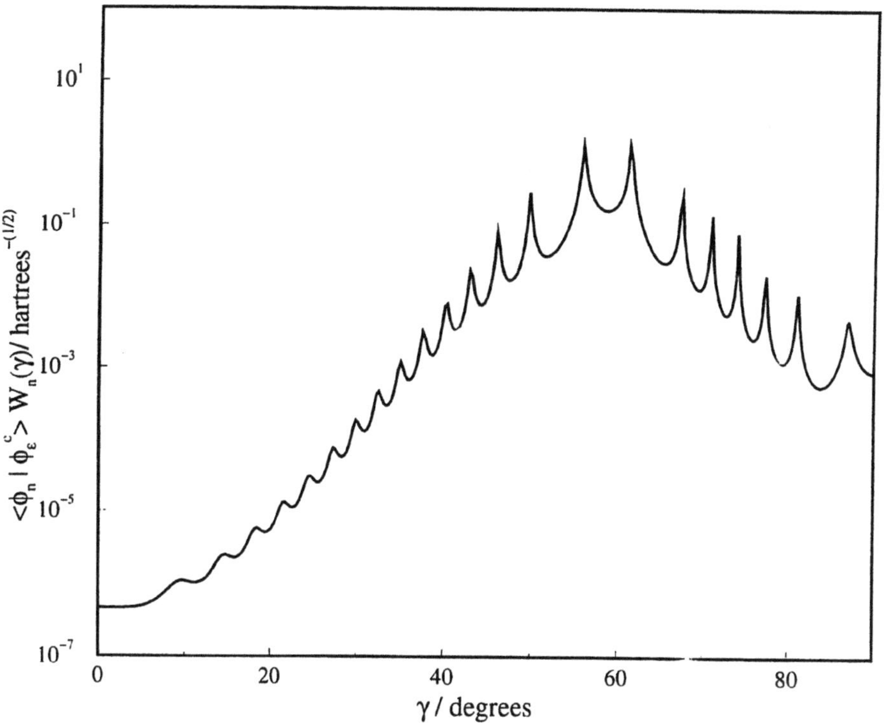

Figure 13. Discrete–continuum potential coupling calculated in the AA approximation for a metastable state with $J = 20$ as a function of γ.

fragmentation process in which the argon atom leaves the cluster from preferential orientations.

The distribution of lifetimes are reported in Figure 14: we note the same trend as in the previous HD calculations, with a very fast fragmentation process at $J = 0$ caused by the strong coupling between the dissociative motion and the hindered rotation of the dimer. Furthermore, as J increases, the results show distributions of lifetimes that spread over a large time scale and a clear tendency of the lifetimes to decreases as the resonance energy increases, reaching the order of magnitude obtained for $J = 0$ at the largest resonance energy. From these calculations, the metastability starts at $J = 20$, where there are some resonances whose lifetimes are comparable to the metastable windows, while the HD results gave before a similar behaviour for $J = 30$. At the same time, we are able to note that the increase of the total angular momentum produces larger energy ranges for which the predissociative states have lifetimes extremely long. Thus, the results previously discussed are substantially confirmed by the more simplified AA approximation.

The different approximate methods discussed in this subsection therefore point to the efficiency of the rotational barrier in producing metastable states with long

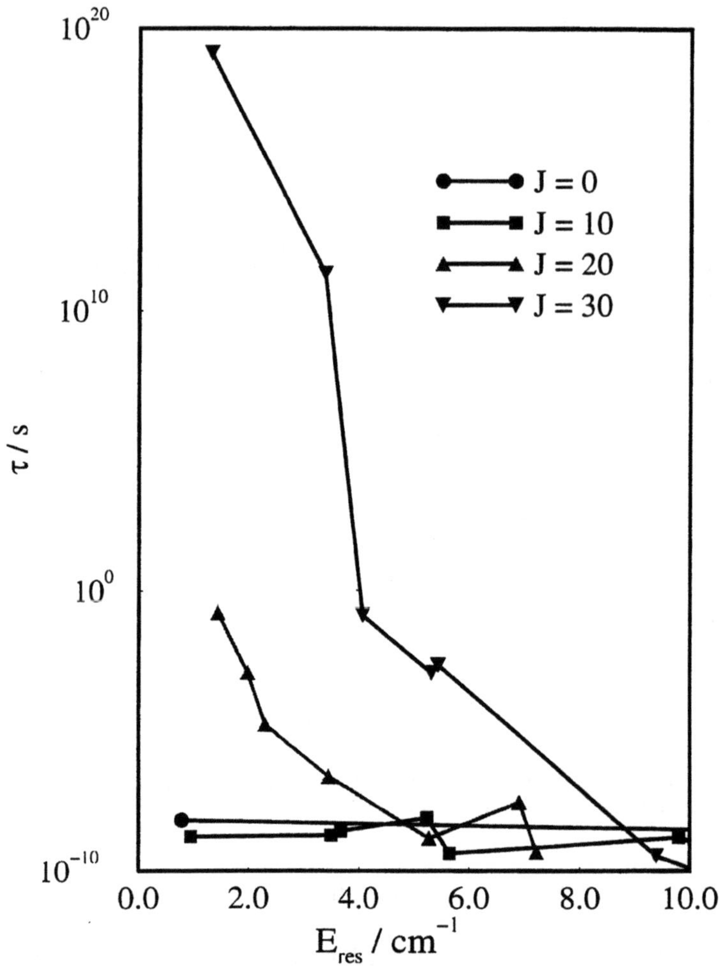

Figure 14. Distributions of lifetimes (in s) as a function of the resonance energy (in cm^{-1}), obtained from AA calculations at three different values of the total angular momentum J.

lifetimes. In fact, the quite strong interaction between the argon atom and the charged dimer Ar_2^+ can support bound and shape resonance states up to $J = 100$, and comparison with the present results shows that a sufficiently large density of predissociative states with lifetimes in the metastable time window can be predicted by this mechanism, taking into account that the total angular momentum distributions have been calculated for a system for which very large J values are to be expected.

5. Conclusions

In the present work we have studied the metastability features of Ar_3^+ in its electronic ground state. In particular, we have analyzed the effects of the total angular momentum of the system on the its fragmentation dynamics, including all the relevant coupling between the dissociative motion and the remaining modes of the clusters. We found with our calculations that the vibrational motion of the charged dimer can be considered almost unchanged by the interaction with the neutral argon atom, allowing the separation between vibrational predissociation and rotational predissociation resonances. The results show that the fragmentation process is very efficient at $J = 0$, due to the strong anisotropy and the large couplings between different vibrational states of Ar_2^+.

In order to extend the calculations to high J values, different theoretical methods have been tested and compared to obtain the most accurate description of the process, to avoid the complexity related to the different projected components of the rotational level that have to be included for large values of J. The results show that the *Helicity Decoupling* approximation is the most suitable for this system when very high total angular momenta are involved in the calculations.

The dependence of the fragmentation process on the total angular momentum has been analyzed and the results show that longer lifetimes can be obtained when J increases, due to at least two different phenomena:

(i) The 'orientational effect', that acts for $J \neq 0$ in the vibrational predissociation process, for the particular angular dependence of the off-diagonal vibrational coupling which is negligible in the perpendicular region when compared to the linear one. Such effect produces a small homogenous increase of the lifetimes and it turns out not to be enough to produce metastability.

(ii) The effect of the centrifugal barrier, that starts at about $J = 30$ and produces lifetimes spanning different orders of magnitude, going from lifetimes of the order of 1 s for energy resonances close to zero to tenths of picoseconds for the larger resonance energies. Both rotational and vibrational redissociation processes are influenced by the rotational barrier; this effect appears to be related to the resonance energy more than to the particular fragmentation process or to the properties of the initial metastable state. The barrier also influences the formation of the final fragments where the tendency to dissociate with the largest available kinetic energy is observed.

According to the reported calculations, the observed metastability can be ascribed to the latter effect, once it is possible to produce the ionized argon trimer with a distribution of total angular momenta that favours high J values. Finally, it is worth to mention that we have also carried out scattering calculations on this systems [42] in which all the contributions coming from different total angular momenta are included. The ensuing cross sections have shown Feshbach resonances in the vibrational inelastic channels, pointing at the role played by the closed channels in this system.

Acknowledgements

We would like to thank G. Delgado-Barrio, P. Villarreal, S. Miret-Atrés and M. P. de Lara-Castells for their help and active collaboration in carrying out this research. The financial support of the Italian National Research Council (CNR) and of the Italian Ministery of University and Research (MURST) is also acknowledged.

References

1. H. Haberland (ed.): *Clusters of Atoms and Molecules*, Vols. I and II, Springer-Verlag, Berlin (1995).
2. T. D. Märk and O. Echt: in H. Haberland (ed.), *Clusters of Atoms and Molecules*, Vol. II, Springer-Verlag, Berlin (1995).
3. K. Stephan, A. Stamatovic, and T. D. Märk: *Phys. Rev. A* **28**, 3150 (1983).
4. K. Stephan, A. Stamatovic, and T. D. Märk: *Phys. Rev. A* **32**, 1447 (1985).
5. B. K. Whitaker, C. A. Woodward, P. J. Knowles, and A. J. Stace: *J. Chem. Phys.* **93**, 376 (1990).
6. K. Stephan and T. D. Märk: *Chem. Phys. Lett.* **90**, 51 (1983).
7. P. Scheier, A. Stamatovic, and T. D. Märk: *J. Chem.. Phys.* **89**, 295 (1988).
8. E. E. Ferguson, C. R. Albertoni, R. Kuhn, Z. Y. Chen, R. G. Keese, and A. W Castleman, Jr.: *J. Chem. Phys.* **88**, 6335 (1988).
9. A. J. Stace: *J. Chem. Phys.* **85**, 5774 (1986).
10. C. A. Woodward and A. J. Stace: *J. Chem. Phys.* **94**, 4234 (1991).
11. P. Scheier and T. D. Märk: *Phys. Rev. Lett.* **59**, 1813 (1987).
12. U. Buck and H. Meyer: *Phys. Rev. Lett.* **52**, 109 (1984). U. Buck and H. Meyer: *Surf. Sci.* **156**, 275 (1985). U. Buck and H. Meyer: *J. Chem. Phys.* **84**, 4854 (1986).
13. A. Bastida, N. Halberstadt, J. A. Beswick, F. X. Gadea, U. Buck, R. Galouska, and C. Lauenstein: *Chem. Phys. Lett.* **289**, 1 (1995).
14. A. J. Stace: *J. Chem. Phys.* **93**, 6502 (1990).
15. W. R. Wadt: *J. Chem. Phys.* **68**, 402 (1978).
16. H. U. Böhmer and S. D. Peyenmhoff: *Z. Phys. D* **3**, 195 (1986).
17. Z. Y. Chen, B. D. May, and A. W. Castleman, Jr.: *Z. Phys. D* **25**, 239 (1993).
18. F. X. Gadea, J. Savrda, and I. Paidarova: *Chem. Phys. Lett.* **233**, 369 (1994)
19. J. Heißlich and P. J. Kuntz: *Z. Phys. D* **2**, 251 (1986). P. J. Kuntz and J. Valldorf: *Z. Phys. D* **8**, 195 (1988). I. Last and T. F. George: *J. Chem. Phys.* **93**, 8925 (1990). T. Ikegami, T. Kondow, and S. Iwata: *J. Chem. Phys.* **98**, 3038 (1993).
20. N. E. Levinger, D. Ray, M. L. Alexander, and W. C. Linenberger: *J. Chem. Phys.* **89**, 5654 (1988).
21. T. Nagata, J. Hirokawa, and T. Kondow:: *Chem. Phys. Lett.* **176**, 526 (1991). J. A. Smith, N. G. Gotts, J. F. Winkel, R. Hallet, C. A. Woodward, A. J. Stace, and B. J. Whitaker: *J. Chem. Phys.* **97**, 397 (1992).
22. M. J. De Luca and M. A. Johnson: *Chem. Phys. Lett.* **162**, 479 (1988).
23. N. E. Levinger, D. Ray, K. K. Murray, A. S. Mullin, C. P. Shultz, and W. C. Linenberger: *J. Chem. Phys.* **89**, 71 (1988).
24. Z. Y. Chen, C. R. Albertoni, M. Hasegawa, R. Kuhn, and A. W. Castleman, Jr.: *J. Chem. Phys.* **91**, 4019 (1980).
25. F. X. Gadea and M. Amarouche: *Chem. Phys* **140**, 385 (1990). F. X. Gadea: *Z. Phys. D* **20**, 25 (1991). F. X. Gadea and F. Le Quéré: *J. Chem. Phys.* **102**, 7830 (1995).
26. A. Bastida and F. X. Gadea: *Chem. Phys.* **209**, 291 (1996).
27. J. T. Mosley, R. P. Saxon, A. Huber, P. C. Cosby, R. Abouof, and M. Tadjeddine: *J. Chem. Phys.* **67**, 2860 (1984).

28. E. Buonomo, F. A. Gianturco, G. Delgado-Barrio, S. Miret-Artés. and P. Villarreal: *J. Chem. Phys.* **67**, 6472 (1994).
29. H. U. Böhmer and S. D. Peyerimhoff: .*Z. Phys. D* **11**, 239 (1989).
30. F.A. Gianturco and M. P. de Lara-Castells: *Chem. Phys.* **208**, 25 (1996). F. A. Gianturco and M. P. de Lara-Castells: *Int. J. Quantum Chem.* **60**, 593 (1996).
31. C. Lee, W. Yang, and R. G. Parr: *Phys. Rev. B* **37**, 785 (1988).
32. M. P. de Lara-Castells, P. Villarreal, G. Delgado-Barrio, S. Miret-Artés, E. Buonomo, and F. A. Gianturco: *Chem. Phys. Lett.* **242**, 336 (1995).
33. M. J. Frisch, G. W. Trunks, M. B. Schlegel, P. M. W. Gill, B. J. Johnson, M. W. Wong, J. B. Foresman, M. A. Robb, M. Head-Gordon, E. S. Replogle, R. Gamperts, J. L. Andres, K. Raghavachari, J. S. Binkley, G. Gonzales, R. L. Martin, D. J. Fox, D. J. De Frees, J. Baker, J. J. P. Stewart, and J. A. Pople: GAUSSIAN92/DFT, Gaussian, Pittsburg (1993).
34. U. Fano: *Phys. Rev.* **12**, 1866 (1961).
35. O. Roncero, J. A. Beswick, N. Halberstadt, P. Villarreal, and G. Delgado-Barrio: *J. Chem. Phys.* **92**, 3348 (1991).
36. J. M. Launay: *J. Phys.* **B9**, 1823 (1976).
37. See, for instance, F. A. Gianturco: *The Transfer of Molecular Energies by Collisions*, Springer-Verlag, Berlin (1979).
38. J. A. Beswick and J. Jortner: *Adv. Chem. Phys.* **47**, 363 (1981).
39. F.A. Gianturco, A. Palma, P. Villarreal, G. Delgado-Barrio, and O. Roncero: *J. Chem. Phys.* **87**,1054 (1987).
40. E. Buonomo, F. A. Gianturco, M. P. de Lara-Castells, G. Delgado-Barrio, S. Miret-Artés and P. Villarreal: *J. Chem. Phys.* **106**, 1718 (1997).
41. E. Buonomo, F. A. Gianturco, M. P. de Lara-Castells, G. Delgado-Barrio, S Miret-Artés, and P. Villarreal: *Chem. Phys.* **218**, 71 (1997).
42. E. Buonomo, F. A. Gianturco, M. P. de Lara-Castells, G. Delgado-Barrio, S. Miret-Artés, and P. Villarreal: *J. Chem. Soc. Faraday Trans.* **93**, 901 (1997).

Molecular Engineering **7**: 219–230, 1997.

Sampling the Initial Conditions for Quasiclassical Trajectory Studies of Vibrational Predissociation Dynamics

G. DELGADO-BARRIO,* A. GARCÍA-VELA, C. GARCÍA-RIZO,
M. I. HERNÁNDEZ and P. VILLARREAL
Instituto de Matemáticas y Física Fundamental, C.S.I.C., Serrano 123, 28006 Madrid, Spain.

(Received: 12 September 1996; in final form: 17 June 1997)

Abstract. A method is proposed for selecting initial conditions in order to study the classical dynamics of vibrational predissociation of van der Waals (vdW) clusters. The method starts from the quantum initial state of the system, which is used to sample and weight the initial positions of the different modes. The initial values of the associated momenta are calculated in such a way that they correspond with the total energy and angular momentum of the system initial state, at which the classical trajectory simulation is to be carried out. An application to the case of Cl_2—He_2 is presented and discussed.

Key words: Sampling initial conditions, vibrational predissociation, van der Waals clusters.

1. Introduction

In the last decade the experimental time-resolved pump-probe techniques have experienced an enormous advance in its application to the field of chemical reaction dynamics [1–11]. This type of experiment makes it possible to follow in real time the course of a reaction occurring in the picosecond or even in the subpicosecond time scale. Detailed information on the time-evolution of a chemical process provides a great deal of physical insight into the dynamics involved and how the reaction proceeds. In addition, time-resolved data may complement other type of dynamical information like that obtained by frequency domain spectroscopic experiments.

Cluster systems composed of several rare-gas atoms weakly bound to a molecule acting as a chromophore present several interesting features. By gradually increasing the cluster size it is possible to establish a bridge between the gas-phase limit and the condensed-matter regime. This allows one to investigate fundamental questions such as the mechanisms of energy transfer from the electronically or vibrationally excited chemical bond (or bonds) to the solvent, the influence of the weak solvation interactions on the molecule dissociation, and the effect of caging and recombination of the chemical impurity induced by the solvent [8].

Clusters with a halogen diatomic molecule weakly bound to one or more rare-gas atoms have been extensively investigated by means of frequency-domain experi-

* Author for correspondence.

ment [2–15]. The pioneering experiments of Levy and co-workers on I_2—He_n (n = 1–3) [16] and I_2—Ne_n (n = 1–7) [17] measured binding energies, predissociation linewidths and lifetimes, and spectral band shifts in order to explore the structure, possible coordination effects, and energy transfer mechanisms in these systems. The dynamics of complexes with diatomic molecules different than I_2, like Br_2—Ne_n (n = 1–3) [18], ICl—Ne_n (n = 1–5) [19], Cl_2—Ne_n (n = 1–3) [20], and Cl_2—He_n (n = 1–2) [21], have also been investigated. The characterization of rare gas–halogen potentials has also been the subject of several experiments [22–24].

The great deal of experimental work mentioned above has generated a rich body of dynamical information in the frequency domain. Consequently, the theoretical analysis of this information has been carried out mostly using time-independent methodologies. Exact close-coupling calculations on several triatomic systems have been reported in the literature [25–27]. Unfortunately, exact quantum methods are limited by the fact that the computational effort required grows very rapidly with the number of system modes. Several approximate methods have been suggested, including Golden Rule approaches [27, 28] and a variety of sudden approximation, [29–32]. The appearance of time-domain experimental data, along with the development of techniques for the efficient solution of the time-dependent Schrödinger equation [33–36] gave a great impetus to the use of time-dependent approaches. Exact [37–40] and approximate [41–47] time-dependent methods have been applied to van der Waals clusters, most of them triatomic and tetraatomic systems. However, a quantum dynamical treatment of a four-atom system including the full dimensionality, even for zero total angular momentum, becomes extremely costly.

Classical Molecular Dynamics (MD) simulations make the problem tractable [48–50], and provide at the very least a qualitative insight into the dynamics. In the classical scheme, prior to the study of the dynamics, an initial state must be specified in the form of initial conditions of the classical trajectories to be integrated. The aim of this work is to propose a technique to select such initial conditions and to apply it to the $Cl_2(B, v)$—He_2 cluster.

The organization of the paper is as follows. In Section 2 we briefly describe the potential surface and the quasiclassical methodology applied. Section 3 describes the model proposed to obtain the initial conditions. Finally, Section 4 presents and discusses distributions of the different modes of $Cl_2(B, v)$—He_2.

2. Quasiclassical Methodology

The VP process in the $Cl_2(B, v)$—He_2 cluster occurs upon laser excitation from the ground state to a well-defined vibrational level v ($v \neq 0$) in the electronic B state of the Cl_2 chromophore. The metastable state $Cl_2^*(B, v)$—He_2 prepared in this way decays to a dissociation continuum giving the products $Cl_2(B, v' = v - m) + 2$He, where $m \geq 2$. Such metastable state is the one to be reproduced by the initial conditions in a classical calculation.

2.1. CLASSICAL HAMILTONIAN AND POTENTIAL-ENERGY SURFACE

In the present work bond coordinates $(\mathbf{r}, \mathbf{R_1}, \mathbf{R_2})$ have been used, where $\mathbf{r}$ is the vector associated with the Cl—Cl bond, and $\mathbf{R_1}$, $\mathbf{R_2}$ are the vectors between the Cl_2 center of mass and the two He atoms, respectively. In these coordinates the classical Hamiltonian can be expressed as

$$H = \frac{\mathbf{P_r^2}}{2\mu_{Cl_2}} + \frac{\mathbf{P_1^2}}{2\mu_{Cl_2\text{—}He}} + \frac{\mathbf{P_2}}{2\mu_{Cl_2\text{—}He}} + \frac{\mathbf{P_1} \cdot \mathbf{P_2}}{2m_{Cl}} + V(\mathbf{r}, \mathbf{R_1}, \mathbf{R_2}), \tag{1}$$

where $\mathbf{P_r}$, $\mathbf{P_1}$, and $\mathbf{P_2}$ are the conjugate momenta associated with $\mathbf{r}$, $\mathbf{R_1}$, and $\mathbf{R_2}$, respectively, and $\mu_{Cl_2} = m_{Cl}/2$, $\mu_{Cl_2\text{—}He} = 2m_{Cl}m_{He}/(2m_{Cl} + m_{He})$ are the corresponding reduced masses. By choosing a body-fixed frame with the z-axis always coinciding with the $\mathbf{r}$ direction, and using a polar coordinate representation where $\mathbf{R_i} = (R_i, \theta_i, \phi_i)$, the Hamiltonian in the case of total angular momentum $\mathbf{J} = 0$ takes the more specific form

$$\begin{aligned} H^{(J=0)} &= \frac{1}{2\mu_{Cl_2}}\left[P_r^2 + \frac{\mathbf{j}^2}{r^2}\right] + \frac{1}{2\mu_{Cl_2\text{—}He}}\left[P_{R_1}^2 + \frac{\mathbf{l}_1^2}{R_1^2}\right] + \\ &\quad + \frac{1}{2\mu_{Cl_2\text{—}He}}\left[P_{R_2}^2 + \frac{\mathbf{l}_2^2}{R_2^2}\right] + \frac{\mathbf{P_1} \cdot \mathbf{P_2}}{2m_{Cl}} + \\ &\quad + V_{Cl_2}(r) + V_{Cl_2\text{—}He}(r, R_1, \theta_1) + V_{Cl_2\text{—}He}(r, R_2, \theta_2) + \\ &\quad + V_{He\text{—}He}(R_1, R_2, \cos\gamma), \end{aligned} \tag{2}$$

$\mathbf{j}$, $\mathbf{l_1}$ and $\mathbf{l_2}$ being the angular momenta associated with $\mathbf{r}$, $\mathbf{R_1}$, and $\mathbf{R_2}$, respectively. Other definitions used in Equation (2) are $\cos\gamma = \sin\theta_1 \sin\theta_2 \cos\phi + \cos\theta_1 \cos\theta_2$, $\phi = \phi_1 - \phi_2$, and the expanded expression of $\mathbf{P_1} \cdot \mathbf{P_2}$ is

$$\begin{aligned} \mathbf{P_1} \cdot \mathbf{P_2} &= P_{R_1} P_{R_2} \cos\gamma - \frac{\cos\phi}{R_1 R_2 \sin\theta_1 \sin\theta_2} P_\phi^2 + \\ &\quad + \frac{\cos\theta_1 \cos\theta_2 \cos\phi + \sin\theta_1 \sin\theta_2}{R_1 R_2} P_{\theta_1} P_{\theta_2} + \\ &\quad + \frac{\sin\theta_1 \cos\theta_2 \cos\phi - \cos\theta_1 \sin\theta_2}{R_2} P_{R_1} P_{\theta_2} + \\ &\quad + \frac{\cos\theta_1 \sin\theta_2 \cos\phi - \sin\theta_1 \cos\theta_2}{R_1} P_{R_2} P_{\theta_1} - \frac{\sin\theta_1 \sin\phi}{R_2 \sin\theta_2} P_{R_1} P_\phi - \\ &\quad - \frac{\sin\theta_2 \sin\phi}{R_1 \sin\theta_1} P_{R_2} P_\phi - \frac{\cos\theta_1 \sin\phi}{R_1 R_2 \sin\theta_2} P_{\theta_1} P_\phi - \frac{\cos\theta_2 \sin\phi}{R_1 R_2 \sin\theta_1} P_{\theta_2} P_\phi. \end{aligned}$$

In order to integrate the trajectories it is necessary to solve the Hamilton equations corresponding to the above Hamiltonian.

Regarding the potential-energy surface for the B state of the Cl_2—He_2 cluster, an addition of atom–atom interactions

$$V = V_{Cl_2} + V_{Cl_2\text{—}He_1} + V_{Cl_2\text{—}He_2} + V_{He_1\text{—}He_2}.$$

Table I. Morse potential parameters used in this work.

	D (cm^{-1})	α (Å^{-1})	R_{eq} (Å)
Cl—Cl	3176.4	2.345	2.435
Cl—He	16.2	1.5	3.8
He—He	7.61	2.126	2.963

was used.

The Cl_2 interaction potential, is described by a Morse analytical form with the parameters reported in Ref. [51]. The $V_{Cl_2\text{—}He_i}$ ($i = 1, 2$) potentials were expressed as a sum of two Cl—He$_i$ interactions,

$$V_{Cl_2\text{—}He_i} = V_{Cl\text{—}He_i} + V_{Cl\text{—}He_i},$$

where each $V_{Cl\text{—}He_i}$ interaction is also described by a Morse function. The corresponding Morse parameters were taken from the potential of Ref. [22]. Finally, a Morse function was also used for the He—He interaction whose parameters were obtained by fitting the more complicate analytical form of Aziz and Slaman [52]. All the Morse potential parameters are listed in Table I.

3. Initial Conditions

As discussed above, the first step in studying the dynamics of the cluster is to obtain the initial conditions for each classical trajectory. From a methodological point of view, a classical distribution for the initial state could be used for the sake of consistency. However, the goal in this work is not to test the applicability of Classical Mechanics, but to describe the dynamics of a system with six coupled degrees of freedom as realistically as possible. It is well known that for the ground states the classical and quantal distributions are very different. It is expected that the quantum description will be the most realistic one. In this sense, to introduce the proper quantification in the initial state we start with a quantum calculation of the Cl_2—He_2 complex, including the full dimensionality (six degrees of freedom) for $\mathbf{J} = 0$. This calculation employs a variational method described elsewhere [53]. A diabatic separation of the Cl_2 vibration is assumed, but the full coupling between all the vdW modes is taken into account in the variational calculation. The diabatic separation is justified because of the frequency mismatch between the diatomic and the vdW vibrations. From the resonance wave function calculated in this way, the quantum-mechanical probability distributions for all the modes are obtained, and used to weight the initial positions selected by random sampling [53(b)].

In the following, all the discussion is referred to the $v = 13$ vibrational level of Cl_2. In the scheme adopted here to select the initial conditions of the r vibration, this mode is separated from the vdW ones, similarly as in the variational calculation

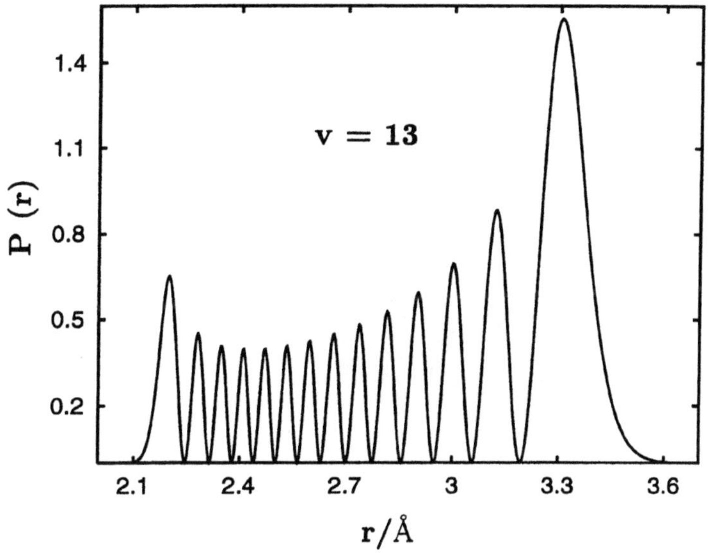

Figure 1. Quantum probability distribution of the r mode.

of the initial resonance state. The quantal probability distribution corresponding with r is displayed in Figure 1. Using this distribution to select and weight initial positions of r, the associated initial momenta can be calculated by means of

$$P_r = \pm\sqrt{2\mu_{Cl_2}[E_v - V_{Cl_2}(r)]},$$

where E_v is the energy of the diatomic vibrational level v, and the corresponding sign of these momenta is further selected at random. The resulting initial positions and corresponding momenta for the Cl_2 vibration are represented in Figure 2. In this drawing, although only the border is meaningful, the non-zero probability regions have been shadowed in order to stress the nodal structure. As can be seen, both figures reflect the same nodal pattern, that of the $v = 13$ level.

Now, initial distributions of positions and momenta for the vdW vibrations still remain to be obtained. Figures 3 and 4 show the quantum probability distributions of the modes R_i, θ_i $(i = 1, 2)$ and γ, and Figure 5 displays the quantum rotational distributions associated with the Cl_2 (**j**) and the two vdW bonds ($\mathbf{l}_i$, i = 1, 2). These distributions are employed to sample initial positions of the vdW modes, values of the quantum numbers j and l_i (i = 1, 2). Only the values fulfilling the constraint $|l_1 - l_2| < j < l_1 + l_2$ are accepted in order to ensure a given total angular momentum (**J** = 0 in the case study here) for the initial conditions.

The initial angular momenta P_{θ_1}, P_{θ_2}, and P_ϕ are then obtained by solving the following system of equations,

$$\mathbf{l}_1^2 = P_{\theta_1}^2 + \frac{P_\phi^2}{\sin^2\theta_1}, \tag{3}$$

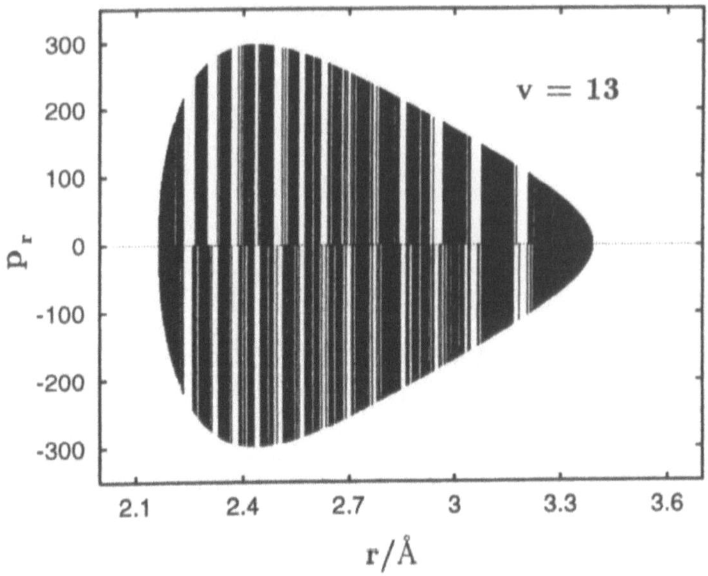

Figure 2. Initial distribution of positions and momenta associated with the r mode.

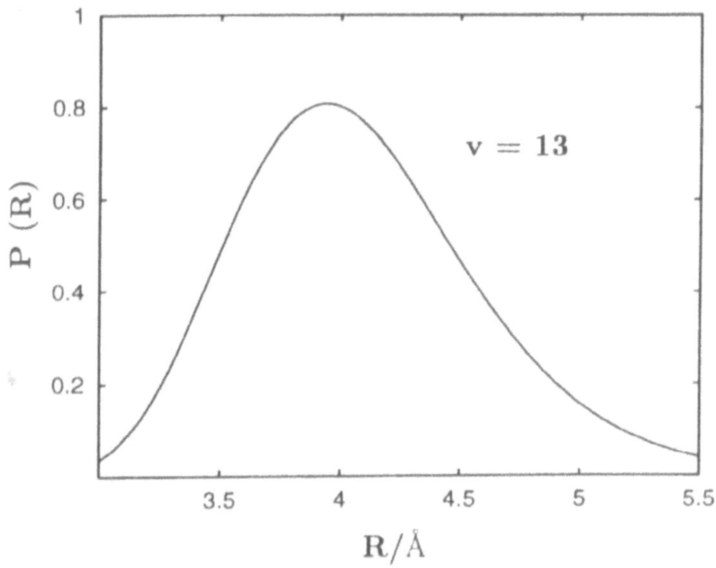

Figure 3. Quantum probability distribution of the R ($R = R_1, R_2$) vdW modes.

$$\mathbf{l}_2^2 = P_{\theta_2}^2 + \frac{P_\phi^2}{\sin^2\theta_2}, \tag{4}$$

$$\mathbf{j}^2 = \mathbf{l}_1^2 + \mathbf{l}_2^2 + \mathbf{l}_1 \cdot \mathbf{l}_2 = P_{\theta_1}^2 + P_{\theta_2}^2 + 2P_{\theta_1}P_{\theta_3}\cos\phi -$$

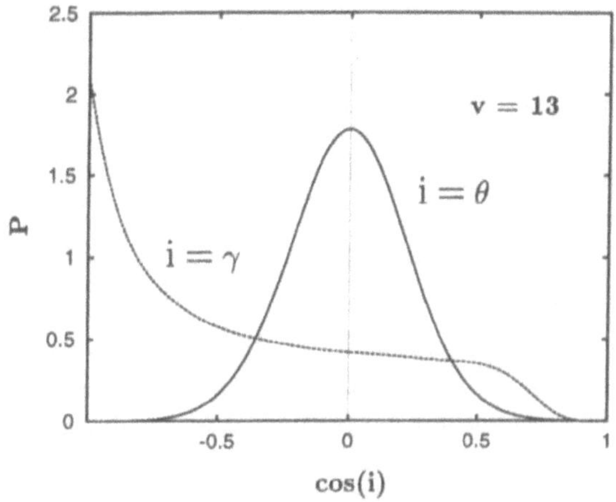

Figure 4. Quantum probability distributions associated with the θ ($\theta = \theta_1, \theta_2$) and γ angles.

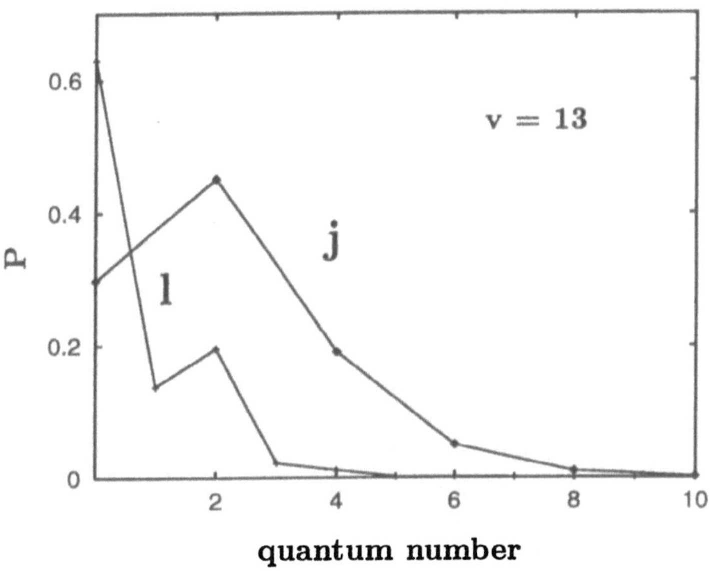

Figure 5. Distributions of the rotational quantum numbers j and l ($l = l_1, l_2$).

$$-2\sin\phi\left(\frac{\cos\theta_2}{\sin\theta_2}P_{\theta_1} + \frac{\cos\theta_1}{\sin\theta_1}P_{\theta_2}\right)P_\phi +$$
$$+\left(\frac{1}{\sin^2\theta_1} + \frac{1}{\sin^2\theta_2} - 2\cos\phi\frac{\cos\theta_1\cos\theta_2}{\sin\theta_1\sin\theta_2} - 2\right)P_\phi^2, \tag{5}$$

By introducing Equations (3) and (4) into Equation (5), a fourth-order polynomial in P_ϕ is obtained. Each root of this polynomial constitutes an initial value of P_ϕ (which already includes the sign). Introducing that value P_ϕ in Equations (3) and (4) gives the initial P_{θ_1} and P_{θ_2}, whose signs are randomly selected. In this way the calculated initial values of P_{θ_1}, P_{θ_2}, and P_ϕ are compatible with a prespecified total angular momentum of the system.

Finally, initial values for P_{R_1} and P_{R_2} are calculated by an iterative method. In a first stage, both momenta are obtained with the expression

$$P_{R_i} = \pm\sqrt{2\mu_{\mathrm{Cl_2—He}}[E_0^{(\nu)} - V_{\mathrm{Cl_2—He}}(\tau, R_i, \theta_i)]}, \quad (i = 1, 2), \tag{6}$$

where $E_0^{(v)}$ is the vdW ground state energy of the triatomic Cl_2—He system in a Cl_2 vibrational level v. The $\pm$ sign of the momenta is selected at random.

At this point initial values for the positions and momenta of all the system modes have been selected. However, the contribution of the $\mathbf{P_1} \cdot \mathbf{P_2}$ and $V_{\mathrm{He—He}}$ terms has so far been neglected in the total energy of the system. This causes the Cl_2—He_2 total energy associated with the initial conditions generated above to be slightly different from that calculated quantum-mechanically. To match the energy of the quantum initial state we must incorporate the two terms previously neglected. The terms $\mathbf{P_1} \cdot \mathbf{P_2}$ and $V_{\mathrm{He—He}}$ are not separable in order to distribute their energy contribution among all the vdW modes. We then chose to distribute such energy difference only among the P_{R_1} and P_{R_2} momenta. To do so, P_{R_1} and P_{R_2} are redefined as

$$P_{R_1} = P_{R_1}^{\mathrm{app}} + \Delta_1,$$

and

$$P_{R_2} = P_{R_2}^{\mathrm{app}} + \Delta_2,$$

where $P_{R_1}^{\mathrm{app}}$ and $P_{R_2}^{\mathrm{app}}$ are the approximate momenta calculated with Equation (6), and Δ_1 and Δ_2 are the variations of P_{R_1} and P_{R_2}, respectively, responsible for absorbing the energy difference. Replacing the new definition of P_{R_1} and P_{R_2} in the expression of the total energy (including the $\mathbf{P_1} \cdot \mathbf{P_2}$ and $V_{\mathrm{He—He}}$ terms) we get a quadratic form in the unknowns Δ_1 and Δ_2. Such an equation is solved iteratively by introducing the constraint that the ratio between Δ_1 and Δ_2 is the same as that between $P_{R_1}^{\mathrm{app}}$ and $P_{R_2}^{\mathrm{app}}$. The final values of Δ_1 and Δ_2 make it possible to fit the energy of the quantum initial state within a desired convergence criterion. In this way, the total energy of the initial conditions now properly matches the quantum resonance energy.

Distributions including 2000 initial conditions for the vdW modes were calculated with the above method. Figure 6 shows the distribution associated with the momenta P_R and the positions R ($R = R_1, R_2$), which corresponds with the quantum distribution of Figure 3. The distributions associated with the angular modes θ ($\theta = \theta_1, \theta_2$) and ϕ are displayed in Figures 7 and 8, respectively. These distributions reflect a strong quantization of the angular momenta, as a result of

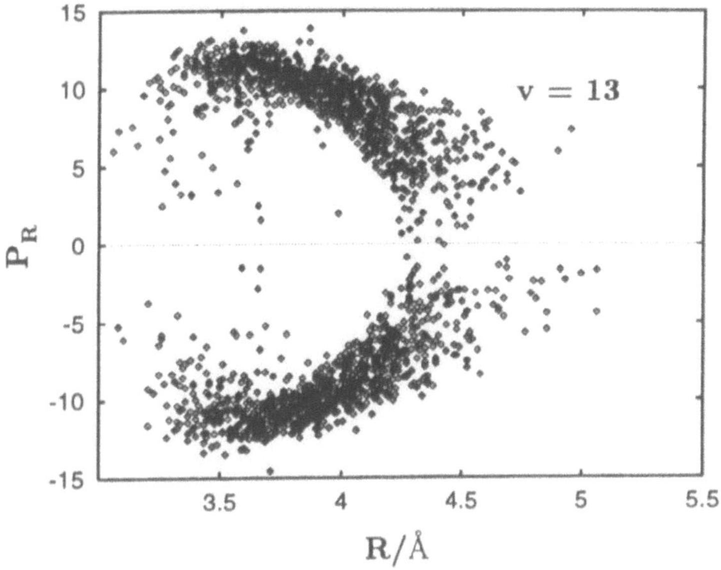

Figure 6. Initial distribution of positions and momenta associated with the R ($R = R_1, R_2$) vdW modes.

using the quantum probability distributions of Figure 5 to select the values of $\mathbf{l}_1$, $\mathbf{l}_2$ and $\mathbf{j}$ that enter in the system of Equations (3)–(5). In fact, the high quantum probability of low rotational numbers $\mathbf{l}_i, \mathbf{j}$ has its counterpart in low values of the angular momenta in the initial conditions.

The initial conditions generated with the method previously described conform an initial state compatible both in total energy and total angular momentum with the quantum initial state from which they come. Such initial conditions can be employed now as the starting point of a classical trajectory simulation of the predissociation dynamics of Cl_2—He_2. Work in this direction is in progress.

4. Conclusions

A method to sample initial positions and momenta to be used in a classical trajectory treatment of the dynamics has been proposed and applied to the Cl_2—He_2 vdW cluster. In the model suggested the quantum probability distributions associated with the initial state of the system are used to sample and weight the initial positions of the modes. Then the initial values of the associated momenta are calculated in such a way that they properly match the total energy and angular momentum of the quantum initial state. In the calculations reported here an initial state corresponding to zero total angular momentum was studied, but the method allows to treat states with total angular momentum different from zero.

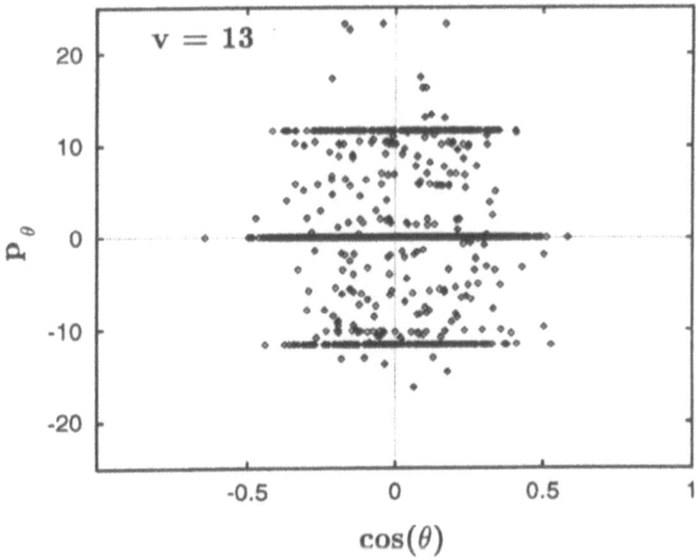

Figure 7. Initial distribution of angles and angular momenta associated with the θ ($\theta = \theta_1, \theta_2$) vdW modes.

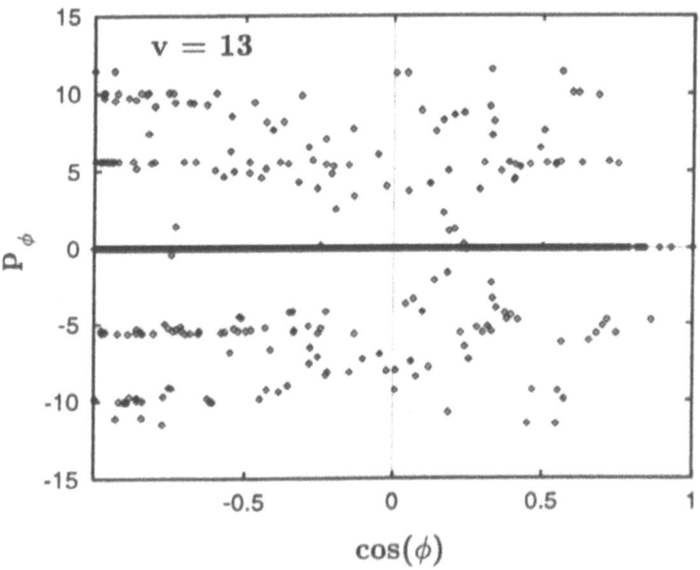

Figure 8. Initial distribution of angles and angular momenta associated with the ϕ mode.

The method proposed in this work can be viewed as a tool to transform an initial state calculated quantum-mechanically into initial conditions for classical trajectories, keeping the main features of the quantum state. Extending the present technique to systems larger than tetraatomic ones still requires more work, but we believe it is feasible. In that sense, the application of this method to many-mode initial states obtained from quantum Monte Carlo approaches [54] appears as a very interesting direction.

Acknowledgement

This work was supported by DGICYT, Spain, under grant no. PB95-0071.

References

1. (a) M. J. Rosker, M. Dantus, and A. H. Zewail: *J. Chem. Phys.* **89**, 6113 (1988). (b) R. M. Bowman, M. Dantus, and A. H. Zewail: *Chem. Phys. Lett.* **156**, 131 (1989). (c) A. H. Zewail: *J. Chem. Soc. Faraday Trans. 2* **85**, 1221 (1989).
2. H. L. Fragnito, J.-Y. Bigot, P. C. Becker, and C. V. Shank: *Chem. Phys. Lett.* **160**, 101 (1989).
3. N. F. Scherer, C. Sipes, R. B. Bernstein, and A. H. Zewail: *J. Chem. Phys.* **92**, 5239 (1990).
4. N. F. Scherer, A. J. Ruggiero, M. Du, and G. R. Fleming: *J. Chem. Phys.* **93**, 856 (1990).
5. (a) D. M. Willberg, M. Gutmann, J. J. Breen, and A. H. Zewail: *J. Chem. Phys.* **96**, 198 (1992). (b) M. Gutmann, D. M. Willberg, and A. H. Zewail: *ibid.* **97**, 8037 (1992). (c) *idem., ibid.* **97**, 8048 (1992).
6. Y. Chen, L. Hunziker, P. Ludöwise, and M. Morgen: *J. Chem. Phys.* **97**, 2149 (1992).
7. (a) S. L. Nikolaisen, H. E. Cartland, and C. Wittig: *J. Chem. Phys.* **96**, 4378 (1992). (b) S. I. Ionov, G. A. Brucker, C. Jaques, L. Valachovic, and C. Wittig: *ibid.* **97**, 9486 (1992).
8. (a) Y. Yan, R. M. Whitnell, K. R. Wilson, and A. H. Zewail: *Chem. Phys. Lett.* **193**, 402 (1992). (b) E. D. Potter, Q. Liu, and A. H. Zewail: *Chem.. Phys. Lett.* **200**, 605 (1992).
9. T. Baumert, J. L. Herek, and A. H. Zewail: *J. Chem. Phys.* **99**, 4430 (1993).
10. (a) S. I. Ionov, G. A. Brucker, C. Jaques, Y. Chen, and C. Wittig: *J. Chem. Phys.* **99**, 3420 (1993). (b) S. I. Ionov, G. A. Brucker, C. Jaques, L. Valachovic, and C. Wittig: *ibid.* **99**, 6553 (1993).
11. R. Zadoyan, Z. Li, P. Ashjian, C. C. Martens, and V. A. Apkarian: *Chem. Phys. Lett.* **218**, 504 (1994).
12. D. H. Levy: *Adv. Chem. Phys.* **47**, 323 (1981).
13. K. C. Janda: *Adv. Chem. Phys.* **60**, 201 (1985), and references therein.
14. A. Weber (ed.): *Structure and Dynamics of Weakly Bound Molecular Complexes*, NATO ASI Series C: Mathematical and Physical Sciences, Vol. 212, Reidel, Dordrecht (1987).
15. N. Halberstadt and K. C. Janda (eds.): *Dynamics of Polyatomic van der Waals Complexes*, NATO ASI Series B: Physics, Vol. 227, Plenum Press, New York (1990).
16. W. Sharfin, K. E. Johnson, L. Warton, and D. H. Levy: *J. Chem. Phys.* **71**, 1292 (1979).
17. J. E. Kenny, K. E. Johnson, W. Sharfin, and D. H. Levy: *J. Chem. Phys.* **72**,1109 (1980).
18. B. A. Swartz, D. E. Brinza, C. M. Western, and K. C. Janda: *J. Phys. Chem.* **88**, 6272 (1984).
19. J. C. Drobits and M. I. Lester: *J. Chem. Phys.* **86**, 1662 (1987).
20. S. R. Hair, J. I. Cline, C. R. Bieler, and K. C. Janda: *J. Chem. Phys.* **90**, 2935 (1989).
21. W. D. Sands, C. R. Bieler, and K. C. Janda: *J. Chem. Phys.* **95**, 729 (1991).
22. L. Beneventi, P. Casavecchia, G. G. Volpi, C. R. Bieler, and K. C. Janda: *J. Chem. Phys.* **98**, 178 (1993).
23. D. G. Jahn, S. G. Clement, and K. C. Janda: *J. Chem. Phys.* **101**, 283 (1994).
24. S. S. Huang, C. R. Bieler, K. C. Janda, F.-M. Tao, W. Klemperer, P. Casavecchia, G. G Volpi, and N. Halberstadt (to be published).
25. J. A. Beswick, G. Delgado-Barrio, and J. Jortner: *J. Chem. Phys.* **70**, 3895 (1979).
26. J. Campos-Martínez, O. Roncero, S. Miret-Artés, P. Villarreal, and G. Delgado-Barrio: *J. Chem. Phys.* **91**,155 (1989).
27. O. Roncero, J. A. Beswick, N. Halberstadt, P. Villarreal, and G. Delgado-Barrio: *J. Chem. Phys.* **92**, 3348 (1990).

28. P. Villarreal, S. Miret-Artés, O. Roncero, S. Serna, J. Campos-Martínez, and G. Delgado-Barrio: *J. Chem. Phys.* **93**, 4016 (1990).
29. S. Miret-Artés, G. Delgado-Barrio, O. Atabek, and J. A. Beswick: *Chem. Phys. Lett.* **98**, 554 (1983).
30. M. Aguado, P. Villarreal, G. Delgado-Barrio, P. Mareca, and J. A. Beswick: *Chem. Phys. Lett.* **102**, 227 (1983).
31. O. Atabek, J. A. Beswick, and G. Delgado-Barrio: *J. Chem. Phys.* **83**, 2954 (1985).
32. G. Delgado-Barrio, P. Mareca, P. Villarreal, A. M. Cortina, and S. Miret-Artés: *J. Chem. Phys.* **84**, 4268 (1986).
33. H. Tal-Ezer and R. Kosloff: *J. Chem. Phys.* **81**, 3967 (1984).
34. (a) M. D. Feit, J. A. Fleck, Jr., and A. Steiger: *J. Comput. Phys.* **47**, 412 (1982). (b) M. D. Feit and J. A. Fleck Jr.: *J. Chem. Phys.* **78**, 301 (1983).
35. (a) J. V. Lill, G. A. Parker, and J. C. Light: *Chem. Phys. Lett.* **89**, 483 (1982). (b) J. C. Light, I. R. Hamilton, and J. V. Lill: *J. Chem. Phys.* **82**, 1400 (1985).
36. R. B. Gerber, R. Korloff, and M. Berman: *Comp. Phys. Rep.* **5**, 59 (1986).
37. S. Das and D. J. Taunor: *J. Chem. Phys.* **92**, 3403 (1990).
38. S. K. Gray and C. E. Wozny: *J. Chem. Phys.* **94**, 2817 (1991).
39. F. Le Quere and S. K. Gray: *J. Chem. Phys.* **98**, 5396 (1993).
40. O. Roncero, G. Delgado-Barrio, M. I. Hernández, J. Campos-Martínez, and P. Villarreal: *Chem. Phys. Lett.* **246**, 187 (1995).
41. G. C. Schatz, V. Buch, M. A. Ratner, and R. B. Gerber: *J. Chem. Phys.* **79**, 1808 (1983).
42. R. H. Bisseling, R. Kosloff, R. B. Gerber, M. A. Ratner, L. Gibson, and C. Cerjan: *J. Chem. Phys.* **87**, 2760 (1987).
43. P. Villarreal, S. Miret-Artés, O. Roncero, G. Delgado-Barrio, J. A. Beswick, N. Halberstadt, and R. D. Coalson: *J. Chem. Phys.* **94**, 4230 (1991).
44. D. H. Zhang and J. Z. H. Zhang: *J. Chem. Phys.* **95**, 6449 (1991).
45. J. Y. Fang and H. Guo: *J. Chem. Phys.* **102**, 1944 (1995).
46. A. García-Vela: *J. Chem. Phys.* **104**, 1047 (1996).
47. J. Campos-Martínez, M. I. Hernández, O. Roncero, P. Villarreal, and G. Delgado-Barrio: *Chem, Phys. Lett.* **246**, 197 (1995).
48. A. García-Vela, P. Villarreal, and G. Delgado-Barrio: *J. Chem. Phys.* **92**, 6504 (1990).
49. A. García-Vela, P. Villarreal, and G. Delgado-Barrio: *J. Chem. Phys.* **94**, 7868 (1991).
50. (a) J. Rubayo-Soneira, A. García-Vela, G. Delgado-Barrio, and P. Villarreal: *Chem. Phys. Lett.* **243**, 236 (1995). (b) A. García-Vela, J. Rubayo-Soneira, G. Delgado-Barrio, and P. Villarreal: *J. Chem. Phys.* **104**, 8405 (1996).
51. K. P. Huber and G. Herzberg: *Molecular Spectra and Molecular Structure*, Vol. IV, *Constants of Diatomic Molecules*, Van Nostrand Reinhold Company, New York (1979).
52. R. A. Aziz and M. J. Slaman: *Chem. Phys.* **130**, 187 (1989).
53. P. Villarreal, O. Roncero, and G. Delgado-Barrio: *J. Chem. Phys.* **101**, 2217 (1994). (b) G. Delgado-Barrio, A. García-Vela, J. Rubayo-Soneira, J. Campos-Martínez, S. Miret-Artés, O. Roncero, and P. Villarreal: in J. Jortner, R. D. Levine, and B. Pullmann (eds.), *Reaction Dynamics in Clusters and Condensed Phases*, Kluwer Academic Publishers, the Netherlands, p. 57 (1994).
54. M. A. Suhm and R. O. Watts: *Phys. Rep.* **204**, 293 (1991).

Molecular Engineering **7**: 231–244, 1997.

Energy Estimates for Local Chemical Processes in Condensed Matter

CESARE PISANI* and SILVIA CASASSA
Department of Inorganic, Physical and Materials Chemistry, University of Torino, via Giuria 5, 10125 Torino, Italy.

(Received: 12 September 1996: accepted 17 June 1997)

Abstract. The problem of local chemical processes in condensed matter is reviewed, with special attention given to the associated changes in energy. The model adopted is based on a partition of the system into a local and an external region: it is assumed that all important changes in the electronic structure and nuclear configuration related with the process considered take place in the local region, while the solution for the unperturbed host system can be taken as a reference in the external region. This Ansatz allows an accurate description of the electronic structure in a neighborhood of the defect to be obtained [10, 16]. However, disregarding long range effects can result in large errors in the estimated value of the defect formation energy, ΔE. A strategy for estimating these long range contributions is outlined, based on the linear response properties of the host system and on the entity of the perturbation created by the defect in the local region. The case of the color center in the MgO monolayer is used to show the importance of one of these contributions to ΔE, the so-called charge balance correction, by comparing embedded-cluster with supercell results.

Key words: Local chemical processes, defects in condensed matter, color centers, magnesium oxide, partition technique.

1. Introduction

Local chemical processes in condensed matter are of interest both from a fundamental viewpoint and in many applications. A few examples can be mentioned: core-hole formation in atoms near a surface in ESCA experiments [1]; migration of a vacancy in a semiconductor [2]; UV adsorption centers in optical fibers [3]; hydrocarbon oxidation at the surface of a catalyst [4]. An atomic-scale description is required in all these cases, based on quantum mechanics. However, straightforward application of the Schrödinger equation is impossible because we are dealing with a macroscopic number of strongly interacting systems:

- which particles should be treated explicitly and which should not?
- how should *external* fields be formulated?
- which boundary conditions should be imposed?

* Author for correspondence.

An accurate analysis of the problem is required not only to obtain sensible theoretical answers, but also to correctly define observable quantities (in particular, energy data) from an experimental point of view.

In the *cluster model* (CM) of defects, one artificially retains a number of strong interactions (those inside the cluster) and disregards or treats approximately those with the rest of the system. The CM is very popular because it allows the use of standard tools from quantum chemistry, but it is far from satisfactory. First of all, there is a large amount of arbitrariness in its formulation, both as concerns shape and size of the cluster, and its termination [5]. Some devices are occasionally adopted for simulating the effect of the surrounding crystalline medium, such as the superposition of an external electrostatic field [6], saturation of dangling bonds [7] or inclusion of pseudoatoms around the cluster [8]. Even so, the CM results cannot converge to a definite limit with increasing cluster size, because boundary effects become energetically much more important than the actual phenomenon of interest.

A different point of view can be adopted when one is interested in processes taking place in ordered or quasi-ordered *host systems* (HS), which is often the case. The electronic structure of such HSs can be obtained with solid state techniques, by imposing periodic boundary conditions. One can then try and reformulate the problem by using a partition scheme, that is, by restricting the calculation to a finite *local region* around the defect (to be defined better in the following), while requiring that the local solution joins smoothly to that prevailing in the host in the infinite *external region*. When a one-electron Hamiltonian is used (Hartree–Fock (HF) or Kohn–Sham, for example) this condition is formulated most conveniently using Green function techniques [9]. This approach is generally referred to as an *embedded cluster model* (ECM).

The quality of ECM results can be checked by comparison with calculations performed with the *supercell model* (SCM), where the defect is repeated periodically within the host. The SCM is conceptually important because it allows an unambiguous formulation of the problem and so represent a bridge between theory and experiment. In particular, the SCM allows us to properly define the *formal charge* and (sometimes) the *spin state* of the system and the *formation energy per individual defect*, ΔE, in such a way that these quantities tend to a definite limit with increasing supercell size. This limit corresponds on the one hand to a sort of reference model for the experiment, on the other hand to an objective for the ECM, which operates with completely different techniques. In this sense, experience has shown that a very accurate description of the electronic structure can be achieved by the ECM, with the use of reasonably small local regions.

A much more delicate problem concerns the evaluation of ΔE. Changes in the electronic structure and in the nuclear configuration in the external region may be undetectable and irrelevant as concerns intensive observables but, due to the macroscopic size of that region, they can contribute a finite amount of energy. Since no brute-force calculation can ever allow a direct evaluation of both short-range

(ΔE^{sr}) and long-range (ΔE^{lr}) contributions to ΔE, the problem is how to estimate the missing terms without unnecessary computational effort. For this purpose, we shall assume that the local region is so large that all short-range, nonlinear effects induced by the defect are confined there. The long-range contribution is then expressed as a linear combination of the form:

$$\Delta E^{\text{lr}} = \sum_i X_i \xi_i,$$

Here, the extensive local quantity X_i measures a given kind of *source* (all X_is are zero in the absence of defects), and the intensive external quantity ξ_i is the associated *linear response coefficient*: the latter depends only on the partition scheme and on the response properties of the host system.

The present work discusses some of the problems just outlined with reference to a specific ECM, based on the perturbed cluster (PC) approach [10]. Section 2 gives a brief account of the PC technique in relation to other ECM approaches and discusses its performance as concerns the determination of the electronic structure. In Section 3, some indications are provided as to the most important contributions to ΔE^{lr} and to the techniques to be adopted for evaluating X_i and ξ_i. The effectiveness of the most important of these contributions, the so-called charge-balance correction, is documented in Section 4, with reference to the case of a color center in MgO, by comparing PC-ECM with SCM results.

2. Perturbed Cluster Determination of the Electronic Structure in the Defect Region

For definiteness, let us consider the host to be a perfect crystal whose general cell, labelled by the translation vector $\mathbf{T}$, contains a few atoms described by p real localized basis functions φ_ν^T. We admittedly know the ground-state wavefunction of the host in the $\mathbf{\Phi} \equiv \{\varphi_\nu^T\}$ basis set for a given one-electron approximation (HF, for instance). This is equivalent to the knowledge of the one-electron Green function represented in this basis set:

$$G^f_{\mu\nu T}(z) = \langle \varphi_\mu^0 | \{\hat{F}^f - z\hat{I}\}^{-1} | \varphi_\nu^T \rangle \tag{1}$$

or of the *projected density of states* (PDOS):

$$\rho^f_{\mu\nu T}(\varepsilon) = \frac{\iota}{2\pi} \lim_{s\to+0} [G^f_{\mu\nu T}(\varepsilon + \iota s) - G^f_{\mu\nu T}(\varepsilon - \iota s)]. \tag{2}$$

In these equations, a superscript f identifies quantities referring to the defect-free HS, $\hat{F}^f$ is the Fock Hamiltonian at self-consistency, while translational invariance has been exploited by making matrix elements depend only on the type of functions involved (μ, ν) and on the translation vector $\mathbf{T}$ which joins the two crystal cells they belong to:

$$Q^f_{\mu\nu T} = \langle \varphi_\mu^{T'} | \hat{Q}^f | \varphi_\nu^{T'+T} \rangle. \tag{3}$$

The advantage of using these quantities instead of making reference to the canonical HF orbitals is that they have *local* properties, that is, their value goes to zero with increasing distance between the centers of the two representative functions at a rate depending on the nature of the HS [11]. The representative basis set of the defect-containing host system (DHS) must differ from the HS one because some atoms may have been displaced or removed, others may have been added, etc. On the other hand, these changes will concern only a finite, hopefully small subset of functions $\alpha' \to \alpha$, while the rest of the basis set comprising an infinite number of basis functions of the $\mathbf{\Phi}$ set will be common to the two systems. We can now define two *cluster* basis sets: $C' \equiv \{\gamma'\}$ and $C \equiv \{\gamma\}$, which include the $\{\alpha'\}$ and $\{\alpha\}$ functions, respectively, plus a finite subset of unchanged crystalline functions (*boundary zone*) in order to accommodate within the cluster all important defect-induced changes in the electronic structure; all other functions, $D \equiv \{\delta\}$, will constitute the *indented crystal* basis set. The representative matrix Q of any one-electron operator $\hat{Q}$ for both the HS and the DHS can be subdivided into a local and an external part:

$$Q^f = Q^f_{\text{loc}} + Q^f_{\text{ext}} \equiv \begin{bmatrix} Q^f_{C'C'} & Q^f_{C'D} \\ Q^f_{DC'} & 0 \end{bmatrix} + \begin{bmatrix} 0 & 0 \\ 0 & Q^f_{DD} \end{bmatrix}$$

$$Q = Q_{\text{loc}} + Q_{\text{ext}} \equiv \begin{bmatrix} Q_{CC} & Q_{CD} \\ Q_{DC} & 0 \end{bmatrix} + \begin{bmatrix} 0 & 0 \\ 0 & Q_{DD} \end{bmatrix}. \tag{4}$$

The external matrices Q^f_{DD} and Q_{DD} can be directly compared to each other, and we will assume that the former is a good approximation for the latter, that is:

$$Q^f_{DD} \approx Q_{DD} \qquad \text{(locality assumption)} \tag{5}$$

provided, of course, that the cluster is large enough. Since Q^f_{DD} is known, equations can be set up concerning the local and unknown part of the problem, Q_{loc}. Most embedding schemes take $\hat{F}$ as the reference quantity in Equation (5) that is, it is assumed that defect-induced contributions to the Hamiltonian are confined to F_{loc} [9]. This choice reflects the fact that, historically, the ECM was first applied to model Hamiltonians of the tight-binding type, where the assumption of locality of the perturbation is strictly justified [12]. Instead, the PC approach uses for this same purpose the PDOS as defined in Equation (2) [10]. The latter choice has been criticized by Fisher [13] who was able to prove, with reference to one-dimensional tight-binding problems, that defect-induced local states or resonances can extend to large distances from the defect. In fact, with the PC approach, the tails of local states cannot be described accurately, and if one is interested in their characterization (for instance, when studying color centers in ionic crystals), one must take care that the local region is large enough to include all the relevant aspects of interest: an example is given below. Apart from this problem, no important changes are usually induced by defects in ρ_{ext}, especially in the case of 2D and 3D host systems. A notable exception is the case of charged centers in insulators, due to the long-range superimposed electrostatic field. To get rid of this problem, a *band-bending correction* can easily be introduced in the PC approach by rigidly shifting $\rho^{f}_{\delta\delta'}(\varepsilon)$ by a quantity $V_{\delta\delta'}$, corresponding to the electrostatic defect-induced field averaged over the $\delta\delta'$ distribution [10]. On the other hand, this is precisely the case where the assumption of locality of the perturbation potential is untenable, when *ab initio* Hamiltonians are used.

Using the locality assumption in the form:

$$\rho_{\delta\delta'}(\varepsilon) = \rho^{f}_{\delta\delta'}(\varepsilon + V_{\delta\delta'}) \qquad \text{(fundamental approximation)} \tag{6}$$

the PC equations are obtained after straightforward algebraic manipulations [14]:

$$P^{(\mathrm{PC})}_{\mathrm{loc}} = P^{\mathrm{clu}}_{\mathrm{loc}}([\rho^{\mathrm{clu}}_{CC}]) + P^{\mathrm{cst}}_{\mathrm{loc}}([S_{CC}, S_{CD}, [\rho^{f}_{DD}]) + P^{\mathrm{cou}}_{\mathrm{loc}}([\rho^{\mathrm{clu}}_{CC}], [\rho^{f}_{DD}]). \tag{7}$$

In this equation, the 'local' one-electron density matrix (P_{loc}) is expressed as a sum of three terms: the 'cluster' density matrix ($P^{\mathrm{clu}}_{\mathrm{loc}}$), built with the cluster eigenvectors of the E_{CC} matrix with eigenvalues below the Fermi energy of the HS; the 'constant' density matrix ($P^{\mathrm{cst}}_{\mathrm{loc}}$), which takes account of the non-orthogonality between the C and D basis sets; and, finally, the 'coupling' density matrix ($P^{\mathrm{cou}}_{\mathrm{loc}}$), which depends on the PDOS in the local and in the external region, and realizes the chemical coupling between the two. The PC equations are exact in the limit of negligible perturbation in the local space, or in the case where the perturbation is restricted to the central part of a large local region.

In spite of their formal simplicity, the actual implementation of the PC equations is a difficult task. First of all, a reliable *ab initio* technique for obtaining the PDOS of the HS must be available; secondly, there must be full compatibility between the algorithms and truncation criteria used for the HS and for the DHS; finally, the calculation of matrix elements for the DHS requires the use of solid state techniques to take full account of the interaction of the cluster region with the infinite surroundings. Our choice is to use the HF-LCAO CRYSTAL program for the HS calculations [15]; and to implement a PC-ECM-HF-LCAO program, EMBED, which uses, wherever possible, the same algorithms as in CRYSTAL [10, 16]. The most recent public versions of the two programs have just been released [17, 18].

ECM-PC versus SCM calculations concerning different types of HSs have proved that the PC approach provides a very accurate description of the density matrix of the DHS [19, 20]. We consider here the case of the neutral (F^0) and charged (F^+) color center in an MgO monolayer, obtained by removing an O^{2-} ion, and replacing it with two and one electrons, respectively. No nuclear relaxation is allowed. A rather sophisticated basis set has been used [21]; the basis functions associated with the removed oxygen ion have been used to describe the electrons at the F center. Though somewhat artificial, this defect system is simple enough to

Table I. Some results of the present computations. The 'size' label identifies the various ECM (cluster composition: OMg_4, O_5Mg_4, O_9Mg_4, O_9Mg_{12}) and SCM calculations (number of MgO units per SC: 2, 4, 8, 16, 32, 64, 128); the central O ion in the cluster, and one (two in the AFM case) O ion in each supercell are removed to create the defect; μ is the electrochemical potential; q^{mis} is the missing change in the embedded cluster, and μq^{mis} is the corresponding charge balance correction (see text); ΔE represents the formation energy per defect calculated as explained in the text, and using the value $E[O^{2-}] = -74.2905$ Ha.: in the ECM case, it includes the charge balance correction. For the F^+ case, two energy data are given: for ECM, data in italics refer to uncorrected energies; for SCM, to the AFM calculation. All data are in atomic units.

Reference HS data for the ECM calculations						
'size'	1	2	3	4		
μ	0.5078	0.6095	1.0663	0.5188		
q^{mis}	0.0005	−0.0007	0.0145	0.0007		
Data for the DHS: F^0 center						
'size'	1	2	3	4		
q^{mis}	0.0022	−0.0092	0.0113	0.0046		
μq^{mis}	0.0011	−0.0057	0.0120	0.0023		
ΔE(ECM)	0.8669	0.8777	0.8824	0.8851		
ΔE(SCM)	0.9022	0.9018	0.9019	0.9019		
Data for the DHS: F^+ center						
'size'	1	2	3	4	5	6
q^{mis}	0.8802	0.5396	0.2730	0.6532		
μq^{mis}	0.4469	0.3288	0.2920	0.3388		
ΔE(ECM)	0.9524	0.9323	0.9404	0.9183		
	0.5056	*0.5993*	*0.6637*	*0.5795*		
ΔE(SCM)	0.7168	0.7870	0.8276	0.8588	0.8809	0.8971
	0.5999	*0.7168*	*0.7867*	*0.8276*		

permit us to explore the effect of supercell and cluster size in SCM and ECM-PC calculations. Restricted Hartree–Fock (RHF) and unrestricted Hartree–Fock (UHF) calculations have been performed for the neutral and charged center respectively. 'in the latter case, a ferromagnetic (FM) or antiferromagnetic (AFM) solution was obtained using the SCM, according to whether all vacancies were populated with α-spin electrons, or alternatively with α- and β-spin. Data concerning these calculations are reported in Table I, which will be commented on in the following. Figure 1 reproduces the spin density around the F^+ center for different EC and SC sizes. It is seen that the three types of calculations converge very rapidly to the same description of the electronic structure up to the most minute details.

3. Long Range Contributions to Reaction Energy

The fact that PC-ECM calculations provide a satisfactory description of the DHS, in the sense that increasing the size of the local region does not introduce visible

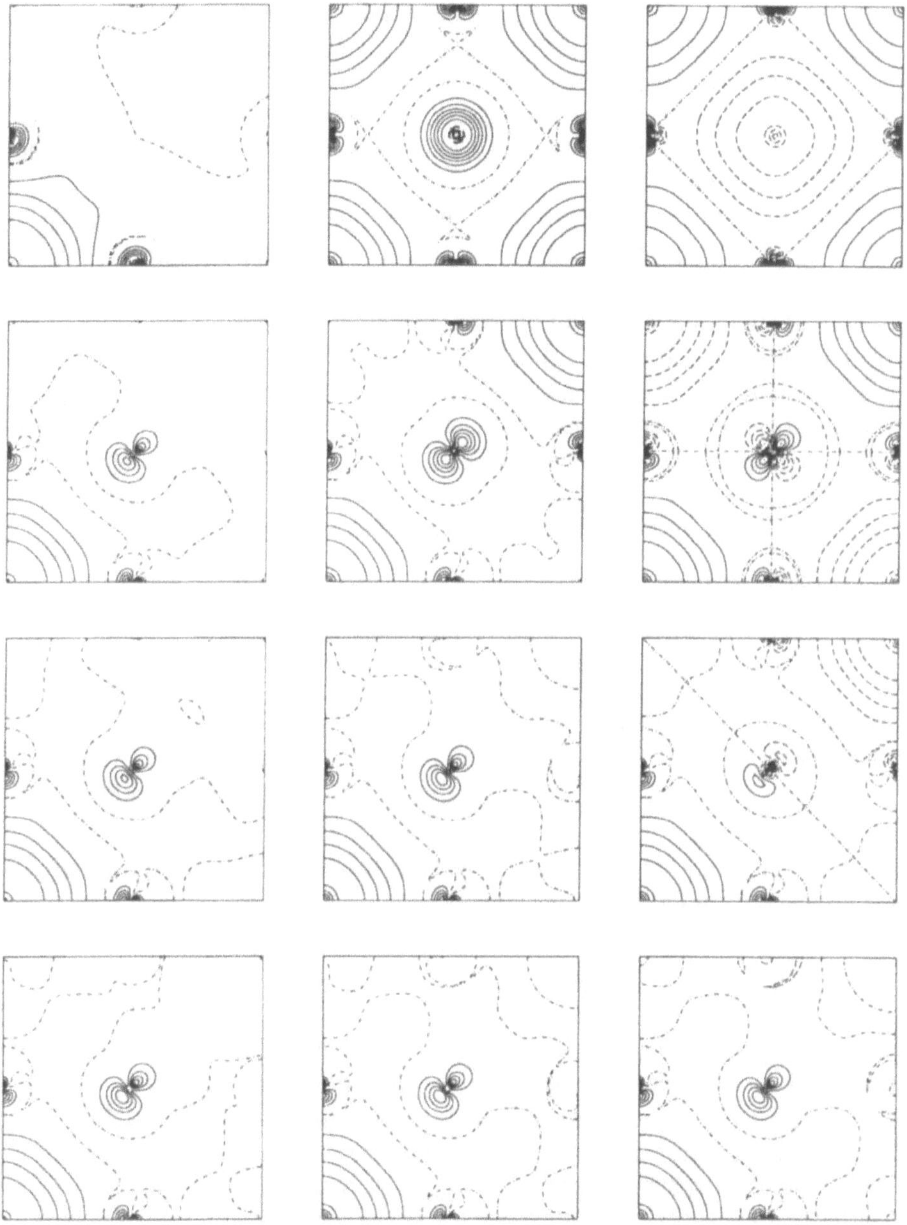

Figure 1. Spin density for the F^+ center in the MgO monolayer. *Left column*: from top to bottom: UHF-PC-ECM results for increasingly larger clusters ('size' from 1 to 4, see Table I). *Center column*: from top to bottom: FM-UHF-SCM results for increasingly larger SCs ('size' from 1 to 4, see Table I). *Right column*: the same as for the center column, but for AFM calculations. Iso-spin-density lines are separated by 0.005 a.u.; continuous, dashed and dot-dash lines correspond to positive, negative and zero spin density, respectively. In all plots, the F center is at the bottom left corner.

changes in the calculated electronic structure, does not imply that all quantities of interest can be obtained accurately from the fundamental approximation. As stated in the Introduction, the defect formulation energy, ΔE, is the most important quantity requiring long-range corrections. To discuss this point, the arguments developed in a recent paper [22] will be followed and generalized, with some changes in notation.

We shall assume that in the DHS the *true* density matrix, P, is given by: $P = P_{\rm loc}^{\rm (PC)} + (P_{\rm ext}^{f} + \delta P_{\rm ext})$, where $P_{\rm loc}^{\rm (PC)}$ is the result of the PC calculation. After attributing the nuclei either to the local or to the external region, we assume that the location of the former ones is fully optimized, while the external ones may exhibit small displacements from $\{\mathcal{Z}_{\rm ext}^{f}\}$ to $\{\mathcal{Z}_{\rm ext}^{f}\} + \{\delta\mathcal{Z}_{\rm ext}\}$, due to elastic strains and/or to electrostatic forces induced by the presence of the defect. On the whole, the rearrangement of charges in the external region, associated with $\delta P_{\rm ext}$ and $\delta\mathcal{Z}_{\rm ext}$, will be denoted as *external polarization charge*, $Q_{\rm ext}^{\rm pol}$. The total HF energy for the system, either perfect or with defects, can be written (using $A \bullet B$ for Trace$[A \times B]$ and $\mathcal{Z} \star \mathcal{Z}'$ for the interaction between two nuclear charge distributions):

$$E[S] = (M_{\rm loc} + B_{\rm loc}/2) \bullet P_{\rm loc} + (M_{\rm ext} + B_{\rm ext}/2) \bullet P_{\rm ext} + + \tfrac{1}{2}\mathcal{Z}_{\rm loc} \star \mathcal{Z}_{\rm loc} + \mathcal{Z}_{\rm loc} \star \mathcal{Z}_{\rm ext} + \tfrac{1}{2}\mathcal{Z}_{\rm ext} \star \mathcal{Z}_{\rm ext} \tag{8}$$

where M and B are the one-electron (kinetic-plus-nuclear attraction) and the two-electron (Coulomb-plus-exchange) parts of the Fock matrix. For a process described by

$$\text{(HS)} + \text{A} \Rightarrow \text{(DHS)} + \text{R} \tag{9}$$

where A and R are atomic, ionic, molecular species or free electrons added to and removed from the HS, respectively, the reaction energy can be expressed as follows:

$$\Delta E = E[\text{DHS}] - E[\text{HS}] + E(\text{R}) - E(\text{A}) \equiv \Delta E^{\rm f.a.} + \Delta E^{\rm lr}. \tag{10}$$

In $\Delta E^{\rm f.a.}$ (f.a. standing for 'fundamental approximation', see Equation (6)), we include all the terms corresponding to changes in the nuclear structure in the local region, and all contributions to ΔE resulting from the use of $Q_{\rm ext}^{\rm pol} = 0$. In order to obtain an explicit expression for $\Delta E^{\rm lr}$, we shall use a prime ($'$) or double prime ($''$) label to indicate contributions to M and B by electrons and nuclei in the local and external region, respectively, and δs are used to indicate changes with respect to the f.a. For instance, $\delta M_{\rm loc}''$ will indicate the change in the local Fock matrix associated with the shift $\delta\mathcal{Z}_{\rm ext}$ of the nuclei in the external region. We can then write:

$$\Delta E^{\rm lr} \equiv \sum_{i=1}^{12} \delta E_i =$$

$$
\begin{aligned}
&= B''^{f}_{\text{ext}}/2 \bullet \delta P_{\text{ext}} + \delta B''_{\text{ext}}/2 \bullet P^{f}_{\text{ext}} + \delta B''_{\text{ext}}/2 \bullet \delta P_{\text{ext}} + \\
&+ M''^{f}_{\text{ext}} \bullet \delta P_{\text{ext}} + \delta M''_{\text{ext}} \bullet P^{f}_{\text{ext}} + \delta M''_{\text{ext}} \bullet \delta P_{\text{ext}} + \\
&+ \delta \mathcal{Z}_{\text{ext}} \star \mathcal{Z}^{f}_{\text{ext}} + \tfrac{1}{2} \delta \mathcal{Z}_{\text{ext}} \star \delta \mathcal{Z}_{\text{ext}} + \\
&+ (M'_{\text{ext}} + B'_{\text{ext}}/2) \bullet \delta P_{\text{ext}} + \mathcal{Z}_{\text{loc}} \star \delta \mathcal{Z}_{\text{ext}} + \\
&+ \delta B''_{\text{loc}}/2 \bullet P_{\text{loc}} + \delta M''_{\text{loc}} \bullet P_{\text{loc}}. \qquad (11)
\end{aligned}
$$

The eight terms in the first three lines give the change in the external energy related to the interaction of the polarization charge with unperturbed external electrons and nuclei, and with itself (polarization self-energy); the subsequent two ones, the change in the external energy related to the interaction of the polarization charge with local electrons and nuclei; finally, those in the last line, the change in the local energy due to modifications in the local Fock matrix induced by the polarization charge.

Following a classical analysis, these terms may be grouped into three main categories; the corresponding contributions to energy are tentatively expressed as follows, for the reasons to be explained shortly:

$$
\begin{aligned}
&(1) \quad \Delta E^{\text{lr}}_{\text{bal}} = q^{\text{mis}} \mu \\
&(2) \quad \Delta E^{\text{lr}}_{\text{rel}} = -F^2 \chi \\
&(3) \quad \Delta E^{\text{lr}}_{\text{pol}} = -(q^{\star})^2 \alpha. \qquad (12)
\end{aligned}
$$

(1) *Electronic charge balance.* This is the most important effect, to be treated explicitly in the following. It is related solely to electron displacement. Due to the different electron affinity of the DHS cluster with respect to the HS one, some electronic charge will be transferred from the local to the external region, and will be missing if no account is taken of changes in the external electron density. One can estimate the corresponding energy contribution by multiplying the *missing charge* (the source) by the *(electro-)chemical potential* of the external region (the associated response coefficient).

(2) *Elastic relaxation.* The insertion, removal of substitution of atoms gives origin to tensions in the surrounding HS structure. If we allow for full electronic and nuclear relaxation in the local region, we can estimate the forces F acting on atoms just outside the cluster boundary, by calculating the derivative of energy with respect to their displacement $(x) : F = -[\partial E/\partial x]_{x=0}$. From the knowledge of the elastic constants of the HS, it is possible to calculate classically the equilibrium structure by minimizing the total energy. This correction corresponds to movement of both electrons and nuclei. In the simplest case where the cluster may be assumed to be a sphere of radius R_0 in a homogeneous elastic medium with bulk modulus B, we have:

$$
E = -Fx + ax^2 \qquad \mathrm{d}E/\mathrm{d}x = 0 \rightarrow x_{\min} = F/(2a)
$$

$$E_{\min} = [-F^2] \times [1/(4a)] \rightarrow \Delta E_{\mathrm{rel}}^{\mathrm{lr}} = X_{\mathrm{rel}} \times \xi_{\mathrm{rel}}, \tag{13}$$

where we have defined: $a = 2\pi B R_0$.

(3) *Dielectric polarization.* In the case of charged defects within an insulator, the position of nuclei and the distribution of electrons in the external zone will be modified to take profit of the long range electrostatic field. In a first approximation, one can express this stabilizing contribution to energy in the form of a product of a squared effective charge (the source) times a polarizability term (the response). A more general expression would include contributions from higher order multipoles of the field created by the defect in the local zone. A lot of information is available on dielectric polarization effects in ionic crystals due to semiclassical calculations based on the so-called shell model [23]. Work is in progress to obtain estimates of q^* and α, by exploiting that kind of experience. Care must be taken when combining this correction with the preceding ones, since there is a risk of counting the same effect twice. As will appear clearly below, a large part of the missing charge is not associated to a real flow of charge, but to the polarization of ions with fixed nuclei. A careful analysis is required to include in this term only the effect of nuclear displacements induced by the charge in the electrostatic field.

Let us now consider the electronic charge balance problem in more detail. Since no nuclear relaxation is allowed in the external region ($\delta \mathcal{Z}_{\mathrm{ext}} = 0$), terms δE_i in Equation (11) will be zero, with i = 5, 6, 10 and 12. To clarify the concept of missing charge in Equation (12), it is expedient to devote a few words to the charge balance problem. In the supercell approach, a definite integer *formal charge* $q^{\mathrm{for}} = q(A) - q(R)$ can be attributed to each unit cell of the DHS (see Equation (9)), if the HS is neutral. Note incidentally that if q^{for} is not zero, one must specify the potential at the location where the isolated species A and R are assumed to be transported. This potential is not uniquely defined in the case of three-dimensional crystals; this problem is discussed elsewhere [24]. Charge conservation requires that $q(\mathrm{DHS}) - q(\mathrm{HS}) = q^{\mathrm{for}}$, or:

$$\delta q_{\mathrm{ext}} = -S_{\mathrm{ext}} \bullet \delta P_{\mathrm{ext}} = -[-S_{\mathrm{loc}} \bullet P_{\mathrm{loc}} + S_{\mathrm{loc}}^f \bullet P_{\mathrm{loc}}^f + \delta \mathcal{Z}_{\mathrm{loc}}] + q(A) - q(R). \tag{14}$$

The δq_{ext} term defined here is called the missing charge, q^{mis}, because it is ignored in the strict PC approximation.

The physical meaning of q^{mis} is worth a comment in the case of charged defects. Consider again the F^+ center in the MgO monolayer, where $q^{\mathrm{for}} = 1$. As seen in Table I, the missing charge is very high, ranging from 0.3 to 0.9 for the various clusters. The departure of the local net charge $q^{\mathrm{net}} = q^{\mathrm{for}} - q^{\mathrm{mis}}$ from the formal charge attributed to this defect, is a measure of dielectric screening: the contribution of the defect to the electrostatic potential around the impurity is found to have the form q^{net}/r which shows that the ratio $q^{\mathrm{for}}/q^{\mathrm{net}}$ should tend to the optical dielectric constant ε_∞ in the limit of very large clusters of quasi-spherical shape and when no nuclear relaxation is allowed. The convergence is usually more rapid for charged defects in bulk: in the present case it is still far apart.

The most important contributions to ΔE associated with this missing charge are those labelled as δE_i in Equation (11), with $i = 1, 2, 4, 5, 9$, corresponding to the interaction of the electronic polarization charge with $Q^{\text{f.a.}}$ and to the change in the external kinetic energy. The overall contribution to energy of these terms will be proportional to q^{mis} for a given structure of δP_{ext}. The proportionality coefficient, μ, is estimated by the program EMBED [18] following the v-technique, amply discussed in previous work [25]. We just remember that this technique consists in superimposing on the HS at equilibrium a perturbative local potential, and then exploiting the fact that we must have $\Delta E^{\text{lr}} = -\Delta E^{\text{sr}}$ to first order in the perturbation, for changes δP which preserve idempotency and number of electrons. The importance and effectiveness of this correction has been demonstrated in a large number of applications, and will be reconsidered in the next section. Certainly, this is not the final answer to the problem of long-range polarization even if no nuclear relaxation is considered. On the one hand, the v-technique sometimes gives unsatisfactory μ data, especially in the case of metals. On the other hand, disregarded terms, especially important in the case of charged defects, require further analysis and more sophisticated corrective terms.

4. A Numerical Example

The case of the color center in the MgO monolayer will help us to show the importance of the charge balance correction to ΔE^{lr}. Instead, no numerical test yet exists of the effectiveness of the nuclear relaxation corrections briefly discussed in the previous section. Table I provides data concerning the present calculations; the corresponding ΔE results are reported in Figure 2. Let us put forward a few comments on these data.

We first consider the SCM results. A very different behavior is observed for the neutral and the charged center. In the former case, convergence of results with respect to SC size is extremely rapid; in the latter case, strong electrostatic interaction of the charged centers between themselves and with the neutralizing charge background are important even for the largest SCs considered. Corrections for these interactions have been suggested [26] but not applied in the present case because no dielectric constant is available for the two-dimensional structure considered. The different FM and AFM results simply reflect the fact that there are twice as many charged centers per supercell in the latter case. From Table I, it is seen that the FM energy for a given SC size practically coincides with the AFM energy for the next larger size.

In the ECM calculations the charge-balance correction is seen to be very large and effective-especially in the F^+ case. Without this correction, the $\Delta E(F^+)$ values would oscillate between 0.50 and 0.65 a.u.: the correction reduces the oscillation by a factor of five. The comparison of the ECM and SCM data for the charged center

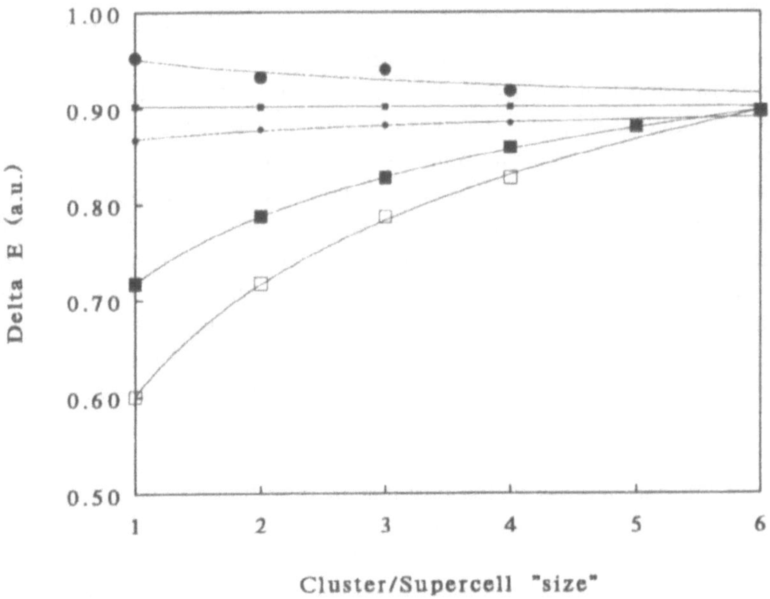

Figure 2. ΔE values for the F^0 (small symbols) and the F^+ center (large symbols) as a function of EC and SC size, for the same four cases as in Figure 1. Circles correspond to PC-ECM, full and open squares to FM-SCM and AFM-SCM results. The lines are a logarithmic interpolation of each series of data.

provides a further confirmation of the fundamental importance of this correction: the three curves appear to tend to the same asymptotic value, about 0.915 a.u. Also note that the ECM data indicate consistently an appreciable electron affinity for the F^+ center, while in the SCM results this behavior appears only asymptotically.

On the whole, the ECM results appear to give essentially correct results even with relatively small clusters, while SCM calculations in many instances need a relatively large distance between defects to reproduce correctly the isolated defect behavior. On the other hand, the energetic trends are more regular for SCM than for ECM calculations, because use is made there of better assessed techniques, where all energy terms are accurately calculated.

5. Conclusions

The *ab initio* study of chemical processes in condensed matter is still in its infancy. A lot of high-quality work is still required in order to formulate approximations based on sound physical arguments, and to test their effectiveness. This paper describes the problem of energy which is probably the most delicate and most important one. A strategy has been outlined for recognizing and estimating the most important long-range contributions to ΔE in the framework of the HF approximation. In this respect, the question is how one is to take correction effects into account. This problem is still unsolved as far as the *local* correlation correction is concerned. Techniques which include correlation in a selected subspace of the system, while treating the rest of the electronic structure as rigid, appear particularly promising for this purpose [27]. In fact, since changes in the density matrix in the external region are both small and featureless, one must expect that the *difference* in the long range correlation energy in the DHS with respect to the HS is again proportional to the missing charge. Once the local correlation energy is known, this long-range term, $\Delta E_{\rm cor}^{\rm lr}$, could then be estimated and incorporated in the charge-balance corrective term by a straightforward modification of the v-technique [25], simply resulting in a change of the electrochemical potential. This kind of problem could be bypassed if a density functional approach were followed, incorporating electronic correlation in the description of both the HS and the DHS.

Acknowledgements

Financial support from the Italian MURST (40%), from the European Community (contrast No. CHRX-CT93-0155), and from CSP (Centro per il Supercalcolo Piemonte) is gratefully acknowledged. The authors are grateful to Uwe Birkenheuer, Erminia Scorza and Roberto Orlando for their help in collecting the results described in the present work.

References

1. W. F. Egelhoff: *Surf. Sci. Rep.* **6**, 253 (1987).
2. J. A. Van Vechten: *Phys. Rev. B* **38**, 9913 (1988).
3. H. Hosono, Y. Abe, D. L. Kinser, R. A. Weeks, K. Muta, and H. Kawazoe: *Phys. Rev. B* **46**, 11445 (1992).
4. O. T. Onsager, R. Lødeng, P. Søraker, A. Anundskaas, and B. Helleborg: *Catal. Today* **4**, 355 (1989).
5. M. C. Zonneyville, J. J. C. Geerlings, and R. A. van Santen: *J. Calal.* **148**, 417 (1994).
6. K. Hermann: in G. Pacchioni, P. S. Bagus and F. Parmigiani (eds.), *Cluster Models for Surface and Bulk Phenomena*, NATO ASI Series B, Vol. 283, Plenum Press, New York, p. 209 (1992).
7. J. Sauer: in G. Pacchioni, P. S. Bagus and F. Parmigiani (eds.), *Cluster Models for Surface and Bulk Phenomena*, NATO ASI Series B, Vol. 283, Plenum Press, New York, p. 533 (1992).
8. M. A. Nygren, L. G. Pettersson, Z. Barandiarán, and L. Seijo: *J. Chem. Phys.* **100**, 2010 (1994).
9. J. E. Inglesfield: *J. Phys. C* **14**, 3795 (1981).
10. C. Pisani, F. Corà, R. Nada, and R. Orlando: *Comp. Phys. Commun.* **82**, 139 (1994).
11. E. N. Economon: *Green Functions in Quantum Physics*, Springer, Berlin (1979).
12. J. Bernholc and S. T. Pantelides: *Phys. Rev. B* **18**, 1780 (1978).
13. A. J. Fisher: *Theor. Chim. Acta* **72**, 319 (1987).
14. C. Pisani: *J. Molec. Catal.* **82**, 229 (1993).
15. C. Pisani, R. Dovesi, and C. Roetti: *Hartree–Fock Ab Initio Treatment of Crystalline Solids*, Lecture Notes in Chemistry, Vol. 48, Springer, Berlin (1988).
16. C. Pisani and U. Birkenheuer: *Comp. Phys. Commun.* **96**, 152 (1996).
17. R. Dovesi, V. R. Saunders, C. Roetti, M. Causà, N. M. Harrison, R. Orlando, and E. Aprà: CRYSTAL95 *User's Manual*, Università di Torino, Torino (1996).
18. C. Pisani, U. Birkenheuer, F. Corà, R. Nada, and S. Casassa: EMBED96 *User's Manual*, Università di Torino, Torino (1996).
19. R. Nada, V. R. Saunders, and C. Pisani: *Chem. Phys.* **169**, 297 (1993).
20. C. Pisani, F. Corà, R. Dovesi, and R. Orlando: *J. Elec. Spectr.* **96**, 1 (1994).

21. M. Causà, R. Dovesi, C. Pisani, and C. Roetti: *Phys. Rev. B* **33**, 1308 (1986); *ibid.* **34**, 2939 (1986).
22. C. Pisani: *Phase. Trans.* **52**, 123 (1994).
23. F. Agullo-Lopez, C. R. A. Catlow, and P. D. Townsend: *Point Defects in Materials*, Academic Press, London (1988).
24. U. Birkenheuer, F. Corà, C. Pisani, E. Scorza, and G. Perego: *Surf. Sci.* **373**, 393 (1997).
25. C. Pisani, and F. Corà: *Comp. Phys. Commun.* **82**, 179 (1994).
26. M. Leslie and M. J. Gillan: *J. Phys. C* **18**, 973 (1985).
27. D. L. Cooper, J. Gerratt, and M. Raimondi: *Chem. Rev.* **91**, 929 (1991).

Molecular Engineering **7**: 245–261, 1997.

Theoretical Study of the TiO_2 and MgO Surface Acidity and the Adsorption of Acids and Bases

ALEXIS MARKOVITS, JAMILA AHDJOUDJ and CHRISTIAN MINOT*
Laboratoire de Chimie Théorique, UPR 9070 CNRS, Tour 23-22, Boîte 137, Université P. et M. Curie, 75252 Paris Cedex05, France.

(Received: 12 September 1996; in final form: 17 June 1997)

Abstract. We present *ab-initio* periodic Hartree–Fock calculations (CRYSTAL program) of small molecules on TiO_2 and MgO. The adsorption of the molecules may be molecular or dissociative. This depends on their acid and basic properties in the gas phase. For the molecular adsorption, the molecules are adsorbed as bases on Ti(+IV) sites, the adsorption energies correlate with the proton affinities. The dissociations on the surface correlate with the gas phase cleavages: thus, the dissociation of MeOH leads to a preferential basic cleavage (the fragment HO^- is adsorbed on a Ti^{+4} ion and the fragment Me^+ is adsorbed on a O^{2-} ion of the oxide). The opposite result is obtained with MeSH. Another important factor is the adsorbate–adsorbate interaction: favorable cases are a sequence of H-bonds for the hydroxyl groups resulting from the water dissociation and the mode of adsorption for the ammonium ions. Lateral interactions also force the adsorbed CO_2 molecules to bend over the surface so that their mutual orientation resembles the geometry of the CO_2 dimer. With respect to water adsorption, MgO appears to be a basic oxide. As experimentally observed, NH_3 adsorbs preferentially on TiO_2 and CO_2 on MgO. However, this difference of reactivity should not be expressed in terms of acid vs. basic behaviour but in terms of hard and soft acidity. The MgO surface is a 'soft' acidic surface that reacts preferentially with the soft base, CO_2.

Key words: Periodic Hartree–Fock calculations, metal oxide, surface acidity, absorption, hard or soft acid or base.

1. The Acid–Base Properties and Adsorption on Metal Oxide Surfaces

The determination of the acidic or basic character of active sites of solid catalysts is of prime interest in determining their adsorption properties and to control the catalytic behaviour [1]. The chemisorption of molecules involves their acidic or basic properties that have to oppose those of the active catalytic site. Metal oxides are amphoteric compounds. On clean surfaces, exposed metallic cations are acidic centers (Lewis acid) and surface oxygen ions are basic centers (Lewis base). In hydrated media, the metallic centers are covered by hydroxyl groups and the surface oxygen atoms are protonated so that the surface is saturated by hydroxyl groups. These groups are also amphoteric; the reactivity through the oxygen atom corresponds to a Brønsted basic species, while that via the proton reveals a Brønsted acidity.

* E-mail: minot@lct.jussieu.fr

Metal oxides are classified as acidic, amphoteric or basic [2, 3]. Experimentally, this classification corresponds to the adsorption of probe molecules. NH_3 is a base probe molecule that reacts with the electron deficient metal atoms (Lewis acid) or the protons adsorbed on the hydrated surface. Because CO_2 is acidic, it is expected to adsorb more strongly on basic sites. According to this classification, TiO_2 belongs to the amphoteric species and MgO to the basic species. Magnesium is more electropositive than titanium (1.23 vs. 1.32 according to Allred [4, 5]); it follows that Mg cations are less acidic than Ti cations; the LUMO (lowest unoccupied molecular orbital) of Ti^{4+} is lower than that of Mg^{2+} and the electron affinities of the ions are 43.2 and 15.0 eV, respectively. It also follows that MgO is more ionic than TiO_2, increasing the basicity of the surface oxygen atoms [2].

In this paper we present the best adsorption modes for these probe molecules and for a series of hydroxylic compounds (water, methanol, carboxylic acids) on well ordered and perfect surfaces. We show that the molecular adsorptions on the clean surfaces imply consideration of the adsorbates as Lewis bases that always bind to the metal cations. The adsorption of water on MgO, where the adsorbate appears to be predominantly an acid, seems to be an exception. Dissociative adsorptions are analyzed with respect to the dissociation in the gas phase (Brønsted acidity or basicity). Finally we discuss the importance of the adsorbate–adsorbate interactions.

2. Calculations and Models

Our calculations are periodic calculations based on first principle, using a restricted Hartree–Fock (RHF) Hamiltonian; the program CRYSTAL [6, 7] is presented in Ref 8. We consider a slab and define a surface unit cell by two translation vectors. The adsorbates are ordered on the surface and the translation vectors are imposed by the translations that relate the adsorbates. The size of the unit cell increases with decreasing coverage or when all the adsorbates are not related by a translation symmetry. The total charge of the system has to be neutral; if not, the periodicity would multiply the charges to infinity. Real surfaces are not usually perfect surfaces and defects are often the most active catalytic sites. This has been assumed on MgO [9–12] and TiO_2 surfaces [13]. Nevertheless, the study of well ordered surfaces corresponds to a considerable amount of experimental work, using specific techniques such as LEED. It provides the richest detailed information on the adsorption modes.

Periodicity introduces two main advantages over cluster models of adsorption.

The first one is to increase the coordination of each atom. Consider the MgO system. A cubic cluster involves four Mg—O units and each atom has only three neighbors. By comparison, the definition of a 2-layer slab requires two Mg—O units only (the other atoms are deduced by translation). Each atom has five neighbors.

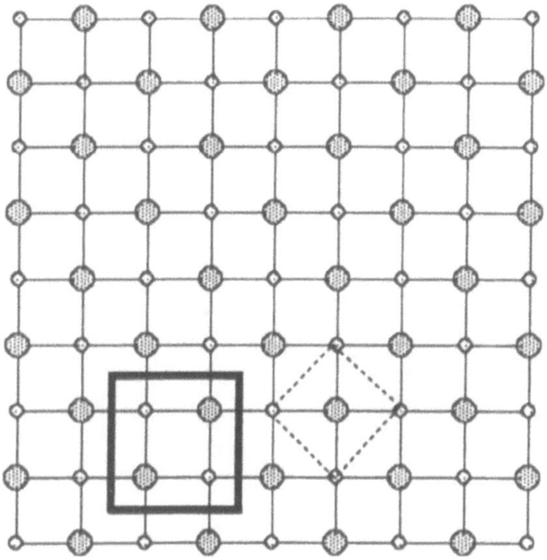

Figure 1. The MgO surface. The large circles represent the oxygen atom and the small ones the metal centers. The dotted lines show the primitive cell, and the bold lines represent the $\sqrt{2}\sqrt{2}R45°$ cell.

The second one is the possibility to take account of the interaction between adsorbates. When adsorbates are all the same, this means a coverage effect. When several adsorbates are present, this means co-adsorption effects.

For MgO(100), we have used a single layer (Figure 1). This surface is non-polar; this is a prerequisite for a truly stable surface [14, 15]; the geometry is the same as in the bulk structure; this surface does not reconstruct [14] as can be concluded from LEED experiments [16]. In the analysis of periodic calculations [17] the electron charge distribution at the surface is found to be similar to that in the bulk.

Our simplest model for representing the TiO_2 surfaces is a polymer that is characteristic of the rutile structure. It is a chain containing a sequence of titanium atoms and the equatorial oxygen atoms attached to them (see Figure 2). It is made of rectangular pieces, the TiO distances are the same as the equatorial distances in the bulk. Rutile is thus decomposed by the cleavages of all the apical Ti—O bonds. It is established that the structure of the TiO_2 surfaces is similar to those of the bulk [14, 18–20]. It presents two sites for adsorption, the cation sites (Lewis acids) and bridging oxygen sites (Lewis bases). The bridging oxygen atoms lie in the polymer plane but similar sites may be found on the outmost atoms in the (110) face (see Figure 3). The polymer is found in the (110) face that corresponds to the easiest cleavage of the crystal. This (110) face may indeed be analysed as made of polymers parallel and perpendicular to the surface plane coupled through apical Ti—O bonds.

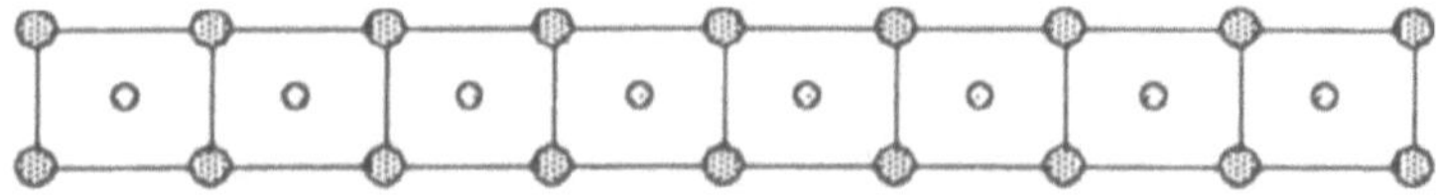

Figure 2. The polymer TiO_2. The large circles represent the oxygen atom and the small ones the metal centers.

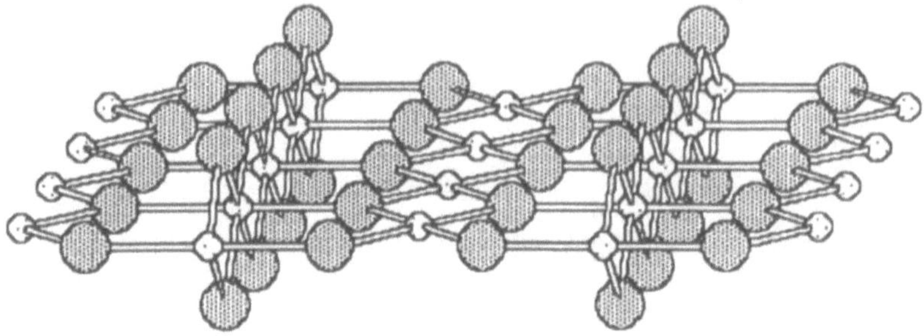

Figure 3. The (110) face of rutile.

The polymer is obviously far from being a complete representation of a crystallographic face. However, as it is a common feature of several faces, differences from this reference emphasize the face specificity of various slabs. By increasing the coordination of the atoms, the use of a slab is supposed to reduce the adsorption energies. The adsorption energy increase which, on the contrary, is found [17, 21, 22] when the (110) surface is fully described is therefore remarkable. It is explained by a polarization of the polymer in the plane and by an enhanced acidity of the exposed metal centers. The (110) face is the most stable face since it is the most dense and since the top layer has no dipole moment. The cohesive energy of the (110) slab, is 118 kcal/mol* per TiO_2 unit (an improvement of 35 kcal/mol relative to the polymer). Usually, a large stability is not an index of reactivity.

Both models contain titanium atoms with the oxidation number +IV and bridging oxygen atoms. The presence of tetracoordinated cations is not enough to create states in the band gap and the electronic structure remains close to that of larger systems. Spectroscopy also supports the notion that the reduction of the ligand coordination of cations is not sufficient to alter the surface electronic structure [14] and all-electron calculations confirm that the features of the density of states and the binding energies of the different surfaces 'agree with a possible dissociation behaviour of the rutile structure into TiO_2 chains, which are calculated to be stable' [23].

Titanium atoms from our polymer model are tetracoordinated. They are more reactive than the pentacoordinated surface atoms would be. To estimate the influence on the heat of adsorption of the increase in coordination, we have also used the 'double polymer model' described in Ref. [24], in which a vertical polymer

* 1 Hartree (1 u.a.) equals 27.2 eV and corresponds to 627.5 kcal per mole.

accounts for the atoms of the first layers beneath the surface. The conclusions on the preferential adsorption modes are preserved in this model even if all the calculated heat of adsorption is decreased. The reduction of heat of adsorption is of the order 25–28% for NH_3 and CO. Based on spectroscopic observations, Heinrich also remarked that ligand coordination was 'not the dominant factor affecting the surface electronic structure' by opposition to the introduction of defects and vacancies [14].

We chose to privilege the description of the valence orbitals and then used effective core pseudopotential calculations. The pseudopotentials are those from Durand and Barthelat [25]. The basis set for titanium atom consists of the d functions contracted to a (4/1) basis set and a single $4sp$ shell with an exponent of 0.484 for polarization purposes [26]. Those for the adsorbates are derived from the PS-31G basis sets [27]. We have retained these basis sets for the study of all the molecules to be able to make a comparison. That for the magnesium atoms is a PS-21G basis set [27]; the exponent of the most diffuse orbitals is 0.11. The description of the $3sp$ (Mg) valence orbitals is improved with respect to the use of the 'ionic' basis set [28] with a unique Gaussian ($\zeta = 0.4$). Usually, one uses more diffuse orbitals for molecular calculations. This is not necessary for the periodic calculations since the diffuse orbitals in a unit cell combine with those of the adjacent cell and would generate extremely diffuse functions. The absence of very diffuse orbitals for the adsorbate, however, contributes to an overestimation of the heat of adsorption since the reference energy involves the molecular calculation of the adsorbate.

The adsorption energies are expressed according to the expression:

$$E_{\text{ads}} = E_{\text{adsorbate}} + E_{\text{substrate}} - E_{(\text{adsorbate+substrate})},$$

where $E_{(\text{adsorbate+substrate})}$ is the energy of the adsorbate/substrate system, $E_{\text{substrate}}$ is the energy of the substrate and $E_{\text{adsorbate}}$ is the energy of the isolated adsorbate in its equilibrium geometry. A positive E_{ads} value corresponds to a stable adsorbate/substrate system.

We have also tested the variations due to the polarization functions in some cases [21]: the adsorption energies are reduced by 7 kcal/mol both for the molecular adsorption and the dissociative adsorption. The comparison between two adsorption modes therefore remains valid. The geometry does not change. We did not correct our calculations to take the electron correlation effects into account. Even though a contribution to the total energy of the system should be large due to the oxygen ions, it should be roughly constant in the different adsorbate + substrate systems where the charge distribution remains comparable. The correlation effects would probably increase the adsorption energies since the adsorbate are more ionic than the gas phase molecule. The energy difference between the various adsorption modes should not be affected.

Table I. Correlation between the heats of adsorption on the polymer model for TiO_2 and the adsorbates [44].

Molecule	Molecular adsorption		Proton affinity
	$\theta = 1/2$	$\theta = 1$	
CO_2	19	10.4	130
CO	20.0	15.2	139
H_2S	21.1	−1.7	157
H_2O	42	36.5	173
MeOH	44	39.6	185
NH_3	47.6	36.2	220

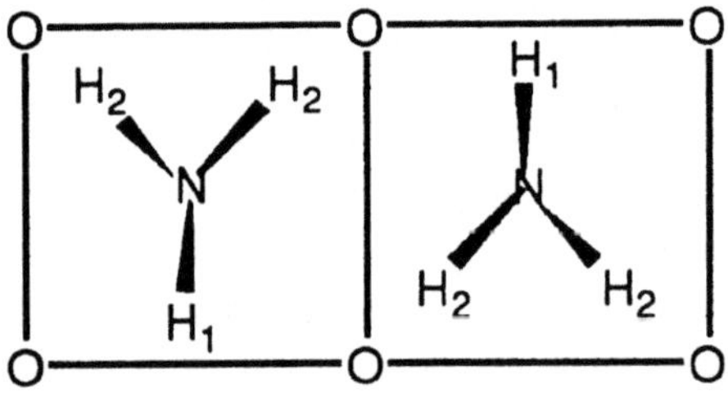

Figure 4. The alternated mode for molecular ammonia adsorption.

3. Molecular Adsorption on Clean Surfaces

On the titanium oxide, the various molecules that are adsorbed without dissociation are always bound to the exposed metal centers. The heat of adsorption correlates with the proton affinities as shown in Table I. Therefore, adsorbates must be considered as bases and the clean surface appears to be acidic. At a $\theta = 1/2$ coverage (one adsorbate per two exposed Ti surface atoms), the most basic adsorbate is ammonia [22]. At $\theta = 1$, a repulsion between adjacent molecules reduces the heat of adsorption. In the best geometry the molecules (Figure 4) alternate to decrease this repulsion that is reduced from 13.1 to 9.4 kcal/mol.

Ammonia behaves as a base that donates its electron pair to the metal generating a dative Ti—N bond perpendicular to the surface polymer. The optimized Ti—N distance, 2.19 Å, is close to the sum of the covalent radii of Ti and N, 2.07 Å; the adsorption is clearly a chemisorption. The charge transfer (0.012 at $\theta = 1$) is very weak. The lack of any appreciable charge transfer is also found with weaker bases [24, 29–32]. The Ti—N bond is very asymmetric (see Figure 5) and, for that reason, the corresponding overlap population is weak, 0.043.

CO_2 is the other extreme and appears as a weak base. It behaves unambiguously as a base since in the best adsorption mode, it interacts with the metal center; it is perpendicular to the surface above the metal [21]. At $\theta = 1/2$, in the adsorption mode parallel to the surface (less favorable than the perpendicular orientation: E_{ads}

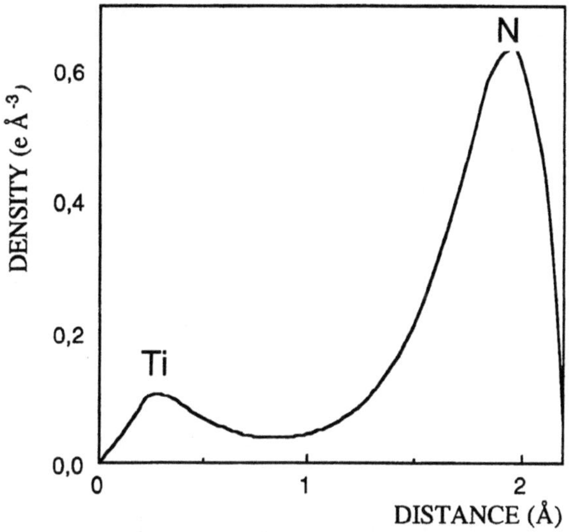

Figure 5. The electronic density along a Ti—N bond.

= 16.4 kcal/mol vs. E_{ads} = 19.0 kcal/mol), CO_2 is also a base. In this orientation, each terminal oxygen atom interacts with one metal center forming a Ti$\cdots$$CO_2$$\cdots$Ti sequence; there is no interaction with the surface oxygen atoms. The frontier orbital for CO_2 is the $1\pi_g$ orbital, the HOMO (highest occupied molecular orbital), while for the perpendicular mode it is the $3\sigma_u$ that is lower in energy. The direct adsorption on the oxygen atom is very weak, E_{ads} = 6.3 kcal/mol and not competitive with the other modes. The optimized O$\cdots$C distance is very large, 2.4 Å. This adsorption mode corresponds to physisorption rather than to chemisorption. The adsorption on a surface hydroxyl group is very weak compared to that on exposed cations.

Water and hydrogen sulfide have intermediate heat of molecular adsorption. The water is in an atop position above the titanium. The four atoms Ti$\cdots$$OH_2$ are in a plane to allow a conjugation of the p electron pair of the water with the vacant d orbitals of the titanium atom of the same symmetry. Since its interaction with the titanium is weaker, H_2S can more easily deviate from the perpendicularity. The sulfur atom is hybridized and H-bonds can be built with the surface oxygen. In the best geometry, H_2S is nearly parallel to the surface plane (the angle is 10°); by this motion toward the surface, E_{ads} is improved from 18.6 kcal/mol to 21.1 kcal/mol).

The best adsorption mode for carboxylic acids involve two adsorbed oxygen atoms. For the formic acid, both the O from the carbonyl group and the O from the hydroxyl group are adsorbed on adjacent metal centers (bidentate adsorption mode). For the oxalic acid, oxygens from each carboxyl group are involved. The adsorption energy for the formic acid and the oxalic acid are 32.1 kcal/mol [33] and 48 kcal/mol, respectively [34].

Table II. Heat of adsorption in kcal/mol for H_2O/MgO for the structures of Figure 6. The first line are our results for $\theta = 1/2$. The second line give the adsorbate-adsorbate repulsion at this coverage. The basic mode is particularly disfavored by this repulsion (the unit cell is $\sqrt{2}\sqrt{2}R45°$ (see Figure 1); for the $p2 \times 1$ cell, the repulsion would be larger, +7.9 kcal/mol leading to a slight endothermicity for the adsorption energy, $E_{ads} = -0.5$ kcal/mol). The two last lines are results for $\theta = 1/4$ from Ref. [35] at the Hartree–Fock level and with the correlation term from Perdew (1991) [55].

Heat of adsorption	A Basic	C 1H-acidic	D 2H-acidic	E amphoteric
Present calculations	+2.1	+7.6	+4.0	+9.7
H_2O—H_2O Repulsions	+4.5	−0.9	+4.4	−2.2
RHF	4.1	–	5.6	–
with correlation	6.3	–	11	–

On MgO(100) at $\theta = 1/2$, the water appears rather as an acid than as a base (see Table II). With our basis set, the heats of adsorption on the oxygen atom (acidic modes, one H-adsorbed (Figure 6C) or 2H-adsorbed (Figure 6D)) are larger than those on the metal (Figure 6A). The best adsorption mode (Figure 6E) uses both properties; it also optimizes the lateral interactions (see Table II). With the 'ionic' basis set CO_2 [28] at $\theta = 1/4$, H_2O is also rather acidic than basic [10, 35]. In these calculations, the two protons of the water are oriented to interact with the surface oxygen (Figure 6D). This is reminiscent of the $H_2O\cdots OH^-$ structure (Figure 6B) with C_{2v} symmetry.

4. Dissociative Adsorptions

If the adsorption is strong enough, the adsorbate may be distorted or dissociated into fragments. Formally considering an heterolytic cleavage of an adsorbate, for instance H_2O, one gets both an acidic and a basic fragments. The former adsorbs on the surface oxygen and the latter on the titanium. Noguera *et al.* [36] have calculated the independent adsorptions of a proton and of an hydroxyl group. For TiO_2, the adsorption of OH^- ($E_{ads}(OH^-) = 8.3$ eV) is larger than that of H^+ ($E_{ads} = 5.9$ eV). In spite of the extreme acidity of the proton, the adsorption of the basic species is the largest. The dissociative adsorption is energetically favorable [37, 38]. For MgO that is more basic, it is the other way round ($E_{ads}(OH^-) = 0.7$ eV vs. $E_{ads}(H^+) = 8.5$ eV). The molecular adsorption remains more favourable on perfect surfaces [10, 12, 36, 39, 40].

It is not possible to investigate the individual adsorption of each charged fragment with the CRYSTAL program since our total system has to remain neutral; therefore we optimize the adsorption of the two moieties. Table III shows that the dissociation on the TiO_2 surface follows the order of the dissociations in the gas phase (see acidic cleavages).

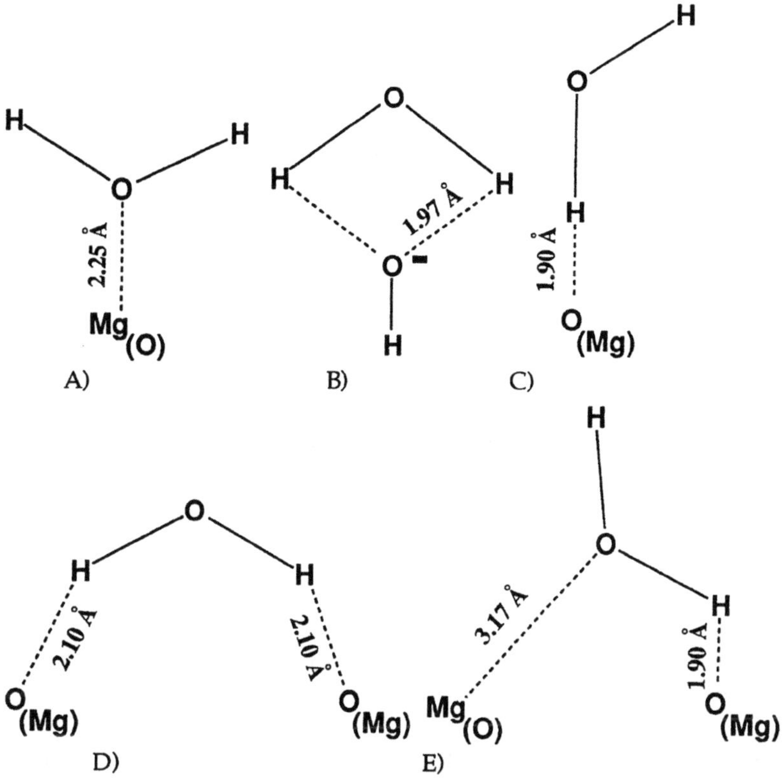

Figure 6. (A) The basic adsorption mode for H_2O/MgO the H atoms are oriented along the Mg—Mg rows so that the 2*p* electron pair of the water molecule is along the perpendicular Mg—Mg row. (B) The $H_2O\cdots OH^-$ structure. (C) And (D) the acidic adsorption modes for H_2O/MgO through one or two H-bonds. (E) The amphoteric adsorption mode; with these parameters, the $O\cdots H$ interaction seems to dominate. It does not seem favorable to make two H-bonds and to have the molecular plane of H_2O parallel to the surface plane.

H_2S cleaves very easily relative to H_2O, which is consistent with gas phase cleavages. For the other adsorbates, two cleavages – acidic or basic – may be distinguished. On the surface, they imply a different orientation of the fragments. When ROH gives a basic cleavage, the fragment R is adsorbed on the bridging oxygen atom and the titanium center is hydroxylated (singly coordinated hydroxyl). By contrast, for an acidic cleavage, the OR fragment goes on the titanium center and a bridging hydroxyl group is formed. For MeOH the basic cleavage is obtained on the surface since, in the gas phase, it corresponds to the least endothermic cleavage. This is the only example. For MeSH, the result is reversed; in the gas phase the acidic cleavage is easier. Surprisingly, carboxylic acids do not appear as Brønsted acids in the gas phase. A basic cleavage is therefore predicted on the surface for the formic acid; it is indeed the best adsorption mode when only one oxygen atom

Table III. Gas phase cleavages and surface dissociations and heat of adsorptions. The polymer model is used for TiO_2. The molecules are Brønsted bases or acids and for the dissociative adsorption both fragments are adsorbed. Values in kcal/mol.

	Gas phase cleavage	Surface dissociation	Dissociative adsorption
Acidic cleavages			
$H_2S''\ H^+ + SH^-$	354.9 (353.4)	−43.3	64.9
$HCOOH''\ H^+ + HCOO^-$	356.7 (345.2)	−58.8*	103*
$MeSH''\ H^+ + MeS^-$	360.6 (359–361.1)	−38	57
$MeOH''\ H^+ + MeO^-$	406.4 (378.2–379.2)	−11.8	51.4
$H_2O''\ H^+ + OH^-$	423.84 (390.8)	−16.9	58.9
$NH_3''\ H^+ + NH_2^-$	447 (403.6)	−10.4	46.6
Basic cleavages			
$HCOOH''\ HCO^+ + OH^-$	290.8 (259)	−7.7*	51.9
$MeOH''\ Me^+ + OH^-$	289.2 (271.4–277.4)	−18.2	62.2
$MeSH''\ Me^+ + SH^-$	455.0	−36	55

* The adsorption involves two adsorbed functions.

is adsorbed. However, the best adsorption involves two oxygen atoms on adjacent titanium centers; this corresponds to an acidic cleavage.

The complete dissociation of ammonia is not favorable since NH_2^- is the strongest base. A dissociation of half of the molecule at $\theta = 1$ is obtained due to H-bonds as shown in Figure 7. The Hard and Soft Acid and Base theory [41, 42] provides an explanation for the molecular adsorption. Two bases NH_2^- (the hard base) and OH^- (the soft base) compete to bind with an hard acid, H^+, and a soft one, Ti^{4+}. The hard acid couples with the hard base, leading to molecular ammonia, while the soft one is hydroxylated. On hydrated surfaces, the formation of NH_4^+ and OH^- on the surface is possible when NH_4^+ is triply bonded to three surface oxygen atoms [22].

As CO_2 is weakly adsorbed, it remains linear in the same geometry as in the gas phase. Under adsorption the central atom becomes much more acidic and can participate in a reaction with other adsorbates. This leads to a consideration of the co-adsorption effects.

5. Lateral Interactions

At high coverages, the adsorbates molecules interact. The main interactions are H-bonds and van der Waals interactions.

The adsorption of H_2O at $\theta = 1$ allows a sequence of H bonds, which does not exist at lower coverage [37]. The heat of adsorption from $\theta = 1/2$ to $\theta = 1$ is larger than that from $\theta = 0$ to $\theta = 1/2$; 67.8 kcal/mol vs. 58.9 kcal/mol.

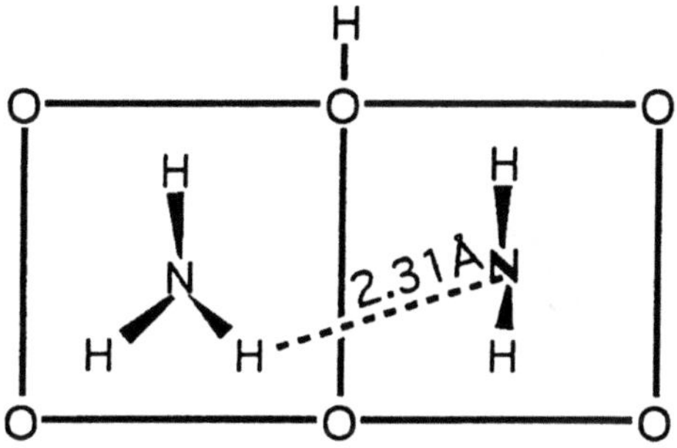

Figure 7. The mixed adsorption mode for ammonia at saturation that combines molecular and dissociative adsorption modes.

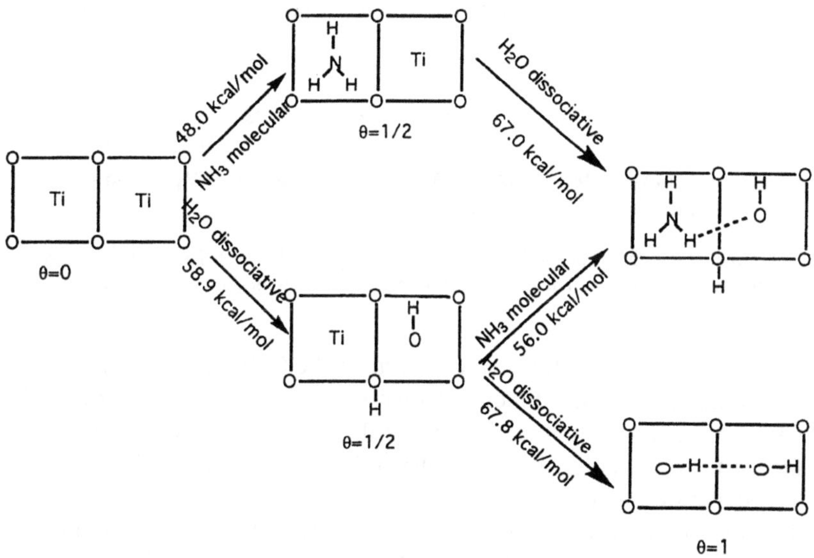

Figure 8. The formation of sequence of adsorbed species at saturation that benefits from sequence of H-bonds.

Since the polymer is a part of the (110) surface, a similar sequence of H-bonds can also be found on the complete (110) surface [37]. When the protons of the hydroxyl groups are rotated by 90° and oriented toward the bridging oxygen out of the surface plane, another hydrogen bonding is found and this explains the nonlinearity of the TiOH sequence in addition to the hybrization of the hydroxyl oxygen atom [39].

A similar effect explains a co-adsorption effect between water and ammonia. The molecular adsorption of ammonia on a hydroxylated surface ($\theta = 1/2$) is more favorable than the adsorption on a bare surface (56 kcal/mol vs. 48 kcal/mol).

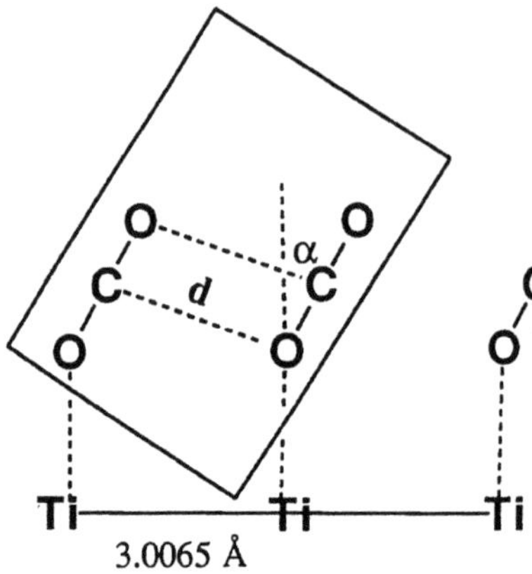

Figure 9. The tilt of the adsorbed CO_2 at saturation; the topology of the CO_2 dimer is indicated in the rectangular frame.

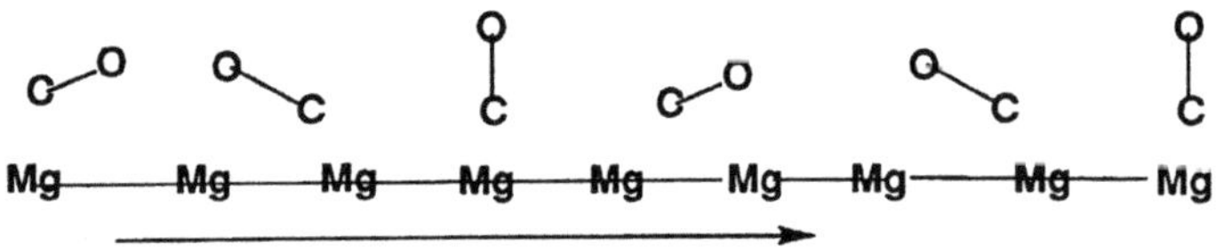

Figure 10. CO/MgO at $\theta = 3/4$. The CO are along the Mg rows. Two of them are symmetrically bent over the surface. The cell vector along the Mg row is shown under the figure.

Van der Waals interactions can be seen on the CO_2 adsorption mode. In Figure 9 the geometry of a CO_2 dimer is shown within the rectangular frame. The optimization of the lateral interactions force the adsorbed molecule to bend over the surface. At low coverage they are independent and normal to the surface. At high coverage, the tilt angle, α, is 29°.

The adsorption of CO on MgO is another example of the importance of lateral interactions between adsorbates [43]. At low coverages, CO is vertical above the surface [44]. It is above a magnesium atom and oriented C down, with the most basic electron pair toward the surface. This forces the adsorbed molecules to be parallel to each other. At higher coverage, $\theta = 3/4$, the molecules move to decrease their mutual repulsion. In the most favorable case they are above the magnesium rows. The adsorption therefore still involves the cations and not the anions of the surface. Two-thirds of the CO molecules are tilted over the surface to decrease the repulsion between the adsorbates.

6. The Clean MgO Surface

Auroux and Cervasini [2, 3] emphasized that the energy adsorption of CO_2 on MgO was larger than that of NH_3, at variance with that on TiO_2. According to these authors, the large heat of adsorption $E_{CO_2/MgO}$ is due to an increased surface basicity, CO_2 behaving as the acid. On the contrary, our calculations show that, in this case, CO_2 behaves as a base being adsorbed on the acidic surface site. We optimized the CO_2 adsorption on the clean MgO(100) surface and found the same adsorption mode as Girardet *et al.* [45–47]. This adsorption mode is parallel to the surface; CO_2 bridging two magnesium atoms. CO_2 therefore binds to the surface acidic sites through the oxygen pairs of the terminal oxygen atoms (Lewis basic centers). The distance to the surface plane is 2.71 Å (MgO = 2.73 Å) and the adsorption energy is 7 kcal/mol. The perpendicular adsorption mode that also corresponds to an acidic surface (at $\theta = 1/2$) leads to a weakly negative E_{ads} (−1.7 Å) at MgO = 2.67 Å whereas the adsorption on the oxygen atom that would imply the basicity of the surface is repulsive by 30 kcal/mol; we conclude that the O atoms from the MgO surface are not basic enough for the CO_2 adsorption, in agreement with an $(OMg_5)^{8+}/CO_2$ cluster calculation [48]. Pacchioni also calculated the parallel adsorption from an *ab initio* calculation on a $Mg_2O_8^{12-}$ cluster [31] and found an energy very similar to that of the perpendicular mode (0.17 eV and 0.18 eV).

The parallel adsorption mode, which is the most favorable case on MgO, has been found slightly less stable than the perpendicular orientation on TiO_2, (16 kcal/mol vs. 19 kcal/mol) according to CRYSTAL calculations [21]. For MgO, we calculated the interaction between the adsorbates in the ordered CO_2 monolayers; for the perpendicular mode the $CO_2 \cdots CO_2$ interactions are repulsive by 2.8 kcal/mol, whereas they are attractive by 2 kcal/mol for the parallel mode shown in Figure 11. Thus, the lateral interactions between the adsorbates contribute to favoring this adsorption at $\theta = 1/2$; this is missing in the cluster calculations. Girardet *et al.* [45] proposed a larger supercell were the relative orientations of the CO_2 alternates, all of them bridging the MgMg first neighbors. The $CO_2 \cdots CO_2$ interaction is attractive by 2.3 kcal/mol, very slightly better than for the uniform orientation shown in Figure 11.

NH_3 on MgO is adsorbed on the magnesium atoms, the acidic centers. It has been shown from periodic HF calculations [49] that the adsorption was exothermic for coverages 1/2 or below. The adsorption on the oxygen atom (basic center) is unstable toward desorption as expected. In Table IV we show the heats of adsorption of CO_2 and NH_3 on TiO_2 and MgO. Results are in agreement with the differential measures of Auroux and Cervasini [2, 3].

The differential adsorption mode may be explained by the HSAB theory [41, 42]. The hard base, NH_3, interacts preferentially on the hard acid, Ti^{4+} (more strongly charged), whereas the soft base, CO_2, interacts on the soft acid, Mg^{2+}. According to Klopman, the hardness is measured by the E^+ energy value, the energy of an orbital under the influence of its partner; those for Ti^{4+} and Mg^{2+} are 4.35 eV and 2.42 eV, respectively.

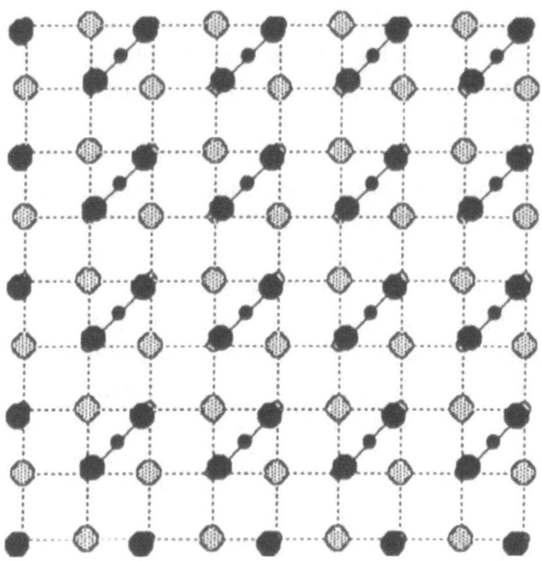

Figure 11. CO_2/MgO. In these calculations all the CO_2 are assumed to be parallel. Better patterns with alternation of orientations have been proposed by Girardet *et al.* [45].

Table IV. Heats of adsorption on TiO_2 (polymer model) and MgO(100) in kcal/mol. Basis sets for the adsorbate are PS31G. For Ti and Mg the basis sets are from [37, 43] except for the value in parentheses where the 'ionic' basis for MgO has been taken [28].

Substrate	E_{ads} (CDO_2)	E_{ads} (NH_3)
TiO_2	19.0	47.6
MgO	7 (8.3)	3.2 (6.3)

It is surprising that MgO appears to be basic for the H_2O adsorption and acidic for the CO_2 adsorption. CO_2 is a stronger acid in the gas phase.

$$CO_2 + OH^- \rightarrow CO_3H^- \qquad E = -82.3 \text{ kcal/mol}$$
$$H_2O + OH^- \rightarrow OH_2 \cdots OH^- \qquad E = -30 \text{ kcal/mol}.$$

OH^- is a much stronger base than MgO; in the gas phase the formation of the bicarbonate corresponds to an activationless chemical reaction [50]. In aqueous phase, solvation effects decrease the nucleophilicity of OH^-_{solvated} and there is an energy barrier. Under adsorption, a different mechanism occurs. The nucleophilic attack would rather look like the formation of a possible $O_2C\cdots OMg$ precursor, which is not confirmed by the calculations. The energy level of the LUMO of CO_2 is high (+0.177 Hartree) and CO_2 is acidic only in a distorted geometry –

bent with elongated CO distances. The adsorption of such species (CO = 1.24 Å, OCO = 120°; the LUMO level is −0.046 Hartree) leads to a stable minimum at $O_{(Mg)}$—$C_{(O_2)}$ = 1.31 Å but corresponding to an endothermic adsorption. Thus, the formation of the bicarbonate should not begin by a nucleophilic attack of the carbon atom of CO_2. The CO_2 molecule needs first to be activated; this is done by the adsorption where CO_2 is the base interacting with a metal cation. An argument supporting this behaviour may be found in biochemically active species. For the mechanism for the hydration of CO_2 by carbonic anhydrase, two mechanisms have been proposed; one [51, 52] corresponds to a nucleophilic attack of M—OH^- groups to the CO_2 molecule while the second one [53] involves the binding of CO_2 to metallic center. In agreement with our surface model, calculations [54] favor the second mechanism.

7. The MgO Hydrated Surface

When water is dissociatively adsorbed, the surface is covered by hydroxyl groups formed either by the OH^- adsorption on the metal centers or by protonation of the surface oxygen atoms. For TiO_2, the oxygen of the hydroxyl groups (MgOH) are less basic than the bridging oxygen from the surface (MgOMg). This is why a proton goes on the latter and does not recombine with the hydroxyl group to give the molecular adsorption. When CO_2 is adsorbed, it is activated and the formation of bicarbonate is possible.

In contrast to TiO_2 (110 face or polymer) the MgO(100) surface does not allow the dissociation of water on the surface. Under dissociation, the adsorption energy is negative, −30 kcal/mol. The surface OH^- groups are very weakly bound to the surface and thus very reactive. This is the case when water is coadsorbed with CO^2. Then, there is a positive overlap population between the carbon atom and the oxygen atom of the neighbouring hydroxyl group. Bicarbonate and carbonate species are stable vs. coadsorption. The adsorption energy for the water (dissociated) and the carbon dioxide leading to the bicarbonate represents 4.9 kcal. This also represents with respect to the hydrated surface at $\theta = 1/2$ (the reference energy is less stable than in the previous case) an adsorption energy for CO_2 of 34.9 kcal/mol. The difference in reactivity between the hydrated surface, E_{ads} (CO_2) = 34.9 kcal/mol, and the clean surface, E_{ads} (CO_2) = 7 kcal/mol, reveals the facility for the bicarbonate formation.

8. Conclusion

The adsorption of molecules on the clean surfaces of the oxide always involve the exposed metal centers whatever the adsorption mode is – perpendicular or flat. These surfaces consequently appear as acidic species reacting with basic adsorbates. H_2O/MgO is the main exception; the driving force for this adsorption is the HOH···OMg bond formation even though a weak H_2O···MgO bond is also

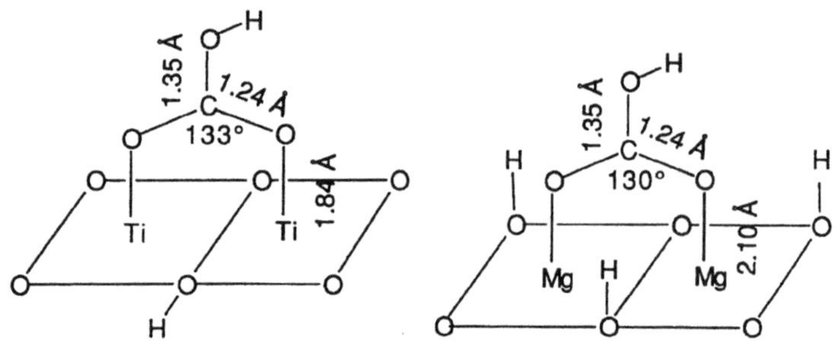

Figure 12. The bicarbonate adsorption on TiO_2 and MgO.

formed. The cleavage of the molecules corresponds to the easiest acidic or basic cleavage in the gas phase. However, in many cases, the fragment adsorbed may involve several atoms. For the carboxylic acid, the acidic cleavage is obtained since the formate ion is adsorbed by its two oxygen atoms. At high coverage, lateral effects are important. Stabilization through H-bonds favors the vicinity of adsorbates on the surface (islands of adsorbates, high coverage and coadsorption with ammonia). Van der Waals interaction explain the change of orientation between low and high coverages. The lateral interactions are more important on MgO than on TiO_2. The difference of reactivity on the clean surfaces between TiO_2 and MgO may be expressed in terms of acid vs. basic behaviour for water adsorption and in terms of hard and soft acidity for CO_2 adsorption.

References

1. C. Noguera: *Chiemie et physique des surfaces d'oxydes*, Eyrolles, Paris (1995).
2. A. Auroux and A. Cervasini: *J. Phys. Chem.* **94**, 6371 (1990).
3. A. Cervasini and A. Auroux: *J. Therm. Analysis* **37**, 1737 (1991).
4. A.L. Allred and E.G. Rochow: *J. Inorg. Nucl. Chem.* **5**, 264 (1958).
5. A.L. Allred: *J. Inorg. Nucl. Chem.* **17**, 734 (1961).
6. R. Dovesi, C. Pisani, C. Roetti, and M. Causà: *Crystal 88*, Bloomington, Indiana, QCPE Program No. 577 (1988).
7. R. Dovesi, V.R. Saunders, and C. Roetti: *Crystal 92, User Documentation*, University of Torino and SERC Daresbury Laboratory (1992).
8. C. Pisani, R. Dovesi, and C. Roetti: *Hartree–Fock Ab Initio Treatment of Crystalline Systems* 48, Lecture Notes in Chemistry, Springer, Heidelberg (1988).
9. A.G. Pelmenschikov, G. Morosi, A. Gamba, S. Collucia, G. Martra, and E.A. Paukshtis: *J. Phys. Chem.* **100**, 5011 (1996).
10. C.A. Scamehorn, N.M. Harrison, and M.I. McCarthy: *J. Chem. Phys.* **101**, 1547 (1994).
11. A.M. Ferrari and G. Pacchioni: *J. Phys. Chem.* **99**, 17010 (1995).
12. J. Goniakowki and C. Noguera: *Surf. Sci.* **330**, 337 (1995).
13. U. Diebold and T.E. Madey: *'Electron Stimulated Desorption (ESD) of Ammonia on TiO_2 (110): The Influence of Substrate Defect Structure' 31*, in A.R. Burn E.B. Stechel and D.R. Jennison (eds.), *Desorption Induced by Electron Transitions DIET V,* Springer-Verlag, Berlin, Heidelberg (1993).
14. V. Heinrich: *Rep. Prog. Phys.* **48**, 1481 (1985).
15. P.A. Cox, F.W.H. Dean, and A.A. Williams: *Vacuum* **33**, 839 (1983).
16. T. Urano, T. Kanaji, and M. Kaburaji: *Surf Sci.* **134**, 109 (1983).

17. M. Causà, R. Dovesi, C. Pisani, and C. Roetti: *Surf. Sci.* **175**, 551 (1986).
18. V. Heinrich and R.L. Kurtz: *Phys. Rev. B* **22**, 6280 (1981).
19. T. Wolfram: *J. Vac. Sci. Technol.* **18**, 428 (1981).
20. M. Tsukada, H. Adachi, and C. Satoko: *Prog. Surf. Sci.* **14**, 113 (1983).
21. A. Markovits, A. Fahmi, and C. Minot: *J. Mol. Struct., Theochem.* **371**, 219 (1996).
22. A. Markovits, J. Ahdjoudj, and C. Minot: *Surf. Sci.* **365**, 649 (1996).
23. P. Reinhardt and B.A. Hess: *Phys. Rev. B* **50**, 12015 (1994).
24. A. Fahmi and C. Minot: *J. Organomet. Chem.* **478**, 67 (1994).
25. P. Durand and J.C. Barthelat: *Theor. Chim. Acta* **38**, 283 (1975).
26. B. Silvi, N. Fourati, R. Nada, and C.R.A. Catlow: *J. Phys. Chem. Solids* **52**, 1005 (1991).
27. Y. Bouteiller, C. Mijoule, M. Nizam, J.C. Barthelat, J.P. Daudey, M. Pelissier, and B. Silvi: *Mol. Phys.* **65**, 295 (1988).
28. M. Causà, R. Dovesi, C. Pisani, and C. Roetti: *Phys. Rev. B* **33**, 1308 (1986).
29. R. Dovesi, R. Orlando, F. Ricca, and C. Roetti: *Surf. Sci.* **186**, 267 (1987).
30. G. Pacchioni, G. Cogliandro, and P. Bagus: *Int. J. Quantum Chem.* **42**, 1115 (1992).
31. G. Pacchioni: *Surf. Sci.* **281**, 207 (1993).
32. M. Causà, E. Kotimin, C. Pisani, and C. Roetti: *J. Phys. C/Solid State Phys.* **20**, 4991 (1987).
33. J. Ahdjoudj and C. Minot: unpublished results (1996).
34. A. Fahmi, C. Minot, P. Fourré and P. Nortier: *Surface Science* **343**, 261 (1995).
35. C.A. Scamehorn, A.C. Hess, and M.I. McCarthy: *J. Chem. Phys.* **99**, 2786 (1994).
36. J. Goniakowki, S. Bouetto-Russo, and C. Noguera: *Surf. Sci.* **284**, 315 (1993).
37. A. Fahmi and C. Minot: *Surf. Science* **304**, 343 (1993).
38. T. Bredow and K. Jug: *Surf. Sci.* **398-408**, 398 (1995).
39. J. Goniakowki and M.J. Gillan: *Surf. Sci.* **350**, 145 (1996).
40. W. Langel and M. Parrinello: *Phys. Rev. Lett.* **73**, 504 (1994).
41. G. Klopman: *J. Am. Chem. Soc.* **90**, 223 (1968).
42. R.G. Pearson: *J. Chem. Ed.* **45**, 581 (1968).
43. C. Minot, M.A. VanHove, and J.P. Biberian: *Surf. Sci.* **346**, 283 (1996).
44. C. Minot, A. Fahmi, and J. Ahdjoudj: *Periodic HF Calculations of the Adsorption of Small Molecules on TiO_2*, Drymen, Scotland.
45. S. Picaud, S. Briquez, A. Lakhlifi, and C. Girardet: *J. Chem. Phys.* in press (1995).
46. V. Panella, J. Suzanne, P.N.M. Hoang, and C. Girardet: *J. Phys. I France* **4**, 905 (1994).
47. S. Briquez, A. Lakhlifi, S. Picaud, and C. Girardet: *Chem. Phys.* **194**, 65 (1995).
48. G. Pacchioni, J.M. Ricart, and F. Illas: *J. Am. Chem. Soc.* **116**, 10152 (1994).
49. Y. Ferro, A. Allouche, F. Corà, C. Pisani, and C. Girardet: *Surf. Sci.* **325**, 139 (1995).
50. Z. Peng and J.K.M. Merz: *J. Am. Chem. Soc.* **115**, 9640 (1993).
51. J. Liang and W.N. Lipscomb: *J. Am. Chem. Soc.* **108**, 5051 (1986).
52. J.-Y. Liang and W.N. Lipscomb: *Int. J. Quantum Chemistry* **XXXVI**, 299 (1989).
53. S. Lindskog: *Zinc Enzymes*, N.Y., Wiley (1983).
54. O. Jacob, R. Cardenas, and O. Tapia: *J. Am. Chem. Soc.* **112**, 8692 (1990).
55. J.P. Perdew: in *Unified Theory of Exchange and Correlation Beyond the Local Density Approximation*, Nova Science, New York, 1991.

Molecular Engineering **7**: 263–281, 1997.

Optical Spectra of High-Temperature Superconductors

MLADEN GEORGIEV and LUBOMIR MIHAILOV
Institute of Solid State Physics, Bulgarian Academy of Sciences, 72 Tsarigradsko Chaussee, 1784 Sofia, Bulgaria.

(Received: 12 September 1996; accepted 17 June 1997)

Abstract. This paper complements our recent survey [1] of experimental and theoretical work on the optical spectra of high-temperature superconductors. The emphasis is laid on experiments involving single crystals and good textured ceramic samples. Data have been collected on chemically doped samples, as well as on samples in which carriers are created by photodoping. One of the most remarkable consequences is the observation of a photoinduced superconductivity and a phase separation of photoexcited carriers. The theory has offered a few main suggestions but they have all failed to produce the magic formula for understanding the optical response. Nevertheless, bipolaron models seem somewhat closer to reality. Future experimental, theoretical, and practical developments are considered.

Key words: Optical spectra, layered perovskites, polarons and bipolarons, photoinduced phenomena, phase segregation

1. Introduction

The optical studies of high-temperature superconductors (HTSC) cover a wide spectral range from FIR to the vacuum UV and beyond at temperatures, T, between 2 K and 300 K [2]. HTSC optical absorptive or conductivity spectra are either obtained by direct measurement or derived from reflectivity data [3, 4]. Missing portions are complemented by a reasonable guess of a spectral dependence outside the direct-access frequency range.

The optical conductivity of a parent insulating phase exhibits a charge-transfer (CT) band in NIR and VIS near 2 eV. Prominent phonon peaks are seen in FIR. Upon doping, a Drude spectrum peaked at zero energy appears in FIR followed by a broad MIR band centered at $\approx$0.5 eV. Phonon bands stand in between the Drude part and MIR.

The MIR band is wide ($\approx$1 eV), grows at the expense of the NIR CT band, and its peak energy decreases as doping increases. It is common for HTSC semimetals from the cuprates La_2CuO_{4+y}, Nd_2CuO_{4-y} (NCO), $La_{2-x}Sr_xCuO_4$ (LSCO), $YBa_2Cu_3O_{7-\delta}$ (YBCO), $Bi_2Ba_2Ca_2CuO_{8-\delta}$ (BBCCO), $Bi_2Sr_2CaCu_2O_{8+y}$ (BSCCO) to lower-T_c $SrTiO_{3-y}$. In La_2CuO_{4+y}, NCO and YBCO single crystals the MIR absorption is resolved into bands at $\approx$0.13 eV and at $\approx$0.5–0.75 eV. The former is T-dependent at $T > 100$ K, the latter is insensitive to T.

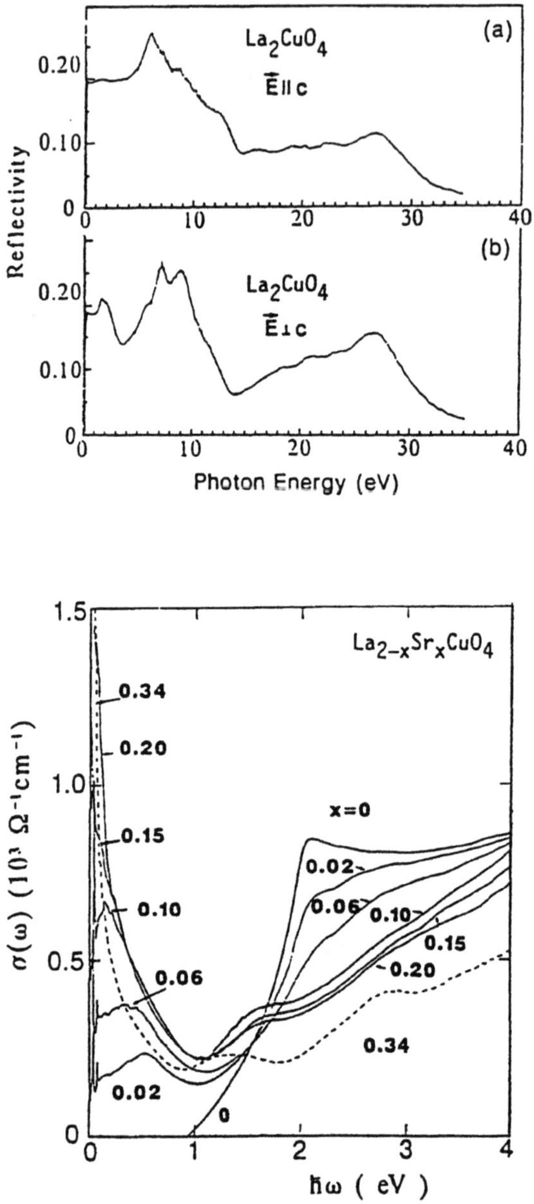

Figure 1. Optical spectra in $La_{2-x}Sr_xCuO_4$ from Ref. [2]. *Top*: reflectivity spectra at $x = 0$ measured with polarisation E parallel to c-axis in (a) and perpendicular to c-axis in (b). *Bottom*: optical conductivity obtained by Kramers–Kronig analysis of E perpendicular to c-axis spectra at various x.

By measuring polarized reflectance spectra, the optical conductivities along the a-, b-, and c-axes are resolved: the CuO_3 chains in YBCO extending along the b-axis, a-axis polarization incites the optical response of CuO_2 planes. The MIR band is a, b-polarized. The c-axis response of YBCO displays interlayer Drude conductivity and unscreened phonon bands.

2. Assignment of Infrared Spectra in Normal State

The optical response is described by the dielectric function

$$\epsilon(\omega) = \epsilon_1(\omega) + \frac{4\pi i}{\omega}\sigma_1(\omega), \tag{1}$$

where $\epsilon_1(\omega)$ is the real dielectric function, $\sigma_1(\omega)$ is the ac-conductivity, $\epsilon_1(0)$ is the static dielectric constant, $\sigma_1(0)$ is the dc-conductivity. These are interrelated by the Kramers–Kronig relations (KKR):

$$\epsilon_1(\omega) = 1 + 8P \int_0^\infty \frac{\sigma_1(\omega')}{\omega'^2 - \omega^2}\, d\omega. \tag{2}$$

The complex refractive index is $N = n + ik$ with KKR between n, the real index, and k, the absorption coefficient. The extinction coefficient is $\alpha = 2(\omega/c)k = (4\pi/nc)\sigma_1$. The reflectance $R(\omega)$ is related to the phase shift ϑ between incident and reflected light:

$$\vartheta(\omega) = \frac{\omega}{\pi} \int_0^\infty \frac{\ln R(\omega') - \ln R(\omega)}{\omega^2 - \omega'^2}\, d\omega. \tag{3}$$

Complementary to the above experimental techniques based on reflection is ellipsometry [5]. One measures the complex reflectarice ratio $\rho(\omega, \phi) = r_p(\omega, \phi)/r_s(\omega, \phi)$, where ϕ is the incident angle while r_s and r_p are the Fresnel reflection coefficients for s- and p-polarized light, respectively. The complex dielectric function $\epsilon(\omega)$ is extracted from $\rho(\omega, \phi)$ by inverting Fresnel's equations, which yields:

$$\epsilon(\omega) = \{[1 - \rho(\omega, \phi)]/[1 + \rho(\omega, \phi)]\}^2 \tan^2\phi \sin^2\phi + \sin^2\phi. \tag{4}$$

Ellipsometry thus gives $\epsilon(\omega)$ directly without requiring any KKR analysis; moreover, it is self-normalizing. Consequently, ellipsometry overcomes two main difficulties with the reflectance technique: the reference and phase problems. The former arises from data accuracy being dependent on one's ability to determine the reference level, the latter from the requirement of an accurate knowledge of reflectance over a wide frequency range for use in Equation (3).

2.1. Drude–Lorentz Models

The normal state HTSC spectrum is fitted by a sum of Drude, damped-Lorentzian, phonon, and high-frequency terms to form the dielectric function:

$$\epsilon(\omega) = \epsilon_\infty - \frac{\omega_{pD}^2}{\omega^2 + i\gamma_D\omega} + \frac{\omega_{pe}^2}{\omega_e^2 - \omega^2 - i\gamma_e\omega} + \sum_j \frac{S_j\omega_j^2}{\omega_j^2 - \omega^2 - i\gamma_j\omega}. \quad (5)$$

Here, ϵ_∞ is a high-frequency dielectric constant, the Drude term is with plasma frequency ω_{pD} and damping coefficient γ_D, the Lorenzians relating to MIR and the phonon bands have intensities ω_{pe}, S_j, center frequencies ω_e, ω_j, and damping coefficients γ_e, γ_j, respectively.

Two theoretical approaches are examined: *all-out model* assuming an overdamped Lorentzian for the MIR band and a Drude term with low plasma frequency, and the *Drude-only model* with high plasma frequency attributing the whole non-phononic part to free carriers.

The Drude model also predicts a relation between dc-resistivity and plasma frequency ω_{pD} and damping γ_D:

$$r_{dc} = \frac{\gamma_D}{\epsilon_0\omega_{pD}^2}. \quad (6)$$

r_{dc} increases linearly with T, and so should γ_D. Thus, knowledge of the T-behavior is essential for choosing between the models. Optical measurements of YBCO powder samples indicate that the integrated MIR absorption, constant below T_c, decreases nearly linearly with T in the normal state [6]. A good fit to the Drude-only model [7] is obtained with damping coefficient γ_D increasing linearly with T within 100–295 K [8].

2.2. *p–d* Models

A model based on in-plane oxygen p and copper d orbitals has been proposed by Mattis *et al.* [9], which exhibits the main features of MIR and NIR bands. In-plane oxygen orbitals are grouped into ones bonding with copper's (α) and others forming a nonbonding band (β). The optical conductivity due to transitions from oxygen p_x, p_y orbitals to copper $d_{x^2-y^2}$ orbitals splits into two components σ_α and σ_β:

$$\left\{ \begin{matrix} \sigma_\alpha \\ \sigma_\beta \end{matrix} \right\}(\omega) = \frac{1}{\omega}\sum_\gamma \left| \left\langle \gamma \left| \left\{ \begin{matrix} H_\alpha \\ H_\beta \end{matrix} \right\} \right| 0_{\rm in} \right\rangle \right|^2 \delta(\omega + E_{0,\rm in} - E_\gamma),$$

$$\left\{ \begin{matrix} H_\alpha \\ H_\beta \end{matrix} \right\} = g t_{pd} \sum_{k,\sigma} \left\{ \begin{matrix} \phi \\ \varphi \end{matrix} \right\}(k, \vartheta)[d_{k,\sigma}^{+}\alpha_{k,\sigma} + \text{h.c.}],$$

$$\left\{ \begin{matrix} \phi \\ \varphi \end{matrix} \right\} (k,\sigma) = \frac{1}{\omega(k)} \times \left\{ \begin{matrix} [\sin k_x \cos\vartheta + \sin k_y \sin\vartheta] \\ [\sin \frac{1}{2}k_x \cos \frac{1}{2}k_y \cos\vartheta + \sin \frac{1}{2}k_y \cos \frac{1}{2}k_x \sin\vartheta] \end{matrix} \right\}. \quad (7)$$

σ_α displays a Drude peak at zero frequency and a broad x-dependent IR plateau. σ_β displays an x-independent NIR peak and a doping-dependent MIR band which increases and shifts to lower energies as x is increased. The resulting absorption, $\sigma_\alpha + \sigma_\beta$, is comparable with experimental results.

The MIR band is an optically-induced transition from β to unoccupied mobile states at Cu^{3+} target above the Fermi level, μ. The MIR peak at $E_{\mathrm{MIR}} = \mu - E_\beta$ shifts to lower energies as x is increased, due to $\mu(x)$. The NIR peak is attributed to an optical CT from β states to Cu^{2+} target at energy $E_{\mathrm{NIR}} = U + E_d - E_\beta$. With $E_{\mathrm{MIR}} \approx 0.5$ eV and $E_{\mathrm{NIR}} \approx 2$ eV at $U \approx 5.5$ eV, we estimate $U + E_d - \mu \approx 1.5$ eV, $\mu - E_d \approx 4$ eV, $E_d - E_\beta \approx -3.5$ eV.

Hartree–Fock calculated ground and extended states by planar *p–d* orbitals are also used for obtaining optical conductivities in papers by Lorenzana *et al.* [10]. The calculations agree with experimental results for peak positions and relative intensities at zero and low doping. The MIR band arises from transitions from states localized at Cu sites to extended states above the Fermi level. A recent extension [11] relates the ground state to Zhang–Rice Singlet (ZRS) and MIR band forms by *p–d* transitions. The Hamiltonian is:

$$\begin{aligned} H = & \sum_{(ij)\sigma} t_{ij} c^+_{i\sigma} c_{j\sigma} + \sum_{i\sigma} e_i c^+_{i\sigma} c_{i\sigma} + \sum_i U_i c^+_{i\sigma} c_{i\sigma} c^+_{i-\sigma} c_{i-\sigma} + \\ & + \sum_{i\neq j,\sigma,\sigma'} U_{ij} c^+_{i\sigma} c_{i\sigma} c^+_{j\sigma'} c_{j\sigma'} + \sum_{i\neq j,\sigma,\sigma'} U_{ij} c^+_{i\sigma} c_{i\sigma} c^+_{j\sigma'} c_{j\sigma'} + \\ & + \sum_{i\neq j,\sigma,\sigma'} K_{ij} c^+_{i\sigma} c_{i\sigma'} c^+_{j\sigma'} c_{j\sigma} \end{aligned} \quad (8)$$

with two sets of LDA parameters in eV (LSCO), as follows: $t_{pd} = 1$, $t_{pp} = 0.4$, $e_p - e_d = 2.2$, $U_d = 5$, $U_p = 2.1$, $U_{pd} = 0.4$, $K_{pd} = -0.27$; or: $t_{pd} = 1.3$, $t_{pp} = 0.65$, $e_p - e_d = 2.6$, $U_d = 7.3$, $U_p = 5$, $U_{pd} = 0.52$, K_{pd} =-0.52.

2.3. PHONON-COUPLED MODELS

Models account for phonon coupling of *p–d* transitions. Rice and Wang [12] assume coupling to an oxygen vibration:

$$\begin{aligned} H_{\mathrm{RW}} = & E_0 a_0^+ a_0 + (t_{pd} + GX_1)(a_0^+ a_1 + \text{h.c.}) - (t_{pd} - GX_2)(a_0^+ a_2 + \text{h.c.}) - \\ & -(t_{pd} + GY_3)(a_0^+ a_3 + \text{h.c.}) + (t_{pd} - GY_4)(a_0^+ a_4 + \text{h.c.}) + \\ & + \tfrac{1}{2}(M(\dot{X}_1^2 + \dot{X}_2^2 + \dot{Y}_3^2 + \dot{Y}_4^2) + K_p(X_1^2 + X_2^2 + Y_3^2 + Y_4^2)). \end{aligned} \quad (9)$$

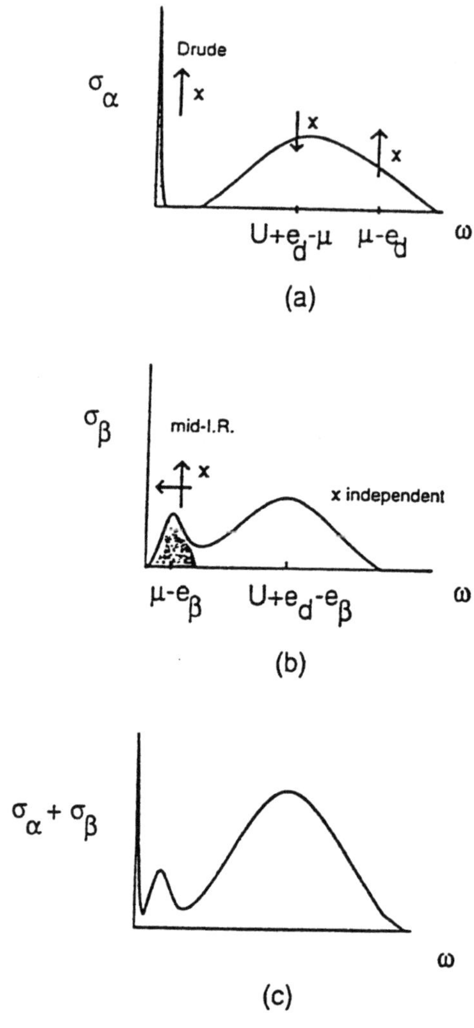

Figure 2. Calculated optical conductivity in $La_{2-x}Sr_xCuO_4$ by in-plane *p–d* orbitals from Ref. [9]. The arrows indicate the motion of a specific component with doping x. See the text in Section 2.2 for further details.

The central Cu ion bears a $3d_{x^2-y^2}$ orbital (a_0) and four nn oxygens bear $2p_{x,y}$ orbitals $(a_{1,2,3,4})$. X_i and Y_i are coordinates, K a force constant, and M the reduced mass of the vibrator, G is electron-mode coupling strength, $E_0 = E_d - E_p$. The extra hole configuration is Cu^{2+} and positive valency P^+ spread over the four O^{2-}. The hole is stabilized by its coupled lattice distortion $\{X_i, Y_i\}$ making the hole self-trapped.

An extension by Mihailov *et al.* [13] applies to MIR, NIR, and VIS optical bands. We assume that phonon-coupled RWS holes move and pair in Cu–O, planes forming bipolarons, the Bose condense of which leads to superconductivity. Each RWS is spreading over an in-plane square of four O around a central Cu; it hops from a square to a neighboring square by virtue of the shared corner oxygen. The binding adherence of another RWS on an nn square induces a pairing perturbation along the centro-connecting line proportional to the binding energy U. The perturbation mixes the two opposite oxygen orbitals; the Hamiltonian reads:

$$H = H_{\mathrm{RW}} - U(a_1^+ a_2 + \mathrm{h.c.}) - U(a_3^+ a_4 + \mathrm{h.c.}). \tag{10}$$

At $U = 0$, a CuO_4 molecule spectrum consists of three nonbonding (NB) oxygen levels $E_{1,2,3} = 0$, one bonding (B) E_5 and one antibonding (AB) E_4 level. In the ground state, all states are doubly-occupied except for Ψ_4, which is empty.

Non-vanishing optical-dipole transitions occur from NB $\Psi_{1,2,3}$ to AB Ψ_4 so that the lowest-lying excitation is at $\hbar\omega = E_4$. At $U \neq 0$ the oxygen reference level, triply degenerate at $U = 0$, splits into an odd-parity singlet (E_1, Ψ_1) at $-U$ and an even-parity doublet $(E_{2,3}, \Psi_{2,3})$ at $+U$. There are three allowed transitions: (i) from $-U$ to $+U$, attributed to the *MIR band*; (ii) from $+U$ to E_4, attributed to the *NIR CT band*; and (iii) from $-U$ to E_4, attributed to the *VIS CT band*.

At $G = 0$, the eigenvalues and the eigenstates are, respectively:

$$E_1 = -U, \quad E_{2,3} = U,$$

$$E_{4,5} = \tfrac{1}{2}\left(-\epsilon_0 - U \pm \sqrt{(\epsilon_0 - U)^2 + 16t_{pd}^2}\right), \tag{11}$$

$$\Psi_1 = \frac{1}{\sqrt{2}}(|1\rangle + |2\rangle + |3\rangle + |4\rangle), \quad \Psi_2 = \frac{1}{\sqrt{2}}(|1\rangle + |2\rangle)$$

$$\Psi_3 = \frac{1}{\sqrt{2}}(|3\rangle + |4\rangle),$$

$$\Psi_4 = \sin\beta|0\rangle + \frac{1}{\sqrt{2}}\cos\beta(|1\rangle - |2\rangle - |3\rangle + |4\rangle),$$

$$\Psi_5 = \sin\gamma|0\rangle + \frac{1}{\sqrt{2}}\cos\gamma(|1\rangle - |2\rangle - |3\rangle + |4\rangle) \tag{12}$$

where

$$\beta, \gamma = \tan^{-1}[(U + E_{4,5})/2t_{pd}].$$

Coupling to an asymmetric mode turns the static IR spectrum into an intense optical band. The optical conductivity derives from electronic polarizability α and linear-response function χ: $\chi F = \alpha F_0$, $F = F_0 + F_{\mathrm{eff}}$. F_{eff} is an effective field by the '*charged-phonon mechanism*':

$$\alpha(p, \omega) = \frac{\chi(p)}{1 - \lambda D \chi(p)},$$

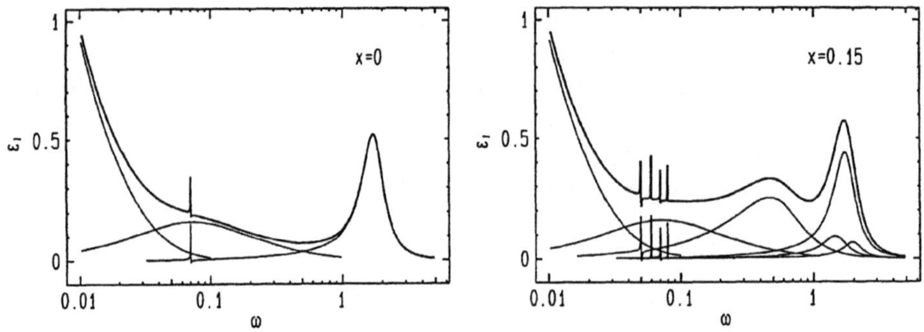

Figure 3. Infrared optical spectrum of $La_{2-x}Sr_xCuO_4$, as calculated by the phonon-coupled model in Ref. [13]. The entire spectrum is resolved into several components as shown.

$$\chi(p,\omega) = \frac{p^2}{\omega_e^2 - \omega^2 - 2i\gamma_e}, \quad D(\omega) = \frac{\omega_0^2}{\omega_0^2 - \omega^2 - 2i\gamma_0}, \tag{13}$$

p is a transition dipole from ground state to excited state at energy $\hbar\omega_e$; γ_e and γ_0 are damping coefficients; D is phonon propagator; ω_0 is bare mode frequency; λ is a dimensionless electron-mode coupling constant.

Using α one derives an optical conductivity σ by way of the Clausius–Mossotti equation:

$$\sigma = \frac{\epsilon - 1}{i}\omega = -\frac{i\omega n\alpha}{1 - n\alpha/3} \simeq -i\omega n\alpha, \quad \epsilon = 1 + \frac{1}{1/n\alpha - 1/3} \tag{14}$$

where n is RWS concentration. The optical conductivity is composed of three parts: $\sigma = \sigma_{\text{VIS}} + \sigma_{\text{NIR}} + \sigma_{\text{dis}}$. They each couple to modes of oxygen sublattice via a *vibronic pseudo Jahn-Teller mechanism.* σ_{VIS} and σ_{NIR} are dynamic CT bands from the phonon coupled $\Psi_1 \to \Psi_4$ and $\Psi_{2,3} \to \Psi_4$ optical transitions at energies $\hbar\omega_{\text{VIS}} \sim 4E_{\text{JT}_1}$ and $\hbar\omega_{\text{NIR}} \sim 4E_{\text{JT}_2}$, respectively ($E_{\text{JT}_{1,2}}$ are Jahn–Teller energies along the respective CT-coupled modes). σ_{dis} is an optical dissociation band, due to $\psi_1 \to \psi_{2,3}$ transition at energy $\hbar\omega_{\text{dis}} = 2U$, the optical dissociation band of vibronic bipolarons to unipolarons.

2.4. Polarons and Bipolarons

Experiments lend support to models where charge carriers are small polarons and bipolarons. The polaron size is determined by Bianconi *et al.* [14] from the statistical distribution of Cu site conformations in the Cu_2O plane: the polaron is intermediate in size, spanning a domain of eight Cu sites. Optical absorption features of large and small polarons (bipolarons) are compared by Emin [15]. A *large-polaron* band, due to photoionization from self-trapped states to extended states, is broad, T-independent, and asymmetric, its high-energy branch being more intense than the low-energy tail. A *small-polaron* band associated with hopping from a self-trapped site to an adjacent site is T-dependent and asymmetric, but its low-energy part is superior to a high-energy tail. Each polaron or bipolaron type, when moving in an ac-field, produces a Drude-like absorption below a characteristic phonon frequency.

Alexandrov and Mott [16] assume that the charge carriers are *small bipolarons*. The IR spectra now include contributions from bipolaron dissociation to small polarons and a low-energy Drude conductivity by small bipolarons itinerant in a narrow band. Subtracting the Drude part, the integrated MIR absorption is:

$$\begin{aligned} I_{\text{opt}}(T) &= \pi e^2 a^2 \Big\{ (\nu^{(2)} - t) n(T) + \\ &+ \int_0^{2t} E N_b(E) \frac{\exp(\mu/k_B T)}{\exp(E/k_B T) - 1} \, \mathrm{d}E \Big\} \end{aligned} \tag{15}$$

$N_b(E)$ is DOS in the bipolaron band, $n(T)$ is bipolaron density determined by chemical potential μ, $\nu^{(2)}$ is interbipolaron repulsive potential. The first term of $I_{\text{opt}}(T)$ is an incoherent light absorption accompanied by multiphonon processes; the second one is a coherent tunneling contribution which couples to the superconducting transition.

Applying (14) to bipolarons at temperature comparable to singlet-to-triplet exchange energy, J, the incoherent part is:

$$\frac{I_{\text{non}}(T)}{I_{\text{opt}}(0)} = 1 - \frac{3AT}{2t} \ln \left\{ \frac{\left[1 - y \exp\left(-\dfrac{2t + J}{k_B T}\right)\right]}{\left[1 - y \exp\left(-\dfrac{J}{k_B T}\right)\right]} \right\} \tag{16}$$

accounting for the temperature dependence in normal the state, and the coherent part is

$$\begin{aligned} \frac{I_{\text{coh}}(T)}{I_{\text{opt}}(0)} &= \delta_c \left(\frac{T}{T_c}\right)^2 \int_0^{2t/T} x \, \mathrm{d}x \left\{ \left(\frac{1}{y} \exp x - 1\right)^{-1} + \right. \\ &\left. +3 \left[\frac{1}{y} \exp\left(x + \frac{J}{kT}\right) - 1\right]^{-1} \right\} \end{aligned} \tag{17}$$

($y = \exp(\mu/k_B T)$, $\delta_c = T_c^2/2nt(\nu^{(2)} - t)$), responsible for the behavior at $T < T_c$. The tendency of absorption to drop at $T < T_c$ is due to the coherent part (16), while the noncoherent part (15) is T-independent. The drop at $T > T_c$ is attributed to singlet-to-triplet conversion.

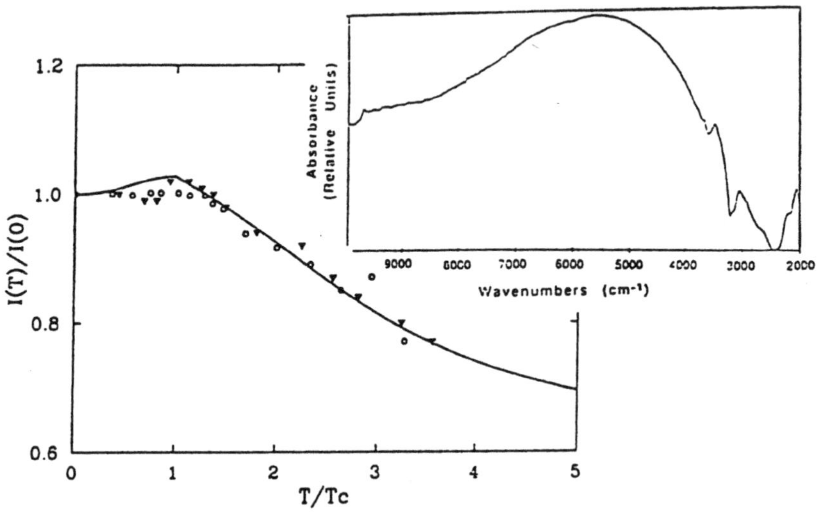

Figure 4. Temperature dependence of integrated optical absorption in the MIR band of $YBa_2Cu_3O_{7-y}$, as calculated by Alexandrov and Mott using small-polaron theory in Ref. [16]. The experiniental data are by Dewing and Salje in Ref. [6]. The insert shows the spectral dependence of NIR absorption.

3. Infrared Spectra in Superconducting State

3.1. TEMPERATURE DEPENDENCE OF SPECTRA

We again refer to the temperature dependence of the MIR bands by Dewing and Salje [6]. Its descending higher-temperature branch may be interpreted as evidencing the formation of bipolarons from unipolarons, since in an Arrhenius plot it yields an activation energy of 10 meV, close to what is found in WO_{3-x}, a classic small-bipolaron material. Now the observed change in temperature slope across T_c follows, since condensed pairs do not absorb light coherently, while noncondensed ones absorb independently of T. The 10% drop in absorption intensity across T_c is indicative of the fraction of condensed pairs.

A careful investigation of the IR reflectivity of thin YBCO films by varying T in small steps across the superconducting transition has been conducted by Van der Marel *et al.* [17]. They obtain the dependence of the superfluid fraction $\propto 1-(T/T_c)^4$. The T-dependence of the absorption edge intensity at 400 cm^{-1} follows the superfluid fraction, while the frequency position of the edge remains virtually constant up to T_c. These experiments can be understood in terms of a two-fluid model with T-independent gap and T-dependent superconducting fraction. The dielectric function of the two-fluid model is:

$$\epsilon(T) = f_s(T)\epsilon_s + [1 - f_s(T)]\epsilon_n \tag{18}$$

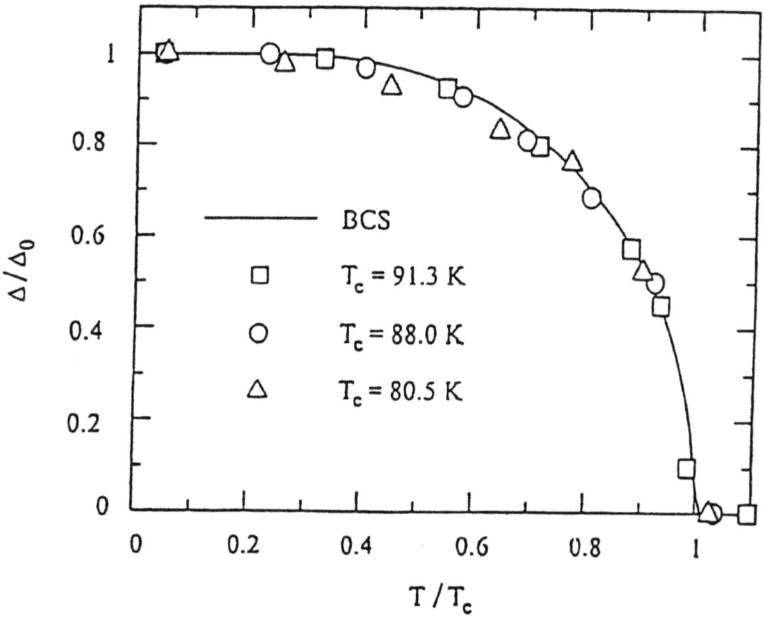

Figure 5. The superconducting gap along the *c*-axis, as manifested by Gervais *et al.* in Ref. [18]. The experimental points are extracted from the IR spectra by fitting to Equation (19), the solid line is the BCS behavior, as explained in Section 3.2.

where $n_s(T)/n_s(0) = f_s(T) = 1 - (T/T_c)^4$. The double fluidity is in the form of electrons (holes) and electron (hole) pairs as the pair condensation at T_c is accompanied by the formation of extra pairs by free electrons.

3.2. SUPERCONDUCTING GAP

Gervais et al. [18] propose a universal Drude–Lorentz model covering the IR spectra of a range of oxide materials:

$$\frac{\epsilon}{\epsilon_\infty} = \prod_{j=1}^{n} \frac{\Omega_{\mathrm{LO}j}^2 - \omega^2 + i\omega\gamma_{\mathrm{LO}j}}{\Omega_{\mathrm{TO}j}^2 - \omega^2 + i\omega\gamma_{\mathrm{TO}j}} - \sum_{k=1}^{m} \frac{\Omega_{pk}^2 + i\omega(\gamma_{pk} - \gamma_{ok})}{\omega(\omega - i\gamma_{ok})}. \qquad (19)$$

The first term gives phonon contributions and the second the contributions of various origins. Within BCS theory an energy gap is expected to occur for $T < T_c$ which imprints itself on IR spectra in that the reflectance is 100% at subgap energies and at overgap energies phonons break Cooper pairs. Due to the complexity of IR spectra, with contributions by phonons, unpaired carriers, and MIR bands, many express doubts about whether the gap, Δ, is observed at all and whether it has the BCS value of $2\Delta = 3.5k_BT_c$.

Gervais *et al.* [18] apply Equation (19) to reflectivity spectra along the c-axis by using $m = 1$ in the normal state and $m = 2$ in the superconducting state.

The temperature dependence below T_c of the plasma frequency Ω_{pSC} ($m = 2$), as derived by fitting to the reflectivity spectra at three T_c, is found to be in agreement with the BCS behavior $\Delta/\Delta_0 = \tanh(\Delta T_c/\Delta_0 T)$. The ratio $2\hbar\Omega_{pSC}/k_B T_c$ varies from 3.8 at $T_v = 80.5$ K to 6.2 at $T_c = 91.3$ K, deviations from the BCS value resulting from non-BCS interference. The applicability to HTSC of the BCS order parameter Δ along the c-axis is confirmed.

4. Photoinduced Phenomena

Photoinduced changes in the phonon and electron parts of IR spectra of insulating parent compounds La_2CuO_4, YBa_2CuO_6, etc. disclose the self-trapping of electrons and holes [19, 20]: photoinduced phonon changes include a bleaching of modes of the parent phase and an appearance of new IR active modes (IRAV). A photoinduced MIR band occurs akin to the doped MIR band. This suggests strong coupling of photoexcited carriers to the lattice and qualifies them as *polarons*. Maximum T_cs of HTSC deriving from the progenitor are found scaling with the dynamic mass $m_d = (I_{MIR}/I_{IRAV})m_e$ and MIR peak position $h\nu_{MIR}$: $T_c \propto m_d^{-1}(h\nu_{MIR})^{-0.6}$ to suggest *quasi-2D dimer carriers* that *Bose condense* to superconductivity [13].

4.1. PHOTOINDUCED PHONON BANDS

Phonon modes strongly coupled to charge carriers are reviewed by Calvani *et al.* [21]. IRAV are responsible for polaron formation in insulating cuprates. They too occur in normal spectra of $Nd_{2-x}Ce_xCuO_{4-y}$ (NCCO) and YBCO. Comparison with NCO–NCCO tunneling data shows that IRAV couple to the superconducting pairs too. IRAV in doped insulators are local modes (LM) arising from the self-trapping of extra charges. IRAV in doped insulators survive across the insulator-to-metal transition and are distinguishable in the metallic phase. Their spectral positions are the same, independent of chemical composition and out-of-plane structures (apex oxygens, etc.) indicating that IRAV are in-plane. This extends to the superconducting phase in that doping-induced LM couple to the pairs.

The 0.13 eV band [22] which forms on chemical doping and in photoinduced experiments is polaronic in origin in both p- and e-doped HTSC cuprates. A two-fluid situation emerges as a small-polaron d-band with a large dynamic mass, $\approx 10m_e$, is superimposed on a free-carrier Drude background. Apparently, band and polaron holes coexist but only the latter are superconducting.

4.2. PHOTOINDUCED ELECTRONIC BANDS

Photoinduced electronic bands at 0.5 eV, 0.13 eV and 0.095 eV in the progenitor phases La_2CuO_4, $YBa_2Cu_3O_{6.25}$ and $Tl_2Ba_2Ca_{(1-x)}Gd_x\,Cu_2O_8$ (TBCGCO) have been reported [23]. The bands in YBCO and TBCGCO resemble the doped-hole photoionization band. Photoinduced bands in semiconducting YBCO form on pumping across the band gap (2.71 eV). Direct absorption bands occur at 0.5, 1.7,

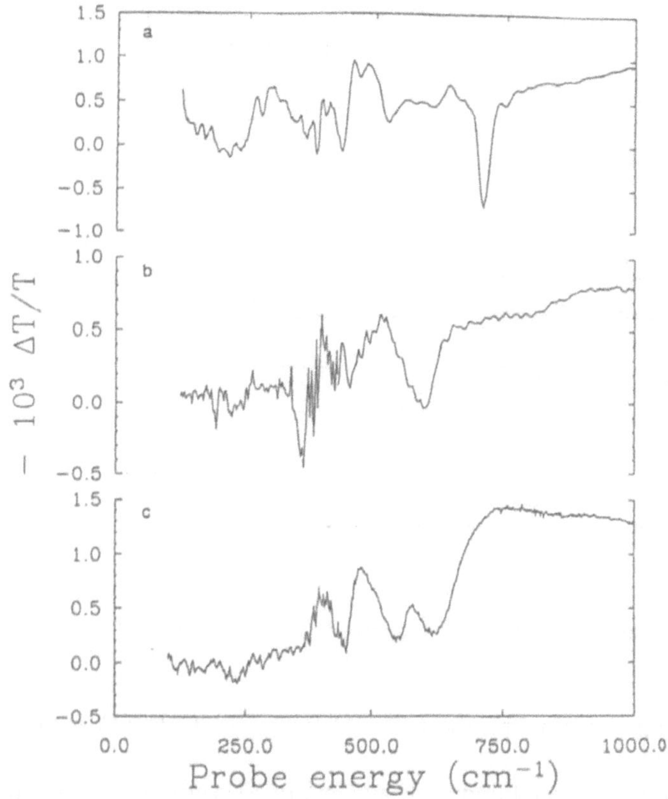

Figure 6. Photoinduced phonon bands in HTSC from Ref. [19]: (a) La_2CuO_4, (b) $YBa_2Cu_3O_{6.25}$, and (c) $Tl_2Ba_2Ca_{1-x}Gd_x cu_2O_8$.

3.0, 4.3, and 4.8 eV ($x = 0$) or 5.2 eV ($x = 0.3$), photoinduced bands at 1.1 eV and 1.4 eV, and photobleached bands at 1.9 eV and 2.6 eV [24]. Models are proposed of photoholes forming polarons itinerant in the Cu–O plane, as free photoelectrons give rise to an absorption continuum. Photoinduced bands at 1.1 eV and 1.4 eV are attributed to transitions from the valence band to a negative-U gap level at the deformed lattice around a hole polaron [25].

4.2.1. *Theories of Photoinduced Electronic Bands*

Nonadiabatic polaron-transport theory has been applied by Mihajlovic *et al.* [26] to photoinduced bands in La_2CuO_4, $YBa_2Cu_3O_{6.25}$, and TBC GCO. The carriers are assumed to couple to both acoustic and optic phonons. The former determine a temperature-dependent contribution to $\sigma(\omega)$ near $\omega = 0$, the latter give rise to a temperature-independent MIR peak at $T < T_{\text{Debye}}$. The spectrum based on small polaron theory accounts for the shakeoff of optical-phonon cloud from polaron

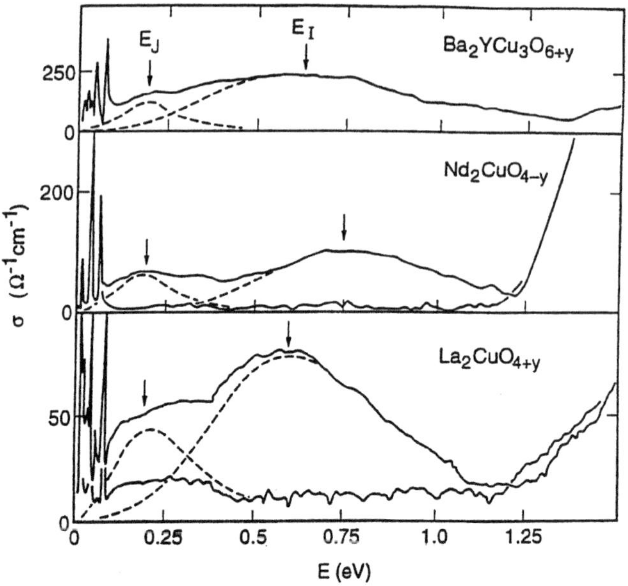

Figure 7. Optical conductivity in HTSC from Ref. [22] showing the *d*-(*J*-) and MIR bands.

due to nonadiabatic intersite transport. The MIR peak is at $2E_b$ (E_b is the polaron binding energy).

A lineshape theory of photoinduced band in YBCO is proposed by Eagles [27]. Transitions to a small-polaron band from an underlying defect level are suggested. There are four fitting parameters: the absorption strength, small-polaron bandwidth W, defect binding energy B, Huang-Rhys factor S. A best fit yields: $W = 75$ meV, $B = 13$ meV, $S = 2.57$. It virtually applies to the absorption to a small-polaron band from a narrow bipolaron band below it.

4.2.2. *Phase Separation of Photoinduced Carriers*

The T-dependence of photoinduced signals in single crystals of YBCO and thin epitaxial $DyBa_2Cu_3O_{7-\delta}$ films at fixed δ and different light intensities, I, are like those of chemically induced conductivities with a different oxygen content, δ [28]. This makes it legitimate to introduce the term 'photodoping' over a range of I. At light levels of 10^{15} photons/cm^2, the thermal activation energy vanishes and a minimum occurs in the T-dependence of resistivity below 100 K. The dip, like the onset of resistivity in inhomogeneous samples, suggests a phase separation of carriers and metallic droplet formation on photoexcitation. The following scenario is plausible: (i) absorbed phonons create electron–hole pairs in the CuO_2 plane. Electrons trap at O vacancies to form F centers along CuO_3 chains.

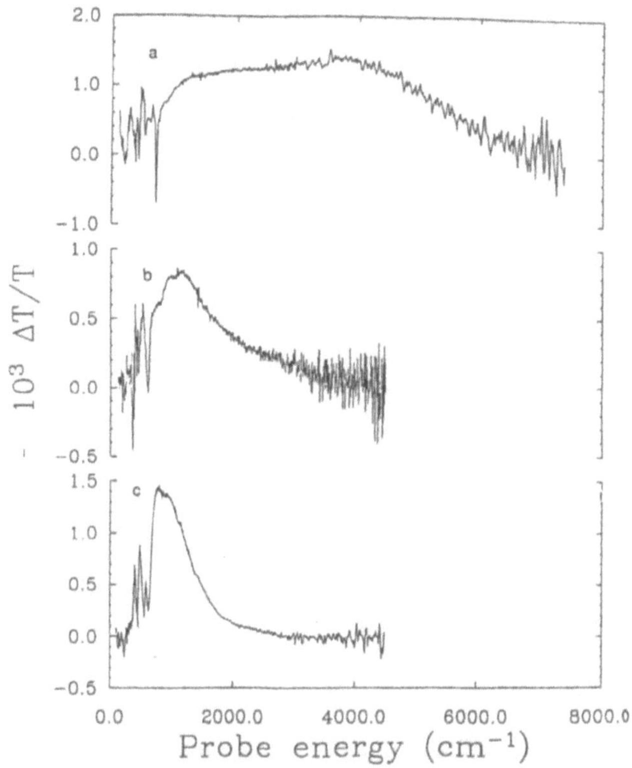

Figure 8. Photoinduced electronic bands in HTSC from Ref. [19]: (a) La_2CuO_4, (b) $YBa_2Cu_3O_{6.25}$, and (c) $Tl_2Ba_2Ca_{1-x}Gd_xCu_2O_8$.

Holes are generated uniformly within illuminated regions. The hole mobility, inhibited by antiferromagnetic phase-separated droplets, leads to local hole-enriched regions of high mobility causing a difflision-limited increase of chemical potential μ. At sufficiently high I, the Fermi level shifts across the mobility edge turning these regions metallic thereby decreasing the total resistivity of the phase-segregated system. (ii) IRAV frequencies observed in photoinduced absorption do not correlate with Raman modes of the insulating phase but do so with Raman modes of the metallic phase in spite of structural differences. The photodoping leads to metallic screening and shifts in phonon frequencies due to droplet formation.

4.3. PHOTOINDUCED SUPERCONDUCTIVITY

Persistent photocurrents are observed in $YBa_2Cu_3O_{6.4}$ films near the semiconductor–metal transition [29, 30]. Transport and magnetic measurements evidence the transition to a metallic state and a *photoinduced superconductivity*. The persistent photoeffects bring about both an enhancement of the normal-state dc-conductivity and an increase of the critical T_c apparently related to the increase in the number of free holes [31]. The enhanced T_c is nearly quadratic with the photohole density as in chemically doped samples. However, the photoinduced changes of critical temperature ΔT_c and of in-plane conductivity $\Delta\sigma_{ab}$ of $YBa_2Cu_3O_{7-\delta}$ films are also reported depending on oxygen content δ as the T_c increases by ordering in oxygen-deficient YBCO [32]. Photoinduced changes are recoverable by RT annealing [33].

The femtosecond quasiparticle dynamics following photogeneration in superconducting YBCO at $T < T_c$ consists of two branches: one $\approx$300 fs is a quasiparticle avalanche production during hot-carrier thermalization and the other is Cooper pair formation within 5 ps [34]. Nonsuperconducting films show only the regular ps response.

Two mechanisms are considered in the literature: (i) *photogeneration of electron–hole pairs* with electrons trapped at oxygen vacancies in the oxygen-depleted material [31, 35], and (ii) *photoassisted oxygen ordering* which creates oxygen-enriched superconducting regions [32, 36]. The actual case may be (i) with phase-segregation effects as in (ii). The decrease of absolute photoinduced change as the oxygen content is increased may relate to the decreasing number of oxygen vacancies along Cu–O chains that trap photoelectrons.

5. Electron–Phonon Coupling

There are signatures of electron–phonon coupling imprinted on the optical spectra, e.g. *Fano-type line-shapes* of phonon bands, *anti-Fano resonance dips* on electronic continua, etc. We shall deal only briefly with the *renormalized phonon frequencies*. The electron coupling modifies the local vibrational potential which evolves from harmonic single well to a double-well type depending on the coupling strength $G = (2E_{\mathrm{JT}}K)^{1/2}$ and T, typically if vibronic mixing is involved. There are critical G_c and T_c at which the potential well flattens, as the renormalized mode frequency ω_{ren} vanishes. Consider the mixing by a vibrational mode of two electronic states separated in energy at E_{sp} and let the local Jahn–Teller energgby be E_{JT}. The free energy, F, of a system of vibronic oscillators coupled weakly through dipole–dipole interactions is considered in detail elsewhere. Here we reproduce the highlights. The critical temperature of single-well to double-well conversion is

$$T_c = (E_{sp}/4k_B)/\tanh^{-1}\{(E_{sp}/4E_{\mathrm{JT}})(1 + [4u(R)/E_{sp}][4E_{\mathrm{JT}}/E_{sp}])\} \quad (20)$$

where $u(R)$ is the pair interaction energy. At weak (single-well) coupling,

$$\omega_{\mathrm{ren}} = \omega_{\mathrm{bare}}\{(4E_{\mathrm{JT}}/E_{sp})[\tanh(E_{sp}/4k_BT_c) - \tanh(E_{sp}/4k_BT)]\}^{1/2}, \quad (21)$$

at strong (double-well) coupling,

$$\omega_{\mathrm{ren}} = \omega_{\mathrm{bare}}[1 - (E_{sp}/4E_{\mathrm{JT}})^2]^{1/2} \quad (22)$$

at low T. It can be shown that $\omega_{\text{ren}} < \omega_{\text{bare}}$ on both sides around the critical temperature T_c, where the anharmonicity is at its strongest, while away from $T_c \omega_{\text{ren}}$ tends to ω_{bare} in either direction. Concommitantly, as the temperature is lowered, the anharmonic model predicts a mode softening at $T > T_c$ and a mode hardening at $T < T_c$. Trends of either type have been observed in the temperature dependences of experimental phonon frequencies. However, no coupling to the superconducting phase has been included in the above model.

6. Final Comments

Doping by donors or acceptors gives rise to two absorption features in the MIR spectrum of insulating cuprates. a band at $\approx$4000 cm^{-1} and a d (or J) band at $\approx$1000 cm^{-1}. The d band, akin to the photoinduced band, is polaronic in origin in p- and e-doped HTSC. The rest of MIR band complements the FIR free-carrier spectrum in an extended Drude absorption at high plasma frequency. A two-fluid image emerges as a small-polaron d-band is superimposed on a free-carrier background indicating the coexistence of band and polaronic carriers.

The p–d CT band in cuprates appears at $\approx$2 eV, the optical electron transfer from $2p_\sigma$ oxygen orbitals to $3d_{x^2-y^2}$ copper orbitals in the CuO_2 plane. If a p-d transition occurs in the insulating phase of p-type so does a hole in the $2p$-conduction band, leading to an extended Drude absorption while free, or to a polaronic band when self-trapped.

Where do photoelectrons absorb, presumably trapped at oxygen vacancies to secure a stable photoinduced state? The vacancies should be doubly charged in the ionic context thereby being able to host two holes forming negative-U centers. Calculations for monovalent ionic crystals at dielectric constants close to HTSC optical values ($\approx$4) give opposite-parity levels spaced at 1.3 eV and an allowed optical transition at $\approx$1.7 eV. If oxygen vacancies in HTSC are partially screened by free carriers, the optical energy will have to be corrected yielding an F band expectancy near $\approx$3 eV. Indeed, three photoinduced absorption bands are observed above the HTSC CT gap.

Another point is the assignment of phonon modes accounting for anharmonic effects, inherent or induced by electron–phonon coupling. Coupled phonon modes should appear in FIR spectra at renormalized, mostly lower, frequencies than will their bare progenitors. Renormalized phonon bands have not been clearly identified in FIR spectra.

As to the theoretical tools, credit should be given to phonon-coupled IR band models. In this respect, models based on the charged phonon mechanism, such as the RWS holes (Rice-Wang-Singlets or other kinds of mode-coupled Zhang-Rice-Singlets) merit attention.

A related matter is the doping dependence of phonon bands. While planar modes may couple to the superconducting transition, out-of-plane modes do so to ipterlayer CT thereby showing a doping dependence. Doping and T-dependences of c-axis polarized ungerade modes are reported from the FIR reflectivity of YBCO [37].

Finally if any part of the MIR band is associated with the photodissociation of composite bosons, as in bipolaron models, then illumination within its spectral portion may be expected to suppress the superconductivity. Observing a photoinduced suppression of superconductivity at MIR frequencies will undoubtedly serve as a check for the bipolaron assignment of the MIR band. An alternative way of optical suppression of the photoinduced superconductivity in YBCO has been demonstrated by NIR illumination near 1 eV, presumably through the optical ionization of electrons trapped at F centers in the Cu–O chain layer and subsequent recombination with holes in the Cu–O plane [38]. Both routes may be used for conductivity control along with the instalment of a photoinduced superconductivity at VIS frequency.

Note Added in Proof

Recently, some doubts have been raised whether the real meaning of the frequency parameter Ω_{pSC}, should be ascribed to the gap along the c-axis or to the density of superconducting carriers. The latter approach gives a penetration depth which follows the $1-(T/T_c)^4$ law [see R. Lobo *et al.*, *SPIE Proc. Spectroscopic Studies of Superconductors*, 2696, 78 (1996)]. The authors are thankful to the Referee for this comment.

References

1. M. Georgiev, M. Borissov, and L. Mihailov: *SPIE Proc. Spectroscopic Studies of Superconductors* **2696** (1996).
2. S. Uchida, T. Ido, H. Takagi, T. Arima, Y. Tokura, and S. Tajima: *Phys. Rev. B* **43**, 7942 (1991).
3. T. Timusk and D. B. Tanner: 'Infrared properties of high T_c superconductors', in D. M. Ginsberg (ed.), *The Physical Properties of High Temperature Superconductors*, World Scientific, Singapore (1989), p. 339.
4. D. B. Tanner and T. Timusk: 'Infrared properties of high-T_c superconductors', in D. M. Ginsberg (ed.), *Physical Properties of High Temperature Superconductors III*, World Scientific, Singapore (1992), p. 363.
5. R. Henn, J. Kircher, M. Cardona, A. Wittlin, V. H. M. Duijn, and A. A. Menovsky: *Phys. Rev. B* **53**, 9353 (1996).
6. H. L. Dewing and E. K. H. Salje: *Supercond. Sci. Technol.* **5**, 50 (1992).
7. I. Bozovic: *Phys. Rev. B* **42**, 1969 (1990).
8. H. L. Dewing and K. Scott: *J. Superconductivity* **7**, 481 (1994).
9. D. C. Mattis and J. M. Wheatley: *Modern Phys. Letters B* **9**, 1107 (1995).
10. J. Lorenzana: *Physica B* **206** / **207**, 675 (1995).
11. J. Lorenzana: *J. Low Temperature Phys.* **99**, 299 (1995).
12. M. J. Rice and Y. R. Wang: *Phys. Rev. B* **36**, 8794 (1987).
13. L. Mihailov, M. Ivanovich, and M. Georgiev: *J Phys. Soc. Japan* **62**, 2431 (1993).
14. A. Bianconi, M. Missori, H. Oyanagi, H. Yamaguchi, D. H. Ha, Y. Nishigara, and S. Della Longa: *Europhys. Letters* **31**, 411 (1995).
15. D. Emin: *Phys. Rev. B* **48**, 13691 (1993).

16. A. S. Alexandrov and N. F. Mott: *Rep. Prog. Phys.* **57**, 1197 (1994). A. S. Alexandrov and N. F. Mott: *High Temperature Superconductors and Other Superfluids*, Taylor & Francis, London (1994).
17. D. van der Marel, M. Bauer, E.H. Brandt, H.-U. Habermeier, D. Heitmann, W. Koenig, and A. Wittlin: *Phys. Rev. B* **43**, 8606 (1991).
18. F. Gervais, R. P. S. M. Lobo, M. Licheron, and F. J. Gotor: *Ferroelectrics* (1995).
19. C. M. Foster, A. J. Heeger, Y. H. Kim, G. Stucky, and N. Herron: *Revs. of Solid State Phys.* **4**, 601 (1990).
20. C. Taliani, R. Zamboni, G. Ruani, A. J. Pal, F. C. Matacotta, Z. Vardeny, and X. Wea: 'Photoinduced optical excitations in the $YBa_2Cu_3O_{7-\delta}$ high T_c superconducting systen', in A. Bianconi and A. Marcelli (eds.), *High T_c Superconductors. Electronic Structure*, Pergamon, Oxford (1989), p. 95.
21. P. Calvani, M. Capizzi, S. Lupi and G. Balestrino: *Europhys. Letters* **31**, 473 (1995).
22. G. A. Thomas: 'Optical properties of insulators and metals with Cu–O planes in High Temperature Superconductivity', in D. P. Turnstall and W. Barford (eds.), *Proc. 39th Scottish Universities Summer School in Physics*, St. Andrews, June 1991, Adam Hilger, Bristol (1991), p. 169.
23. C. M. Foster, A. J. Heeger, Y. H. Kim, and G. Stucky: *Synthetic Metals* **33**, 171 (1989).
24. J. M. Leng, J. M. Ginder, W. E. Farneth, S. I. Shah, and A. J. Epstein: *Phys. Rev. B* **43**, 10582 (1991).
25. H. J. Ye, R. P. McCall, W. E. Farneth, E. M. McCarron III, and A. J. Epstein: *Phys. Rev. B* **43**, 10574 (1991).
26. D. Mihailovic, C. M. Foster, K. Voss, and A. J. Heeger: *Phys. Rev. B* **42**, 7989 (1990).
27. D. M. Eagles: *Solid State Commun.* **76**, 715 (1990).
28. G. Yu, C. H. Lee, A. J. Heeger, N. Herron, E. M. McCarron, Lin Cong, G. C. Spalding, C. A. Nordman, and A. M. Goldman: *Phys. Rev. B* **45**, 4964 (1992).
29. G. Yu, A. J. Heeger, G. Stucky, N. Herron, and E. M. McCarron: *Solid State Commun.* **72**, 345 (1989).
30. V. I. Kudinov, A. I. Kirilyuk, N. M. Kreines, R. Laiho, and E. Laehderanta: *Phys. Letters A* **151**, 358 (1990).
31. K. Tanabe, S. Kubo, F. Hosseini Teherani, H. Asano, and M. Suzuki: *Phys. Rev. Letters* **72**, 1537 (1994).
32. E. Osquiguil, M. Maenhoudt, B. Wuyts, Y. Bruynseraede, D. Lederman, and I. Schuller: *Phys. Rev. B* **49**, 3675 (1994).
33. W. Goeb, W. Lang, W. Markowitsch, V. Schlosser, W. Kula, and R. Sobolewski: *Solid State Commun.* **86**, 431 (1995).
34. S. G. Han, Z. V. Vardeny, K. S. Wong, O. G. Symko, and G. Koren: *Phys. Rev. Letters* **65**, 2708 (1990).
35. S. L. Bud'ko, H. H. Feng, M. F. Davis, J. C. Wolfe and P. H. Hor: *Phys. Rev. B* **48**, 16707 (1993).
36. K. Kawamoto and I. Hirabayashi: *Phys. Rev. B* **49**, 3655 (1994).
37. J. Schutzmann, S. Tajima, S. Miyamoto, Y. Sato, and R. Hauff: *Phys. Rev. B* (1995).
38. D. C. Chew, J. F. Federici, J. Guiterrez-Solana, G. Molina, W. Savin, and W. Wilber: *SPIE Proc. Spectroscopic Studies of Superconductors* **2696**, 654 (1996).

R. McWeeny et al. (eds.), Quantum Systems in Chemistry and Physics, 283–308.

Multireference Brillouin–Wigner Coupled-Cluster Theory: Hilbert Space Approach

JOZEF MÁŠIK and IVAN HUBAČ
Department of Chemical Physics, Faculty of Mathematics and Physics, Comenius University, 842 15 Bratislava, Slovakia.

(Received and accepted: 17 June 1997)

Abstract. Recently developed Brillouin–Wigner coupled-cluster theory (I. Hubač and P. Neogrády: *Phys. Rev. A* **50**, 4558 (1994)) is extended to the multireference case using the Hilbert space approach. We present two computational strategies, namely the 'multi-root' as well as 'single-root' formulations of the multireference Brillouin–Wigner coupled-cluster (MR BWCC) theory and we discuss their relations to existing multireference coupled-cluster (MR CC) theories. We show that the multi-root formulation of the MR BWCC theory is equivalent to existing MR CC theories while the single-root formulation does not have such an analogy. Using the Hilbert space approach to the wave operator we present basic equations of the MR BWCCSD approximation, i.e. a coupled-cluster method with the inclusion of singly and doubly excited cluster components.

Key words: Multireference, coupled clusters, perturbation theory, Brillouin-Wigner, single-root

1. Introduction

In the past two decades, the single-reference coupled-cluster (CC) method for closed-shell systems has become one of the most efficient methods to include correlation effects in many-body systems [1–9]. Nevertheless, its multireference version is far from being the standard method to account for correlation effects of generally open-shell systems. Among the reasons why the multireference coupled cluster (MR CC) theory as well as related multireference Rayleigh–Schrödinger many-body perturbation theory (MBPT) are not used as standard methods is the fact that both theories suffer from the intruder state problem or generally from the convergence problems. It is well-known that both MR MBPT/CC theories are size-extensive due to the fact that they employ the linked cluster theorem which results in a full cancellation of the so-called disconnected diagrams. We note that this property is achieved by the use of the Rayleigh–Schrödinger type of resolvent in perturbation theories and using the Baker–Campbell–Hausdorff (BCH) formula in the case of CC theories.

In our recent articles [10–12] we have studied the Brillouin–Wigner type of the coupled cluster (BWCC) theory in the nondegenerate case. We have shown that the BWCC theory is fully equivalent to the standard CC theory, but it differs in the fact that it does not employ the BCH formula and therefore it does not employ the linked cluster theorem. Nevertheless, the nondegenerate BWCC theory is a size-extensive method since the disconnected diagrams are cancelled out by the iterative procedure. In this article we study the multireference formulation of the BWCC theory. The question we ask is whether the multireference Brillouin–Wigner coupled cluster (MR-BWCC) theory can bring more advantages over the standard MRCC methods; especially with respect to the intruder state problem. Needless to say, the Brillouin–Wigner perturbation theory (BWPT) is not often used in quantum molecular physics because it is not size-extensive and its convergence is believed to be slow. Moreover, it has to be solved iteratively. On the other hand, the nondegenerate BWCC is size-extensive and has to be solved iteratively, anyway. In the past, BWPT was intensively studied by other authors; we should mention the series of articles by Löwdin [13] and also Brandow [14] who used BWPT to develop the size-extensive multireference Rayleigh–Schrödinger perturbation theory.

In this article we discuss two versions of the MR BWCC theory. First, we formulate the 'multi-root' version of the MR BWCC method and we show its connection with standard MRCC theories. Then, we formulate the so-called 'single-root' (or state-specific or one-state) version of the MR-BWCC method and exploiting the Hilbert space approach to the wave operator we present the basic equations for the MR-BWCCSD approximation, i.e. coupled-cluster method with the inclusion of singly and doubly excited cluster components. Finally, we compare our MR-BWCCSD results in the case of the trapezoidal H4 model system with corresponding full configuration interaction (FCI) values as well as other correlated multireference techniques.

2. Multireference Brillouin–Wigner Perturbation Theory

As is usual in perturbation theory, let us split the exact Hamiltonian H into two parts, namely

$$H = H_0 + V \tag{1}$$

where H_0 is the zeroth-order Hamiltonian and V is a perturbation. Let us further assume that we know the solution of the characteristic problem of the zeroth-order Hamiltonian H_0

$$H_0 \Phi_\mu = E_\mu \Phi_\mu \tag{2}$$

and our task is to find a solution of the Schrödinger equation for the exact Hamiltonian H

$$H \Psi_\alpha = \mathcal{E}_\alpha \Psi_\alpha. \tag{3}$$

In general, we do not need to know the whole energy spectrum, but we are interested in several lowest lying states, say, Ψ_α, where $\alpha = 1, 2, \ldots, d$. Let us suppose that the most important contributions to d exact wave functions Ψ_α come from d configurations Φ_μ represented by Slater determinants in the spin-orbital formalism, where $\mu = 1, 2, \ldots, d$. Given dominant configurations span the so-called model or reference space. To simplify our derivation, we will use Greek indices α, β to denote exact wavefunctions, indices μ, ν for configurations spanning the model space and the Roman index q for configurations from the complementary space. If we separate the complete configuration space into the d-dimensional model space P and its orthogonal complement Q, the projection operators associated with these subspaces will have the form

$$P = \sum_{\mu=1}^{d} |\Phi_\mu\rangle\langle\Phi_\mu| = \sum_{\mu\in P} |\Phi_\mu\rangle\langle\Phi_\mu| \tag{4}$$

$$Q = \sum_{q>d} |\Phi_q\rangle\langle\Phi_q| = \sum_{q\in Q} |\Phi_q\rangle\langle\Phi_q| \tag{5}$$

$$P + Q = 1. \tag{6}$$

Projections of the exact wave functions onto the model space (sometimes denoted as model functions)

$$\Psi_\alpha^P = P\Psi_\alpha; \quad \alpha = 1, \ldots, d \tag{7}$$

are, in general, nonorthogonal but they are assumed linear independent. Therefore, they can be expressed as a linear combination of configurations spanning the model space

$$\Psi_\alpha^P = \sum_{\nu\in P} c_{\nu\alpha}\Phi_\nu \tag{8}$$

and, in turn, the model space configurations can be expressed using the inverse transformation

$$\Phi_\mu = \sum_{\alpha=1}^{d} c_{\alpha\mu}^{-1}\Psi_\alpha^P. \tag{9}$$

The expansion coefficients $c_{\nu\alpha}$ obey the following identity

$$\sum_{\alpha=1}^{d} c_{\nu\alpha}c_{\alpha\mu}^{-1} = \delta_{\nu\mu}. \tag{10}$$

Within the multireference Brillouin–Wigner perturbation theory [15], the exact wave functions Ψ_α for $\alpha = 1, \ldots, d$ can be expanded in the Brillouin–Wigner (BW) perturbation series as follows

$$\Psi_\alpha = (1 + B_\alpha V + B_\alpha V B_\alpha V + \cdots)\Psi_\alpha^P \tag{11}$$

where Ψ_α^P is the projection of the exact wave function onto the model space and B_α is the BW type of propagator

$$B_\alpha = \sum_{q \in Q} \frac{|\Phi_q\rangle\langle\Phi_q|}{\mathcal{E}_\alpha - E_q}. \tag{12}$$

At this point it should be mentioned that the BW expansion (11) at the same time prescribes the normalization of the exact wave functions, i.e.

$$\langle\Psi_\alpha^P|\Psi_\alpha\rangle = \langle\Psi_\alpha^P|\Psi_\alpha^P\rangle = 1 \tag{13}$$

which is referred to as the intermediate normalization. If we introduce a new operator Ω_α to denote the operator expansion inclosed in parentheses

$$\Omega_\alpha = 1 + B_\alpha V + B_\alpha V B_\alpha V + \cdots \tag{14}$$

Equation (11) will now read

$$\Psi_\alpha = \Omega_\alpha \Psi_\alpha^P. \tag{15}$$

As one can see, the operator Ω_α has the property of a wave operator (it transforms projection of the exact wave function into the exact wave function), however, it should be stressed that one wave operator Ω_α converts just one projected wave function into the corresponding exact wave function; so in order to avoid any confusion we will denote it as a 'partial' or 'state-specific' wave operator. From definition (14) it is immediately seen that the state-specific wave operators obey the following system of equations for $\alpha = 1, \ldots, d$

$$\Omega_\alpha = 1 + B_\alpha V \Omega_\alpha \tag{16}$$

which may be viewed as a Bloch equation in the Brillouin–Wigner form.

3. Multi-root Formulation of the Multireference Brillouin–Wigner Coupled-Cluster Theory

3.1. MULTI-ROOT VERSION OF MR-BWPT

If one is interested in all d exact energies, we construct a 'multi-root' wave operator Ω, acting on states from the model space, in the following way

$$\Omega = \sum_{\alpha=1}^{d} \Omega_\alpha P_\alpha \tag{17}$$

where P_α is the projection operator

$$P_\alpha = \sum_{\beta=1}^{d} |\Psi_\alpha^P\rangle S_{\alpha\beta}^{-1} \langle\Psi_\beta^P| \tag{18}$$

and $S^{-1}_{\alpha\beta}$ is the inverse matrix of the overlap matrix $S_{\alpha\beta}$

$$S_{\alpha\beta} = \langle \Psi^P_\alpha | \Psi^P_\beta \rangle. \tag{19}$$

The multi-root wave operator Ω has the property

$$\Psi_\alpha = \Omega \Psi^P_\alpha = \Omega_\alpha \Psi^P_\alpha \tag{20}$$

which implies that our multi-root wave operator represents a 'true' wave operator. In fact, we should speak about the Bloch wave operator [16] because the projected wave functions Ψ^P_α are, in general, nonorthogonal which results in a non-Hermitian wave operator in contrast to Kato's approach [17] leading to a Hermitian wave operator. Using the multi-root wave operator (17), the system of Equations (15) can be replaced by one Equation (20) and, likewise, the intermediate normalization (13) can be replaced by one equation in the operator form

$$P\Omega = P. \tag{21}$$

However, a similar statement does not come into consideration in the case of the system of equations for state-specific wave operators (16) since we are not able to replace this system just by one equation in the operator form and, at the same time, to keep its BW character. To some extent, this can be achieved within the matrix formalism. Using the multi-root wave operator (17), the system of Equations (16) is equivalent to

$$\Omega \Psi^P_\alpha = (1 + B_\alpha V \Omega) \Psi^P_\alpha. \tag{22}$$

Projecting this equation onto configurations from the Q space, one obtains the system of equations for the multi-root wave operator Ω

$$(\mathcal{E}_\alpha - E_q) \langle \Psi_q | \Omega | \Psi^P_\alpha \rangle = \langle \Phi_q | V \Omega | \Psi^P_\alpha \rangle \tag{23}$$

which may be viewed as a matrix formulation of the Bloch equation in the Brillouin–Wigner form. As concerns determination of exact energies, substituting Equation (20) into the Schrödinger Equation (3) we obtain

$$H(\Omega \Psi^P_\alpha) = \mathcal{E}_\alpha \Psi_\alpha \tag{24}$$

and projecting this equation onto the model space gives us

$$PH\Omega\Psi^P_\alpha = \mathcal{E}_\alpha P \Psi_\alpha = \mathcal{E}_\alpha \Psi^P_\alpha \tag{25}$$

so the exact energies of interest can be obtained as eigenvalues of the operator

$$H_{\text{eff}} = PH\Omega P \tag{26}$$

which acts within the model space and which can be referred to as a 'multi-root' effective Hamiltonian. The eigenfunctions of the effective Hamiltonian are projected wavefunctions Ψ_α^P; as these functions are not necessarily orthogonal, the effective Hamiltonian is generally non-Hermitian. Of course, its eigenvalues are always real since they correspond to true energies.

In order to obtain the wave operator Ω in a form suitable for practical calculations, we substitute the expansion (8) into the Schrödinger equation in the BW form (23), which brings us the following system of equations

$$(\mathcal{E}_\alpha - E_q) \sum_{\nu \in P} \langle \Phi_q | \Omega | \Phi_\nu \rangle c_{\nu\alpha} = \sum_{\nu \in P} \langle \Phi_q | V\Omega | \Phi_\nu \rangle c_{\nu\alpha} \tag{27}$$

where the coefficients $c_{\nu\alpha}$ are obtained by diagonalization of the effective Hamiltonian. Now, we can proceed in two ways: either we eliminate summation on the left-hand side (lhs) or on the right-hand side (rhs). The first way is to put the denominator back on the rhs and multiply both sides by $c_{\alpha\mu}^{-1}$ followed by the summation over all α. This provides us with a set of equations

$$\langle \Phi_q | \Omega | \Phi_\nu \rangle = \sum_{\nu \in P} \langle \Phi_q | V\Omega | \Phi_\nu \rangle d_{\nu\mu} \tag{28}$$

where

$$d_{\nu\mu} = \sum_{\alpha=1}^{d} \frac{c_{\nu\alpha} c_{\alpha\mu}^{-1}}{\mathcal{E}_\alpha - E_q} \tag{29}$$

that can be used for the calculation of the wave operator Ω. Since the coefficients $d_{\nu\mu}$ are explicitly dependent on exact energies as well as expansion coefficients within the model space, the wave operator as well as effective Hamiltonian also become energy dependent and the system of Equations (26) and (28) must be solved simultaneously in an iterative manner. The second possibility, when casting Equation (27), is to 'glue' operators H_0 and V together; i.e. the terms containing the zeroth-order energy E_q are added to the rhs

$$\sum_{\nu \in P} \langle \Phi_q | \Omega | \Phi_\nu \rangle c_{\nu\alpha} \mathcal{E}_\alpha = \sum_{\nu \in P} \langle \Phi_q | H\Omega | \Phi_\nu \rangle c_{\nu\alpha}. \tag{30}$$

Multiplying both sides by $c_{\alpha\mu}^{-1}$ followed by the summation over all α leads us to the following set of equations

$$\sum_{\nu \in P}^{d} \langle \Phi_q | \Omega | \Phi_\nu \rangle \left\{ \sum_{\alpha=1}^{d} c_{\nu\alpha} \mathcal{E}_\alpha c_{\alpha\mu}^{-1} \right\} = \langle \Phi_q | H\Omega | \Phi_\mu \rangle. \tag{31}$$

If we realize that the term in curly brackets is the matrix element of the effective Hamiltonian (see Appendix), we obtain the Bloch equation in the matrix form

$$\langle \Phi_q | H\Omega | \Phi_\mu \rangle = \sum_{\nu \in P}^{d} \langle \Phi_q | \Omega | \Phi_\nu \rangle H^{\text{eff}}_{\nu\mu}. \tag{32}$$

The corresponding operator form of the previous equation is the well-known Bloch equation [15, 16, 18–20]

$$H\Omega P = \Omega P H \Omega P \tag{33}$$

or the so-called generalized Bloch equation [15, 16, 18, 20–22]

$$[\Omega, H_0]P = V\Omega P - \Omega P V \Omega P \tag{34}$$

which determines the wave operator Ω in the MR case. As one can see, the wave operator does not depend on the exact energies any more, which represents much more suitable basis for practical calculations. One can thus conclude, that within the multireference BWPT we were able to construct a multi-root wave operator as well as the effective Hamiltonian H_{eff} which formally possess the same properties as those employed in the multireference theories based on the Bloch equation; for this reason the adjective 'multi-root' is not necessary here. Second, we derived the Bloch equation starting from the BW perturbation expansion, but in contrast to Brandow's approach [14] we did not expand the exact energy dependence out of the denominators. Finally, two sets of equations for the wave operator (28) and (32) are equivalent, so the first approach represented by the set (28) may be viewed as a Bloch equation in the BW form. Likewise, in terms of perturbation theory, the generalized Bloch Equation (34) may be viewed as a Bloch equation in the Rayleigh–Schrödinger form.

3.2. MULTI-ROOT MR-BWCC THEORY: HILBERT SPACE APPROACH

So far, we have specified the wave operator Ω by means of the multireference Brillouin–Wigner perturbation expansion (16) and (17). If we adopt an exponential ansatz for the wave operator Ω, we can speak about the multireference Brillouin–Wigner coupled-cluster (MR BWCC) theory. The simplest way how to accomplish the idea of an exponential expansion is to employ the so-called state-universal or Hilbert space exponential ansatz by Jeziorski and Monkhorst [23]

$$\Omega = \sum_{\mu \in P} e^{T^\mu} |\Phi_\mu\rangle\langle\Phi_\mu| \tag{35}$$

where T^μ is a cluster operator defined with respect to the μth configuration involving, in general, one-body (T_1^μ), two-body (T_2^μ) up to the N-body (T_n^μ) cluster components

$$T^\mu = T_1^\mu + T_2^\mu + \cdots + T_N^\mu \tag{36}$$

with N being the total number of electrons. Comparing two forms for the wave operator (17) and (35), the Hilbert space approach appears as a more straightforward way of extension of an exponential ansatz to the multireference case in contrast to the so-called valence-universal or Fock space approach [22, 25–33] which employs a single (valence universal) cluster operator. Substituting the cluster ansatz (35) into (20) leads to the following cluster expansion for the exact wave function

$$\Psi_\alpha = \sum_{\mu \in P} c_{\mu\alpha} e^{T^\mu} |\Phi_\mu\rangle \tag{37}$$

where the coefficients $c_{\mu\alpha}$ are obtained by diagonalization of the effective Hamiltonian. Needless to say, the resulting wave function represents a generalization of the cluster expansion previously introduced by Silverstone and Sinanoğlu for open-shell wave functions with an empty set of core orbitals [24].

Although in the past ten years we have witnessed a great deal of progress in the Hilbert space MR CC methods (see e.g. [23, 33–45]), only a few applications of this method have been reported, mostly oriented to the simple H_4 and H_8 model systems exploiting a two-dimensional model space. Among the reasons are the choice of an appropriate model space, convergence difficulties (intruder state problem, multiple solutions of nonlinear CC equations) and also spin-contamination when handling open-shell systems. From the computational point of view, excessively large model spaces require a great number of cluster amplitudes (each configuration in the model space is associated with its own set of cluster amplitudes) and, even worse, they are more likely to be plagued by intruder states. On the other hand, in order to guarantee the size-extensivity property, most of the Hilbert space CC developments employ a complete model space. Although recently formulated incomplete model space formulations also provide us with size-extensive results [44], the use of a complete model space is still conceptually simpler. In the following, we limit ourselves to a complete model space which implies that amplitudes corresponding to internal excitations (i.e. excitations within the model space) are equal to zero.

Now, let us start with the calculation of exact energies within the multi-root MR-BWCC theory. In our next derivations we will work with the Hamiltonian in the normal-ordered-product form, i.e.

$$H = \langle \Phi_\mu | H | \Phi_\mu \rangle + H_N(\mu) = H_{\mu\mu} + H_N(\mu) \tag{38}$$

where the μth configuration plays the role of a Fermi vacuum. In general, when calculating matrix elements containing the wave operator Ω, we work with different Fermi vacua for different reference configurations; in our derivations the role of the Fermi vacuum is always played by the configuration with the μth subscript. Using the exponential ansatz (35), the off-diagonal matrix elements of the effective Hamiltonian (26) take the form

$$H^{\text{eff}}_{\nu\mu} = \langle \Phi_\nu | H \Omega | \Phi_\mu \rangle = \langle \Phi_\nu | H_N(\mu) e^{T^\mu} | \Phi_\mu \rangle \tag{39}$$

and, analogously, diagonal elements can be written as

$$\begin{aligned} H^{\text{eff}}_{\mu\mu} &= \langle\Phi_\mu|H\Omega|\Phi_\mu\rangle = H_{\mu\mu} + \langle\Phi_\mu|H_N(\mu)e^{T^\mu}|\Phi_\mu\rangle \\ &= H_{\mu\mu} + \langle H_N(\mu)e^{T^\mu}\rangle_\mu. \end{aligned} \tag{40}$$

We recall that all matrix elements of the effective Hamiltonian are expressed by means of connected diagrams only; in the case of diagonal elements just connected vacuum diagrams may come into consideration and in the case of off-diagonal elements at least one part of a disconnected diagram would correspond to an internal excitation.

As concerns the determination of unknown cluster amplitudes, there are at least three possibilities, how to implement the Hilbert space exponential ansatz (35). Here, we discuss them in more detail. The first possibility is to keep the original BW character of the calculation; i.e. formal dependence of the wave operator on exact energies. We start from the Bloch equation in the BW form (28) and using the exponential ansatz (35) we get

$$\langle\Phi_q(\mu)|e^{T^\mu}|\Phi_\mu\rangle = \sum_{\nu\in P}\langle\Phi_q(\mu)|Ve^{T^\nu}|\Phi_\nu\rangle d_{\nu\mu} \tag{41}$$

where $\Phi_q(\mu)$ is an excited configuration with respect to the μth configuration. This approach may be viewed as a 'net' or 'pure' MR BWCC approach that is manifested by the presence of typical BW-like denominators in (29). As regards the summation on the right-hand size of Equation (41), beside the direct term $\nu = \mu$, the structure of which is the same as in the nondegenerate case, we also recognize the so-called coupling terms ($\nu \neq \mu$), the structure of which differs from the coupling terms used in the MR CC Hilbert space approach by the presence of a perturbation. Hence, the coupling among various cluster operators T^ν is provided in two ways: directly through the so-called coupling terms and indirectly through the presence of exact energies in denominators. From the computational point of view, such an approach is not advantageous due to the complicated structure of coupling terms. Although the exponential expansion $\exp(T^\nu)$ terminates at some level of excitation; this level is too high. For instance, if we confine ourselves to a two-dimensional model space spanned by two closed-shell-type configurations (inter-related as biexcited configurations), the level of excitation is 6 in the case of the CCSD approximation because our Hamiltonian is at most a two-particle operator. Moreover, the explicit splitting of our Hamiltonian into the zeroth-order Hamiltonian and a perturbation brings another formal complications when using various vacua.

The second possibility is to employ the standard Bloch Equation (35) in its matrix form. Substituting the exponential ansatz (35) into Equation (32), we obtain a set of equations

$$
\begin{aligned}
(H^{\text{eff}}_{\mu\mu} - H_{\mu\mu})\langle\Phi_q(\mu)|e^{T^\mu}|\Phi_\mu\rangle &= \langle\Phi_q(\mu)|H_N(\mu)e^{T^\mu}|\Phi_\mu\rangle \\
&\quad - \sum_{\nu\neq\mu}\langle\Phi_q(\mu)|e^{T^\nu}|\Phi_\nu\rangle H^{\text{eff}}_{\nu\mu}
\end{aligned}
\tag{42}
$$

which can be used for calculation of cluster amplitudes. Now, in contrast to the previous alternative the explicit dependence on exact energies is removed, so the adjective 'Brillouin–Wigner' is no longer appropriate and we should speak about the MR CC method. The coupling among various cluster operators is provided just through the coupling terms; i.e. terms on the rhs of the previous equation. As one can see, their structure is very simple because the exponential expansion $\exp(T^\nu)$ terminates at some level of excitation and, moreover, the matrix element $\langle\Phi_q(\mu)|e^{T^\nu}|\Phi_\nu\rangle$ does not contain internal summations. For instance, in the case of our model space spanned by two closed-shell-type configurations the level of excitation is 4 for the CCSD approximation. It should be stressed, however, that such an approach is manifested by the presence of disconnected diagrams as well, so one must be very careful when doing lower approximations in order not to destroy the size-extensivity property. This approach is very suitable for practical calculations and, so far, it has being intensively developed and tested in Bartlett's group [39–44], especially at the level of the CCSD approximation using the Slater determinant based spin-orbital formulation. Nevertheless, there is a formal difference between our derivation and those employed by Bartlett's group since we do not split Hamiltonian in contrast to Bartlett's group who started from the Bloch equation in the 'Rayleigh–Schrödinger' form (34).

The third possibility is to take advantage of the Baker–Campbell–Hausdorff formula. As mentioned above, this approach was initiated by Jeziorski and Monkhorst [23] and, so far, it has been intensively developed and tested by Paldus's group [5, 34–38]; in particular at the level of the CCSD approximation using the spin-adapted formulation. We start from the standard Bloch equation in the operator form (33), however, before projecting this equation against configurations spanning the Q and P spaces, we multiply both sides from the left by $\exp(-T^\mu)$, which gives us

$$
\langle\Phi_q(\mu)|e^{-T^\mu}He^{T^\mu}|\Phi_\mu\rangle = \sum_\nu\langle\Phi_q(\mu)|e^{-T^\mu}e^{T^\nu}|\Phi_\nu\rangle H^{\text{eff}}_{\nu\mu}. \tag{43}
$$

The use of the BCH formula causes that only connected diagrams will survive

$$
\langle\Phi_q(\mu)|H_N(\mu)e^{T^\mu}|\Phi_\mu\rangle_C = \sum_{\nu\neq\mu}\langle\Phi_q(\mu)|e^{-T^\mu}e^{T^\nu}|\Phi_\nu\rangle_C H^{\text{eff}}_{\nu\mu} \tag{44}
$$

and this set of equations is used for calculation of cluster amplitudes (to be more precise, the BCH lemma just guarantees the connectedness property of the direct term, the connectedness of coupling terms has to be proved *a posteriori*). Apparent advantage of such an approach is freedom in doing various approximations without the loss of size-extensivity, but this freedom is paid by the complicated structure of coupling terms. The product of exponentials $\exp(-T^{\mu})\exp(T^{\nu})$ results in the presence of internal summations followed by the enormous growth of diagrams – which is considered as 'cumbersome' or 'unwieldy' [40, 41]. For this reason, the full expansion of a coupling term, even in the simplest case of the model space spanned by two closed-shell-type at the level of the CCSD approximation practically does not come in consideration and one must resort to further approximations.

3.3. BASIC APPROXIMATIONS EMPLOYED IN THE MR-BWCC METHOD

In the previous section we presented three different computational schemes of determination of the 'multi-root' wave operator within the MR-BWCC approach. While the first approach characterized by the set of equations (41) may be looked upon as a pure MR-BWCC one (it must be solved simultaneously with the equation for the effective Hamiltonian); the remaining two approaches characterized by the set of Equations (42) and (44), respectively, represent standard MR-CC approaches and the adjective 'Brillouin–Wigner' is useless in this case (the wave operators do not depend on exact energies). As concerns the computational complexity, there is no doubt that the most advantageous approach is the second one (employed by Bartlett's group) followed by the third one. Of course, all three approaches are equivalent when the full expansion of exponentials is carried out; however, this can be done only in very limited cases of small electron systems. For this reason, some 'reasonable' truncation schemes for the wave operator have to be adopted. The most common truncation is the so-called CCSD approximation with the cluster operators T^{μ} being approximated by their singly and doubly excited components, i.e.

$$T^{\mu} = T_1^{\mu} + T_2^{\mu}. \tag{45}$$

So far, a great deal of the Hilbert space MR-CCSD approximation schemes have been formulated which are known under various names. First of all, we should mention the linear MR-CCSD (MR-L-CCSD) approximation by Jeziorski and Paldus [35] which was first applied to the H_4 model system by Paldus *et al.* [37] exploiting a two-dimensional reference space. This approximation neglects all terms nonlinear in T^{μ} when calculating the effective Hamiltonian (39, 40) as well as in equations for cluster amplitudes (44). Although such a linear theory is size-extensive and works well in highly quasidegenerate regions, it undergoes singular behaviour in the nondegenerate region (singularities appear whenever the energies of low-lying excited configurations become close to energies of the reference configurations). It is remarkable, that while the lowest root of the effective Hamiltonian always describes the ground state, the second root successively approximates higher and higher excited states. Moreover, the MR-LCCSD method fails to produce real energies in some regions near singularities; it is worthwhile to note that such a breakdown does not have an analogue in a linear single-reference theory which may become singular but never yield complex energies.

The next step in the implementation of the MR-CCSD method is the inclusion of quadratic and, in general, nonlinear terms. Here, we should mention orthogonally spin-adapted MR-CCSD-1, MR-CCSD-2 and MR-CCSD-3 approximations developed by Paldus *et al.* [38] and tested on the H_4 system. The first two approximations were designed just for testing purposes in order to better assess the importance of various nonlinear terms. All three approximations are size-extensive and they differ by the presence of quadratic and bilinear terms in the direct as well as coupling terms in the equation for cluster amplitudes (44). To be more precise, in addition to absolute and all linear terms, the MR-CCSD-1 method contains the quadratic $(T_2^{\mu})^2$ term involved in the direct term; the MR-CCSD-2 method contains the quadratic $(T_2^{\mu})^2$ term involved in both the direct as well as coupling terms and, finally, the MR-CCSD-3 method represents a fully quadratic MR-CCSD approximation considering all bilinear terms. The main conclusions to be drawn here are that the inclusion of quadratic terms eliminates singular behaviour of the linear MR-CCSD approximation (even at the MR-CCSD-1 level) and the inclusion bilinear components usually further improves the results.

At the same time, Meissner, Kucharski and others [39, 40] developed the so-called quadratic MR-CCSD method which is nowadays amply employed by Bartlett's group. As already stated, they start from Equation (42) and the word 'quadratic' denotes that only linear and quadratic (bilinear) terms are retained in coupling terms while the full expansion of the direct term is taken into account. This method is size-extensive and its applications to molecular systems as well as corresponding equations for monoexcited and biexcited amplitudes can be found in [39–43]. The quadratic MR-CCSD method is very similar to the MR-CCSD-3 approximation and one may expect somewhat better results due to the presence of cubic and quartic terms involved in the direct term. Moreover, the spin-orbital formulation also enables us to obtain some open-shell (although spin-contaminated) energies in a special case of the model space spanned by two open-shell Slater determinants that combine into a singlet and triplet. This method was first proposed by Balková and Bartlett [42] and applied to a study of open-shell singlet states of ozone, singlet-triplet separation in methylene etc., see [43]. Needless to say, all nonlinear approximations presented so far employ the full expansion of the effective Hamiltonian (39–40).

In spite of a great progress achieved in development of nonlinear approximations, some problems still persist. Although the description of the ground state is very accurate in the quasidegenerate region and remains satisfactory in the nondegenerate one; excited states are often described qualitatively poorer and it is probable that a worsened description of an excited state may also affect the ground state energy through the effective Hamiltonian. Another problem is the fact that nonlinear MR-CCSD equations possess multiple solutions capable of describing not only d states of interest but also various other d states as far as they contain a significant contribution from the model space (some of them may not represent physically meaningful eigenstates). Depending on the starting approximation, the Newton–Raphson iterative procedure converges to different solutions, so the initial approximation of cluster amplitudes becomes a nontrivial task, in particular, when highly accurate CI results are not available. For this reason, a very useful approximation seems to us the linearized version of the second approach represented by Equation (42) and we denote it as a linear MR-BWCCSD (MR-L-BWCCSD) method. The effective Hamiltonian (39–40) in this case reads

$$H^{\mathrm{eff}}_{\nu\mu} = H_{\mu\mu}\delta_{\nu\mu} + \langle\Phi_\nu|H_N(\mu)(1+T_1^\mu+T_2^\mu)|\Phi_\mu\rangle \tag{46}$$

and the monoexcited and biexcited cluster amplitudes are given by

$$\begin{aligned}(H^{\mathrm{eff}}_{\mu\mu} - H_{\mu\mu})t^{A(B)}_{I(J)}(\mu) &= \langle\Phi^{A(B)}_{I(J)}(\mu)|H_N(\mu)(1+T_1^\mu+T_2^\mu)|\Phi_\mu\rangle - \\ &- \sum_{\nu\neq\mu}\langle\Phi^{A(B)}_{I(J)}(\mu)|T_1^\nu+T_2^\nu|\Phi_\nu\rangle H_{\nu\mu}.\end{aligned} \tag{47}$$

It is apparent that such an approach is not size-extensive because the lhs is diagrammatically represented by unlinked diagrams which have no counterpart on the rhs (product of a closed vacuum diagram and the cluster amplitude); but it may serve as an initial guess when convergence difficulties with the size-extensive formulations may be expected. This approximation is very similar to MR-CISD (configuration interaction limited to singles and doubles), but not equivalent because the so-called semi-internal triples and quadruples are missing (i.e. those configurations from the Q space that are at most doubly excited with respect to one reference configuration but triply and quadruply excited with respect to another one). Application of this method to the ground state and the first excited state of the H_4 model system as well as comparison with other multireference techniques is provided in the section "Results and discussion".

4. Single-Root Formulation of the Multireference Brillouin–Wigner Coupled-Cluster Theory

4.1. SINGLE-ROOT VERSION OF MR-BWPT

As mentioned in the previous section, the simultaneous calculation of all d states within the multi-root MR-CC approach may be computationally very demanding or even impossible in view of the intruder state problems, multiple solutions, convergence difficulties and, to a lesser extent, the choice of the basis set (different electronic states may have different orbital requirements). Likewise, a worsened description of one state (due to the lack of significant configurations in the reference space or truncation of the cluster operator) may have a negative effect on the convergence of other states. Therefore, it is highly desirable to develop such a theory which would deal with one state at a time, while employing a multiconfigurational reference. Such approaches are often referred to as 'one-state' or 'state-selective' or 'single-root' ones and were proposed by Harris [46], Paldus *et al.* [47], Nakatsuji and Hirao [48], Banerjee and Simons [49], Laidig and Bartlett [50], Hoffmann and Simons [51] and, very recently, by Li and Paldus [52–54], who formulated size-extensive open-shell CC theory based on the unitary group approach (UGA) formalism. It is beyond the scope of this article to give a detailed survey of all these methods, but it is worth noting that most of them must face the size-extensivity problem owing to their conceptual simplicity.

In this article we formulate a single-root version of the multireference BWPT. To this end, let us assume that we are interested in one exact state, say, for simplicity, the ground state Ψ_0. As was shown in Section 2, within the multireference BWPT we can introduce a state-specific or single-root wave operator $\tilde{\Omega}$, acting on states from the model space, in the following way

$$\tilde{\Omega} = 1 + B_0 V + B_0 V B_0 V + \cdots \tag{48}$$

or, equivalently

$$\tilde{\Omega} = 1 + B_0 V \tilde{\Omega} \tag{49}$$

where B_0 is the BW type of propagator

$$B_0 = \sum_{q \in Q} \frac{|\Phi_q\rangle\langle\Phi_q|}{\mathcal{E}_0 - E_q} \tag{50}$$

and $\mathcal{E}_0$ is the exact energy of the ground state. It should be stressed, however, that Equations (48) and (49) are equivalent only in the case when the BW perturbation expansion converges. Then, in accordance with (11), the wave operator $\tilde{\Omega}$ has the property

$$\Psi_0 = \tilde{\Omega}(P\Psi_0) = \tilde{\Omega}\Psi_0^P. \tag{51}$$

This means that it transforms the projection Ψ_0^P of the ground state back into the exact wavefunction. In contrast to the multi-root wave operator Ω introduced in Section III, our single-root wave operator $\tilde{\Omega}$ is a state-specific wave operator (14) and does not convert projections of other exact wavefunctions $P\Psi_\alpha$ into corresponding exact states; so in order to avoid any confusion we use a tilde for a single-root wave operator.

If we denote the 'action' of the wave operator $\tilde{\Omega}$ on the model space configurations as $\tilde{\Psi}_\mu$, i.e.

$$\tilde{\Psi}_\mu = \tilde{\Omega}\Phi_\mu \tag{52}$$

it is very easy to show that the operator $\tilde{\Omega}$ is a bijection (one-to-one mapping). Using the BW perturbation expansion (48) gives us

$$\tilde{\Psi}_\mu = \Phi_\mu + B_0(V + VB_0V + \cdots)\Phi_\mu \tag{53}$$

which implies that the resulting states $\tilde{\Psi}_\mu$ are linear independent and the operator $\tilde{\Omega}$ represents a bijection (it transforms d linear independent states into d linear independent states). Since the states $\tilde{\Psi}_\mu$ are not necessarily orthogonal, the single-root wave operator $\tilde{\Omega}$ is, in general, non-Hermitian (like in the multi-root theory). The only drawback to such an approach is the existence of the wave operator $\tilde{\Omega}$ alone, or, in other words, the convergence of the BW perturbation expansion (48). In the case of the ground state, it is quite reasonable to assume that the convergence can be achieved in view of expected larger differences $(\mathcal{E}_0 - E_q)$, but in general, we are not able to assure convergence for excited states, when the exact energy $\mathcal{E}_\alpha$ of our interest becomes close to some zeroth-order energy E_q. In that case one can start from Equation (49) which prescribes an explicit form of the operator $\tilde{\Omega}$ to be the inverse of $(1 - B_0V)$; i.e.

$$\tilde{\Omega} = (1 - B_0V)^{-1}. \tag{54}$$

The inverse operator may exist even in the case when the corresponding power series in terms of B_0V does not converge. Therefore, we will start next from Equation (49) instead of (48). Now, we can introduce the 'effective' Hamiltonian, acting within the model space, in the same way as in the multi-root theory, i.e.

$$\tilde{H}_{\text{eff}} = PH\tilde{\Omega}P. \tag{55}$$

Employing Equation (51) one can write

$$\tilde{H}_{\text{eff}}\Psi_0^P = PH\Psi_0 = \mathcal{E}_0\Psi_0^P \tag{56}$$

which implies that the exact energy of the ground state can be obtained as an eigenvalue of the effective Hamiltonian $\tilde{H}_{\text{eff}}$ and, likewise, the projection of the ground state onto the model space Ψ_0^P can be found as eigenvector of $\tilde{H}_{\text{eff}}$. As concerns remaining eigenvalues and eigenvectors of $\tilde{H}_{\text{eff}}$, these are uniquely determined by definition of the single-root wave operator (48) even though they do not represent any physical meaningful solution. Likewise in the multi-root approach, the effective Hamiltonian $\tilde{H}_{\text{eff}}$ is a non-Hermitian operator (due to the non-Hermicity of $\tilde{\Omega}$). For a better distinction from the multi-root theory, it would be possibly desirable to put some adjective in front of the term 'effective Hamiltonian' but we think that the use of a tilde is fully sufficient. We recall that in both 'multi-root' as well as 'single-root' approaches the effective Hamiltonian acts within the same d-dimensional reference space, but within the 'multi-root' approach all eigenvalues (roots) of the effective Hamiltonian are physically meaningful in contrast to the 'single-root' approach when just one eigenvalue (root) possesses physical meaning. And this was also the reason why we decided to employ the names 'multi-root' approach and 'single-root' approach.

In order to obtain the wave operator $\tilde{\Omega}$ in a form suitable for practical calculations, we project Equation (49) onto configurations from the Q and P subspaces

$$\langle\Phi_q|\tilde{\Omega}|\Phi_\mu\rangle = (\mathcal{E}_0 - E_q)^{-1}\langle\Phi_q|V\tilde{\Omega}|\Phi_\mu\rangle \tag{57}$$

which brings us

$$(\mathcal{E}_0 - E_q)\langle\Phi_q|\tilde{\Omega}|\Phi_\mu\rangle = \langle\Phi_q|V\tilde{\Omega}|\Phi_\mu\rangle. \tag{58}$$

If we use the Schrödinger equation for the zeroth-order Hamiltonian (2), we get the system of coupled equations for $\mu = 1, 2, \ldots, d$

$$\mathcal{E}_0\langle\Phi_q|\tilde{\Omega}|\Phi_\mu\rangle = \langle\Phi_q|H\tilde{\Omega}|\Phi_\mu\rangle \tag{59}$$

which determines the wave operator $\tilde{\Omega}$ in the single-root case. Note, that the splitting of the exact Hamiltonian H is now eliminated (which is advantageous from the computational point of view) but the wave operator $\tilde{\Omega}$ still depends on the exact energy. One can thus conclude, that within the multireference BWPT, we were able to construct a single-root wave operator $\tilde{\Omega}$ as well as the effective Hamiltonian $\tilde{H}_{\text{eff}}$ that determine the exact wavefunction as well as energy just for one state exploiting a multiconfigurational reference. Second, we derived an analogue of the Bloch equation for the single-root wave operator (59), but in contrast to the multi-root theory, such an equation is dependent on the exact energy of our interest and it must be solved simultaneously with the eigenvalue problem for the effective Hamiltonian.

4.2. SINGLE-ROOT MR-BWCC THEORY: HILBERT SPACE APPROACH

So far, we have specified the wave operator $\tilde{\Omega}$ in the BW form (49). If we adopt an exponential ansatz for the wave operator $\tilde{\Omega}$, we can speak about the single-root multireference Brillouin–Wigner coupled-cluster (MR-BWCC) theory. The simplest way to accomplish the idea of an exponential expansion is again to employ the exponential ansatz of Jeziorski and Monkhorst [23]

$$\tilde{\Omega} = \sum_{\mu\in P} e^{T^\mu}|\Phi_\mu\rangle\langle\Phi_\mu| \tag{60}$$

exploiting a complete model space. If we employ the exact Hamiltonian in the normal-ordered-product form (38) with the μth configuration being a Fermi vacuum, the basic equation for the single-root wave operator (59) takes the form

$$(\mathcal{E}_0 - H_{\mu\mu})\langle\Phi_q|\tilde{\Omega}|\Phi_\mu\rangle = \langle\Phi_q|H_N(\mu)\tilde{\Omega}|\Phi_\mu\rangle. \tag{61}$$

If we substitute the Hilbert space exponential ansatz (60) for the wave operator $\tilde{\Omega}$, we obtain the system of equations

$$(\mathcal{E}_0 - H_{\mu\mu})\langle\Phi_q|e^{T^\mu}|\Phi_\mu\rangle = \langle\Phi_q|H_N(\mu)e^{T^\mu}|\Phi_\mu\rangle \tag{62}$$

that can be used for the calculation of cluster amplitudes in the single-root MR-BWCC theory. As one can see, these equations do not mix various sets of cluster amplitudes (i.e. amplitudes belonging to various reference configurations) and the coupling among them is provided indirectly through the exact energy $\mathcal{E}_0$; that means through the off-diagonal matrix elements of the effective Hamiltonian. This fact dramatically simplifies the theory as a whole because, in contrast to the multi-root approach, there is no need to calculate coupling terms and, even more, the structure of matrix elements in (62) is precisely the same as in the nondegenerate CC theory (the only modification is that the cluster amplitudes corresponding to internal excitations are equal to zero). As concerns the calculation of the effective Hamiltonian, using the Hilbert space exponential ansatz (60), the off-diagonal elements of the effective Hamiltonian (55) can be expressed in the form

$$\tilde{H}^{\text{eff}}_{\nu\mu} = \langle\Phi_\nu|H_N(\mu)e^{T^\mu}|\Phi_\mu\rangle \tag{63}$$

and its diagonal elements are given by

$$\tilde{H}^{\text{eff}}_{\mu\mu} = H_{\mu\mu} + \langle\Phi_\mu|H_N(\mu)e^{T^\mu}|\Phi_\mu\rangle = H_{\mu\mu} + \langle H_N(\mu)e^{T^\mu}\rangle_\mu. \tag{64}$$

Obviously, we gain precisely the same matrix elements as in the multi-root theory, cf. (39–40), because within the single-root strategy we postulated the same expression for the effective Hamiltonian as in the multi-root theory. Again, all matrix elements of $\tilde{H}_{\text{eff}}$ can be expressed by means of connected diagrams only.

For the sake of completeness, we should recall that the idea of the single-root formalism exploiting the Hilbert space approach was also proposed by Banerjee and Simons [49] and Laidig and Bartlett [50]. In both approaches they started from the multiconfigurational self-consistent-field (MC SCF) wave function, however, in order to eliminate redundant cluster amplitudes they approximated the wave operator by

$$\Omega \simeq e^T \tag{65}$$

with the exact energy being

$$\mathcal{E} = \langle\Psi_{\text{MCSCF}}|e^{-T}He^T|\Psi_{\text{MCSCF}}\rangle. \tag{66}$$

While Banerjee and Simons take into account only the valence correlations (obtained by exciting valence electrons into particle orbitals), which may not be justified in many cases, Laidig *et al.* also incorporate core excitations including semi-internal ones with all contributions being considered at the linear level in T. The latter approach is referred to by its authors as the MR linearized coupled-cluster method (MR-LCCM); however, it should not be confused with the linear MR-CC method by Paldus *et al.* [35–38] which is size-extensive and does not employ 'prediagonalization' of the effective Hamiltonian. Later on, Hoffmann and Simons developed a UGA-based formalism [51]. To our knowledge, the MR-LCC method by Laidig *et al.* has been applied to study potential energy curves of F_2 and N_2 diatomics; in view of approximations involved it is not surprising that it yields spectroscopic constants very similar to the MR-CISD method.

4.3. SINGLE-ROOT MR-BWCCSD APPROXIMATION

In this section, we derive basic equations for the monoexcited and biexcited cluster amplitudes at the CCSD level of approximation, i.e. with the cluster operators T^{μ} being approximated by their singly and doubly excited cluster components (45). To this end, let us assume that the qth configuration is a monoexcited configuration with respect to the μth configuration, i.e.

$$\Phi_q(\mu) = \Phi_I^A(\mu). \tag{67}$$

In that case, Equation (62) reduces to

$$(\mathcal{E}_0 - H_{\mu\mu}) t_I^A(\mu) = \langle \Phi_I^A(\mu) | H_N(\mu) e^{T^{\mu}} | \Phi_{\mu} \rangle \tag{68}$$

where we used the well-known fact

$$\langle \Phi_I^A(\mu) | e^{T^{\mu}} | \Phi_{\mu} \rangle = t_I^A(\mu). \tag{69}$$

In accordance with [12, 55], the direct term on the rhs of Equation (68) can be split into connected (subscript C) and disconnected terms as follows

$$\begin{aligned}\langle \Phi_I^A(\mu) | H_N(\mu) e^{T^{\mu}} | \Phi_{\mu} \rangle &= \langle \Phi_I^A(\mu) | H_N(\mu) e^{T^{\mu}} | \Phi_{\mu} \rangle_C + \\ &+ t_I^A(\mu) \langle H_N(\mu) e^{T^{\mu}} \rangle_{\mu}.\end{aligned} \tag{70}$$

If we take into account the expression for diagonal elements of the effective Hamiltonian (64), we get a system of nonlinear equations

$$(\mathcal{E}_0 - \tilde{H}^{\text{eff}}_{\mu\mu}) t_I^A(\mu) = \langle \Phi_I^A(\mu) | H_N(\mu) e^{T^{\mu}} | \Phi_{\mu} \rangle_C \tag{71}$$

which can be used for the calculation of singly excited cluster amplitudes within the single-root MR-BWCCSD approximation.

In the case of biexcited amplitudes we will proceed in a similar way. Assuming the qth configuration to be a biexcited configuration with respect to μth configuration, i.e.

$$\Phi_q(\mu) = \Phi_{IJ}^{AB}(\mu). \tag{72}$$

Equation (62) can be expressed in the form (for simplicity, we omit subscript μ at individual amplitudes):

$$(\mathcal{E}_0 - H_{\mu\mu})(t_{IJ}^{AB} + t_I^A t_J^B - t_J^A t_I^B)_\mu = \langle \Phi_{IJ}^{AB}(\mu)|H_N(\mu)e^{T^\mu}|\Phi_\mu\rangle, \tag{73}$$

where we used the well-known fact

$$\langle \Phi_{IJ}^{AB}(\mu)|e^{T^\mu}|\Phi_\mu\rangle = (t_{IJ}^{AB} + t_I^A t_J^B - t_J^A t_I^B)_\mu. \tag{74}$$

Then, according to [12, 55], the direct term on the rhs of Equation (73) can be split into connected and disconnected terms as follows

$$\begin{aligned}\langle \Phi_{IJ}^{AB}(\mu)|H_N(\mu)e^{T^\mu}|\Phi_\mu\rangle &= \langle \Phi_{IJ}^{AB}(\mu)|H_N(\mu)e^{T^\mu}|\Phi_\mu\rangle_C + \\ &+(t_{IJ}^{AB} + t_I^A t_J^B - t_J^A t_I^B)_\mu \langle H_N(\mu)e^{T^\mu}\rangle_\mu + \\ &+P_{IJ}P_{AB}[t_I^A \langle \Phi_J^B(\mu)|H_N(\mu)e^{T^\mu}|\Phi_\mu\rangle_C]\end{aligned} \tag{75}$$

where P_{IJ} and P_{AB} are antisymmetrizers of the form

$$P_{IJ} = 1 - P(I \leftrightarrow J). \tag{76}$$

If we take into account the expression for diagonal elements of the effective Hamiltonian (64), we get

$$\begin{aligned}&(\mathcal{E}_0 - \tilde{H}_{\mu\mu}^{\text{eff}})(t_{IJ}^{AB} + t_I^A t_J^B - t_J^A t_I^B)_\mu \\ &\quad = \langle \Phi_{IJ}^{AB}(\mu)|H_N(\mu)e^{T^\mu}|\Phi_\mu\rangle_C + \\ &\quad\quad +P_{IJ}P_{AB}[t_I^A \langle \Phi_J^B(\mu)|H_N(\mu)e^{T^\mu}|\Phi_\mu\rangle_C]\end{aligned} \tag{77}$$

and, finally, if we substitute for the connected part corresponding to projections onto singly excited configurations from Equation (71), we obtain a system of nonlinear equations

$$\begin{aligned}(\mathcal{E}_0 - \tilde{H}_{\mu\mu}^{\text{eff}})t_{IJ}^{AB}(\mu) &= \langle \Phi_{IJ}^{AB}(\mu)|H_N(\mu)e^{T^\mu}|\Phi_\mu\rangle_C \\ &+(\mathcal{E}_0 - \tilde{H}_{\mu\mu}^{\text{eff}})(t_I^A t_J^B - t_J^A t_I^B)_\mu\end{aligned} \tag{78}$$

which can be used for the calculation of doubly excited cluster amplitudes in a special case of a two-determinant model space corresponding to two active orbitals of different spatial symmetry.

Equations (71) and (78) can be used for the calculations of monoexcited and biexcited cluster amplitudes in the single-root MR-BWCCSD approximation. As already stated, these equations do not mix various sets of cluster amplitudes and the coupling among amplitudes is provided through the off-diagonal elements of the effective Hamiltonian (if these off-diagonal elements were zeroes, the coupling would vanish). On the other hand, the equations explicitly depend on the exact energy of our interest and they must be solved simultaneously with the eigenvalue problem for the effective Hamiltonian. Hence, the set of Equations (71) and (78) is solved iteratively and after each iteration we compute the effective Hamiltonian (63–64). Diagonalization of the effective Hamiltonian provides us with several eigenvalues; the lowest one is taken as a new exact energy $\mathcal{E}_0$, while the remaining ones (not physically meaningful) are discarded.

Finally, let us pay attention to the convergence of presented single-root Hilbert space MR-BWCCSD approximation. When calculating the ground state energy, denominators appearing in Equations (71) and (78) are always negative, so one could not expect problems with vanishing denominators or intruder states. Moreover, as the coupling between amplitudes is provided through the off-diagonal elements of $\tilde{H}_{\rm eff}$, one may expect good convergence over the whole potential energy curve, because the off-diagonal elements enable us 'to switch' in a continuous manner between a single-reference and multireference approach. To be more precise, in the nondegenerate case, when the coupling between reference configurations is weak, the off-diagonal elements of $\tilde{H}_{\rm eff}$ are negligible (in fact, the off-diagonal elements would correspond to amplitudes of internal excitations divided by corresponding denominators), which implies that the difference $(\mathcal{E}_0 - \tilde{H}^{\rm eff}_{\mu\mu})$ approaches zero for the ground state configuration Φ_μ and Equations (71) and (78) reduce to standard nondegenerate CCSD equations for the ground state. In the quasi-degenerate case, when the coupling between the model space configurations is getting large, the coupling between Equations (71) and (78) becomes stronger and these equations correspond to the multireference case.

From the above discussion one may expect that the single-root MR-BWCCSD approximation is well suited to the calculation of the ground state energy. As concerns the calculation of excited states, the convergence problems can not be excluded in view of unknown behaviour of denominators. Moreover, chosing the correct eigenvalue from several eigenvalues of the effective Hamiltonian may not seem as unambiguous as in the case of the ground state. Application of this method to the ground state and the first excited state of the H_4 model system as well as comparison with other multireference techniques is provided in the next section.

5. Results and Discussion

In order to get a better idea about the behaviour of presented multi-root and single-root MR-BWCC approaches, we carried out multi-root linear MR-BWCCSD as well as single-root MR-BWCCSD calculations for the ground state and the first excited state of the H_4 model system. This model system was introduced by Jankowski and Paldus [56] and nowadays is being used to explore the performance of various single-reference and MR approaches in quasidegenerate cases. In this work we confine ourselves to the trapezoidal H4 model consisting of two stretched hydrogen molecules in an isosceles trapezoidal arrangement with all

Table I. Correlation energies for the ground state of the H4 DZP model obtained using various multireference correlation techniques within a two-dimensional model space spanned by the ground state configuration and the first biexcited configuration. For simplicity, we present only differences from corresponding FCI values ($\Delta = X - \text{FCI}$) in mhartree.

alpha	FCI	MR MBPT3[a]	MR CISD[b]	MR L-CCSD[c]	MR CCSD[d]	MR L-BWCCSD	Single-root MR BWCCSD
0.0	−131.361	2.172	–	−3.685	−0.687	3.942	−0.100
0.005	−123.831	2.234	1.794	−3.670	−0.664	3.973	−0.074
0.01	−117.956	2.295	1.771	−3.610	−0.594	3.961	−0.012
0.015	−113.412	2.348	1.741	−3.527	−0.503	3.918	0.065
0.02	−109.870	2.386	1.706	−3.436	−0.408	3.863	0.146
0.05	−98.647	2.373	1.507	−3.120	−0.084	3.524	0.454
0.1	−91.006	2.155	1.277	−5.027	−0.172	3.212	0.582
0.12	−89.169	2.106	–	−13.442	−0.322	3.153	0.591
0.15	−87.121	2.073	–	NC	−0.714	3.102	0.590
0.2	−84.953	2.079	1.039	NC	−1.186	3.071	0.579
0.3	−83.042	2.136	–	NC	−1.995	3.092	0.559
0.4	−82.460	2.161	–	NC	−2.305	3.136	0.551
0.5	−82.333	2.166	0.895	NC	−2.375	3.151	0.548

[a]Third-order MR MBPT calculation; taken from [58]. Intruder state problem is obviated by averaging of active orbital energies.
[b] MR CISD approximation; taken from [57]. Missing values were not present.
[c]Linear MR CCSD approximation. Symbol 'NC' indicates that no convergence is achieved when the ground and first biexcited states are calculated simultaneously.
[d]Quadratic MR CCSD approximation; taken from [41].

nearest-neighbour internuclear separations being fixed at 2 a.u. The geometry of the H4 model is uniquely specified by a single parameter α (defining the trapezoidal angle $\phi = \alpha * \pi$) that enables us to vary the degree of quasidegeneracy in a continuous manner from a non-degenerate case (α = 0.5) to a fully degenerate case (α = 0.0). Molecular orbitals are taken from restricted Hartree–Fock (RHF) calculations for the ground state; for the sake of comparison, we employ the same DZP basis set as Paldus *et al.* [57].

As is well known, the H4 model shows a configurational degeneracy in the square arrangement; i.e. the weights of the ground state configuration and the first biexcited configuration in the FCI expansion are equal. This implies that the model space should be spanned by the two closed-shell-type configurations. Such a two-dimensional model space is complete since the monoexcited configurations that combine into a singlet and triplet are of different spacial symmetry. In Table I we compare correlation energies for the ground state of the H4 DZP model obtained by the MR-BWCCSD methods with other multireference techniques exploiting the same model space; namely the third-order MR Rayleigh–Schrödinger MBPT [58] (RHF orbital energies of two active orbitals had to be averaged in order to eliminate

Table II. Correlation energies for the first biexcited state of the H4 DZP model obtained using various multireference correlation techniques within a two-dimensional model space spanned by the ground state configuration and the first biexcited configuration. For simplicity, we present only differences from corresponding FCI values ($\Delta = X - \text{FCI}$) in mhartree.

alpha	FCI	MR MBPT3[a]	MR L-CCSD[b]	MR CCSD[c]	MR L-BWCCSD	Single-root MR BWCCSD
0.0	−49.683	5.975	−1.499	3.370	3.221	3.115
0.005	−40.119	6.148	−1.730	3.356	3.321	3.109
0.01	−29.192	6.351	−2.044	3.299	3.475	3.078
0.015	−17.275	6.600	−2.429	3.228	3.668	3.037
0.02	−4.729	6.899	−2.879	3.160	3.889	3.004
0.05	71.152	10.071	−7.724	3.060	5.506	3.085
0.1	168.217	21.819	−51.385	3.491	9.096	3.744
0.12	194.947	28.667	−132.149	3.585	10.551	3.919
0.15	224.624	39.942	NC	3.462	12.233	3.903
0.2	254.831	57.217	NC	2.767	13.273	3.329
0.3	279.104	76.350	NC	1.721	12.004	2.329
0.4	284.609	81.839	NC	1.652	10.578	2.207
0.5	285.262	82.797	NC	1.723	10.087	2.250

[a]Third-order MR MBPT calculation; taken from [58]. Intruder state problem is obviated by averaging of active orbital energies.
[b] Linear MR CCSD approximation. Symbol 'NC' indicates that no convergence is achieved when the ground and first biexcited states are calculated simultaneously.
[c]Quadratic MR CCSD approximation; taken from [41].

intruder states), MR-CISD method [57], linear MR CCSD (MR-L-CCSD) method and 'quadratic' MR-CCSD [41]. Analogous results for the first biexcited state are presented in Table II. For simplicity, instead of correlation energies we present only their differences from corresponding FCI values, i.e. $\Delta = X - \text{FCI}$.

Let us begin our comparison with the ground state. It is undisputable, that the superior approximation to the FCI method, in spite of the size-extensivity error, is provided by the single-root MR-BWCCSD approach. As might be expected, the results of the multi-root linear MR-BWCCSD are somewhat worse than the MR-CISD approximation (probably in view of missing semi-internal excitations). Needless to say, both MR-BWCC approaches do not encounter the intruder state problem in contrast to MR-L-CCSD approximation which undergoes a singular behaviour and fails in the nondegenerate region (no convergence is achieved when the ground and first biexcited states are to be solved simultaneously). As concerns the first biexcited state, the single-root MR-BWCCSD approach remains at the level of the quadratic MR CCSD method, whereas the behaviour of the linear MR-BWCCSD is not satisfactory in the nondegenerate region. Although the presented results are limited, it is apparent that the single-root MR-BWCC method seems to be a viable approach for the calculation of the ground state energy over the whole potential energy surface.

6. Summary and Conclusion

In this article we have analyzed the coupled-cluster theory based on the Brillouin–Wigner perturbation expansion. To our knowledge, BWPT is not often used in quantum molecular physics because it is not size-extensive and its convergence is believed to be slow. Moreover, the BW perturbation series has to be solved iteratively. With the coupled-cluster formulation of BWPT the situation is different. Although the BWCC equations must be solved iteratively anyway, the single-reference BWCC theory is size-extensive and provides us with equivalent results as the standard CC theories [10–12]. This is due to the fact that the disconnected diagrams which appear in the single-reference BWCC theory are fully cancelled out in an iterative procedure; we recall that in contrast to standard CC theories we do not use the BCH expansion (i.e. we do not employ the linked cluster theorem).

In this article we focused on the multireference formulation of the BWCC theory within the Hilbert space approach. We discuss two different formulations of the MR-BWCC theory, namely the 'multi-root' and 'single-root' versions. The multi-root version of the MRBWCC theory provides us with three different computational schemes which are equivalent to the use of the effective Hamiltonian and the Bloch equation known in standard MR-CC theories. While the first scheme, which may be looked upon as a pure Brillouin–Wigner-like approach is not advantageous from the computational point of view (cluster amplitudes are dependent on exact energies), the second one leads to the standard MR-CC method employed by Bartlett's group [39–44], and the third one is equivalent to that employed by Paldus's group [34–38].

Another formulation of the MR-BWCC is presented by its single-root version. The single-root MR-BWCC method employs a state-specific wave operator which leads to only one root of the quasidegenerate energy spectrum. Such an approach has many advantages over the multi-root approaches; it does not contain the so-called coupling terms and it is relatively very simple (all necessary diagrams are also present in the standard single-reference CC theories). Equations for cluster amplitudes do not mix various sets of amplitudes and the coupling among amplitudes is provided just through the presence of the exact energy in denominators. Of course, one must employ a typical Brillouin–Wigner-like scheme of calculation: equations for amplitudes have to be solved simultaneously with the eigenvalue problem for the effective Hamiltonian. Diagonalization of the effective Hamiltonian provides us with several eigenvalues; only one is taken as a new exact energy, the remaining ones are thrown away (they do not represent any physical meaningful solution). On the other hand, the method is not fully size-extensive in view of the fact that it does not work with the 'genuine' Bloch equation for the wave operator. A question of the size-extensivity will be discussed in a subsequent paper.

In order to better assess the performance of presented multi-root and single-root MR-BWCC approaches, we carried out multi-root linear MR-BWCCSD as well as single-root MR-BWCCSD calculations for the case of the H_4 model system using a DZP basis set and compared them with other multireference techniques. We have found that in the case of the ground state the single-root MR-BWCCSD method provides the best approximation to the FCI energies of all presented methods and, moreover, it does not suffer from the intruder state problem; what was another reason of our work. In spite of the fact, that our results are preliminary, one can still conclude that the single-root MR-BWCC method appears as a very promising approach which deserves our future attention.

Acknowledgements

The authors would like to thank Dr. L. Meissner for very valuable discussions and comments as well as the organizers of European Workshop 'Quantum Systems in Chemistry and Physics' in San Miniato, 1996 for their kind invitation. This work was supported by the grant 1/1334/94 of the Slovak Grant Agency and the COST Action D3 project.

Appendix

In order to prove

$$H^{\text{eff}}_{\nu\mu} = \sum_{\alpha=1}^{d} c_{\nu\alpha}\mathcal{E}_{\alpha}c^{-1}_{\alpha\mu} \tag{79}$$

we start from the definition of the effective Hamiltonian (25)

$$H_{\text{eff}}\Psi^P_{\alpha} = \mathcal{E}_{\alpha}\Psi^P_{\alpha}. \tag{80}$$

Projecting this equation onto the νth configuration from the model space, we obtain

$$\langle\Phi_{\nu}|H_{\text{eff}}|\Psi^P_{\alpha}\rangle = \mathcal{E}_{\alpha}\langle\Phi_{\nu}|\Psi^P_{\alpha}\rangle. \tag{81}$$

Substituting for the projected wave function Ψ^P_{α} its expansion within the model space (8) gives us

$$\sum_{\kappa\in P}\langle\Phi_{\nu}|H_{\text{eff}}|\Phi_{\kappa}\rangle c_{\kappa\alpha} = \mathcal{E}_{\alpha}c_{\nu\alpha}. \tag{82}$$

If we multiply this equation by $c^{-1}_{\alpha\mu}$ and sum over all α, we finally get

$$\sum_{\kappa\in P}\langle\Phi_{\nu}|H_{\text{eff}}|\Phi_{\kappa}\rangle\left\{\sum_{\alpha=1}^{d} c_{\kappa\alpha}c^{-1}_{\alpha\mu}\right\} = \langle\Phi_{\nu}|H_{\text{eff}}|\Phi_{\mu}\rangle = \sum_{\alpha=1}^{d} c_{\nu\alpha}\mathcal{E}_{\alpha}c^{-1}_{\alpha\mu} \tag{83}$$

since the term in curly brackets is the identity (10).

References

1. F. Coester: *Nucl. Phys.* **7**, 421 (1958). F. Coester and H. Kümmel: *Nucl. Phys.* **17**, 477 (1960).
2. J. Čížek: *J. Chem. Phys.* **45**, 4256 (1966). *Adv. Chem. Phys.* **14**, 35 (1969).
3. J. Čížek and J. Paldus: *Int. J. Quantum Chem.* **5**, 359 (1971).
4. J. Paldus, J. Čížek and I. Shavitt: *Phys. Rev. A* **5**, 50 (1972).
5. J. Paldus: in *Methods in Computational Molecular Physics*, NATO ASI Series, S. Wilson and G. H. F. Diercksen (eds.), Plenum, New York (1992), pp. 99–194; in *Relativistic and Correlation Effects in Molecules and Solids*, NATO ASI Series, G. L. Malli (ed.), Plenum, New York (1994), pp. 207–282.
6. R. J. Bartlett: *Ann. Rev. Phys. Chem.* **32**, 359 (1981). *J. Chem. Phys.* **93**, 1697 (1989).
7. R. J. Bartlett, C. E. Dykstra, and J. Paldus: in *Advanced Theories and Computational Approaches for the Electronic Structure of Molecules*, C. E. Dykstra (ed.), Reidel, Dordrecht (1984), pp. 127–159.
8. K. Jankowski: in *Methods in Computational Chemistry*, Vol. 1, S. Wilson (ed.), Plenum, New York (1987), pp. 1–116. M. Urban, I. Černušák, V. Kellö, and J. Noga: *ibid.*, pp. 117–250.
9. S. Wilson: *Electron Correlation in Molecules*, Clarendon, Oxford (1984).
10. I. Hubač and P. Neogrády: *Phys. Rev. A* **50**, 4558 (1994).
11. I. Hubač; in *New Methods in Quantum Theory*, NATO ASI Series, C. A. Tsipis, V. S. Popov, D. R. Herschbach, and J. S. Avery (eds.), Kluwer, Dordrecht (1996), pp. 183–202.
12. J. Mášik and I. Hubač: *Coll. Czech. Chem. Commun.* **62**, 829 (1997).
13. P.-O. Löwdin: *Perturbation Theory and its Applications in Quantum Mechanics*, C. H. Wilcox (ed.), Wiley, New York, (1966). *Int. J. Quantum Chem.* **2**, 867 (1968).
14. B. H. Brandow: *Rev. Mod. Phys.* **39**, 771 (1967).
15. I. Lindgren: *J. Phys. B* **7**, 2441 (1974).
16. C. Bloch: *Nucl. Phys.* **6**, 329 (1958).
17. T. Kato: *Perturbation Theory for Linear Operators*, Springer, Berlin (1966).
18. V. Kvasnička: *Czech. J. Phys. B* **24**, 605 (1974).
19. V. Kvasnička: *Czech. J. Phys. B* **27**, 599 (1977). *Adv. Chem. Phys.* **36**, 345 (1977).
20. I. Lindgren: *Int. J. Quantum Chem. Symp.* **12**, 33 (1978).
21. V. Kvasnička: *Chem. Phys. Lett.* **79**, 89 (1981).
22. I. Lindgren and J. Morrison: *Atomic Many-Body Theory*, Springer-Verlag, Berlin (1982).
23. B. Jeziorski and H. J. Monkhorst: *Phys. Rev. A* **24**, 1668 (1981).
24. H. J. Silverstone and O. Sinanoğlu: *J. Chem. Phys.* **44**, 3608 (1966).
25. F. Coester: in *Lectures in Theoretical Physics*, Vol. 11B, K. T. Mahanthappa and W. E. Brittin (eds.), Gordon and Breach, New York (1966), pp. 157–186.
26. D. Mukherjee, R. K. Moitra, and A. Mukhopadhyay: *Pramana* **4**, 247 (1975). *Mol. Phys.* **30**, 1861 (1975). *Mol. Phys.* **33**, 955 (1977). A. Haque and D. Mukherjee: *J. Chem. Phys.* **80**, 5058 (1984).
27. R. Offermann, W. Ey, and H. Kümmel: *Nucl. Phys. A* **273**, 349 (1976). R. Offermann: *Nucl. Phys. A* **273**, 368 (1976). W. Ey: *Nucl. Phys. A* **296**, 189 (1978).
28. I. Lindgren: *Int. J. Quantum Chem. Symp.* **12**, 33 (1978).
29. L. Z. Stolarczyk and H. J. Monkhorst: *Phys. Rev. A* **32**, 725 (1985). *Ibid.* 743 (1985). *Phys. Rev. A* **37**, 1908 (1988). *Ibid.* 1926 (1988).
30. W. Kutzelnigg: *J. Chem. Phys.* **80**, 822 (1984).
31. A. Haque and U. Kaldor: *Chem. Phys. Lett.* **117**, 347 (1985). *Chem. Phys. Lett.* **120**, 261 (1985). *Int. J. Quantum Chem.* **29**, 425 (1986).
32. U. Kaldor: *J. Chem. Phys.* **87**, 467 (1987). *Theor. Chim. Acta* **80**, 427 (1991), and references therein.
33. D. Mukherjee and S. Pal: *Adv. Quantum Chem.* **20**, 292 (1989), and references therein.
34. J. Paldus: in *New Horizons of Quantum Chemistry*, P.-O. Löwdin and B. Pullman (eds.), Reidel, Dordrecht (1983), pp. 31–60.
35. B. Jeziorski and J. Paldus: *J. Chem. Phys.* **88**, 5673 (1988).

36. K. Jankowski, J. Paldus, and J. Wasilewski: *J. Chem. Phys.* **95**, 3549 (1991).
37. J. Paldus, L. Polypow, and B. Jeziorski: in *Many-Body Methods in Quantum Chemistry*, Lecture Notes in Chemistry, Vol. 52, U. Kaldor (ed.), Springer, Berlin (1989), pp. 151–170.
38. J. Paldus, P. Piecuch, L. Polypow, and B. Jeziorski: *Phys. Rev. A* **47**, 2738 (1993).
39. L. Meissner, K. Jankowski and J. Wasilewski: *Int. J. Quantum Chem.* **34**, 535 (1988).
40. S. A. Kucharski and R. J. Bartlett: *J. Chem. Phys.* **95**, 8227 (1991). S. A. Kucharski, A. Balková, P. G. Szalay, and R. J. Bartlett: *J. Chem. Phys.* **97**, 4289 (1992).
41. A. Balková, S. A. Kucharski, L. Meissner, and R. J. Bartlett: *Theor. Chim. Acta* **80**, 335 (1991).
42. A. Balková and R. J. Bartlett: *Chem. Phys. Lett.* **193**, 364 (1992).
43. A. Balková, S. A. Kucharski, and R. J. Bartlett: *Chem. Phys. Lett.* **182**, 511 (1991). A. Balková, S. A. Kucharski, L. Meissner, and R. J. Bartlett: *J. Chem. Phys.* **95**, 4311 (1991). A. Balková and R. J. Bartlett: *J. Chem. Phys.* **99**, 7907 (1993). *Ibid.* **101**, 4936 (1994). *Ibid.* **102**, 7116 (1995).
44. L. Meissner, S. A. Kucharski and R. J. Bartlett: *J. Chem. Phys.* **91**, 6187 (1989). L. Meissner and R. J. Bartlett: *J. Chem. Phys.* **92**, 561 (1990).
45. S. Berkovic and U. Kaldor: *Chem. Phys. Lett.* **199**, 42 (1992). *J. Chem. Phys.* **98**, 3090 (1993).
46. E. F. Harris: *Int. J. Quantum Chem. Symp.* **11**, 403 (1977).
47. J. Paldus, J. Čížek, M. Saute, and A. Laforgue: *Phys. Rev. A* **17**, 805 (1978). M. Saute, J. Paldus, and J. Čížek: *Int. J. Quantum Chem.* **15**, 463 (1979).
48. M. Nakatsuji and K. Hirao: *J. Chem. Phys.* **68**, 2053 (1978). *Ibid.* 4279 (1978).
49. A. Banerjee and J. Simons: *Int. J. Quantum Chem.* **19**, 207 (1981). *J. Chem. Phys.* **76**, 4548 (1982).
50. W. D. Laidig and R. J. Bartlett: *Chem. Phys. Lett.* **104**, 424, (1984). W. D. Laidig, P. Saxe and R. J. Bartlett: *J. Chem. Phys.* **86**, 887 (1987).
51. M. R. Hoffman and J. Simons: *J. Chem. Phys.* **88**, 993 (1988).
52. J. Paldus and X. Li: in *Symmetries in Science* VI, B. Gruber (ed.), Plenum, New York (1993), pp. 573–591.
53. X. Li and J. Paldus: *Int. J. Quantum Chem. Symp.* **27**, 269 (1993). *J. Chem. Phys.* **101**, 8812 (1994).
54. B. Jeziorski, J. Paldus, and P. Jankowski: *Int. J. Quantum Chem.* **56**, 129 (1995).
55. J. Paldus, J. Čížek, and B. Jeziorski: *J. Chem. Phys.* **90**, 4356 (1989).
56. K. Jankowski and J. Paldus: *Int. J. Quantum Chem.* **18**, 1243 (1980).
57. J. Paldus, P. E. S. Wormer, and M. Bénard: *Coll. Czech. Chem. Commun.* **53**, 1919 (1988).
58. J. Mášik, I. Hubač, and P. Mach: *Int. J. Quantum Chem.* **53**, 207 (1995).

R. McWeeny et al. (eds.), Quantum Systems in Chemistry and Physics, 309–321.

Core-Valence Separation for an Open-Shell Atom in the LST-DFT

L. S. GEORGIEV, YA. I. DELCHEV and R. L. PAVLOV
Institute for Nuclear Research and Nuclear Energy, Tsarigradsko Chaussee 72, 1784 Sofia, Bulgaria.

J. MARUANI
Laboratoire de Chimie Physique, CNRS and UPMC, 11, rue Pierre et Marie Curie, 75005 Paris, France.

(Received and accepted: 17 June 1997)

Abstract. We apply the local-scaling tranformation (LST) method to core-valence split model wave-functions (CVS-MWF) for open-shell atoms described by the LS coupling scheme. We derive an explicit form for the total energy in terms of single-particle, repulsion and exchange functionals of the LST scalar fimction $f(\mathbf{r})$, and write the corresponding Euler–Lagrange integro-differential equation, which involves derivatives of $f(\mathbf{r})$ up to the fourth order. Solving numerically this equation would yield the density $\rho_{LS}(\mathbf{r})$ and energy E_{LS} within the orbit described by the CVS-MWF. A formal illustration is given with the lithium atom.

Key words: Core-valence separation, open-shell atom, local-scaling transformation, density functional theory.

Introduction

Point transformations, i.e., mappings of coordinates of the three-dimension vector space R^3 onto itself, have not been used very often in quantum chemistry [1–7]. Except for the direct application of this type of transformation, the idea that they could be of importance in a rigorous formulation of density functional theory (DFT) has been stated quite early [8–18].

Petkov and Stoitsov [19–20] have studied the type of transformation that keeps the direction of the original vector unchanged. Because of its similarity with the scaling transformation $\mathbf{r} \to f\mathbf{r}$ it was called the *local scaling transformation* (LST). On the basis of the developed LST variational method a novel version of DFT was formulated (LST-DFT) ([20–22] and references therein).

In the frame of the LST method one can express the energy as a rigorous functional of the LST function: $E = E[f]$. Then, due to the one-to-one correspondence between the LST function $f(\mathbf{r})$ and the particle density $\rho(\mathbf{r})$, the energy can be written as a functional of the particle density: $E = E[\rho]$. The main advantages of this approach are its rigorousness, constructivity and n- and v-representability. However, because the functional $f[\rho]$ is implicit, the numerical minimization of $E[\rho]$ has to be realized through a suitable expansion (parameterization) of $\rho(\mathbf{r})$.

The solution of the corresponding Euler–Lagrange equation for $f(\mathbf{r})$ is equivalent to parameterization of $\rho(\mathbf{r})$ with an infinite number of parameters in the orbit.

In the present article, we are concerned with the derivation of an Euler–Lagrange equation for a special model of atom in the LST-DFT framework, as a first step towards a strict implementation of this approach. The main features of the LST-DFT method are recalled in the next section. The third section is devoted to the concretization of the model wavefunction and derivation of the corresponding energy functional. It is built in the λLS coupling frame using the angular momenta summation model. In the fourth section the energy density functional is derived with respect to the LST function, and the Euler–Lagrange equation is obtained for the LST function through variation of this functional. In the last section an explicit form of this equation is given for the lithium atom. The possibilities for further development of this approach are discussed in the conclusion.

1. The Local-Scaling Transformation for Multifermionic Distributions

The LST is a reversible continuous transformation in 3D space R^3 defined as follows:

$$\mathbf{f} : \mathbf{r} \rightarrow \mathbf{f}(\mathbf{r}) = f(\mathbf{r})\mathbf{r}/r, \tag{1}$$

where $f(\mathbf{r})$ is a scalar function. Transformations of this type form a group $\mathcal{F}$ with respect to their composition law if the functions $f(\mathbf{r})$ satisfy the following conditions [19]:

(i) $\forall(\theta,\phi) \quad r_1 < r_2 \Rightarrow f(r_1,\theta,\phi) < f(r_2,\theta,\phi),$

$$\text{(ii)} \quad J[\mathbf{f}(\mathbf{r}),\mathbf{r}] \equiv \frac{D[\mathbf{f}(\mathbf{r})]}{D[\mathbf{r}]} = \frac{f^2(\mathbf{r})}{r}\frac{\partial f(\mathbf{r})}{\partial r} \neq 0, \tag{2}$$

where $J[\mathbf{f}(\mathbf{r}),\mathbf{r}]$ is the Jacobian of the LST. The unitary representation $\mathcal{U}_\mathbf{f}$ of the LST group $\mathcal{F}$ in the Hilbert space $\mathcal{H}^N$ of N-particle wavefunctions $\Psi(\mathbf{r}_1,\ldots,\mathbf{r}_N)$ is given by:

$$\mathcal{U}_\mathbf{f}\Psi(\mathbf{r}_1,\ldots,\mathbf{r}_N) = J[\mathbf{f}(\mathbf{r}_1),\mathbf{r}_1]^{1/2}\cdots J[\mathbf{f}(\mathbf{r}_N),\mathbf{r}_N]^{1/2}\Psi[\mathbf{f}(\mathbf{r}_1),\ldots,\mathbf{f}_N)]. \tag{3}$$

A given *model* wavefunction (MWF) $\bar{\Psi} \in \mathcal{H}^N$ induces a corresponding *orbit* $\mathcal{O}_{\bar{\Psi}} \subset \mathcal{H}^N$ with respect to the group $\mathcal{U}_\mathbf{f} = \{\mathcal{U}_\mathbf{f}\,;\, \mathbf{f} \in \mathcal{F}\}$:

$$\mathcal{O}_{\bar{\Psi}} = \{\mathcal{U}_\mathbf{f}\Psi\,;\, \mathcal{U}_\mathbf{f} \in \mathcal{U}_\mathcal{F}\}. \tag{4}$$

Through this procedure the Hilbert space $\mathcal{H}^N$ can be split into a set of nonintersecting orbits.

1.1. THE LST VARIATIONAL METHOD

The local-scaling transformation can be used in a variational process for determining the ground-state wavefunction of a multifermionic system. This can be done in the following steps.

1. Select a MWF $\bar{\Psi} \in \mathcal{H}^{\mathcal{N}}$ which induces an orbit $\mathcal{O}_{\bar{\Psi}} \subset \mathcal{H}^{\mathcal{N}}$ with respect to $\mathcal{U}_{\mathcal{F}}$. The function
$$\Psi_{\mathbf{f}} = \mathcal{U}_{\mathbf{f}}\bar{\Psi} \tag{5}$$
is called the *trial* wavefunction.
2. Construct the energy functional
$$E[\mathbf{f}] = \frac{\langle \Psi_{\mathbf{f}}|\hat{H}|\Psi_{\mathbf{f}}\rangle}{\langle \Psi_{\mathbf{f}}|\Psi_{\mathbf{f}}\rangle} = E[f] \tag{6}$$
or any approximation of it.
3. Solve the variational equation
$$\frac{\delta E[f]}{\delta f(\mathbf{r})} = 0, \quad \mathbf{f} \in \mathcal{F} \tag{7}$$
within the orbit $\mathcal{O}_{\bar{\Psi}}$. The solution $f(\mathbf{r})$ of this equation yields the approximation to the ground-state wavefunction through Equation (5).

1.2. THE LST VERSION OF DFT

The model density distribution (MDD) $\bar{\rho}$; associated with the MWF $\bar{\Psi}$ is

$$\bar{\rho}(\mathbf{r}) = N \int |\bar{\Psi}(\mathbf{r}, \mathbf{r}_2, \ldots, \mathbf{r}_n)|^2 \, d\mathbf{r}_2 \ldots d\mathbf{r}_N. \tag{8}$$

For a given LST function $f(\mathbf{r})$ the density distribution $\rho(\mathbf{r})$ associated with the trial wavefunction (5) can be written as

$$\rho(\mathbf{r}) = \frac{f^2(\mathbf{r})}{r} \frac{\partial f(\mathbf{r})}{\partial r} \bar{\rho}[\mathbf{f}(\mathbf{r})]. \tag{9}$$

This equation depends parametrically on $\mathbf{r}_0 = \mathbf{r}/r$. For any fixed direction $\mathbf{r}_0$ defined by angles (θ, ϕ) one can invert Equation (9) to obtain $f(\mathbf{r})$ as a functional of $\rho(\mathbf{r})$:

$$\frac{\partial f(r, \mathbf{r}_0)}{\partial r} = \frac{r^2 \rho(r, \mathbf{r}_0)}{f^2(r, \mathbf{r}_0)\bar{\rho}[f(r, \mathbf{r}_0)]}, \quad \forall(\theta, \phi) \qquad f(0, \theta, \phi) = 0. \tag{10}$$

For all values of (θ, ϕ) the solution of this ordinary differential equation always exists and is unique: $f = f[\rho]$. Substiting this solution in Equation (6) we then obtain the energy as a functional of the density: $E[\rho] = E[f[\rho]]$.

We thus see that the variational equation for $f(\mathbf{r})$ is equivalent to a variational equation for $\rho(\mathbf{r})$:

$$\frac{\delta E[f]}{\delta f(\mathbf{r})} = 0 \Leftrightarrow \frac{\delta E[\rho]}{\delta \rho(\mathbf{r})} = 0, \qquad \frac{\partial f(\mathbf{r})}{\partial r} > 0 \Leftrightarrow \int \rho(\mathbf{r})\, \mathrm{d}r = N. \tag{11}$$

This is a local-density formulation of the many-body problem. However we must notice that the density functional is explicit only in terms of the LST function.

2. The Energy Density Functional in the LS-Coupling Scheme

Consider the Schrödinger Hamiltonian of an N-electron atom (or ion):

$$\hat{H} = \sum_{i=1}^{N} \hat{h}(\mathbf{r}_i) + \frac{1}{2} \sum_{i \neq j=1}^{N} V(\mathbf{r}_i, \mathbf{r}_j), \tag{12}$$

where $\hat{h}(\mathbf{r}) = -\frac{1}{2}\nabla^2 + U(\mathbf{r})$ is the one-electron Hamiltonian and $U(\mathbf{r})$ and $V(\mathbf{r}_i, \mathbf{r}_j)$ designate the nucleus–electron and electron–electron Coulomb interactions.

2.1. THE 'CORE-VALENCE' SPLIT, MODEL WAVEFUNCTION (DVS-MWF)

As a model wavefunction for an atomic system with one shell and several closed shells, which is an eigenfunction of $\widehat{\mathbf{L}^2}$, $\widehat{L_z}$, $\widehat{\mathbf{S}^2}$, $\widehat{S_z}$ in the L–S coupling scheme, we take an antisyininetrized projuct of wavefunctions for the closed shells and for the open shell [23]:

$$\Psi(\zeta_1, \ldots, \zeta_N) = C\mathcal{A}[\Phi_1(\zeta_1, \ldots, \zeta_p)\Phi_2(\zeta_{p+1}, \ldots, \zeta_N)], \tag{13}$$

where $\mathcal{A}$ is the antisymmetrizer for N particles and $\zeta_i = (\mathbf{r}_i, \sigma_i)$ is the set of space and spin coordinates for the ith electron. The first p electrons form the system of closed shells and their wavefunction can be taken as a single Slater determinant. The total orbital and spin momenta of these electrons are $L = S = 0$ and this is the reason the total wavefunction is factorized (13). One can write here

$$\begin{aligned}\Phi_1(\zeta_1, \ldots, \zeta_p) &= \frac{1}{\sqrt{p!}} \det[\phi_1(\mathbf{r}_1)\alpha(\sigma_1) \cdots \phi_{\frac{p}{2}}(\mathbf{r}_{\frac{p}{2}})\alpha(\sigma_{\frac{p}{2}}) \times \\ &\quad \times \phi_1(\mathbf{r}_{\frac{p}{2}+1})\beta(\sigma_{\frac{p}{2}+1}) \cdots \phi_{\frac{p}{2}}(\mathbf{r}_p)\beta(\sigma_p)].\end{aligned} \tag{14}$$

The last $q = N - p$ electrons are *equivalent* and form an s, p, d...-type open shell (however the formalism can be extended to non-equivalent orbitals involved in a bond). Their wavefunction is taken as an antisymmetrized product of single-electron orbitals [23]:

$$\Phi_2(\zeta_{p+1}, \ldots, \zeta_N) = \sum_{i_{p+1}, \ldots, i_N = p+1}^{p+q} C_{i_{p+1}, \ldots i_N} \phi_{i_{p+1}}(\zeta_{p+1}) \ldots \phi_{i_N}(\zeta_N). \tag{15}$$

This representation of the open-shell wavefunction is useful because of the factorization property of the total energy resulting from Equation (13).

A concrete realization of Equation (15) [24] is the sum of all possible products of '$q-1$ ion' wavefunctions with the wavefunction of the 'detached electron', coupled to be eigenfunctions of $\widehat{\mathbf{L}^2}$, $\widehat{L_z}$, $\widehat{\mathbf{S}^2}$, $\widehat{S_z}$ with eigenvalues L, M_L, S, M_S in order to have the same orbital and spin momenta L and S as the total wavefunction (13):

$$\Phi_{\gamma LSM_LM_S}(nl)^q = \sum_{\gamma',L',S'} (l^{q-1}[\gamma' L'S']lLS) \times \\ \times \Psi_{\gamma LSM_LM_S}(l^{q-1}[\gamma' L'S']l_q), \tag{16}$$

where γ is a seniority number which distinguishes between different terms with same L, S and the $(l^{q-1}[\gamma' L'S']lLS)$ are *coefficients of fractional parentage* which give the weights of the 'initial ions' and guarantee the antisymmetry of $\Phi_{\gamma LSM_LM_S}$. The $\Psi_{\gamma LSM_LM_S}$ in Equation (16) are built according to the vector coupling rule notwithstanding antisymmetry ([24–24]):

$$\Psi_{\gamma LSM_LM_S}(l^{q-1}[\gamma' L'S']l_q) = \sum_{m',\mu',M'_L,M'_S} (L'M'_L lm'|L'lLM_L) \times \\ \times (S'M'_S \tfrac{1}{2}\mu'|S'\tfrac{1}{2}SM_S)\Psi^{(q-1)}_{\gamma' L'S'M'_LM'_S} \times \\ \times(\zeta_{p+1},\ldots,\zeta_{N-1})\phi_{nlm'\mu'}(\zeta_N), \tag{17}$$

where $(L'M'_L lm'|L'lLM_L)$ and $(S'M'_S\frac{1}{2}\mu'|S'\frac{1}{2}SM_S)$ are the Clebsch–Gordan coefficients for vector coupling to L, S.

2.2. THE ENERGY AS A FUNCTION OF THE CVS-MWF

It will be shown that the structure of the total wavefunction (13) does not change in the variational process provided that the single-electron orbitals retain their central-symmetry form:

$$\phi_i(\zeta) = R_{n_il_i}(r)Y_{l_im_i}(\theta,\varphi)\chi_{s_i}(\sigma), \tag{18}$$

and that the variation concerns only their radial part:

$$R_{n_il_i}(r) = \frac{P_{n_il_i}(r)}{r}. \tag{19}$$

Using standard techniques for the symmetric one-electron and two-electron parts of the Hamiltonian (12), the expansion of the electron–electron interaction in terms of spherical harmonics, the properties of the Slater determinant (14) and the orthogonality of the single-electron orbitals and coefficients in Equation (15), we obtain the following factorization of the total energy for the wavefunction (13) in kinetic energies for the closed and open shells, Coulomb repulsion and exchange terms within the closed shells, Coulomb repulsion and exchange terms between the closed shells and the open shell, and electron–electron interactions within the open shell:

$$\begin{aligned} E[\Phi] &= \sum_{i=1}^{p} h_{ii} + q h_{NN} + \frac{1}{2}\sum_{i,j=1}^{p}(V_{ijji} - V_{jiji}) + q\sum_{\alpha}(2l_\alpha + 1) \times \\ &\times \left\{ 2F^0(n,l;n_\beta,l_\beta) - \sum_{k=|l_\alpha - l_\beta|}^{l_\alpha + l_\beta} \begin{pmatrix} l_\alpha & k & l_\beta \\ 0 & 0 & 0 \end{pmatrix}^2 G^k(n_\alpha,l_\alpha;n_\beta,l_\beta) \right\} + \\ &+ \langle \Phi_2(\zeta_{p+1},\ldots,\zeta_N) | \frac{1}{2}\sum_{i\neq j=1}^{N} V(\mathbf{r}_i,\mathbf{r}_j) | \Phi_2(\zeta_{p+1},\ldots,\zeta_N) \rangle, \end{aligned} \tag{20}$$

wherc the matrix elements and the Slater integrals are given by:

$$h_{ii} = \int \mathrm{d}\mathbf{r} \sum_{\sigma} \phi_i^*(\mathbf{r},\sigma)\left[-\frac{1}{2}\Delta_{\mathbf{r}} + U(\mathbf{r})\right]\phi_i(\mathbf{r},\sigma) \tag{21}$$

$$V_{ijkl} = \int \mathrm{d}\zeta \int \mathrm{d}\zeta' \phi_i^*(\zeta)\phi_j^*(\zeta')V(\mathbf{r},\mathbf{r}')\phi_k(\zeta')\phi_l(\zeta) \tag{22}$$

$$F^k(n,l;n',l') = \int_0^\infty \int_0^\infty \mathrm{d}r\,\mathrm{d}r' \frac{r_<^k}{r_>^{k+1}} P_{n,l}^2(r) P_{n',l'}^2(r') \tag{23}$$

$$G^k(n,l;n',l') = \int_0^\infty \int_0^\infty \mathrm{d}r\,\mathrm{d}r' \frac{r_<^k}{r_>^{k+1}} P_{n,l}^2(r) P_{n,l}(r') P_{n',l'}(r) P_{n',l'}(r'), \tag{24}$$

$$G^k(n,l) \equiv G^k(n,l;n,l) = F^k(n,l;n,l) \equiv F^k(n,l), \tag{25}$$

and

$$\begin{pmatrix} j_1 & j_2 & j_3 \\ m_1 & m_2 & m_3 \end{pmatrix}$$

denote the $3j$-symbols.

The term $\frac{1}{2}\sum_{i,j=1}^{p}(V_{ijji} - V_{jiji})$ in Equation (20) is also expressible in terms of the integrals $F^k(n,l;n',l')$ and $G^k(n,l;n',l')$. To compute the last term in Equation (20) we use Equations (15)–(17). We expand the electron–electron interaction for the eqiuvalent electrons in the open shell in terms of irreducible tensor operators, use the Wigner–Eckart theorem for their matrix elements, and introduce the standard notation [24] for irreducible tensor operators U^k. Thus we obtain the final partition for the total energy:

$$\begin{aligned}
E[\Phi, \gamma LS] &= qI(n,l) + \sum_\alpha 2(2l_\alpha+1)I(n_\alpha,l_\alpha) + \\
&+\sum_\alpha (2l_\alpha+1)\Big\{(4l_\alpha+1)F^0(n_\alpha,l_\alpha) - (2l_\alpha+1)\times \\
&\times \sum_{k=0}^{2l_\alpha} \begin{pmatrix} l_\alpha & k & l_\alpha \\ 0 & 0 & 0 \end{pmatrix}^2 F^k(n_\alpha,l_\alpha)\Big\} + \\
&+2\sum_{\alpha<\beta}(2l_\alpha+1)(2l_\beta+1)\Big\{2F^0(n_\alpha,l_\alpha;n_\beta,l_\beta) - \\
&- \sum_{k=|l_\alpha-l_\beta|}^{l_\alpha+l_\beta} \begin{pmatrix} l_\alpha & k & l_\beta \\ 0 & 0 & 0 \end{pmatrix}^2 G^k(n_\alpha,l_\alpha;n_\beta,l_\beta)\Big\} + \\
&+q\sum_\alpha (2l_\alpha+1)\Big\{2F^0(n,l;n_\alpha,l_\alpha) - \\
&- \sum_{k=|l-l_\alpha|}^{l+l_\alpha} \begin{pmatrix} l & k & l_\alpha \\ 0 & 0 & 0 \end{pmatrix}^2 G^k(n,l;n_\alpha,l_\alpha)\Big\} + \\
&+\frac{(2l+1)^2}{2}\sum_{k=0}^{2l}\Big\{\frac{1}{(2L+1)(2S+1)}\times \\
&\times \sum_{\gamma' L'} |(l^q\gamma LS \parallel \tilde{U}^k \parallel l^q\gamma' L'S')|^2 - \frac{q}{2l+1}\Big\}\times \\
&\times \begin{pmatrix} l & k & l \\ 0 & 0 & 0 \end{pmatrix}^2 F^k(n,l),
\end{aligned} \tag{26}$$

where γLS is the spectroscopic term, q is the number of (equivalent) electrons in the open shell (with orbital momentum l) α, β run over the number of closed shells ($\alpha < \beta$ if $n_\alpha < n_\beta$ or if $l_\alpha < l_\beta$ when $n_\alpha = n_\beta$), the one-electron integrals $I(n,l)$ are given by

$$I(n,l) = \int_0^\infty dr P_{n,l}(r)\left[-\frac{1}{2}\frac{\partial^2}{\partial r^2} + \frac{Z}{r} + \frac{l(l+1)}{2r^2}\right]P_{n,l}(r), \tag{27}$$

and the matrix elements of $\tilde{U}^k$ are connected to the standard ones (given in tables [26]) by:

$$(l^q\gamma LS \parallel \tilde{U}^k \parallel l^q\gamma' L'S') = \sqrt{2S+1}(l^q\gamma LS \parallel U^k \parallel l^q\gamma' L'S'). \tag{28}$$

In Equation (26) the first term represents the one-particle energies of the q open-shell electrons in the field of the nucleus, the following summation bears on the one-particle energies of the electrons filling the closed shells, the following two lines refer to the Coulomb repulsion and exchange interactions of electrons within and between closed shells, the fourth line gives the interactions between the closed-shell and the open-shell electrons, and the last summations refer to the interactions of electrons within the open shell (which reduce to 0 if a single unpaired electron is present).

2.3. THE ENERGY AS A FUNCTIONAL OF THE LST FUNCTION

The energy (26) will be an explicit functional of the scalar function $f(\mathbf{r})$ after the LST is performed in the Slater integrals $I(n,l)$, $F^k(n,l;n',l')$, $G^k(n,l;n',l')$. The LST bears only on the radial variable of the one-electron orbitals and thus preserves the structure and symmetry of the total wavefunction (13):

$$R_{n_il_i}(r) \to \frac{f(\mathbf{r})}{r}\sqrt{f'(\mathbf{r})}R_{n_il_i}[f(\mathbf{r})], \quad f(\mathbf{r}) = f(r,\theta,\phi) = f(r)\forall(\theta,\phi). \tag{29}$$

The LST transforms the Slater integrals as follows:

$$\begin{aligned} I(n,l)[f] &= \frac{1}{2}\int_0^\infty \left\{\frac{\partial}{\partial r}\left[f\sqrt{\frac{\partial f}{\partial r}}R_{n,l}(f)\right]\right\}^2 \mathrm{d}r + \\ &+ \int_0^\infty \left[\frac{Z}{r}+\frac{l(l+1)}{2r^2}\right] f^2\frac{\partial f}{\partial r}R_{n,l}^2(f), \end{aligned} \tag{30}$$

$$\begin{aligned} F^k(n,l;n'l')[f] &= \int_0^\infty\int_0^\infty \mathrm{d}r\,\mathrm{d}r'\frac{r_<^k}{r_>^{k+1}}f^2(r)\frac{\partial f}{\partial r}R_{n,l}^2[f(r)] \times \\ &\times f^2(r')\frac{\partial f}{\partial r'}R_{n',l'}^2[f(r')], \end{aligned} \tag{31}$$

$$\begin{aligned} G^k(n,l;n',l')[f] &= \int_0^\infty\int_0^\infty \mathrm{d}r\,\mathrm{d}r'\frac{r_<^k}{r_>^{k+1}}f^2(r)\frac{\partial f}{\partial r}R_{n,l}[f(r)]R_{n',l'}[f(r)] \times \\ &\times f^2(r')\frac{\partial f}{\partial r'}R_{n,l}[f(r')]R_{n',l'}[f(r')]. \end{aligned} \tag{32}$$

The latter two expressions become identical for $n' = n$, $l' = l$: $F^k(n,l)[f] = G^k(n,l)[f]$.

2.4. THE EULER–LAGRANGE EQUATION FOR THE ENERGY FUNCTIONAL

The Euler–Lagrange Equation (11) is constructed from Equation (26) with the variational derivatives of the Slater integrals entering it:

$$\frac{\delta I(n,l)[f]}{\delta f(r)} =$$

$$\frac{f^{(iv)}}{4}\frac{f^2 R^2_{n,l}(f)}{f'} + \frac{f'''}{2}\left[\frac{\{f^2 R^2_{n,l}(f)\}'}{f'} - \frac{f^2 R^2_{n,l}(f) f''}{f'^2}\right] + \frac{f''^3}{4}\frac{f^2 R^2_{n,l}(f)}{f'^3} +$$

$$+ f''[2ff'\{R'_{n,l}(f) + fR''_{n,l}(f)\}R_{n,l}(f) - f'\{R^2_{n,l}(f) + f^2 R'^2_{n,l}(f)\}] +$$

$$+ f'^3\left[fR_{n,l}(f)R''_{n,l}(f) + \frac{f^2}{2}R_{n,l}(f)R'''_{n,l}(f) - fR'^2_{n,l}(f) - \right.$$

$$\left. - R_{n,l}(f)R'_{n,l}(f) - \frac{f^2}{2}R'_{n,l}(f)R''_{n,l}(f)\right] + f^2 R^2_{n,l}(f)\left[\frac{Z}{r^2} + \frac{l(l+1)}{r^3}\right], \quad (33)$$

$$\frac{\delta F^k(n,l;n',l')[f]}{\delta f(r)} =$$

$$\frac{k+1}{r^{k+2}} f^2(r) R^2_{n,l}[f(r)] \int_0^r \mathrm{d}r' f^2(r')(r')^k f'(r') R^2_{n',l'}[f(r')] -$$

$$- k r^{k-1} f^2(r) R^2_{n,l}[f(r)] \int_r^\infty \mathrm{d}r' \frac{f^2(r')}{(r')^{k+1}} f'(r') R^2_{n',l'}[f(r')] +$$

$$+ \frac{k+1}{r^{k+2}} f^2(r) R^2_{n',l'}[f(r)] \int_0^r \mathrm{d}r' f^2(r')(r')^k f'(r') R^2_{n,l}[f(r')] -$$

$$- k r^{k-1} f^2(r) R^2_{n',l'}[f(r)] \int_r^\infty \mathrm{d}r' \frac{f^2(r')}{(r')^{k+1}} f'(r') R^2_{n,l}[f(r')], \quad (34)$$

$$\frac{\delta G^k(n,l;n',l')[f]}{\delta f(r)} = 2\frac{k+1}{r^{k+2}} f^2(r) R_{n,l}[f(r)] R_{n',l'}[f(r)] \times$$

$$\times \int_0^r \mathrm{d}r' f^2(r')(r')^k f'(r') R_{n,l}[f(r')] R_{n',l'}[f(r')] -$$

$$- 2k r^{k-1} f^2(r) R_{n,l}[f(r)] R_{n',l'}[f(r)] \times$$

$$\times \int_r^\infty \mathrm{d}r' \frac{f^2(r')}{(r')^{k+1}} f'(r') R_{n,l}[f(r')] R_{n',l'}[f(r')]. \quad (35)$$

Equation (33) involves derivatives of $f(r)$ up to the 4th order due to the kinetic-energy contributions to the single-particle terms.

3. An Illustration: Lithium

The lithum atom has $N = Z = 3 \rightarrow p = 2$, $q = 1$, $L = 0$, $S = 1/2$, $n_1 = 1$, $n_2 = 2$, $l_1 = l_2 = 0$, and the spectral term for the ground state is ^2S. We can choose hydrogenoid orbitals:

$$R_{n_1,l_1}(r) = R_{n_1,l_2}(r) = R_{1,0}(r) = 6\sqrt{3}\exp(-3r),$$

$$R_{n_2,l_2}(r) = R_{2,0}(r) = 3\sqrt{3/2}\left(1 - \frac{3}{2}r\right)\exp\left(-\frac{3}{2}r\right). \tag{36}$$

In this relatively simple case the integrals in the variational derivatives (33–35) can be computed using the incomplete gamma function $\gamma(n+1, z)$ [27], and the variational Equation (11) can be written:

$$0 = \frac{\delta E(Li[^2S])[f]}{\delta f(r)}$$

$$= 2\frac{\delta I(n_1,0)}{\delta f(r)} + \frac{\delta I(n_2,0)}{\delta f(r)} + 2\frac{\delta F^0(n_1,0;n_2,0)}{\delta f(r)} + \frac{\delta F^0(n_1,0)}{\delta f(r)} -$$

$$- \frac{\delta G^0(n_1,0)}{\delta f(r)} = f^{(iv)}\frac{f^2[2R_1^2(f) + R_2^2(f)]}{4f'} +$$

$$+ f'''\left\{f[2R_1^2(f) + R_2^2(f)] + f^2[2R_1(f)R_1'(f) + R_2(f)R_2'(f)] -\right.$$

$$\left.- f''\frac{f^2[2R_1^2(f) + R_2^2(f)]}{2f'^2}\right\} + f''^3\frac{f^2[2R_1^2(f) + R_2^2(f)]}{4f'^3} +$$

$$+ f''\{2ff'[2R_1(f)[R_1'(f) + fR_1''(f)] + R_2(f)[R_2'(f) + fR_2''(f)]] -$$

$$- f'[2R_1^2(f) + R_2^2(f) + f^2[2R_1'^2(f) + R_2'^2(f)]]\} +$$

$$+ f'^3\left\{\frac{f^2}{2}[2[R_1(f)R_1'''(f) - R_1'(f)R_1''(f)] + [R_2(f)R_2'''(f) - R_2'(f)R_2''(f)]] +\right.$$

$$+ f[2[R_1(f)R_1''(f) - R_1'^2(f)] + [R_2(f)R_2''(f) - R_2'^2(f)]] -$$

$$\left.- [2R_1(f)R_1'(f) + R_2(f)R_2'(f)]\right\} + \frac{3f^2}{r^2}[2R_1^2(f) + R_2^2(f)] +$$

$$+\frac{18f^2}{r^2}\left\{\left[12\exp(-6f)+\frac{3}{2}\left(1-\frac{3}{2}f\right)^2\exp(-3f)\right]\times\right.$$

$$\times\left[1-(18f^2+6f+1)\exp(-6f)\right]+12\exp(-6f)\times$$

$$\times\left[1-\left(\frac{81}{8}f^4+\frac{9}{2}f^2+3f+1\right)\exp(-3f)\right]-$$

$$\left.-54\left(1-\frac{3}{2}f\right)f^3\exp(-9f)\right\}. \qquad (37)$$

As we can see, the Euler–Lagrange equation for the LST function appears to be a *nonolinear differential equation*, which may be solved numerically. This equation has singular points $\{r_s \mid f'(r_s) = 0,\ f(r_s) \neq 0\}$. The initial conditions can be determined from Equation (37) and its asymptotic equation in the neighborhood of the singular points:

$$0=\frac{f''^2}{2f'^3}-\frac{f'''}{f'^2},\quad r\to r_s(f'\to 0). \qquad (38)$$

The non-trivial solution of the above equation is given by:

$$f(r)=b(r-r_s)^3+d,\quad r\to r_s, \qquad (39)$$

where b, d are constants. There is another singular point: $r_0 = 0$, where $f'(r_0) \neq 0$. It can be shown [23] that the asymptotic behaviour is there:

$$f(r)=cr\exp(3r),\quad r\to r_0. \qquad (40)$$

Using this latter asymptotic and its derivatives we find the initial conditions:

$$f(r_{in})=cr_{in},\quad f'(r_{in})=c(1+3r_{in}),$$

$$f''(r_{in})=3c(2+3r_{in}),\quad f'''(r_{in})=27c(1+r_{in}). \qquad (41)$$

There remains one free parameter: $c = f'(0)$. If we impose the condition that $c > 0$, then we have, due to the asymptotic from Equation (39), $f'(r) > 0\forall r$, which is precisely the constraint shown in Equation (11). The energy determined from Equations (26–32) must be a minimum with respect to parameter c.

We are presently investigating specific numerical procedures aimed at solving the exact Euler–lagrange equation in $f(\mathbf{r})$ for the lithium and other alkali metal atoms. Various types of parameterized forms for $f(\mathbf{r})$ ([1, 2, 28]) are also being considered.

4. Conclusion

In this paper we have applied the local-scaling transformation method to core-valence split, model wavefunctions for open-shell atoms described by the LS coupling scheme. We have derived an explicit form for the total energy in terms of single-particle, repulsion and exchange functionals of the LST scalar function $f(\mathbf{r})$. We have written the corresponding Euler–Lagrange integro-differential equation, which involves derivatives of $f(\mathbf{r})$ up to the fourth order. Solving numerically this equation would in principle constitute a very powerful way of deriving the density $\rho_{LS}(\mathbf{r})$ and energy E_{LS} within the orbit described by the CVS-MWF. After this optimization is achieved one could also consider performing orbit jumping by moving through the transverse direction [22] in order to reach the eigenorbit, from which the exact density $\rho_0(\mathbf{r})$ and energy E_0 would be obtained, A preliminary approach to this may be to vary the one-electron orbitals with the same symmetry, i.e., to introduce k-dependent LST f functions. However, even the simple case of lithium reveals some specific divergency problems, which can be solved only by resorting to sophisticated mathematical procedures. A way out of these analytical difficulties would be to use a convenient parameterized form for $f(\mathbf{r})$ [28]. The problem would then reduce to finding the stationay values of the density $\rho^k_{LS}(\mathbf{r})$ and energy E^k_{LS} with respect to the parameters k defining the LST function $f(\mathbf{r})$. Further work along this line is now in progress.

References

1. G. G. Hall: *Proc. Roy. Soc. (London)* **75**, 575 (1960).
2. M. G. T. Hoor: *Int. J. Quantum Chem.* **33**, 563 (1968).
3. N. M. Witriol: *Int. J. Quantum Chem. S* **6**, 145 (1972).
4. K. B. Wolf: *Rev. Mex. Fis.* **22**, 45 (1973).
5. B. I. Zhilinsky and N. F. Stepanov: *Teor. Mat. Fiz.* **15**, 146 (1973).
6. S. T. Epstein: *The Variational Method in Quantum Chemistry*, Academic Press, New York (1974).
7. F. M. Fernandez and E. A. Castro: *Rev. Braz. Fiz.* **11**, 145 (1981).
8. D. I. Blokhintsev: *Zh. Exp. Teor. Fiz.* **9**, 1166 (1939).
9. W. Macke: *Phys. Rev.* **100**, 992 (1955).
10. W. Macke and P. Rennert: *Ann. Phys. (Leipzig)* **22**, 18 (1958).
11. N. H. March and W. H. Young: *Proc. Roy. Soc. (London)* **72**, 182 (1958).
12. J. R. Percus: *Int. J. Quantum Chem.* **13**, 89 (1978).
13. P. W. Payne: *Proc. Nat. Acad. Sci. (USA)* **77**, 6293 (1980).
14. H. K. H. Siedentorp and A. M. K. Muller: *Phys. Lett. A* **81**, 5 (1981).
15. V. Yu. Kolmanovich and I. M. Reznik: *Dokl. Acad. Nauk SSSR* **258**, 1100 (1981).
16. J. E. Harriman: *Phys. Rev. A* **24**, 680 (1981). *Phys. Rev. A* **27**, 632 (1983). *Int. J. Quantum Chem.* **83**, 6283 (1985).
17. G. Zumbach and K. Maschke: *Phys. Rev. A* **28**, 544 (1983). *Erratum A* **29**, 1585 (1984).
18. E. V. Ludeña: *J. Chem. Phys.* **79**, 6174 (1983).
19. I. Zh. Petkov and M. V. Stoitsov: *Teor. Mat. Fiz.* **55**, 407 (1983). I. Zh. Petkov, M. V. Stoitsov and E. S. Kryachko: *Int. J. Quantum Chem.* **29**, 149 (1986).
20. I. Zh. Petkov and M. V. Stoitsov: *Nuclear Density Functional Theory*, Clarendon, Oxford (1991).
21. E. S. Kryachko and E. V. Ludeña: *Energy Density Functional Theory of Many-Electron Systems*, Kluwer, Doredrecht (1989).

22. E. V. Ludeña, R. Lopez-Boada, J. E. Maldonado, E. S. Kryachko, T. Koga and J. Hinze: *Int. J. Quantum Chem.* **56**, 285 (1995).
23. L. S. Georgiev: Diploma thesis, Sofia (1994).
24. I. I. Sobelman: *Introduction to the Theory of Atomic Spectra* (in Russian), Fizmatgiz, Moscow (1963).
25. L. C. Biedenharn and J. D. Louck: *Angular Momentum in Quantum Physics*, Addison-Wesley, Boston (1981).
26. R. Karazia, Y. Vizbaraite, Z. Rudzias and A. Yucis: *Tables for the Calculation of the Matrix Elements of Atomic Quantities* (in Russian), Nauka, Moscos (1986).
27. M. Abramowitz and I. Stegun: *Handbook of Mathematical Functions*, National Bureau of Standards, Washigton (1964).
28. E. S. Kryachko, E. V. Ludeña and T. Koga: in *METECC-89B*, E. Clementi (ed.), Cagliari (1993).

R. McWeeny et al. (eds.), Quantum Systems in Chemistry and Physics, 323–341.

On the Accuracy of the Algebraic Approximation in Molecular Electronic Structure Studies: VII. Matrix Valence Bond Calculations for the Hydrogen Molecule Ground State

D. MONCRIEFF
Supercomputer Computations Research Institute, Florida State University, Tallahassee, Florida, U.S.A.

S. WILSON
Rutherford Appleton Laboratory, Chilton, Oxfordshire OX11 0QX, U.K.

(Received and accepted: 17 June 1997)

Abstract. The accuracy of generalized valence bond calculations made within the algebraic approximation (that is by making finite basis set expansions) is investigated by using systematic sequences of distributed even-tempered Gaussian basis sets for the ground state of the hydrogen molecule. In contrast to the matrix Hartree–Fock model, the valence bond model approach afford a qualitatively correct description of molecular dissociation processes and provides a useful reference with respect to which dynamic electron correlation effects can be described. A basis set is developed which supports an accuracy approaching the sub-μhartree level in energy calculations using the matrix valence bond and matrix Hartree–Fock approaches. A comparison is made with the results of second-order many-body perturbation theory calculations. An interpretation of the utility of bond-centred basis functions is provided in terms of generalized valence bond theory.

Key words: Algebraic approximation, finite basis sets, Gaussian basis sets, even-tempered basis sets, distributed basis sets, valence bond theory.

1. Introduction

Over the past decade it has been demonstrated that the algebraic approximation, that is the use of finite basis set expansions, can be implemented in matrix Hartree–Fock calculations for diatomic molecular systems so as to afford an accuracy approaching that achieved in finite difference and finite element calculations [1–8]. More recently, it has been demonstrated that the algebraic approximation facilitates high precision applications to polyatomic systems [9], to the description of electron correlation effects [10, 11], and to the relativistic electronic structure problem [12–15].

In this work, we seek to establish a basis set capable of supporting a sub-μhartree level of accuracy for the energy corresponding to a generalized valence bond wave function of the form

$$\Psi_{vb} = \sqrt{N!}\mathcal{A}(\Phi\Theta) \tag{1}$$

where the spatial function, Φ, is a product of non-orthogonal orbitals, ϕ_i, that is

$$\Phi = \prod_i^N \phi_i \tag{2}$$

and Θ is an N-electron spin function. This wave function, unlike the molecular orbital functions studied in our previous work, provides a qualitatively correct description of molecular dissociation processes and thus affords an appropriate reference for electron correlation studies at geometries away from equilibrium.. Wave functions of the type (1) were considered, independently, in the late 1960s and early 1970s by Goddard and coworkers [16–18], by Gerratt and coworkers [19–23], and by Kaldor and Harris [24]. For more recent discussions of valence bond models see, for example, Gallup *et al.* [25], Raimondi *et al.* [26], Cooper *et al.* [27], McWeeny [28–30], and Wilson [31].

Specifically, in this work, we systematically develop an even-tempered basis set of Gaussian primitive functions capable of supporting sub-μhartree accuracy for the ground state energy of the hydrogen molecule within the framework of matrix valence bond theory; that is, generalized valence bond theory formulated within the algebraic approximation. Such a formalism affords a qualitatively correct description of the dissociative process

$$H_2(X^1\Sigma_g^+) \to 2H(^2S). \tag{3}$$

The wave function employed in the present study has the form

$$\Psi = \sqrt{2}\mathcal{A}(\phi_1(\mathbf{r}_1)\phi_2(\mathbf{r}_2)\Theta) \tag{4}$$

where ϕ_1 and ϕ_2 are two distinct non-orthogonal orbitals each having σ symmetry and related by the reflection operation $\hat{\sigma}_h : \phi_1 = \hat{\sigma}_h\phi_2$. $\mathbf{r}_1$ and $\mathbf{r}_2$ are the spatial coordinates of electrons 1 and 2, respectively. Θ is the two-electron singlet spin function: $\sqrt{\frac{1}{2}}(\alpha_1\beta_2 - \beta_1\alpha_2)$ and $\mathcal{A}$ is the antisymmetrizing operator. Applications of the generalized valence bond formalism to the hydrogen molecule ground state were described by (C.W.) Wilson and Goddard [32] and by (S.) Wilson and Gerratt [33]. Such formulations can be regarded as a generalization of the Coulson–Fischer wave function [34, 46] for the hydrogen molecule which is not only a generalization of both the simple molecular orbital and valence bond treatments but also serves to emphasize the equivalence of the two methods when suitably refined. (See also the work of Mueller and Eyring [35].)

The basis sets developed in this work are distributed sets of Gaussian functions [36] [37]. Molecular basis sets are usually constructed from functions centred on each of the atoms in the system. Distributed basis sets may also contain functions located on other centres and, in particular, at the mid-point of the bond. A number of classical valence bond studies [38–41] of the hydrogen molecule ground state using basis sets of atom-centred functions have been reported. Some early valence bond calculations [42–44] explored the use of floating functions but only the recent classical valence bond study by Harcourt [45] has made use of mid-bond functions.

In Section 2, our strategy for systematically designing a basis set is described first for the case of atom-centred basis sets and then for distributed basis sets containing both atom-centred and bond-centred primitive functions. The interpretation of the effects of bond-centred functions with the context of generalized valence bond theory are considered. The computational details of the calculations reported in the present work are given in Section 3. Section 4 contains our results and Section 5 a discussion. A summary is given in Section 6 together with our concluding remarks.

2. Universal Systematic Sequence of Distributed Even-Tempered Basis Sets for Matrix Valence Bond Calculations

The orbitals $\phi_i(\mathbf{r})$ in the molecular wave function (1) are approximated by means of a linear expansion in terms of some basis set $\{\chi_p(\mathbf{r})\}$ which in this work will be taken to be spherical harmonic Gaussian functions [47]; thus

$$\phi_i(\mathbf{r}) = \sum_{p=1}^{N} \chi_p(\mathbf{r}) c_{ip}, \quad i = 1, 2. \tag{5}$$

Formally, a complete set of Gaussian functions centred on any arbitrary point may be employed. In practice, this approach is found to be poorly convergent and a distributed basis set is most often employed with different subsets centred on different points in space. In the widely used LCAO (Linear Combination of Atomic Orbitals) approximation, these points are taken to be coincident with the nuclei. We consider first the development of atom-centred basis sets.

2.1. ATOM-CENTRED BASIS SETS

A basis set containing functions centred on each of the atoms in the system is designated

$$S = S_A \oplus S_B \tag{6}$$

where S_A (S_B) denotes the basis subset centred on nucleus A (B) containing N_A (N_B) primitive Gaussian functions. For a homonuclear diatomic species $N_A = N_B$.

Now the approximate orbital (5) for the hydrogen molecule ground state can be written

$$\phi_1(\mathbf{r}) = \sum_{p=1}^{N_A} \chi_p^{(A)}(\mathbf{r})c_{1p} + \sum_{p=N_A+1}^{N} \chi_p^{(B)}(\mathbf{r})c_{1p} \tag{7}$$

where $\chi_p^{(A)}(\mathbf{r})$ $[\chi_p^{(B)}(\mathbf{r})]$ is a function centred on nucleus A $[B]$ and $N = N_A + N_B$. Putting

$$\phi_1(\mathbf{r}) = \mathcal{N}[\phi_1^{(A)}(\mathbf{r}) + \lambda\phi_1^{(B)}(\mathbf{r})] \tag{8}$$

then the second orbital has the form

$$\phi_2(\mathbf{r}) = \mathcal{N}[\phi_1^{(B)}(\mathbf{r}) + \lambda\phi_1^{(A)}(\mathbf{r})] \tag{9}$$

where

$$\phi_i^{(A)}(\mathbf{r}) = \mathcal{N}_A \sum_{p=1}^{N_A} \chi_p^{(A)}(\mathbf{r})c_{ip} \tag{10}$$

and

$$\phi_i^{(B)}(\mathbf{r}) = \mathcal{N}_B \sum_{p=N_A+1}^{N} \chi_p^{(B)}(\mathbf{r})c_{ip} \tag{11}$$

$\mathcal{N}$, $\mathcal{N}_A$ and $\mathcal{N}_B$ are normalization constants. The valence bond orbitals have $C_{\infty v}$ symmetry so that the total spatial wave function

$$\Phi = \phi_1(\mathbf{r}_1)\phi_2(\mathbf{r}_2) + \phi_2(\mathbf{r}_1)\phi_1(\mathbf{r}_2) \tag{12}$$

belongs to the molecular point group symmetry ($D_{\infty h}$).

It is well known that this wave function, which is a generalization of that first discussed by Coulson and Fischer [34, 46], is entirely equivalent to a multistructure valence bond function of the type

$$\Phi = \Phi_{\text{covalent}} + \Phi_{\text{ionic}} \tag{13}$$

which may be written explicitly as

$$\begin{aligned}\Phi = {} & \mathcal{M}^2\{(1+\lambda^2)(\phi^{(A)}(\mathbf{r}_1)\phi^{(B)}(\mathbf{r}_2) + \phi^{(B)}(\mathbf{r}_1)\phi^{(A)}(\mathbf{r}_2)) + \\ & +2\lambda(\phi^{(A)}(\mathbf{r}_1)\phi^{(A)}(\mathbf{r}_2) + \phi^{(B)}(\mathbf{r}_1)\phi^{(B)}(\mathbf{r}_2))\}\end{aligned} \tag{14}$$

where $\mathcal{M}$ is a normalization factor. Φ_{covalent} describes a covalent structure which is usually represented by H—H and Φ_{ionic} describes the ionic structures H^+H^- and H^-H^+.

The wave function (12) can also be written as a multiconfiguration self-consistent-field function

$$\Phi = C_g\phi_g(\mathbf{r}_1)\phi_g(\mathbf{r}_2) + C_u\phi_u(\mathbf{r}_1)\phi_u(\mathbf{r}_2). \tag{15}$$

Both of the molecular orbitals have $D_{\infty h}$ symmetry

$$\phi_g(\mathbf{r}) = \mathcal{N}_g[\phi^{(A)}(\mathbf{r}) + \phi^{(B)}(\mathbf{r})] \tag{16}$$

$$\phi_u(\mathbf{r}) = \mathcal{N}_u[\phi^{(A)}(\mathbf{r}) - \phi^{(B)}(\mathbf{r})]. \tag{17}$$

The generalized valence bond function (12) is equivalent both to a multistructure valence bond function and a multiconfiguration self-consistent field wave function.

2.2. ADDITION OF BOND-CENTRED BASIS FUNCTIONS

We turn now to the case of a particular distributed basis set in which atom-centred functions are supplemented with bond-centred functions. Such a basis set can be designated

$$S = S_A \oplus S_{AB} \oplus S_B \tag{18}$$

where S_{AB} labels a set of primitive Gaussian functions centred at the mid-point of the 'bond' AB. The generalized Coulson–Fischer orbitals may now be written

$$\begin{aligned} \phi_i(\mathbf{r}) &= \sum_{p=1}^{N_A} \chi_p^{(A)}(\mathbf{r})c_{ip} \sum_{p=N_A+1}^{p=N_A+N_B} \chi_p^{(AB)}(\mathbf{r})c_{ip} + \\ &+ \sum_{p=N_A+N_{AB}+1}^{N} \chi_p^{(B)}(\mathbf{r})c_{ip}, \quad i=1,2 \end{aligned} \tag{19}$$

with

$$\phi_1(\mathbf{r}) = \mathcal{N}[\phi^{(A)}(\mathbf{r}) + \mu\phi^{(AB)}(\mathbf{r}) + \lambda\phi^{(B)}(\mathbf{r})] \tag{20}$$

and

$$\phi_2(\mathbf{r}) = \mathcal{N}[\phi^{(B)}(\mathbf{r}) + \mu\phi^{(AB)}(\mathbf{r}) + \lambda\phi^{(A)}(\mathbf{r})] \tag{21}$$

where

$$\phi^{(A)}(\mathbf{r}) = \mathcal{N}_A \sum_{p=1}^{N_A} \chi_p^{(A)}(\mathbf{r})c_{ip} \tag{22}$$

$$\phi^{(AB)}(\mathbf{r}) = \mathcal{N}_{AB} \sum_{p=N_A+1}^{N_A+N_{AB}} \chi_p^{(AB)}(\mathbf{r})c_{ip} \tag{23}$$

and

$$\phi^{(B)}(\mathbf{r}) = \mathcal{N}_B \sum_{p=N_A+N_{AB}+1}^{N} \chi_p^{(B)}(\mathbf{r})c_{ip} \tag{24}$$

$\mathcal{N}, \mathcal{N}_A, \mathcal{N}_B$ and $\mathcal{N}_{AB}$ are normalization constants.

Substituting the explicit form for the valence bond orbitals in the spatial wave function, Φ (Equation (12)), we have

$$\begin{aligned}\Phi = \; & \mathcal{M}^2\{(1+\lambda^2)(\phi^{(A)}(\mathbf{r}_1)\phi^{(B)}(\mathbf{r}_2) + \phi^{(B)}(\mathbf{r}_1)\phi^{(A)}(\mathbf{r}_2)) + \\ & +2\lambda(\phi^{(A)}(\mathbf{r}_1)\phi^{(A)}(\mathbf{r}_2) + \phi^{(B)}(\mathbf{r}_1)\phi^{(B)}(\mathbf{r}_2)) + \\ & +\mu(1+\lambda)((\phi^{(A)}(\mathbf{r}_1)\phi^{(AB)}(\mathbf{r}_2) + \phi^{(AB)}(\mathbf{r}_1)\phi^{(A)}(\mathbf{r}_2)) + \\ & +(\phi^{(AB)}(\mathbf{r}_1)\phi^{(B)}(\mathbf{r}_2) + \phi^{(B)}(\mathbf{r}_1)\phi^{(AB)}(\mathbf{r}_2))) + \\ & +2\mu^2\phi^{(AB)}(\mathbf{r}_1)\phi^{(AB)}(\mathbf{r}_2)\}\end{aligned} \tag{25}$$

which may be rewritten as a multistructure valence bond function

$$\Phi = \Phi_{\text{covalent}} + \Phi_{\text{ionic}} + \Phi_{\text{atom—bond}} + \Phi_{\text{bond}} \tag{26}$$

where the first two terms, Φ_{covalent} and Φ_{ionic}, are the familiar covalent and ionic terms which arose in (13) and (14) and the two new terms have the form

$$\begin{aligned}\Phi_{\text{atom—bond}} \simeq \; & (\phi^{(A)}(\mathbf{r}_1)\phi^{(AB)}(\mathbf{r}_2) + \phi^{(AB)}(\mathbf{r}_1)\phi^{(A)}(\mathbf{r}_2)) + \\ & +(\phi^{(AB)}(\mathbf{r}_1)\phi^{(B)}(\mathbf{r}_2) + \phi^{(B)}(\mathbf{r}_1)\phi^{(AB)}(\mathbf{r}_2))\end{aligned} \tag{27}$$

and

$$\Phi_{\text{bond}} \simeq \phi^{(AB)}(\mathbf{r}_1)\phi^{(AB)}(\mathbf{r}_2). \tag{28}$$

If we introduce the symbol b to represent a pseudo-atom at the bond centre then $\Phi_{\text{atom—bond}}$ describes the strutures (H—b)$^-$ H^+ and H^+ (b—H)$^-$ while the function Φ_{bond} describes the structure $H^+\ b^{2-}\ H^+$.

The wave function with the orbitals (20) and (21) may also be re-written as a multiconfiguration self-consistent-field function. The orbitals must have $D_{\infty h}$ symmetry and therefore have the form

$$\phi_g(\mathbf{r}) = \mathcal{N}_g[\phi^{(A)}(\mathbf{r}) + \nu\phi^{(AB)}(\mathbf{r}) + \phi^{(B)}(\mathbf{r})] \tag{29}$$

and

$$\phi_u(\mathbf{r}) = \mathcal{N}_u[\phi^{(A)}(\mathbf{r}) - \phi^{(B)}(\mathbf{r})]. \tag{30}$$

The bond-centred basis set only contributes to the orbital with gerade symmetry.

3. Computational Details

The calculations reported in this note were carried out with the GAUSSIAN94 [48] program on the CRAY J-932 computer at the Rutherford Appleton Laboratory and a Silicon Graphics PowerChallenge computer at the Florida State University. Integrals over spherical harmonic Gaussian-type functions [47] were evaluated to an accuracy of 10^{-10} and the molecular orbital self-consistent field iterations were continued until an absolute accuracy of 10^{-8} was achieved in the elements of the density matrix. The so-called direct approaches were employed in both the self-consistent field [49] and the perturbation theory [50–52] stages of the computation so as to avoid the need to store the two-electron integrals over the Gaussian basis functions. The generalized valence bond calculations were executed within the GAUSSIAN94 program using the software described by Bobrowicz and Goddard [53].

When using large basis sets it is important to monitor any near computational linear dependence in the basis set. All of the basis sets employed in the present work were monitored using the approach that we have employed in our previous studies by examining the eigenvalues of the overlap matrix and determining the number of eigenvalues less than 10^{-5} and the smallest eigenvalue. Where necessary diffuse functions in the bond-centred basis subset were deleted so as to control any near linear dependence effects..

4. Results

4.1. MATRIX HARTREE–FOCK ENERGIES

4.1.1. *Atom-Centred Basis Sets*

The matrix Hartree–Fock energies calculated by employing atom-centred basis sets with $n = 10$ and function of s, s and p, s, p and d, and s, p, d and f symmetries are presented in Table I. An energy decrease of 5026.5 μhartree is obtained on adding functions of p symmetry to the 20s set, and a further decrease of 94.6 μhartree on adding functions of d symmetry. The addition of atom-centred functions of f symmetry leads to a further energy lowering of 6.3 μhartree. The results of two further calculations for basis sets containing functions of s, p, d and f type with $n = 11, 12$ are also presented in Table I. The first of these achieves an additional energy lowering of 2.0 μhartree and the second a lowering of 0.8 μhartree.

Table I. Matrix Hartree–Fock calculations using atom-centred basis sets for the ground state of the hydrogen molecule with a nuclear separation of 1.4 bohr*.

Basis set	E_{mHF}	ΔE_{mHF}
$20s$	$-1.128\,498\,601\,5$	
$20s10p$	$-1.133\,525\,065\,1$	5026.463 6
$2010p10d$	$-1.133\,619\,668\,9$	94.603 8
$20s10p10d10f$	$-1.133\,625\,991\,1$	6.322 2
$22s11p11d11f$	$-1.133\,628\,009\,5$	2.018 4
$24s12p12d12f$	$-1.133\,628\,844\,4$	0.834 9

* Energies are given in Hartree.

4.1.2. *Atom-Centred and Bond-Centred Basis Sets*

The results of matrix Hartree–Fock calculations using basis sets containing both atom-centred and bond centre functions are collected in Table II.

Adding a $20s$ set at the bond centre yields a decrease of 5010.4 μhartree with respect to the $20s$ atom-centred basis set, a decrease which should be compared with that of 5026.5 μhartrec observed when the $20s$ atom-centred basis set is supplemented by $10p$ atom-centred sets. 99.7% of the energy decrease obtained by adding atom-centred p functions to the $20s$ atom-centred basis set is obtained by a $20s$ set at the bond centre. The $20s10p$ basis set contains a total of 100 functions whereas the $20s$; $20s$ bc set contains only 60 functions.

Adding a $20s$ set at the bond centre yields a decrease of 61.7 μhartree in the matrix Hartree–Fock energy with respect to the $20s10p$ atom-centred basis set, a decrease which should be compared with that of 94.6 μhartree observed when the $20s10p$ atom-centred basis set is supplemented by $10d$ atom-centred sets. 65.2% of the energy decrease obtained by adding atom-centred d functions to the $20s10p$ atom-centred basis set is obtained by a $20s$ set at the bond centre. Symmetry precludes any further decrease in the matrix Hartree–Fock on adding bond-centred functions of p symmetry to the $20s10p$; $20s$ set.

The addition of a $10f$ atom-centred set to the $20s10p10d$ set gives an energy lowering of 6.3 μhartree. The addition of a $20s$ bond-centred set to the $20s10p10d$ set gives an energy lowering of 6.9 μhartree; 8.7% more than that resulting from the addition of a $10f$ atom-centred set. The addition of a further $6d$ bond-centred set to the $20s10p10d$; $20s$ set gives a further energy lowering of 0.02 μhartree.

Adding a $17s$ set at the bond centre yields a decrease of 0.08 μhartree in the matrix Hartree–Fock energy with respect to the $20s10p10d10f$ atom-centred basis set. Adding a $18s7p6d$ set at the bond centre yields a decrease of 1.03 μhartree in the matrix Hartree–Fock energy with respect to the $20s10p10d10f$ atom-centred basis set.

Table II. Matrix Hartree–Fock calculations using basis sets containing both atom-centred and bond-centred functions for the ground state of the hydrogen molecule with a nuclear separation of 1.4 bohr*.

Basic set	E_{mHF}	ΔE_{mHF}
$20s$; $20s$ bc	−1.133 508 957 3	
$20s10p$; $20s$ bc	−1.133 586 768 0	77.810 7
$20s10p$; $20s10p$ bc	−1.133 586 768 0	0.000 0
$20s10p10d$; $20s$ bc	−1.133 626 539 7	39.771 7
$20s10p10d$; $20s7p$ bc	−1.133 626 539 7	0.000 0
$20s10p10d$; $10s7p6d$ bc	−1.133 626 556 8	0.017 1
$20s10p10d10f$; $17s$ bc	−1.133 626 069 1	0.487 7
$20s10p10d10f$; $18s7p$ bc	−1.133 626 997 0	0.927 9
$20s10p10d10f$; $18s7p6d$ bc	−1.133 627 021 5	0.024 5
$24s12p12d12f$; $21s$ bc	−1.133 628 890 8	1.869 3
$24s12p12d12f$; $1s8p$ bc	−1.133 628 890 8	0.000 0
$24s12p12d12f$; $21s8p9d$ bc	−1.133 628 900 4	0.009 6
$24s12p12d12f$; $21s8p9d8f$ bc	−1.133 628 900 4	0.000 0

* Energies are given in Hartree.

Extending the sequence $2nsnpndnf$ from n = 10 to n = 11 yields an energy decrease of 2.02 μhartree with a further decrease of 0.8 μhartree for the n = 12 case. Adding bond centre functions to the $24s12p12d12f$ atom-centred set gives energy decreases of 0.05 μhartree and 0.06 μhartree on adding s and d functions, respectively.

4.2. SECOND-ORDER MANY-BODY PERTURBATION THEORY ENERGIES

4.2.1. *Atom-Centred Basis Set*

The results of second-order correlation energy calculations using atom-centred basis sets are collected in Table III. An energy decrease of 17278.8 μHartree is obtained on adding functions of p symmetry to the $20s$ set (a decrease which is over three times the size of the corresponding decrease for the matrix Hartree–Fock calculations), and a further decrease of 2455.8 μHartree (almost 26 times the corresponding decrease for the matrix Hartree–Fock calculations) on adding d type functions. A further decrease of 622.8 μHartree is measured on adding atom-centred f functions; a decrease which is 99 times that observed in the matrix Hartree–Fock calculations. For the basis set of s, p, d and f functions there are energy decreases of 9.7 and 7.0 μHartree, respectively, for n = 11, and n = 12.

Table III. Second order many-body perturbation theory correlation energies obtained by using atom-centred basis sets for the ground state of the hydrogen molecule with a nuclear separation of 1.4 bohr*.

Basis set	$E_{[2/0]}$	$\Delta E_{[2/0]}$
$20s$	−1.146 994 093	
$20s10p$	−1.164 272 935	17278.842
$20s10p10d$	−1.166 728 768	2455.833
$20s10p10d10f$	−1.167 351 607	622.839
$22s11p11d11f$	−1.167 361 317	9.710
$24s12p12d12f$	−1.167 368 359	7.042

* Energies are given in Hartree.

4.2.2. *Atom-Centred and Bond-Centred Basis Sets*

The results of second-order correlation energy treatments using basis sets containing both atom-centred and bond-centred functions are collected in Table IV. An energy decrease of 6489.4 μHartree is obtained on adding a $20s$ bond-centred set to the $20s$ atom-centred sets. This is only ~38% of the decrease measured when atom-centred sets of p-type functions were added. Adding a $20s$ bond-centred set to the $20s10p$ atom-centred sets gives an energy decrease of 1004.6 μHartree; ~41% of the energy increment recovered by adding atom-centred functions of d symmetry. Adding bond-centred basis sets to the $20s10p10d$ atom-centred sets results in successive energy decreases of ~32.4, ~104.9 and ~117.6 μHantree for the $20s$, $20s10p$ and $20s10p10d$ sets, respectively. The addition of atom-centred $10f$ sets to the $20s10p10d$ atom-centred sets results in an energy change of ~622.8 μHartree compared with a total lowering of ~254.8 μHartree obtained on adding the $20s10p10d$ bond centred set. Adding bond-centred basis sets to the $20s10p10d10f$ atom-centred sets results in successive energy decreases of ~8.4, ~26.4 and ~35.1 μHartree for the $20s$, $20s10p$ and $20s10p10d$ sets, respectively. A total energy lowering of ~69.9 μHartree obtained on adding the $20s10p10d$ bond centred set to the $20s10p10d10f$ atom-centred sets. Extending the sequence $2nsnpndnf : 2nsnpnd$ bc from $n = 10$ to $n = 12$ yields an energy decrease of ~24.4 μhartree. There is a further decrease of ~60.7 μHartree on adding functions of f-type at the bond centre.

4.3. MATRIX VALENCE BOND ENERGIES

4.3.1. *Atom-Centred Basis Sets*

The results of the matrix valence bond calculations using atom-centred basis sets for the hydrogen molecule ground state are presented in Table V. A decrease of 3409.1 μhartree was obtained on supplementing the $20s$ sets with $10p$ sets. This corre-

Table IV. Second order many-body perturbation theory correlation energies obtained by using basis sets containing both atom-centred and bond-centred functions for the ground state of the hydrogen molecule with a nuclear separation of 1.4 bohr*.

Basis set	$E_{[2/0]}$	$\Delta E_{[2/0]}$
$20s$; $20s$ *bc*	$-1.153\,483\,504$	
$20s10p$; $20s$ *bc*	$-1.164\,494\,473$	11010.969
$20s10p$; $20s10p$ *bc*	$-1.165\,499\,086$	1004.613
$20s10p10d$; $20s$ *bc*	$-1.166\,761\,134$	1262.048
$20s10p10d$; $20s10p$ *bc*	$-1.166\,866\,018$	104.884
$20s10p10d$; $20s10p10d$ *bc*	$-1.166\,983\,574$	117.556
$20s10p10d10f$; $20s$ *bc*	$-1.167\,359\,994$	376.420
$20s10p10d10f$; $20s10p$ *bc*	$-1.167\,386\,407$	26.413
$20s10p10d10f$; $20s10p10d$ *bc*	$-1.167\,421\,545$	35.138
$24s12p12d12f$; $21s$ *bc*	$-1.167\,376\,217$	45.328
$24s12p12d12f$; $21s8p$ *bc*	$-1.167\,398\,819$	22.602
$24s12p12d12f$; $21s8p9d$ *bc*	$-1.167\,445\,936$	47.117
$24s12p12d12f$; $1s8p9d8f$ *bc*	$-1.167\,506\,631$	60.695

* Energies are given in Hartree.

Table V. Marix valence bond calculations using atom-centred basis sets for the ground state of the hydrogen molecule with a nuclear separation of 1.4 bohr*.

Basic set	E_{mVB}	ΔE_{mVB}
$20s$	$-1.148\,666\,732\,4$	
$20s10p$	$-1.152\,075\,816\,8$	3409.0844
$20s10p10d$	$-1.152\,151\,891\,6$	76.0748
$20s10p10d10f$	$-1.152\,157\,032\,2$	5.1406
$24s12p12d12f$	$-1.152\,159\,269\,6$	2.2374

* Energies are given in Hartree.

sponds to 67.8% of the 5026.5 μhartree decrease measured for the corresponding matrix Hartrer–Fock calculations. A further decrease of 76.1 μhartree was obtained on supplementing the $20s10p$ atom-centred sets with a $10d$ set which should be compared with a change of 94.6 μhartree in the matrix Hartree–Fock case. An energy decrease of 5.1 μhartree was observed on adding $10f$ sets which should be compared with an energy lowering of 6.3 μhartree in the matrix Hartree–Fock case.

Extending the sequence $2nsnpndnf$ from $n = 10$ to $n = 12$ yields an energy decrease of 2.2 μhartree compared with a decrease of 2.8 μhartree for the matrix Hartree–Fock case.

Table VI. Matrix valence bond calculations using basis sets containing both atom-centred and bond-centred functions for the ground state of the hydrogen molecule with a nuclear separation of 1.4 bohr*.

Basis set	E_{mVB}	ΔE_{mVB}
$20s$; $20s$ *bc*	−1.152 051 067 7	
$20s10p$; $20s$ *bc*	−1.152 126 832 0	75.7643
$20s10p$; $20s10p$ *bc*	−1.152 126 944 2	0.1122
$20s10p10d$; $20s$ *bc*	−1.152 157 418 0	30.4738
$20s10p10d$; $20s7p$ *bc*	−1.152 157 420 6	0.0026
$20s10p10d$; $20s7p6d$ *bc*	−1.152 157 794 9	0.3743
$20s10p10d10f$; $17s$ *bc*	−1.152 157 091 9	0.703
$20s10p10d10f$; $18s7p$ *bc*	−1.152 157 776 1	0.6842
$20s10p10d10f$; $18s7p6d$ *bc*	−1.152 157 794 9	0.0188
$24s12p12d12f$; $21s$ *bc*	−1.152 159 309 1	1.5142
$24s12p12d12f$; $21s8p$ *bc*	−1.152 159 309 2	0.0001

* Energies are given in Hartree.

4.3.2. *Atom-Centred and Bond-Centred Basis Sets*

The results of the matrix valence bond calculations carried out with basis sets containing both atom-centred and bond-centred functions are presented in Table VI. An energy decrease of ~3384.3 μHartree is obtained upon supplementing $20s$ atom-centred basis sets by a $20s$ bond-centred set. This decrease amounts to 99.3% of the energy increment measured when $10p$ atom-centred sets are added to the $20s$ sets. The addition of bond-centred sets to the $20s10p$ atom-centred sets results in successive energy changes of ~75.8 and ~0.1 μHartree for the $20s$ and $20s10p$ sets, respectively, whereas adding sets of atom-centred d functions reduces the energy by ~76.1 μHartree.

The successive addition of bond-centred s, p and d sets to the $20s10p10d$ set gives energy increments of ~5.5, ~0.002 and ~0.4 μHartree. The addition of atom-centred sets of f-type functions gives an energy change of ~5.1 μHartree. The successive addition of bond-centred s, p and d sets to the $20s10p10d10f$ set gives energy increments of ~0.06, ~0.69 and ~0.002 μHartree.

Extending the sequence $2nsnpndnf : 2ns$ from $n = 10$ to $n = 12$ yields an energy decrease of 2.2 μhartree compared with a decrease of 2.8 μhartree for the matrix Hartree–Fock case.

5. Discussion

The convergence of the matrix valence bond energies for the sequence of atom-centred basis sets $20s$, $20s10p$, $20s10p10d$, $20s10p10d10f$ is displayed in Figure 1 where it is compared with that of the matrix Hartree–Fock energies and of the

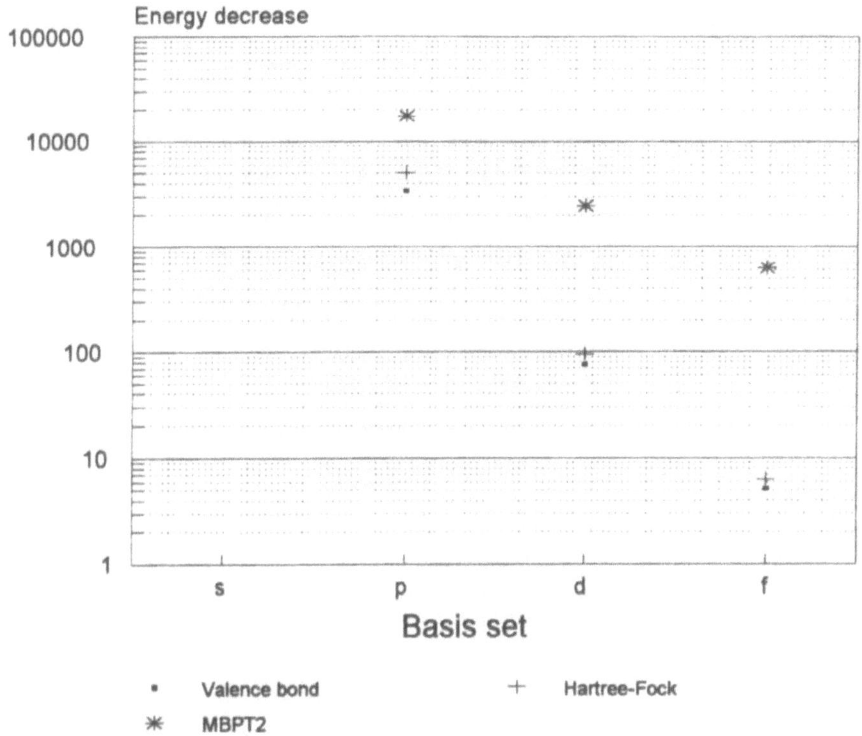

Figure 1. Energy decreases for matrix valence bond, matrix Hartree–Fock and second-order many-body perturbation theory correlation energy calculations for the sequence of atom-centred basis sets designated $20s$, $20s10p$, $20s10p10d$, $20s10p10d10f$.

second order correlation energies. In this figure, the energy decrease measured on adding higher symmetries is plotted on a logarithmic scale against the highest included symmetry. It can be seen that the convergence behaviour of the matrix valence bond calculations mirrors that of the matrix Hartree–Fock energies whereas the second-order correlation energies exhibit a less rapid convergence.

Figure 2 displays the measured energy decreases for the matrix Hartree–Fock and the matrix valence bond calculations. For the calculations using basis sets which contain only atom-centred functions the decreases are measured with respect to the previous members of the sequence *s*, *sp*, *spd*, *spdf*. For the calculations using both atom-centred and bond-centred sets the decrease in energy is measured with respect to the corresponding calculation using only atom-centred functions. For example, the bars labelled *sp*; *sp* display the energy decreases for the matrix Hartree–Fock and matrix valence bond energies with respect to those obtained with the atom-centred *sp* sets. In both the matrix Hartree–Fock and the matrix valence bond studies the addition of bond centre functions of *s* symmetry to a basis set of atom-centred functions of *s*-type only is seen to have an effect on the energy comparable

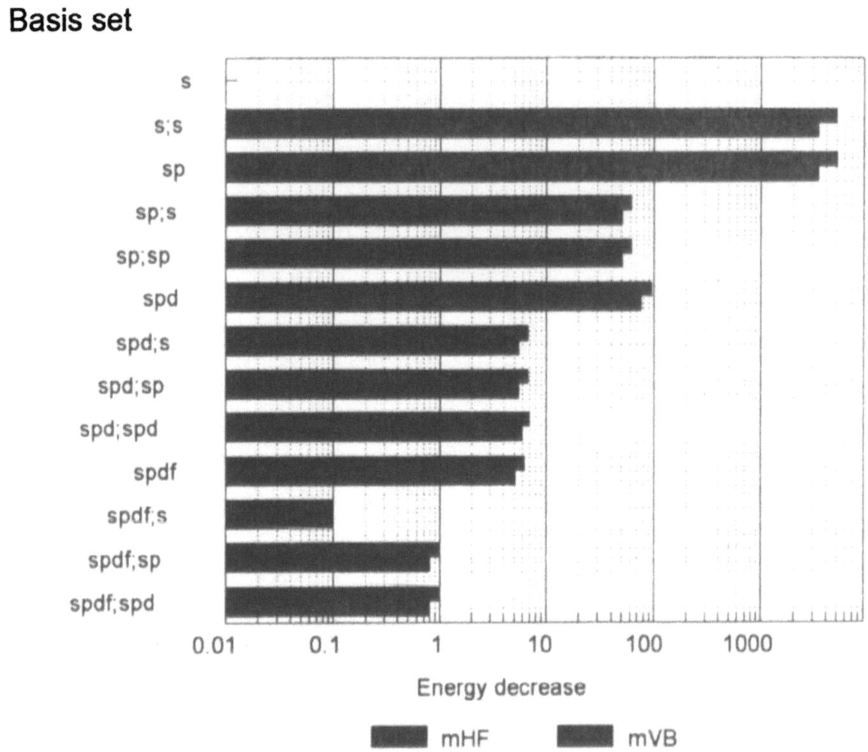

Figure 2. Energy decreases meaured for matrix Hartree–Fock and matrix valence bond calculations. See text for details.

with that observed upon adding atom-centred p functions. Similarly, the addition of bond centred functions to a set of atom-centred functions of s and p type functions has an effect comparable with that achieved by adding atom-centred functions of d symmetry. When bond centred functions are added to atom-centred sets of functions with s, p and d symmetry the energy lowering observed is very similar to that resulting from the addition of f type atom-centred functions. Furthermore, most of the energy decrease is attributable to the bond-centred functions of s symmetry.

In Table VII, the ground state matrix Hartree–Fock energy obtained in the present work for the hydrogen molecule with an equilibrium nuclear separation of 1.4 bohr is compared with previously reported values. The fully numerical finite difference value reported by Laaksonen, Pyykkö and Sundholm [2] defines the Hartree–Fock limit for the total electronic energy. It agrees with the finite element result reported by Yang, Heinemann and Kolb [56]. The calculation reported in 1989 by Wells and Wilson [1] using an extensive basis set of elliptical functions and labelled B in Table VII agrees with this value to the 0.01 μHartree level of

Table VII. Comparison of matrix Hartree–Fock energies for a nuclear separation of 1.4 bohr*.

Authors	$E_{\rm HF}$	Reference
Kolos and Roothaan	−1.133 630	[54]
Laaksonen, Pyykkö and Sundholm	−1.133 629 57	[2]
Yang, Heinemann and Kolb	−1.133 629 571 7(2)	[56]
Wells and Wilson, A**	−1.133 629 44	[1]
Wells and Wilson, B	−1.133 629 57	[1]
Present work (20s10p10d10f; 18s7p6d bc)	−1.133 627 02	
Present work (24s12p12d12f; 21s8p9d8f bc)	−1.133 628 90	
	$E_{\rm DF}$	
Parpia and Mohanty	−1.133 643 833 4	[55]
Yang, Heinemann and Koib	−1.133 643 970 08(1)	[56]

* Energies are given in Hartree.
** This energy was re-calculated for the present study.

Table VIII. Comparison of matrix valence bond energies for a nuclear separation of 1.4 bohr*.

Authors	$E_{\rm vb}$	Reference
Wilson (S.) and Gerratt	−1.51 751	[33]
Wilson (C.W.) and Goddard	−1.51 887	[32]
Present work (20s10p10d10f; 18s7p6d bc)	−1.52 157 79	

* Energies are given in Hartree.

accuracy. Their calculation using basis sets of atom-centred primitive Gaussian functions, labelled A in Table VII, was repeated in the present study using. the GAUSSIAN94 program and was found to differ by 0.07 μHartree from the previously reported value. This difference is almost certainly attributable to differences in the integral evaluation stage of the calculations. The new matrix Hartree–Fock energy obtained with the 23s9p4d4f basis set of Wells and Wilson [1] now differs from the limiting value of Laaksonen, Pyykkö and Sundholm [2] by only 0.13 μhartree. The calculation carried out as part of the present study using a 20s10p10d10f; 18s7p6d bc basis set of even-tempered Gaussian primitive functions yields a total electronic energy which differs from the limiting value by 2.42 μhartree. Use of the 24s12p12d12f; 21s8p9d8f bc basis set reduces this difference to only 0.67 μhartree.

The Dirac–Fock energies reported by Parpia and Mohanty [55] from finite basis set calculations and by Yang Heinemann and Kolb [56] from finite element studies are also recorded in Table VII. The difference between the (non-relativistic) Hartree–Fock energy and the (relativistic) Dirac–Fock energy for the hydrogen

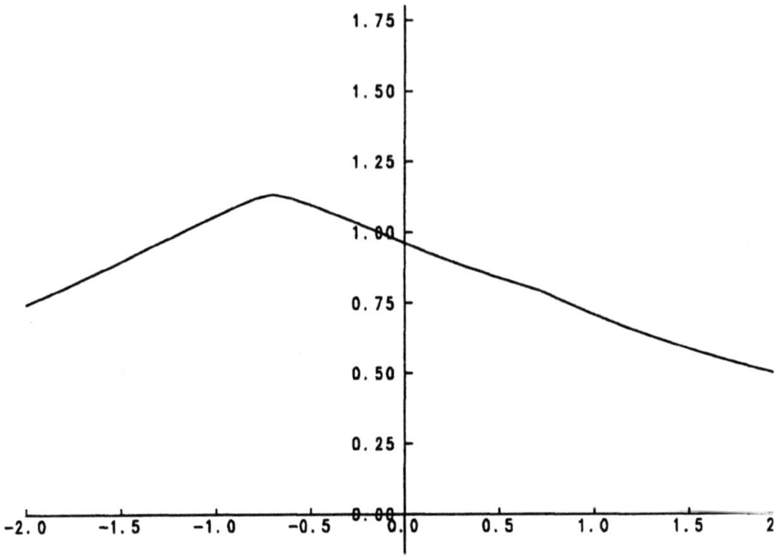

Figure 3. Orbital amplitude for the valence bond function obtained using atom-centred basis sets.

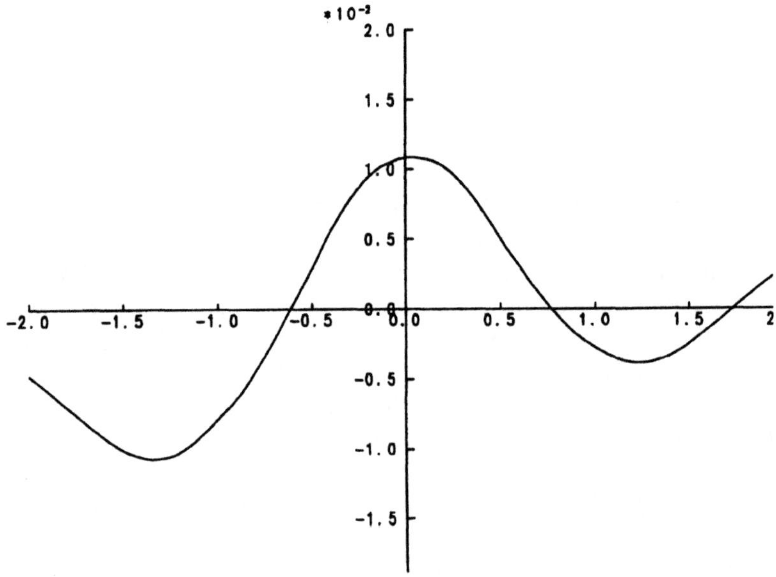

Figure 4. Orbital amplitude difference functions for the valence bond model obtained using the 20*s* atom-centred basis sets with and without the 20*s* bond-centre sets.

molecule at its equilibrium nuclear geometry is 14.40 μhartree; an order of magnitude larger than the estimated error in the present calculations.

The ground state matrix valence bond energy obtained in the present work for the hydrogen molecule with an equilibrium nuclear separation of 1.4 bohr is compared with previously reported values in Table VIII. The $20s10p10d10f$; $18s7p6d$ *bc* basis set of even-tempered primitive Gaussian function used in the largest matrix valence bond calculation supported an accuracy of 2.42 μhartree in our matrix Hartree–Fock calculation and a comparable level of accuracy is expected here. The energy reported by Wilson and Gerratt [33] lies 406.8 μhartree above the present value while that of Wilson and Goddard [32] is 270.8 μhartree above our present result. The best matrix valence bond energy obtained in the present study lies 18528.22 μhartree below the Hartree–Fock limit. This should be compared with the Dirac–Fock energy which lies 14.40 μhartree below the Hartree–Fock limit and the lowest second-order correlation energy recorded in Table IV which lies 33877.06 μhartree below the Hartree–Fock limit. The matrix valence bond calculation therefore accounts for $\sim$54.7% of the difference between the Hartree–Fock limit and the second-order correlation energy limit. The exact non-relativistic electronic energy of the hydrogen molecule ground state is -1.17447 μhartree [57] for a nuclear separation of 1.4 bohr. The matrix valence bond calculation recovers $\sim$38.8% of the correlation energy as it is usually define, i.e. the difference between the Hartree–Fock energy and the exact non-relativistic energy.

The amplitude of one of the valence bond orbitals obtained in the present study from a calculation using an atom-centred basis set of $20s$ Gaussian primitive functions is plotted in Figure 3 along the axis (z-axis) passing through the nuclei. The nuclei are located at $z = \pm 0.7$ bohr. The orbital can be seen to have two maxima; the global maximum is situated on the left-hand nucleus and the second maximum is located on the right-hand nucleus. As the nuclei are drawn apart the left-hand maximum increases until in the separated atom limit it become a hydrogen atom whilst the right-hand maximum decreases and disappears in the limit. The amplitude of the other orbital obtained from this calculation is obtained by a reflection in the plane perpendicular to the internuclear axis and passing through its mid-point.

In Figure 4, the orbital amplitude difference function is shown for the matrix valence bond model implemented by using a $20s$ atom-centred basis set with and without the $20s$ bond centre set. The addition of bond centred functions facilitates an increase in the orbital amplitude in the region near to the bond centre and a decrease in the amplitude in the regions beyond the atoms. Notice that the decrease in amplitude along the negative z-axis is greater than that along the positive z-axis.

6. Summary and Prospects

It has been shown that systematic sequences of distributed even-tempered basis sets of primitive Gaussian functions can be employed in matrix valence bond calculations to approach the exact valence bond total electronic energy with an accuracy approaching the sub-μhHartree level. The convergence pattern observed for the valence bond approach mirrors that seen previously in matrix Hartree–Fock calculations.

The atom-centred even-tempered basis sets of Wells and Wilson [1] for the hydrogen molecule ground state have recently, been employed by Parpia and Mohanty [55] in a Dirac–Hartree–Fock study. The present study suggests that the basis sets developed will prove useful in accurate relativitistic valence bond approaches, which unlike the Dirac–Hartree–Fock ansatz supports a qualitatively correct description of the dissociation of the hydrogen molecule.

Acknowledgement

One of us (DM) acknowledges the support of the US Department of Energy through contract No. DE-FC05-85ER2500000 and US Department of Energy Chemical Science Contract No. DF-FG05-95ER-14523. Support by Florida State University through the allocation of supercomputer resources on the SGI Power Challenge computer is acknowledged.

References

1. B. H. Wells and S. Wilson: *J. Phys. B: At. Mol. Opt. Phys.* **22**, 1285 (1989).
2. L. Laaksonen, P. Pyykko and D. Sundholm: *Int. J. Quantum Chem.* **23**, 319 (1983).
3. J. W. Thompson and S. Wilson: *J. Phys. B: At. Mol. Opt. Phys.* **23**, 2205 (1990).
4. D. Moncrieff and S. Wilson: *J. Phys. B: At. Mol. Opt. Phys.* **26**, 1605 (1993).
5. J. Kobus, D. Moncrieff and S Wilson: *J. Phys. B: At. Mol. Opt. Phys.* **27**, 2867 (1994).
6. J. Kobus, D. Moncrieff and S Wilson: *J. Phys. B: At. Mol. Opt. Phys.* **27**, 5139 (1994).
7. J. Kobus, D. Moncrieff and S Wilson: *Mol. Phys.* **86**, 1315 (1995).
8. D. Moncrieff, J. Kobus and S Wilson: *J. Phys B: At. Mol. Opt. Phys.* **28**, 4555 (1995).
9. D. Moncrieff and S. Wilson: *J. Phys B: At. Mol. Opt. Phys.* **28**, 4007 (1995).
10. D. Moncrieff and S. Wilson: *J. Phys. B: At. Mol. Opt. Phys.* **29**, 2425 (1996).
11. D. Moncrieff and S. Wilson: *J. Phys. B: At. Mol. Opt. Phys.* **29**, 6009 (1996).
12. H. M. Quiney, I. P. Grant and S. Wilson: in *Many-Body Methods in Quantum Chemistry*, U. Kaldov (ed.), Lecture Notes in Chemistry, Springer Verlag (1989).
13. H. M. Quiney, I. P. Grant and S. Wilson: *J. Phys. B: At. Mol. Opt. Phys.* **22**, L15 (1989).
14. S. Wilson: in *Methods in Computational Molecular Physics*, Proceedings of NATO Advanced Study Institute, Bad Windsheim, Germany, July–August, 1991, S. Wilson and G. H. F. Diercksen (eds.), Plenum (1992).
15. S. Wilson: in *Problem Solving in Computational Molecular Science: Molecules in Different Environments*, Proceedings of NATO Advanced Study Institute, Bad Windsheim, Germany, August, 1996, S. Wilson and G. H. F. Diercksen (eds.), Kluwer (1997).
16. W. A. Goddard: *Phys. Rev.* **157**, 73 (1967).
17. W. A. Goddard: *Phys. Rev.* **157**, 81 (1967).
18. R. C. Ladner and W. A. Goddard: *J. Chem. Phys.* **51**, 1073 (1969).
19. J. Gerratt and W. N. Lipscomb: *Proc. Nat. Acad. Sci. U.S.A.* **59**, 332 (1968).
20. J. Gerratt: *Adv. At. Mol. Phys.* **7**, 141 (1971).
21. S. Wilson and J. Gerratt: *Mol. Phys.* **30**, 765, 777 (1975).
22. N. C. Pyper and J. Gerratt: *Proc. Roy. Soc.* **A355**, 407 (1977).
23. J. Gerratt and M. Raimondi: *Proc. Roy. Soc.* **A371**, 525 (1980).
24. U. Kaldor and F. E. Harris: *Phys. Rev.* **183**, 1 (1969).

25. G. A. Gallup, R. L. Vance, J. R. Collins and J. M. Norbeck: *Adv. Quantum Chem.* **16**, 229 (1982).
26. M. Raimondi, M. Simonetta and G. F. Tantardini: *Comput. Phys. Rept.* **2**, 171 (1985).
27. D. L. Cooper, J. Gerratt and M. Raimondi: *Adv. Chem. Phys.* **69**, 319 (1987).
28. R. McWeeny: *Methods of Molecular Quantum Mechanics*, Academic Press (1989).
29. R. McWeeny: *Pure and Appl. Chem.* **61**, 2087 (1989).
30. R. McWeeny: in *Methods in Computational Molecular Physics*, Proceedings of NATO Advanced Study Institute, Bad Windsheim, Germany, July–August, 1991, S. Wilson and G. H. F. Diercksen (eds.), Plenum (1992).
31. S. Wilson: *Electron Correlation in Molecules*, Clarendon Press (1984).
32. C. W. Wilson and W. A. Goddard: *Theor. Chim. Acta.* **26**, 195 (1972).
33. S. Wilson and J. Gerratt: *Mol. Phys.* **30**, 777 (1975).
34. C. A. Coulson and I. Fischer: *Phil. Mag.* **40**, 386 (1949).
35. C. R. Mueller and H. Euring: *J. Chem. Phys.* **19**, 1495 (1951).
36. S. Wilson: in *New Methods in Quantum Theory*, Proceedings of NATO Advanced Research Workshop held in Halkidiki, Greece, May 1995, C. A. Tsipis, V. S. Popov, D. R. Herschbach and J. S. Avery (eds.), Kluwer, Dordrecht (1996).
37. S. Wilson and D. Moncrieff: *Adv. Quantum Chem.* **28**, 47 (1997).
38. S. Weinbaum: *J. Chem. Phys.* **1**, 593 (1933).
39. L. J. Schaad and I. R. Thomas: *Theor. Chim. Acta* **21**, 115 (1971).
40. B. Freihaut and L. M. Raff: *J. Chem. Phys.* **58**, 1202 (1973).
41. J. M. Norbeck and P. R. Certain: *J. Chem. Phys.* **63**, 4127 (1975).
42. E. F. Gurney and J. L. Magee: *J. Chem. Phys.* **18**, 142 (1950).
43. A. C. Hurley: *Proc Roy. Soc.* **A226**, 179 (1954).
44. H. Shull and D. D. Ebbing: *J. Chem. Phys.* **28**, 866 (1958).
45. R. D. Harcourt: *Theor. Chim. Acta.* **78**, 267 (1991).
46. R. McWeeny: *Coulson's Valence*, pp. 129–131, Oxford University Press (1979).
47. V. R. Saunders: in *Methods in Computational Molecular Physics*, Proceedings of NATO Advanced Study Institute Bad Windsheim, Germany, July–August, 1982, G. H. F. Diercksen and S Wilson (eds.), Plenum (1983).
48. GAUSSIAN 94, Revision C.3, M. J. Frisch, G. W. Trucks, H. B. Schlegel, P. M. W. Gill, B. G. Johnson, M. A. Robb, J. R. Cheeseman, T. Keith, G. A. Petersson, J. A. Montgomery, K. Raghavachari, M. A. Al-Laham, V. G. Zakrzewski, J. V. Ortiz, J. B. Foresman, J. Cioslowski, B. B. Stefanov, A. Nanayakkara, M. Challacombe, C. Y. Peng, P. Y. Ayala, W. Chen, M. W. Wong, J. L. Andres, E. S. Replogle, R. Gomperts, R. L. Martin, D. J. Fox, J. S. Binkley, D. J. Defrees, J. Baker, J. P. Stewart, M. Head-Gordon, C. Gonzalez, and J. A. Pople, Gaussian, Inc., Pittsburgh PA (1995).
49. J. Almlöf, K. Faegri and K. Korsell: *J. Comput. Chem.* **3**, 385 (1982).
50. M. Head-Gordon, J. A. Pople and M. J. Frisch: *Chem. Phys. Lett.* **153**, 503 (1988).
51. S. Saebo and J. Almlöf: *Chem. Phys. Lett.* **154**, 83 (1989).
52. S. Wilson: in *Electron Correlation in Atoms and Molecules* (Meth. Comput. Chem. 1), S. Wilson (ed.), Plenum (1987).
53. F. W. Bobrowicz and W. A. Goddard III: in *Methods of Electronic Structure Theory*, H. F. Schaefer III (eds.), Plenum, New York (1977).
54. W. Kolos and C. C. J. Roothaan: *Rev. Mod. Phys.* **32**, 219 (1960).
55. F. A. Parpia and A. K. Mohanty: *Phys. Rev. A* **52**, 962 (1995).
56. L. Yang, D. Heinemann and D. Kolb: *Phys. Rev. A* **48**, 2700 (1993).
57. W. Kolos and L. Wolniewicz: *J. Chem. Phys.* **49**, 404 (1968).

R. McWeeny et al. (eds.), Quantum Systems in Chemistry and Physics: 343 - 347.

On the Evaluation of the Matrix Elements Between Non-Orthogonal Slater Determinants

CLAUDIO AMOVILLI*
Dipartimento di Chimica e Chimica Industriale, Via Risorgimento 35, 56126 Pisa, Italy.

(Received and accepted: 17 June 1997)

Abstract. An alternative procedure is presented for the calculation of Hamiltonian matrix elements between non-orthogonal Slater determinants. Using the singular-value decomposition theorem a simple and general expression is derived for the cofactors needed in the Löwdin formula. The algorithm proposed has a cost in terms of CPU time proportional to N^4, N being the number of electrons, and does not present any problem of data storage.

Key words: VB calculations, non-orthogonal Slater determinants.

1. Introduction

In recent years there has been considerable interest in developing new methodologies for studies of configuration interaction problems using non-orthogonal orbitals [1–5]. Non-orthogonality, however, leads to well known problems in the calculation of the Hamiltonian matrix elements arising from the consideration of, in principle, $N!$ products of non-zero, one-particle matrix elements. In the limit of a complete expansion, the quality of the wavefunctions does not depend on the nature of the orbitals (orthogonal or not) but for a truncated expansion, the use of non-orthogonal orbitals can lead to a very good level of accuracy even with a relatively small number of terms. For a spin free approach and in the usual Born–Oppenheimer approximation many efficient techniques that avoid the $N!$ problem have been developed by many authors (see, for example [6–9]). In this note an alternative simple procedure for the calculation, by the direct use of the Löwdin formula, of the Hamiltonian matrix elements between non-orthogonal Slater determinants is proposed. The method is closely related to the parameter method of Gallup *et al.* [1] but it offers a different approach.

2. Theory

The most general expression for the matrix elements between non-orthogonal Slater determinants is due to Löwdin [10]. Considering a system of N electrons and two

* Author for correspondence.

general Slater determinants Φ_a and Φ_b, built up from non-orthogonal spin-orbitals, for the overlap, 1- and 2-electron Hamiltonian matrix elements, one can write

$$\langle \Phi_a | \Phi_b \rangle = D_{ab} \tag{1}$$

$$\left\langle \Phi_a \left| \sum_{i=1}^{N} \hat{h}(i) \right| \Phi_b \right\rangle = \sum_{i=1}^{N} \sum_{j=1}^{N} \langle a_i | \hat{h}(1) | b_j \rangle D_{ab}(a_i | b_j) \tag{2}$$

$$\left\langle \Phi_a \left| \sum_{i,j}{}' g(i,j) \right| \Phi_b \right\rangle = \sum_{i=1}^{N} \sum_{j=1}^{N} \sum_{k=1}^{N} \sum_{l=1}^{N} \langle a_i a_k | g(1,2) | b_j b_l \rangle \times \\ \times \epsilon_{ik} \epsilon_{jl} D_{ab}(a_i a_k | b_j b_l) \tag{3}$$

where a_i and b_i denote respectively the spin-orbitals of Φ_a and Φ_b, D_{ab} the determinant of the matrix of the overlap integrals between the two sets of spin-orbitals $\{a\}$ and $\{b\}$ and finally $D_{ab}(a_i|b_j)$ and $D_{ab}(a_i a_k|b_j b_l)$ refer respectively to the cofactors of order $N-1$ and $N-2$. The coefficient ϵ_{ik} is a unimodular factor with value ± 1; it is -1 when the order of the indices ik is reversed with respect to the order in the starting Slater determinant. The most difficult expression to evaluate is, of course, Equation (3) because it involves a four-fold summation and, in principle, for each term the calculation of a cofactor. A direct evaluation of (3) requires then N^7 operations, as is well known. In the 'cofactor-driven algorithms' [6–9] this calculation is broken down into separate steps, the evaluation of all the unique cofactors, that are then stored in the memory of the computer, and separately the summations. The efficiency of such processes depends largely on the generation of the cofactors and on the assumption that the number of cofactors is relatively small. In practice, only in the case where the matrix of the overlap integrals is non-singular it is possible to avoid the generation of the cofactors. In this case, if $\mathbf{S}_{ab}$ is the matrix whose cofactors appear in Equations (1)–(3), Jacobi's ratio theorem [11] shows that

$$D_{ab}(a_i|b_j) = S^{-1}_{ab,ji} D_{ab} \tag{4}$$

$$\epsilon_{ik}\epsilon_{jl} D_{ab}(a_i a_k|b_j b_l) = [S^{-1}_{ab,ji} S^{-1}_{ab,lk} - S^{-1}_{ab,li} S^{-1}_{ab,jk}] D_{ab}. \tag{5}$$

When $\mathbf{S}$ is singular it has been shown [6], by using the unsymmetrical Choleski decomposition, that this result can be applied to non-singular minors in order to compute only a reduced number of cofactors. Similar developments have been made more recently by several authors (see, for example, [12–14]).

The purpose of the present work is to show how to avoid any cofactor analysis, even for a singular $\mathbf{S}$ matrix, by means of the singular-value decomposition theorem

[15] combined with the parameter method of Gallup *et al.* [1]. The theorem states that any $\mathbf{S}$ matrix can be reduced as follows

$$\mathbf{U}^{\dagger}\mathbf{S}\mathbf{W} = \mathbf{D} \tag{6}$$

where $\mathbf{U}$ and $\mathbf{W}$ are two unitary matrices and $\mathbf{D}$ is diagonal. Whenever $\mathbf{S}$ is singular there will be k zero diagonal elements (for simplicity $D_{11} = D_{22} = \cdots = D_{kk} = 0$). In terms of $\mathbf{U}$, $\mathbf{W}$ and $\mathbf{D}$, the matrix $\mathbf{S}$, being square, can be written as

$$\mathbf{S} = \mathbf{U}\mathbf{D}\mathbf{W}^{\dagger}. \tag{7}$$

It is then convenient to substitute $\mathbf{D}$ by $\mathbf{D} + \lambda\mathbf{X}$, where λ is an arbitrary small parameter and $\mathbf{X}$ a matrix whose elements X_{ij} are δ_{ij} when $i,\ j \leq k$ and 0 in all other cases. The new matrix

$$\mathbf{S}_{\lambda} = \mathbf{U}(\mathbf{D} + \lambda\mathbf{X})\mathbf{W}^{\dagger} \tag{8}$$

is then no longer singular and can be inverted. The inverse $\mathbf{S}_{\lambda}^{-1}$ can always be written in the following form

$$\mathbf{S}_{\lambda}^{-1} = \frac{1}{\lambda}\mathbf{A} + \mathbf{B}. \tag{9}$$

This expression can be used finally to evaluate the cofactors of Equations (4) and (5) simply by taking the limit $\lambda \to 0$ of the following expressions

$$P_{ji}^{ab} = \lim_{\lambda \to 0} S_{\lambda,ji}^{-1} \det(\mathbf{S}_{\lambda}) \tag{10}$$

$$\Pi_{jl,ik}^{ab} = \lim_{\lambda \to 0} [S_{\lambda,ji}^{-1} S_{\lambda,lk}^{-1} - S_{\lambda,li}^{-1} S_{\lambda,jk}^{-1}] \det(\mathbf{S}_{\lambda}) \tag{11}$$

which are, in fact, elements of 1- and 2-electron (transition) density matrices. From Equation (8) it is very easy to show that $\mathbf{A}$ and $\mathbf{B}$ are given by

$$\mathbf{A} = \sum_{j=1}^{k} \mathbf{W}_j \mathbf{U}_j^{\dagger} \quad (D_{jj} = 0) \tag{12}$$

$$\mathbf{B} = \sum_{j=k+1}^{N} \frac{1}{D_{jj}} \mathbf{W}_j \mathbf{U}_j^{\dagger} \quad (D_{jj} \neq 0) \tag{13}$$

where $\mathbf{U}_i$ and $\mathbf{W}_i$ are respectively the ith columns of $\mathbf{U}$ and $\mathbf{W}$. Assuming that the determinants of $\mathbf{U}$ and $\mathbf{W}$ are equal to unity,[1] the determinant of $\mathbf{S}_{\lambda}$ is then

$$\det(\mathbf{S}_{\lambda}) = \lambda^k \prod_{j=k+1}^{N} D_{jj} \tag{14}$$

[1] If the determinant of $\mathbf{U}$, for example, is -1 it is possible to change the sign of one column $\mathbf{U}_i$ and the sign of the corresponding diagonal element D_{jj} without changing the meaning of Equation (6).

and on combining the limits (10, 11) with Equations (12)–(14) one gets explicitly

$$
\begin{aligned}
P_{ji}^{ab} &= \left(\prod_{m=1}^{N} D_{mm}\right) B_{ji} \quad (k=0)\\
&= \left(\prod_{m=2}^{N} D_{mm}\right) A_{ji} \quad (k=1)\\
&= 0 \quad (k>1)
\end{aligned}
\tag{15}
$$

$$
\begin{aligned}
\Pi_{jl,ik}^{ab} &= \left(\prod_{m=1}^{N} D_{mm}\right) [B_{ji}B_{lk} - B_{li}B_{jk}] \quad (k=0)\\
&= \left(\prod_{m=2}^{N} D_{mm}\right) [A_{ji}B_{lk} + B_{ji}A_{lk} - A_{li}B_{jk} - B_{li}A_{jk}] \quad (k=1)\\
&= \left(\prod_{m=3}^{N} D_{mm}\right) [A_{ji}A_{lk} - A_{li}A_{jk}] \quad (k=2)\\
&= 0 \quad (k>2).
\end{aligned}
\tag{16}
$$ [2]

This result is identical to that obtained by Gallup *et al.* [1] but the construction of the matrices **A** and **B** is substantially different. Turning to the decomposition (6) the proposed method for the calculation of **A** and **B** involves the solution of the following secular equations

$$\mathbf{S}\mathbf{S}^{\dagger}\mathbf{U} = \mathbf{U}\mathbf{D}^2 \tag{17}$$

$$\mathbf{S}^{\dagger}\mathbf{S}\mathbf{W} = \mathbf{W}\mathbf{D}^2. \tag{18}$$

This procedure, by itself, is not sufficient because it leaves the uncertainty in the sign of D_{ii} and, when D_{ii}^2 is degenerate, **U** and **W** obtained by Equations (17) and (18) are not unique and they could not satisfy the condition (6). All these problems can be easily avoided simply writing the matrix **B** and the product (14) in terms of **S**. Combining (17, 18) and Equations (7), (13), (14) one finally has

$$\mathbf{B} = \mathbf{S}^{\dagger} \sum_{j=k+1}^{N} \frac{1}{D_{jj}^2} \mathbf{U}_j \mathbf{U}_j^{\dagger} \tag{19}$$

$$\prod_{j=k+1}^{N} D_{jj} = \det(\mathbf{U}^{\dagger}\mathbf{S}\mathbf{W} + \mathbf{X})(-1)^{\nu}(-1)^{\mu}. \tag{20}$$

[2] When $k = 1$ it should be noted that $A_{ji}A_{lk} - A_{li}A_{jk} = W_{j1}U_{i1}W_{l1}U_{k1} - W_{l1}U_{i1}W_{j1}U_{k1} = 0$.

Equation (19) does not depend on the matrix **W** and on the sign of D_{ii} while Equation (20) does not require $\mathbf{U}^{\dagger}\mathbf{SW}$ to be diagonal and then they can work even for any solutions of Equations (17), (18).[3] The determinant of Equation (20) can be determined very quickly because $\mathbf{U}^{\dagger}\mathbf{SW}$ is a block-diagonal matrix where the dimension of each block is given by the degeneracy of D_{ii}^2. ν and μ are (respectively) the number of eigenvector permutations in **U** and **W** needed to have D_{ii}^2 in the same ascending order. Similar considerations are not required for the parameter subspace because λ is externally imposed.

3. Conclusion

The main result of the present work is represented by the relations (15, 16), which give explicitly all the cofactors involved in the Löwdin formulas (1)–(3) without any complicated analysis. In the case of non-singular overlap matrices all the cofactors are determined by a single matrix inversion: here it is shown that even in a general case it is sufficient to determine the matrices **A** and **B** in Equations (12) and (13). The cost of this operation is only slightly greater than the inversion of a matrix and it does not affect significantly the total cost of the complete calculation because the latter is determined almost completely by the fourfold summation of Equation (3). The total CPU time required is thus proportional to N^4. Work at present in progress includes the analytical development of the energy gradient for a wavefunction of VB type, for the variation of orbital parameters. Analytical gradients require 3-electron densities and consequently cofactors of order $N - 3$ with the inclusion of the overlap matrices with three singularities ($k = 3$).

References

1. G. A. Gallup, R. L. Vance, J. R. Collins and J. Norbeck: *Adv. Quantum Chem.* **16**, 229 (1982).
2. J. H. van Lenthe and G. G. Balint-Kurti: *J. Chem. Phys.* **78**, 5699 (1983).
3. M. Raimondi, M. Simonetta and G. F. Tantardini: *Comput. Phys. Rep.* **2**, 171 (1985).
4. D. L. Cooper, J. Gerratt and M. Raimondi: *Adv. Chem. Phys.* **69**, 319 (1987).
5. R. McWeeny: *Int. J. Quantum Chem.* **34**, 25 (1988).
6. F. Prosser and S. Hangstrom: *Int. J. Quantum Chem.* **2**, 89 (1968).
7. M. Raimondi, W. Champion and M. Karplus: *Mol. Phys.* **34**, 1483 (1977).
8. M. Raimondi and E. Gianinetti: *J. Phys. Chem.* **92**, 899 (1988).
9. S. Rettrupp, T. Thorsteinsson and C. R. Sarma: *Int. J. Quantum Chem.* **40**, 709 (1991).
10. P. O. Löwdin: *Phys. Rev.* **97**, 1474 (1955).
11. J. W. Archbold: *Algebra*, Pitman, London, 338 (1961).
12. I. C. Hayes and A. J. Stone: *Mol. Phys.* **53**, 69 (1984).
13. V. Magnasco and G. Figari: *Mol. Phys.* **55**, 319 (1985).
14. J. Verbeek and J. H. van Lenthe: *J. Mol. Struct. (Theochem)* **75**, 115 (1991).
15. A. T. Amos and G. G. Hall: *Proc. R. Soc. (London)* **A263**, 483 (1961); see also, H. F. King, R. E. Stanton, H. Kim, R. E. Wyatt and R. G. Parr: *J. Chem. Phys.* **47**, 1936 (1967).

[3] Using Jacobi's algorithm to diagonalize $\mathbf{SS}^{\dagger}$ and $\mathbf{S}^{\dagger}\mathbf{S}$ one always gets det(**U**) = det(**W**) = 1, but the two separate diagonalizations do not guarantee the same order for the eigenvalues.

R. McWeeny et al. (eds.), Quantum Systems in Chemistry and Physics, 349–371.

The Calculation of Some Processes Involving Transitions to the Continuum

R. MOCCIA,* S. MENGALI and R. MONTUORO
Dipartimento di Chimica e Chimica Industriale, Università di Pisa Via Risorgimento 35, I-56126 Pisa, Italy.

(Received and accepted: 17 June 1997)

Abstract. The calculation of the photoionization cross section for processes involving one or more photons entails a satisfactory representation of the electronic continuum. Here, after a brief exposition of a technique developed and tested in recent years, which yields accurate values of the properties pertaining to the continuum by using L^2 basis, two applications of this approach are reported. The first one deals with the continuum spectrum of Mg up to the third ionization threshold, where the two active electrons are allowed to interact with an *a priori* evaluated polarization potential of the core. The second one is the first application of the method to a Floquet-type Hamiltonian for the case of a hydrogen atom subjected to a rather intense laser field.

Key words: Continuum spectrum of Mg, photoionization cross section, Floquet-type Hamiltonian, hydrogen atom in intense laser field.

1. Introduction

The representation of the continuum by using L^2 basis is of considerable interest since, if the computation may be brought into the domain of the discrete states calculations, the ample expertise gained in this area by theoretical chemistry may be successfully exploited. Among several theoretical methods developed on this subject [1–4], the discretization of the system of integral equations for the half-on-the-shell transition T–matrix, or the reaction K-matrix, upon the grid points yielded by a variational calculation, performed with L^2 basis functions [5, 6], has proved to be quite useful. It has been successfully applied to compute the photoionization cross sections of a variety of atoms and molecules at quite different levels of approximation [5, 7–9, 11].

After a brief summary of the theory in Section 2, calculations for the Mg atom are presented in Section 3. The results were obtained by using an *a priori* approach to consider more rigorously the effects of the polarization of the Ne-like core, differently from previous work were it was approximated by a semi-empirical polarization potential [11]. Furthermore, the study is extended for energies up to the third ionization threshold.

* Author for correspondence.

As a novel application of our technique we have considered the case of the ionization rate of an H atom subjected to intense radiation. The approach, based upon a Floquet Hamiltonian, has required the introduction of several adaptations, both theoretical and numerical, in order to have a working algorithm. The results of this last application are presented in Section 5.

2. Theory

Since a full account of the method may be found in previous works [5, 6], here only a synthetic outline is presented with the aim to emphasize the main points. The application to Mg reported in Section 3 should be helpful in clarifying several details. Assuming the LS coupling to be valid, the N-electron basis, for each manifold, comprises different sets of functions obtained by the appropriate coupling of orbitals. The first one, called the 'localized channel' (lc), includes functions Φ_{C,E_j} built from localized orbitals ϕ. The other sets, called 'partial wave channels' (pwc), include functions $\Phi_{\gamma,E'}$ built as products of some well defined localized ion state $\Phi^{N-1}_{E_{I_\gamma}L_{I_\gamma},S_{I_\gamma}}$, times orbitals $\varphi_{\epsilon,l_\gamma,s_\gamma}$, representing the one-electron continuum, coupled to give states of the selected manifold. The Greek index γ collects all relevant quantum numbers, including those which specify the manifold.

These multielectron basis functions are chosen such as to diagonalize the atomic Hamiltonian separately for each channel, i.e.:

$$\langle \Phi_{C,E_j}|\hat{H}|\Phi_{C,E_k}\rangle = \delta_{j,k}E_j; \qquad \langle \Phi_{C,E_j}|\Phi_{C,E_k}\rangle = \delta_{j,k} \tag{1}$$

$$\langle \Phi^{N-1}_{I'}|\hat{H}^{N-1}|\Phi^{N-1}_{I}\rangle = \delta_{I',I}E_I; \qquad \langle \Phi^{N-1}_{I'}|\Phi^{N-1}_{I}\rangle = \delta_{I',I} \tag{2}$$

$$\langle \Phi_{\beta,E''}|\hat{H}|\Phi_{\beta,E'}\rangle = \delta_{E'',E'}E''; \qquad \langle \Phi_{\beta,E''}|\Phi_{\beta,E'}\rangle = \delta_{E'',E'}. \tag{3}$$

It must be pointed out that, besides continuum states, discrete states may also occur in the pwc's. Thus the $\delta_{E'',E'}$ stands for a Dirac $\delta(E''-E')$ for the continuum and for a Kroneker symbol for the discrete part of the spectrum. It should also be obvious that, by this procedure, the multielectron basis functions belonging to different channels of the same manifold will not be orthogonal, i.e.:

$$\text{for} \quad \theta \neq \gamma, \quad \langle \Phi_{\theta E''}|\Phi_{\gamma,E'}\rangle \neq 0; \quad \langle \Phi_{\theta,E''}|\hat{H}|\Phi_{\gamma,E'}\rangle \neq 0 \tag{4}$$

thus matrix elements $V_{\theta E'',\gamma E'}(E)$ may be defined by

$$\langle \Phi_{\theta,E''}|\hat{H}-E|\Phi_{\gamma,E'}\rangle = \delta_{\theta,\gamma}\delta_{E'',E'}(E''-E) + V_{\theta E'',\gamma E'}(E)$$

$$V_{\theta E'',\gamma E'}(E) = (1-\delta_{\theta,\gamma})\langle \Phi_{\theta,E''}|\hat{H}-E|\Phi_{\gamma,E'}\rangle. \tag{5}$$

The asymptotic behaviour of the pwc states $|\Phi_{\gamma,E'}\rangle$ lying in the continuum will be determined by the φ orbitals, whose radial part for large values of r will behaves as

$$\text{for} \quad r \to \infty, \quad \varphi_{\epsilon,l_\gamma,s_\gamma}(x) \propto \sin[\Theta_{l_\gamma}(r,k) + \delta_\gamma(k)]/r$$

where the total energy $E' = E_{I_\gamma} + \epsilon$, $\epsilon = k^2/2$; $\Theta_{l_\gamma}(r,k)$ is the purely Coulombic phase shift and $\delta_\gamma(k)$ the additional phase shift due to the other electrons. With the above defined multielectron basis, the final states $|\Psi^{(\pm)}_{\beta,E}\rangle$ are obtained by formally diagonalizing the total Hamiltonian matrix as in a standard, time-honored, variational calculation first proposed by Fano [12], to obtain a system of integral equations. In a more scattering oriented terminology, the same equations may be derived from a Newton variational principle which ensures a stationary value for the reaction K-matrix on the energy shell [13]. Thus by expressing $|\Psi^{(\pm)}_{\beta,E}\rangle$ as

$$|\Psi^{(\pm)}_{\beta,E}\rangle = |\Phi_{\beta,E}\rangle + \sum_\gamma \sum\!\!\!\!\!\!\int \mathrm{d}E' |\Phi_{\gamma,E'}\rangle \frac{1}{E - E' \pm i0+} T^{(\pm)}_{\gamma E',\beta E} \tag{6}$$

and by requiring that

$$\langle \Phi_{\theta,E''}|\hat{H} - E|\Psi^{(\pm)}_{\beta,E}\rangle = 0 \tag{7}$$

the following system of integral equation is obtained for the reaction $T^{(\pm)}_{\gamma E',\beta E}$ matrix elements

$$\sum_\gamma \sum\!\!\!\!\!\!\int \mathrm{d}E' \left\{ \delta_{\theta,\gamma}\delta_{E'',E'} - V_{\theta E'',\gamma E'} \frac{1}{E - E' \pm i0+} \right\} T^{(\pm)}_{\gamma E',\beta E} = V_{\theta E'',\beta E} \tag{8}$$

where the symbol $\sum\!\!\!\!\!\!\int$ stands for summation upon the discrete states and integration upon the energy in the continuum.

In some cases it may be convenient to work with stationary waves since they will be determined by the real reaction K-matrix. All the formulae written so far are readily transformed by making the substitution

$$\frac{1}{E - E' \pm i0+} \rightarrow \frac{\mathcal{P}}{E - E'}.$$

The K-matrix, the T-matrix and the S-matrix are connected by the well known relations

$$T^{(\pm)}_{\gamma E',\beta E} = \sum_\theta K_{\gamma E',\theta E}[\mathbf{1} \pm i\pi \mathbf{K}(E)]^{-1}_{\theta,\beta}$$

$$\mathbf{S}(E) = \frac{\mathbf{1} - i\pi\mathbf{K}(E)}{\mathbf{1} + i\pi\mathbf{K}(E)} = \mathbf{1} - 2i\pi\mathbf{T}^{(+)}(E)$$

where $\mathbf{K}(E)$ and $\mathbf{T}^{(+)}(E)$ are the matrices upon the energy shell.

2.1. DISCRETIZATION PROCEDURE

The use of L^2 orbitals implies that our multielectron basis functions cannot yield the continuum, therefore the system of integral Equations (8) is discretized using as grid points the energy values provided by separate channel variational calculations, according to Equation (1)–(3). Thus, by using, for instance, a Lagrange interpolation, the integration in Equation (6) may be approximated as

$$\int \mathrm{d}E' V_{\theta E'',\gamma E'} \frac{1}{E - E' \pm i0^+} T^{(\pm)}_{\gamma E',\beta E}$$

$$\simeq \sum_{k,l} V_{\theta E''_j,\gamma E'_k}(E) P^{(\pm)}_{\gamma E'_k,\gamma E'_l}(E) T^{(\pm)}_{\gamma E'_l,\beta E} \tag{9}$$

where the $P^{(\pm)}_{\gamma E'_k,\gamma E'_l}$ is an integration matrix as discussed in [5, 6]. By this device the matrix representation of Equation (6) upon a finite L^2 basis may, in fact, be recast as

$$[\mathbf{1} - \mathbf{V}(E)\mathbf{P}^{(\pm)}(E)]\mathbf{T}^{(\pm)}_{\bullet\bullet,\beta E} = \mathbf{V}_{\bullet\bullet,\beta E} \tag{10}$$

where the double dot stands for the channel and energy indices.

2.2. Resonances: Positions and Widths; Branching Ratios

When it happens that some states $|\Phi_a\rangle$ of the lc with energy E_a correspond to the localized part of an autoionizing state, it is convenient to write down the expression (6) as

$$\begin{aligned} |\Psi^{(+)}_{\beta,E}\rangle &= |\Phi_{\beta,E}\rangle + \sum_{\gamma} \not\!\!\int \mathrm{d}E' |\Phi_{\gamma,E'}\rangle \frac{1}{E - E' + i0^+} T^{(+)}_{\gamma E',\beta E} + \\ &\quad + |\Phi_a\rangle \frac{T^{(+)}_{a,\beta E}}{E - E_a + i0^+} \end{aligned} \tag{11}$$

making explicit the contribution of $|\Phi_a\rangle$. Equation (10) may then be recast as

$$\begin{pmatrix} E - E_a - V_{aa} & \mathbf{V}_{a,\bullet\bullet}\mathbf{P}^{(+)}(E) \\ \mathbf{V}_{\bullet\bullet,a} & [\mathbf{1} - \mathbf{V}(E)\mathbf{P}^{(+)}(E)] \end{pmatrix} \begin{pmatrix} \frac{T^{(+)}_{a,\beta E}}{E - E_a + i0^+} \\ \mathbf{T}^{(+)}_{\bullet\bullet,\beta E} \end{pmatrix} = \begin{pmatrix} V_{a,\beta E} \\ \mathbf{V}_{\bullet\bullet,\beta E} \end{pmatrix} \tag{12}$$

which yields

$$\frac{T^{(+)}_{a,\beta E}}{E - E_a + i0^+} = \frac{V_{a,\beta E} + \mathbf{V}_{a,\bullet\bullet}\mathbf{P}^{(+)}(E)[\mathbf{1} - \mathbf{V}(E)\mathbf{P}^{(+)}(E)]^{-1}\mathbf{V}_{\bullet\bullet,a}}{E - E_a - \mathbf{V}_{a,\bullet\bullet}\mathbf{P}^{(+)}(E)[\mathbf{1} - \mathbf{V}(E)\mathbf{P}^{(+)}(E)^{-1}\mathbf{V}_{\bullet\bullet,a}}. \tag{13}$$

Thus the pole of the S-matrix in the neighbourhood of E_a may be found, by analytical continuation, as the root E_{Res} of

$$E - E_a - \mathbf{V}_{a,\bullet\bullet}\mathbf{P}^{(+)}(E)[\mathbf{1} - \mathbf{V}(E)\mathbf{P}^{(+)}(E)]^{-1}\mathbf{V}_{\bullet\bullet,a} = 0. \tag{14}$$

Once determined,the position ReE_{Res} and the width $\Gamma = 2\text{Im}\, E_{\text{Res}}$ of the resonance, the autoionizing state $|\Psi^r_{a,E}\rangle$ at $E = E_{\text{Res}}$ will be

$$|\Psi^r_{a,E}\rangle = |\Phi_a\rangle + |\{\Phi_{\bullet\bullet}\}\rangle\mathbf{P}^{(+)}(E)[\mathbf{1} - \mathbf{V}(E)\mathbf{P}(E)]^{-1}\mathbf{V}(E)_{\bullet\bullet,a}.$$

The branching ratios will readily be obtained from the second term in the rhs of the last equation.

3. Mg Atom

The continuum spectrum of Mg up to the second ionization threshold has been the subject of some previous works by one of the present authors. The study was carried out by considering the Ne-like core either frozen [9] or polarizable according to a semiempirical model [11]. Furthermore, with a single open channel, a simpler method [8] was sufficient to obtain the continuum properties from an L^2 basis set variational calculation. Here we report some results of a calculation [14, 15] performed up to the third ionization threshold by the method described in Section 2. A non-empirical description of the core polarization effects is obtained by including in the C.I. calculation also single excitations of the core. The calculations are much more demanding than those faced with the model polarization potential since, now, four quasi-particles are involved. In addition,it should be remembered that in our method a complete diagonalization of all the individual channels, lc and pwc's, is required. The following description of the selection of the bases will elucidate several of the features required for a successful application of the method.

3.1. Orbital Basis

All the orbitals are orthonormal. Three different group of orthogonal orbitals are employed: (i) $\{\psi_1\}$ localized basis, expressed as linear combinations of *STO*; (ii) $\{\psi_2\}$, with a hydrogenic character, is obtained from a double zeta basis $\{\phi_2\}$ of hydrogenic functions; (iii) $\{\psi_3\}$, with a diffuse character, is obtained from a set $\{\phi_3\}$ of *STO* multiplied by a cosine (*STOCOS*) [16]

$$\chi_{\zeta klm}(\mathbf{r}) = N_{\zeta kl} r^l e^{-\zeta r}\cos(kr) Y_{lm}(\hat{r}) \tag{15}$$

with the k values chosen according to

$$k_j = (j + 1/2)\pi/R_0 \tag{16}$$

where R_0 represents, roughly, the radial dimension of the obtainable wave packets lying in the continuum spectral region.

3.2. The Multielectron Basis: Bound States and Localized Channel

Configurations describing bound states, the localized part of the auto ionizing states and short range correlation are used to build the lc basis which, if $|core\rangle$ is a closed shell single determinant for the core, are

$$|\Phi^{\mathrm{loc}}_{\beta,E_J}\rangle = \sum_{\lambda,\mu,\nu,j} C_{\lambda,\mu,\nu,j;\beta,E_J}(\hat{a}^\dagger_\lambda \hat{a}^\dagger_\mu)(\hat{a}^\dagger_\nu \hat{a}_j)|\mathrm{core}\rangle \tag{17}$$

λ and μ are elements of $\{\psi_1\}$ or $\{\psi_2\}$, and refer to valence electrons, where ν and j belong only to $\{\psi_1\}$ and are concerned with the electron excitation from the core occupied orbital j to the valence orbital ν. The coefficients C are such as to ensure that $|\Phi^{\mathrm{loc}}_{\beta,E_J}\rangle$ belongs to the selected manifold β.

The (localized) polarized-core bound states of the ion $|\Phi^{N-1}_{I_\gamma}\rangle$, employed subsequently to build the pwc, are generally expanded on 400–500 configurations of the type

$$|\Phi^{N-1}_{I_\gamma}\rangle = \sum_{\lambda,\nu,j} C_{\lambda,\nu,j;I_\gamma}\hat{a}^\dagger_\lambda(\hat{a}^\dagger_\nu \hat{a}_j)\,|\mathrm{core}\rangle. \tag{18}$$

New functions for the neutral system may now be constructed as

$$|\Phi^{N-1}_I \phi_w\rangle_\beta = \sum_v C_{v,\beta}\hat{a}^\dagger_v|\Phi^{N-1}_{I_\beta}\rangle. \tag{19}$$

With the orbitals ϕ_v belonging to the ψ_2 set, the $|\Phi^{N-1}_I \phi_v\rangle$'s will provide a particularly suitable basis for the description of Rydberg series.

3.3. The Partial Wave Channels

A pwc describes asymptotic states in the continuum specified by: (1) the state I of the ion; (ii) the orbital angular momentum l of the external electron; and (iii) all the labels defining the manifold. Using, as before, a single index β which collects all the labels of a pwc, the multielectron functions

$$|\Phi^{pwc}_\beta\rangle = \sum_w C_{\beta,w}|\Phi^{N-1}_{I_\beta}\chi_{w,l_\beta}\rangle, \tag{20}$$

with χ_{w,l_β} belonging to the $\{\psi_3\}$ set with given value of l_β, are used to represent the partial wave channels pwc. The diagonalization of the atomic Hamiltonian upon these bases will provide the pwc basis functions $|\Phi_{\beta,E_n}\rangle = |\Phi^{N-1}_{I_\beta}\varphi_{\epsilon_n,l_\beta}\rangle$ where ϵ_n specify the energy of the wave packet related to the total energy E_n of the variational calculation (made for each *pwc* separately) and to the wave-vector k_n by $\epsilon_n = E_n - E_I$, $\epsilon_n = k_n^2/2$. It is useful to add localized $|\Phi_I^{N-1}\phi_v\rangle$ in the expansion, or otherwise the CI calculation will try to combine the diffused $|\Phi_{I_\beta}^{N-1}\chi_{w,l_\beta}\rangle$ to reproduce bound states rather than continuum states. For large values of the radial coordinate, the orbital $|\varphi_{\epsilon_n,l}\rangle$ behaves as a shifted Coulomb wave in a wide region of space. The analysis of its behavior in this region yields its phaseshift $\delta_\beta(\epsilon)$ and a normalization constant C such that the C^2-normalized state $C|\Psi_{\beta,\epsilon_n}\rangle$ for $r < R_0$ for almost coincides with the true continuum state. Thus it may be continued to infinity as an energy-normalized shifted Coulomb wave. We emphasize here that the addition of localized functions cause the *pwc* to be non-orthogonal to the localized channel. In fact, if χ_v is a localized orbital, $|\Phi_{I_\beta}^{N-1}\chi_{w,l_\beta}\rangle$ may overlap with those configurations $|\Phi_{\lambda\mu\nu j}^{\text{loc}}\rangle$ having one of the labels λ, μ or ν equal to v. In the construction of a *pwc* one typically puts roughly 80 $|\Phi_{I_\beta}^{N-1}\chi_{w,l_\beta}\rangle$, as many as the number of diffused plus localized orbitals employed. On the other hand, since the ion states are expanded on ~500 configurations, ~40 000 configurations are actually involved. Furthermore, since one has several *pwc* in a typical calculation above the third ionization threshold, the full K matrix calculation is based upon more than 10^5 basic configurations, contracted to give the 2000–2500 unperturbed states of the localized channel plus the partial wave channels.

3.4. MG RESULTS

Since this article is focused on the calculation of the continuum spectrum and a more thorough report may be found elsewhere [14, 15], here only the results more pertinent to the illustration of the method's possibilities are presented.

3.4.1. *3pnl Resonances*

An energy range of roughly 0.16 au from the first ionization threshold ($3s$ ion) up to almost the second threshold ($3p$ ion) has been explored. The energy difference $E_{3p} - E_{3s} = 0.1627$ au which we have computed is in excellent agreement with the experimental value $E_{3p}^{\text{exp}} - E_{3s}^{\text{exp}} = 0.1628$ au, where E_{3p}^{exp} is obtained as the barycenter of the $J = 1/2$ and $J = 3/2$ levels reported by Bashkin and Stoner (1975) [17]. This agreement allows us to express our levels directly with respect to the ion $3s$, and in the following, when speaking of energies, we mean photoelectron energies. In Table I our positions and widths are obtained from the real part, E_r, and from the imaginary part, $\Gamma/2$, of the S matrix poles. These same quantities are also computed through the analysis of one-photon ionization patterns when marked peaks occur according to the Fano profile formula [12, 18]. A good agreement between the two methods has always been obtained. The results reported in Table I together with the recent data available in the literature, show that, on the whole, our calculated values are satisfactory and eliminate almost all the large discrepancies between experimental data and theoretical computations, as is the case with the

$3p3d\ H^1P^0$ resonance. In the case of the $3p^2\,{}^1S^e$ we find: $E_r(S$ pole$) = 0.0301$ au, E_r (Fano profile) = 0.0301 au, $E_{\sigma_{\max}} = 0.0305$ au; Γ (S pole) = 1.80×10^{-3} au, Γ (Fano profile) = 1.70×10^{-3} au. On the other hand, the analysis of the two-photon cross section [14] seems to lead to the smaller width of 1.65×10^{-3} au. Two-photon and stepwise experiments all put the width of the $3p^2\,{}^1S^e$ resonance within the range $1.26 - 1.28 \times 10^{-3}$ au [19–21]. Theoretical calculations instead produce larger values, which generally become even larger when core polarization is taken into account both rigorously (present work) and semiempirically [11, 23]. It is not sure if the result obtained by Chang (1987) [25], closer to the experiment (1.43×10^{-3} au), may be due to the use of too limited a basis [11]. It may be observed that experimental spectra are generally affected by noticeable noise and one wonders how Shao and coworkers can report as many as three significant digits for their value of Γ, without specifying either the procedure employed to obtain Γ from the cross section or the error to be expected from this analysis. Even so, theoretical values are probably too large, and this might indicate that electron correlation in the $3p^2$ resonance should be treated at a more accurate level than single core excitations. On the other hand it is interesting to note that in a recent paper [26], coauthored by two of the authors of [21], a value of 1.70×10^{-3} au is reported for the width of the $3p^2\,{}^1S^e$ resonance.

When the widths of the other resonances are considered, experimental investigations are far less systematic; apart for a few data for the $3pnp\,{}^1S^e$ series, there is only the result by Shao and coworkers for the $3p4f\,{}^1D^e$ resonance. While the agreement for the former series may be considered satisfactory, our result for $3p4d\,{}^1D^e$, though larger than the frozen-core calculation by Moccia and Spizzo [9], is still far smaller than the experiment at value. The agreement between our values and frozen core width values is fairly good in general, and good when the $3pns\,{}^1P^o$ and $3pnp\,{}^1D^e$ series are concerned.

3.4.2. *4snl and 3dnl Resonances*

All these resonances lie above the second ionization threshold. Here we deal with the lowest members of several different series, in an energy range of roughly 0.15 au. The relative positions of the states $4s$ and $3d$ of Mg(II) and the ground state of Mg(III) (hereafter referred to as Mg^{++}) are well reproduced by our calculations: $E_{4s} - E_{Mg^{++}} = -0.2248$ au (-0.2345 au exp.); $E_{3d} - E_{Mg^{++}} = -0.2267$ au (-0.2268 au exp.). On the other hand, our calculations give a value for $E_{Mg^{++}} - E_{3s}$ which is roughly 1.5×10^{-3} au too large. As in the case of the ionization potential, this fact may be associated with the different amounts of core correlation for Mg^{++} and for $3s\ Mg^{++}$ as accounted by the single core excitations coupled to the single and double valence-shell excitations. The autoionizing states, formally ascribed to the $4snl$ and $3dnl$ series, have their localized parts mostly external to the core. Thus, it is reasonable to expect that in such a case the single excitations of the core would account only for polarization effects without introducing perceptible

Table I. *Wave energies* and widths of the $3pnl$ levels. Neglecting spin-orbit interactions, ${}^1P^e$ do not autoionise. Numbers in parentheses are powers of ten.

	Present	a	b	c*	Expt.
$3p^2\ {}^1S^e$	0.0301 1.80(−3)	0.0306 1.72(−3)	0.0327 1.56(−3)	0.0294 1.43(−3)	0.0301[d] 1.27(−3)
$3p4p$	0.1113 3.64(−5)	0.1116	0.1121 3.09(−5)	0.1108 3.00(−5)	0.1114[d] 3.18(−5)
$3p5p$	0.1334 7.21(−6)	0.1336	0.1338 2.08(−6)	0.1326 2.38(−6)	0.1334[d] 7.75(−6)
$3p6p$	0.1437 1.39(−6)	0.1439	0.1440 2.50(−7)	0.1428 9.75(−8)	0.1438[d] 2.73(−6)
$3p7p$	0.1494 2.90(−7)		0.1497 1.05(−9)		
$3p8p$	0.1528 2.55(−8)		0.1531		
$3p9p$	0.1551				
$3p4s\ {}^1P^0$	0.0774 1.19(-2)	0.0760	0.0768 1.36(−2)	0.0738 1.96(−2)	0.0774[e]
$3p5p$	0.1201 3.70(−3)	0.1205	0.1207 3.72(−3)	0.1195 4.66(−2)	0.1202[e]
$3p6s$	0.1373 1.61(−3)	0.1375	0.1376 1.63(−3)	0.1371 1.84(−3)	0.1377[e]
$3p7s$	0.1458 8.48(−)		0.1461 1.05(−3)	0.1457 9.90(−4)	0.1458[e]
$3p8s$	0.1506 6.23(−4)		0.1512	0.1506 5.12(−4)	0.1508[e]
$3p9s$	0.1536 3.22(−4)			0.1537	
$3p3d\ {}^1P^0$	0.1106 5.70(−5)	0.1112	0.1116 9.2(−5)	0.1117 6.16(−5)	0.1105[e]
$3p4d$	0.1328 1.58(−5)	0.1330	0.1331 2.6(−5)	0.1334 7.25(−6)	0.1326[e]
$3p5d$	0.1433 6.32(−6)	0.1435	0.1436 1.11(−5)	0.1437 2.36(−7)	0.1434[e]
$3p6d$	0.1492 3.14(−6)	0.1493	0.1493 5.84(−6)	0.1494 4.23(−7)	0.1492[e]
$3p7d$	0.1527 1.81(−6)		0.1529		
$3p8d$	0.1550 1.13(−6)			0.1552	
$3p4p\ {}^1D^e$	0.1045 1.32(−2)	0.1058	0.1058 1.29(−2)		0.1056[f]
$3p5p$	0.1306 4.79(−3)	0.1314	0.1312 4.21(−3)		0.1313[f]
$3p6p$	0.1422 2.30(−3)	0.1422	0.1424 1.86(−3)		0.1427[f]
$3p7[$	0.1485 1.27(−3)		0.1487 1.10(−3)		0.1490[f]
$3p8p$	0.1523 7.78(−4)				
$3p9p$	0.1547 5.10(−4)				
$3p4f\ {}^1D^e$	0.1322 1.71(−6)		0.1323 1.10(−6)		0.1318[d] 7.3(--4)
$3p5f$	0.1431 1.31(−6)		0.1432 1.11(−6)		
$3p6f$	0.1490 1.22(−6)		0.1491 9.65(−7)		
$3p7f$	0.1526 1.09(−6)				
$3p8f$	0.1549 9.56(−7)				
$3p9f$	0.1566 8.59(−7)				
$3p4p\ {}^1P^e$	0.0917				0.0917[f]
$3p5p$	0.1262				0.1258[f]
$3p6p$	0.1402				0.1398[f]
$3p7p$	0.1474				
$3p8p$	0.1517				

* When reference states different from the $3s$ ion are used in the original paper, experimental data are employed for the conversion.

[a] Moccia and Spizzo [11], model potential *CI; phase shift* analysis.

[b] Moccia and Spizzo [9], *frozen-core CI; phase shift* and Fano analysis.

[c] Chang [24, 25]; *phase shift* analysis.

[d] Shao *et al.* [21] two-photon experiment.

[e] Baig and Connerade [28]; photoabsorption spectroscopy.

[f] Rassi *et al.* [29]; ejected electron spectroscopy.

amounts of energy correlation. It is not surprising then that the choice of Mg^{++} as the reference level for the energies leads to errors not larger than 3×10^{-4} au. In Table II our results are compared with the few recent data available in the literature.

Most resonances had never before been reported, and we had to classify them. The character of these resonances was obtained through the analysis of the eigenstates (6) obtained from the K matrix in the neighbourhood of the S matrix poles. The analysis was performed at the frozen core level since polarized-core results are very difficult to rationalize and, on the other hand, the range of energies investigated should exclude any heavy involvement of core excitations. The narrow resonances $3dng\,{}^1D^e$ and $3dnd\,{}^1P^e$ have very marked dg and dd characters, respectively, and no doubt arises about their classification. Most other resonances have no well defined character, for instance the lowest resonances of the ${}^1D^e$ manifold are clearly of the sd type, but one cannot distinguish whether they belong to the $3dns$ or to $4snd$ series since their localized parts are, more or less, an equivalent mixture of the two. No assignment is given in Table II when it is doubtful.

In general, positions and widths have been evaluated only through the S matrix poles, since most cross section peaks are too weak and asymmetric to permit an accurate analysis of their profiles.

There is a good agreement on the position of the $4s4p\,{}^1P^o$ resonance, but things are not so good for the $4s5p$ and, in particular, for the $4s6p\,{}^1P^o$. For this resonance, differences reach almost 2×10^{-3} au between the measurements of Pejcev and Baig [27], while our result lies in between. On the average, our results are slightly closer to those by Baig and Connerade (28].

3.4.3. *Photoionization Spectrum of the* $3s3p\,{}^1P^o$ *Excited State*

The lower lying states yielded by the lc calculation with a basis including a large number of the functions (18) may be used to compute accurate values of several properties [14, 15]. For instance, the photoionization cross section which involves the transition matrix elements between one of these low lying states and the $|\Psi^{(-)}_{\beta,\mathrm{E}}\rangle$ expressed as Equation (6), satisfying the boundary conditions appropriate for a photoionization process, is a routine calculation of our programs. Here, as an example, the photoionization cross section of the first excited state $3s3p\,{}^1P^o$ is presented in Figures 1 and 2. The calculations were made for the two more common gauges to show how much our results satisfy the requirement of gauge invariance. Concerning the region below the second ionization threshold, Figure 1 shows the calculated total absorption cross section σ_1 and the asymmetry parameter β. It is of some interest to compare the present results for β with those obtained by a frozen-core CI calculation [9]. Both the spectra, at the threshold, start with a value of $\beta \sim 1$ and rise smoothly up to the $3s^2\,{}^1S^e$ resonance. Here, while the frozen core calculation gives a weak and flat peak, we find a well pronounced minimum ($\beta = 0.6$). Since in this region the ${}^1S^e$ contribution to the cross section is dominant,

Table II. Positions and widths in au of the $4snl$ and $3dnl$ levels; Mg(III) ground state is the reference state for the energies. Numbers in parentheses are powers of ten.

		Present		Other positions			
	$E(\sigma_{max})$	E_r (S pole)	Γ	[a]	[b]	[c]	[d]
$4s^2\ ^1S^e$		−0.3556	6.0(−3)				
$3d^2$		−0.2988	4.6(−3)				
$4s5s$		−0.2850	2.9(−3)				
$4s6s$		−0.2637	6.7(−4)				
$3d4d$		−0.2594	1.5(−3)				
$4s7s$		−0.2536	5.3(−4)				
$3d5d$		−0.2483	1.7(−4)				
$4s8s$		−0.2473	7.9(−4)				
$4s9s$		−0.2441	1.1(−4)				
$3d6d$		−0.2422	7.2(−5)				
$4s4p\ ^1P^o$	−0.3129	−0.3126	4.3(−3)	−0.3124	−0.3120	−0.3117	−0.3125
$3d4p$		−0.2877	5.5(−3)				
$4s5p$	−0.2726	−0.2725	1.3(−3)	−0.2716		−0.2727	−02733
$3d5p$		−0.2629	1.9(−3)				
$4s6p$	−0.2590	−0.2590	2.6(−4)	−0.2587		−0.2580	−0.2597
$3d4f$		−0.2561	1.3(−3)				
$4s7p$	−0.2512	−0.2511	1.1(−4)	−0.2507		−0.2518	−0.2515
$3d6p$		−0.2492	9.4(−4)				
$4s8p$		−0.2468	2.1(−5)	−0.2459		−0.2473	−0.2461
$3d5f$		−0.2460	8.8(−4)				
$4s9p$		−0.2439				−0.2451	−0.2432
$3d7p$		−0.2423	5.6(−4)				
$^1D^e$		−0.3303	1.2(−2)				
		−0.3056	9.7(−3)				
(ds)		−0.2836	7.9(−4)				
(ds)		−0.2713	9.3(−3)				
		−0.2666	2.3(−3)				
		−0.2615	2.7(−3)				
(ds)		−0.2583	5.4(−4)				
(ds)		−0.2538	1.6(−3)				
		−0.2509	8.8(−3)				
(ds)		−0.2467	1.4(−4)				
$3d5g$		−0.2462	1.0(−5)				
		−0.2444	7.5(−4)				
(ds)		−0.2435	6.3(−4)				
$3d4d\ ^1P^e$	−0.2699	−0.2699	3.39(−5)		−0.2697		
$3d5d$	−0.2518	−0.2518	1.69(−5)		−0.2523		
$3d6d$	−0.2433	−0.2433	9.60(−6)				
$3d7d$	−0.2386	−0.2386	5.83(−6)				
$3d8d$	−0.2356	−0.2356	3.81(−6)				
$3d9d$	−0.2336	−0.2336	2.64(−6)				

[a] Lengyel *et al.* [22], frozen core CI.
[b] Mendoza and Zeippen [23], 8 target close-coupling.
[c] Pejcev *et al.* [27], electron impact spectroscopy.

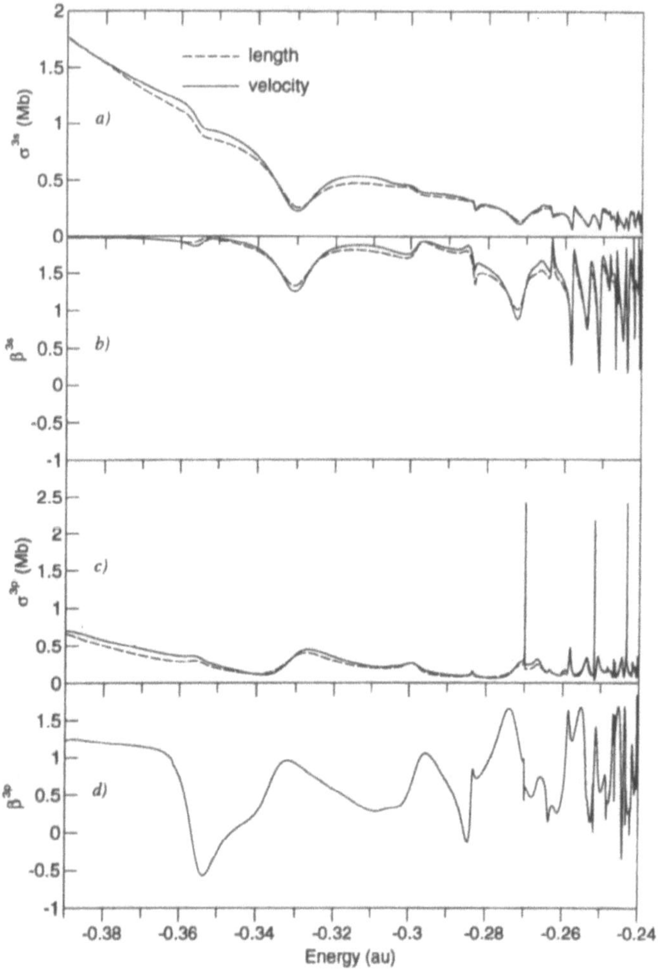

Figure 1. $3s3p\,^1P^o$ state one-photon ionization below the $3p$ threshold: (a) total cross section σ and(b) asymmetry parameter β.

the present result seems to emphasize more the $^1S^e$ component. In fact, if only the $^1S^e$ contribution were to be considered, β should be equal to zero.

Two cross sections σ^{3s} and σ^{3p} have been considered above the second threshold. In the present case we have $\sigma^{3s} = \sigma^{3sks\,^1S^e} + \sigma^{3skd\,^1D^e}$, $\sigma^{3p} = \sigma^{3pkp\,^1S^e} + \sigma^{3pkp\,^1P^e} + \sigma^{3p\,^1D^e}$. So as not to encumber this paper with too many figures, we report only σ^{3s} and σ^{3s} and σ^{3p} in Figure 2 together with their asymmetry parameters. Mendoza and Zeippen [23] report partial cross sections for the transitions $3s3p \rightarrow\ ^1S^e$, $3s3p \rightarrow\ ^1P^e$ and $3s3p \rightarrow\ ^1D^e$. Their spectra are quite different from ours; in particular, they find a narrow peak of intensity ~ 2 Mb in the $^1S^e$ manifold at the apparent energy ~ -0.283 au below Mg^{++}, and classify it as due

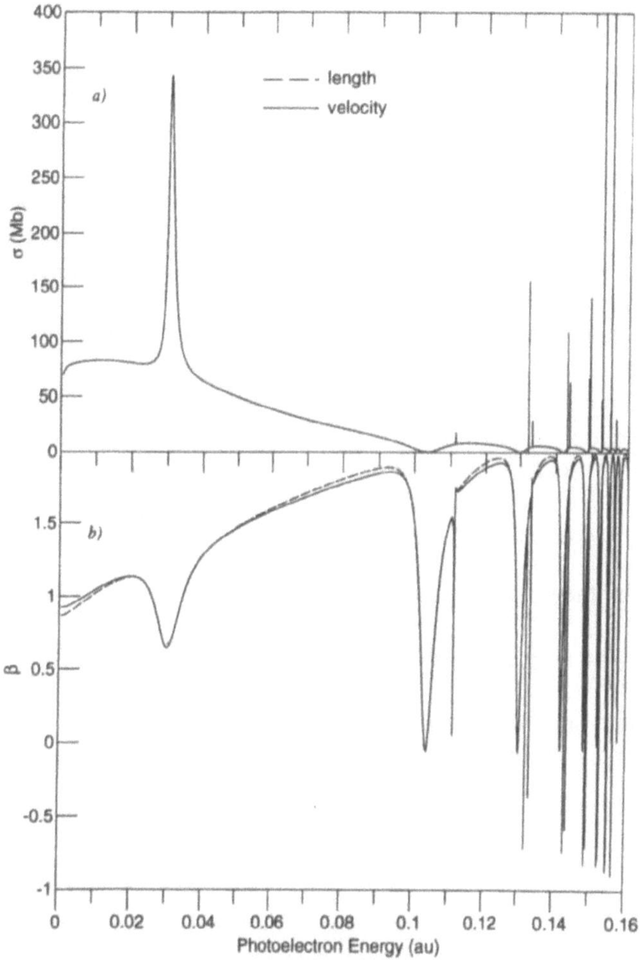

Figure 2. $3s3p\,^1P^o$ state one-photon ionization beyond the $3p$ threshold: (a) cross section σ^{3s} and (b) asymmetry parameter β^{3s} for processes leaving the ion in the $3s$ state; (c) cross section σ^{3p} and (d) asymmetry parameter β^{3p} for processes leaving the ion in the $3p$ state. For β^{3p} only VG results are presented. Energies are relative to Mg^{++}.

to the $3d^2$ resonance. We associate instead the $3d^2\,^1S^e$ resonance with a broad and weak peak at -0.2988 au, while no marked peak is found at -0.283 au, and the background is below 0.5 Mb. Furthermore, our spectra give narrow peaks of intensity $\sim$2.5 Mb due to $3dnd\,^1P^e$ resonances, while these same peaks, according to Mendoza and Zeippen, are $\sim$0.05 Mb high and as wide as $\sim$0.002 au.

4. The Floquet Hamiltonian

The effects of intense laser fields upon atoms involve very high order processes that are best studied by using non-perturbative approaches. Among the several methods proposed [30], those based upon the Floquet theory have been successfully applied to a variety of cases [31–37]. Our interest in this method rests upon the fact that the Floquet theory allows one to treat time-dependent processes formally in the same way as time-independent ones [38, 39]. For the sake of simplicity, all the following exposition refers to a one-electron system, its extension to the multielectron cases being formally straightforward.

If a quantized radiation state may be described by a coherent state $|\alpha\rangle$ of a single frequency ω, Shirley [38] has shown that the total Hamiltonian, in the limit of a very large average photon number $\bar{n}$, corresponds closely to the Floquet Hamiltonian. In fact, in the dipole approximation of the minimal coupling interaction, the time-dependent operator associate with a monochromatic radiation field

$$\hat{h}(t) = \frac{1}{c}\mathbf{A}_0 \cdot \hat{\mathbf{p}} \cos(\omega t) + \frac{1}{2c^2}\mathbf{A}_0^2 \cos^2(\omega t) \tag{21}$$

may be simplified by the unitary transformation

$$\hat{U}_1 = \exp\left[-\frac{i}{2c^2}\int^t \mathrm{d}t' A_0^2 \cos^2(\omega t')\right] \tag{22}$$

of the time-dependent Schrödinger equation. The transformation will eliminate the second term of the rhs of Equation (21) yielding the new Hamiltonian

$$\hat{H}_1 = \hat{H}_0 + (1/c)\mathbf{A}_0 \cdot \hat{\mathbf{p}} \cos(\omega t) \tag{23}$$

and modifying the energy levels by the amount $E_P = A_0^2/(4c^2)$, the ponderomotive energy.

The eigenstates $|\Phi_{\nu\gamma E_f^0}\rangle\rangle$ of the unperturbed Floquet Hamiltonian

$$\hat{\mathcal{H}}_0^f = \hat{H}_0 - i\frac{\partial}{\partial t} \tag{24}$$

corresponding to an energy $E_f^0 = E + \nu\omega$, can be built as direct products of eigenstates with energy E of the unperturbed matter Hamiltonian $\hat{H}_0$ times $\exp(i\nu\omega t)$, and provide an appropriate basis set for the matrix representation of the full Floquet Hamiltonian

$$\hat{\mathcal{H}}^f = \hat{H}_0 + \frac{1}{c}\mathbf{A}_0 \cdot \hat{\mathbf{p}} \cos(\omega t) - i\frac{\partial}{\partial t}. \tag{25}$$

The matrix elements are evaluated in the extended Floquet space $\mathcal{R}^3 \otimes [0, \tau]$ [39, 40], according to

$$\langle\langle \ \{\ldots\} \ \rangle\rangle = \frac{1}{\tau}\int_{-\tau/2}^{\tau/2} \mathrm{d}\mathbf{r}\{\ldots\} \tag{26}$$

where $r = 2\tau/\omega$.

The elements of the Floquet matrix $\mathbf{H}^f$ coincide with the elements of the matrix $\mathbf{H}^Q_\omega$, obtainable with a radiation quantized Hamiltonian $\hat{\mathcal{H}}^Q_\omega$, upon a basis of states $|\Phi_{\gamma E}\rangle|\bar{n}+\nu\rangle$ expressed as direct product of the $|\Phi_{\gamma E}\rangle$ times the number states $\bar{n}+\nu\rangle$ of the monochromatic radiation. Therefore, these states can be made to correspond to those of Floquet, allowing one to associate their Fourier components to a photon number within neighbourhood of $\bar{n}$, specifying a quantum state of radiation.

In other words [41], the time-independent Hamiltonian

$$\hat{\mathcal{H}}^Q_\omega = \hat{H}_0 + \omega(\hat{b}^\dagger_\omega \hat{b} - \hat{\mathbf{I}}\bar{n}) + \frac{\mathbf{A}_0 \cdot \hat{\mathbf{p}}}{2c}(\hat{b}^\dagger_\omega + \hat{b}_\omega) \tag{27}$$

is equivalent to $\hat{H}_0 + (1/2c)\mathbf{A}_0 \cdot \hat{\mathbf{p}}\cos(\omega t)$, in the sense that the correct cycle-averaged time evolution operator may be expressed as $\exp(-i\hat{\mathcal{H}}^f t)$. Thus, for a system irradiated by a monomode laser field it appears legitimate to interpret the Floquet states as linear combinations of matter eigenstates times different radiation number states. The analogy between the time-dependent Floquet Hamiltonian and a time-independent Hamiltonian, where the radiation is considered quantized, allows us to apply the method of section 2 to the calculation of the resonances arising from the interaction of an unperturbed discrete state of the matter, say $|a\rangle$, coupled to $\bar{n}$-photons, $|a\rangle|\bar{n}\rangle$, with all the other matter-radiation unperturbed states including, obviously, the continua. The expression (6) may now be written as

$$\begin{aligned} |\Psi_{\nu\beta E_f}\rangle\rangle &= |\Phi_{\nu\beta E_f}\rangle\rangle + \sum_\mu \sum_\gamma \not{\int} \mathrm{d}E' |\Phi_{\mu\gamma E'_f}\rangle\rangle \times \\ &\times \frac{1}{E_f - (E' + \mu\omega) + i0+} T^{(+)}_{\mu\gamma E'_f, \nu\beta E_f} \end{aligned} \tag{28}$$

where a channel is now identified by the pair (ν, β), the photon number and the collective index of the matter state.

In this study we have considered only the case of a linear polarized radiation field, so the resulting Hamiltonian has a cylindric symmetry. This fact, together with the lack of electronic structure in the H^+ ion, makes the collective index β coincide with the electron orbital angular momentum l, and the channel index becomes simply (ν, l).

The spatial integration of the matrix elements $\langle\langle\Phi_{\lambda\theta E''_f}|\hat{\mathcal{H}}^f - E_f|\Phi_{\nu\gamma E'_f}\rangle\rangle$ will involve integrals $\langle\Phi_{\theta E''}|\hat{\mathbf{p}}|\Phi_{\gamma E'}\rangle$ between continuum states which, according to the standard way of treating the dipole approximation [42], will assume the general form

$$\langle\Phi_{\theta E''}|\sum_j \hat{\mathbf{p}}_j|\Phi_{\gamma E'}\rangle = B(E'', E')\frac{\mathcal{P}}{E' - E''} + C(E')\delta(E' - E''). \tag{29}$$

These integrals, evaluated between wave-packets that mimic continuous functions in a large region of space, display large oscillation for small values of $E' - E''$, undermining the interpolation procedure and thus introducing serious numerical instabilities. In order to avoid this shortcoming we adopt the 'accelerated frame' representation introduced by Henneberger [43]. By applying the transformation

$$\hat{U}_2 = \exp\left[-\frac{i}{c}\mathbf{A}_0 \cdot \hat{\mathbf{p}} \int^t dt' \cos(\omega t')\right] \tag{30}$$

the Hamiltonian $\hat{H}_1$ (Equation (23) will become

$$\hat{H}_2 = \hat{H}_0 + \left[\frac{1}{r} - \frac{1}{|\mathbf{r} - \mathbf{a}(t)|}\right] \tag{31}$$

with

$$\mathbf{a}(t) = \frac{\mathbf{A}_0}{\omega c} \sin(\omega t'). \tag{32}$$

The integrals evaluated with the new perturbing operator, which is now a function of $\sin(\omega t)$, do not present divergencies and the method becomes numerically stable.

By the use of a spatial Fourier transform, one may easily show that the time-dependent perturbation operator in (31) may be expressed as a power series of $|\mathbf{a}(t)|$, whose s term is proportional to

$$\{i[\exp(i\omega t) - \exp(-i\omega t)]\}^s = i^s \sum_{j=0}^{s} (-1)^j \binom{s}{j} \exp[i\omega(s - 2j)t]. \tag{33}$$

It is clear then that the matrix elements of the Floquet Hamiltonian corresponding to $\hat{H}_2$ will be obtainable without difficulty.

It is not obvious that the Floquet matrix $\mathbf{H}^f$, corresponding to the accelerated frame, is closely related to a radiation quantized Hamiltonian. Fortunately, in the limit of a very large value of $\bar{n}$, the equivalence may be proved by the following argument. By applying to the Hamiltonian (27) a Pauli–Fierz type transformation [44]

$$\hat{U}_{PF} = \exp\left[-i\left(\frac{2\pi}{V\omega^3}\right)^{1/2} (\hat{b}_\omega^\dagger - \hat{b}_\omega)\hat{\epsilon}_\omega \cdot \hat{\mathbf{p}}\right], \tag{34}$$

closely related to that of Henneberger, and neglecting smaller terms, the resulting time-independent interaction operator $\hat{h}^Q$ may be written as

$$\hat{H}^Q = \frac{1}{r} - \frac{1}{\left|\mathbf{r} + \hat{\epsilon}_\omega i \left(\frac{2\pi}{V\omega^3}\right)^{1/2} (\hat{b}_\omega^\dagger - \hat{b}_\omega)\right|} \tag{35}$$

where V is the volume of the quantization box. This operator will have a power series expression whose s term will be proportional to $[i(\hat{b}_\omega^\dagger - \hat{b}_\omega)]^s$. Applying the boson commutation rules and retaining only the dominant term, it may be seen that

$$\langle \bar{n} + \mu | [i(\hat{b}_\omega^\dagger - \hat{b}_\omega)]^s | \bar{n} + \nu \rangle$$

$$\simeq i^s \sum_{j=0}^{s} (-1)^j \binom{s}{j} \langle \bar{n} + \mu | (b_\omega^\dagger)^{s-j} \hat{b}_\omega^j | \bar{n} + \nu \rangle, \tag{36}$$

the contribution of the neglected terms being smaller at least by an order $1/\bar{n}$.

Thus, since the spatial integration will be identical both for the perturbation operator of Equation (31) and for $\hat{h}^Q$, the following relation will hold between the matrix elements of $\mathbf{H}_2^f$ and $\mathbf{H}_\omega^Q$

$$\langle\langle \Phi_{\beta\mu} | \hat{H}_2 - i\frac{\partial}{\partial t} | \Phi_{\gamma\nu} \rangle\rangle \simeq \langle \Phi_\beta | \langle \bar{n}+\mu | \hat{H}_0 + \omega(\hat{b}_\omega^\dagger \hat{b}_\omega - \hat{\mathbf{I}}\bar{n}) + \hat{h}^Q | \Phi_\gamma \rangle | \bar{n}+\nu \rangle. \tag{37}$$

The conclusion is that, even in the accelerated frame, we are dealing formally with a time-independent problem and the resonance poles will arise from the matter-radiation interaction.

The choice of sine or cosine functions to describe classically the time evolution of vector potential $\mathbf{A}(t)$ depends on initial conditions. Because our calculations refer to a stationary regime, the dependence upon initial values must be absent. This is confirmed by noting that the matrix elements of the perturbation operators, using sine or cosine, are the same, up to a phase factor. Moreover, the two forms of the operator are related by an unitary transformation, which obviously leaves the poles unchanged.

5. A Calculation for the H Atom

The unperturbed basis $|\Phi_{\gamma E}\rangle$ for a hydrogen atom are well known but, in order to test the approach based upon the use of finite set of L^2 functions (which should be applicable to more complex atoms), a basis set comprising *STO*'s, hydrogenic Rydberg, and *STOCOS*'s has been employed. The selection of the *STOCOS* basis was first made as indicated by (15) with $R_0 \simeq$ 70–90 au. To reduce the basis dimension, a new set of *STOCOS* functions was generated using the following expression for the k's:

$$k_{j+1} = k_j + a(1 + bk_j) \tag{38}$$

where the parameters a and b were chosen to obtain 40 functions, for each manifold, having $k_{\min}$ = 0.2 au, $k_{\max}$ = 3.7 au. The ζ orbital exponents were calculated iteratively by imposing a constant value of the overlap between two contiguous *STOCOS*.

With these bases the resulting wave-packet in the continuum were such as to mimic quite well the Coulomb wave of the same energy in a wide region of space, roughly for $0 \leq r \leq 40$ au. On the average the phase shift with respect to the true Coulomb wave was ≤ 0.001 Rad.

The evaluation of the matrix elements of the perturbing operator in Equation (32) for a linearly polarized beam, namely

$$\hat{h}_a(t) = \frac{1}{r} - \frac{1}{\left|\mathbf{r} - \frac{\mathbf{A}_{0_z}}{\omega c} \sin \omega t\right|} \tag{39}$$

was performed by first computing the spatial integrals for different values of the variable $z(t) = A_0/(\omega c)\sin(\omega t)$, corresponding to a regular grid of 20 points for the variable t built over the interval $[0, \tau/4]$. These points are sufficient to describe the temporal variation of the integrals in a complete cycle, the others being evaluated by symmetry. The time integration in (26) can now be readily carried out as a discrete Fourier transform.

By carcful handling of the formulae and of the numerical procedures the final matrix elements are quite accurate, as verified by several tests.

5.1. Selection Rules

The computational effort required to calculate the photoionization parameters can be greatly reduced by noting some simple symmetry properties. A space-time inversion operation in the extended Floquet space applied to the matrix element of $\hat{h}_a(t)$ provides the following relation

$$\langle\langle \Phi_{\nu\gamma E}|\hat{h}_a|\Phi_{\nu'\gamma' E'}\rangle\rangle = (-1)^{(l+\nu)+(l'+\nu')}\langle\langle \Phi_{\nu\gamma E}|\hat{h}_a|\Phi_{\nu'\gamma' E''}\rangle\rangle. \tag{40}$$

The only non-zero elements are consequently those between states having the same parity of $l + \nu$. The total Hamiltonian matrix reduces then to a two-block diagonal matrix, corresponding to odd and even parity, and these two blocks lead to the same eigenvalues. The calculations can therefore be made, using indifferently either one of the two blocks.

6. Results

The aim of this section is to discuss the results of some preliminary calculations of resonances of the H atom irradiated by various laser fields, as a function of their intensity and frequency. The computational schemes used, in order to compare their efficiency, are derived by implementing both $\hat{T}^{(+)}$ and $\hat{K}$ scattering matrix techniques.

We attempt to reproduce some values reported in literature, calculated by use of a specially adapted R-matrix technique [45, 46] or complex scaling of spatial coordinates [47]. Furthermore, we have compared our results with the exact calculations for one photon processes, made using an analytical formula [48], and we have verified the correct behaviour in the limit of low intensity. With respect to the cited schemes, our techniques can be more simply applied to complex atomic systems: in particular, they do not suffer from problems due to a bad representation of inner shells, as occurs in the complex-coordinate method. On the other hand, the use of the Henneberger transformation leads to a slow convergence of the results with respect to the number of channels employed. This is not unexpected, looking at the wavefunction transformed by the operator (30):

$$\Psi'[\mathbf{r} - \mathbf{a}(t), t] = \hat{U}_2 \Psi(\mathbf{r}, t) \tag{41}$$

The time dependence appears now in the spatial coordinate also, making more difficult the representation of the wavefunction by a limited basis set.

All the reported calculations, if not otherwise stated, have been performed using a basis set including spatial functions up to $l = 7$. Channels are built using a number of photon ν in the interval $(-n, n')$, centered around $\bar{n}$ so that the absorption processes are described by $\nu < 0$ and emission ones by $\nu > 0$.

We report here the cases of two different radiation frequencies, corresponding to several ionization regimes. First, we examine the behaviour of H in presence of an electromagnetic field with angular frequency $\omega = 0.650$ au ($\lambda = 70$ nm). In this case a single photon absorption is enough to ionize the atom, so the physics is dominated essentially by one-photon processes. Calculations have been performed at increasing intensities, using a set of spatial functions having a maximum value of angular momentum $l_{\rm max} = 6$. The maximum number of channels employed was 28, corresponding to about 1400 states for the $(-4, +3)$ photon interval. As reported in Table III, in the cases of lower intensities, the results agree well with LOPT previsions reported for single photon ionization. Increasing the values of intensity the contribution of multiphoton processes is more and more important, so remarkable deviations from LOPT results appear. Figure 3 shows the characteristic dependence of the total rate of ionization of $1s$ state upon the laser intensity, which describes the well known phenomenon of 'stabilization' [45, 49–51]. Second, a true multiphoton regime is achieved by considering a field of frequency $\omega = 0.184$ au ($\lambda = 248$ nm), having intensity $I = 10^{14}$ W/cm^2. This particular photon energy allows the states $|1s, 0\rangle$ and $|2s, -2\rangle$ to be quasi-degenerate ($E_{1s,0} = -0.5$ au, $E_{2s,-2} = -0.493$ au), so, instead of Equation (14), a different relation derived from (10) has been used in order to find the poles. A succession of calculations was made to study the convergence of the results with respect to the extension of the basis of channels, characterized by the maximum value of the orbital angular momentum of the spatial functions employed ($l_{\rm max}$) and by the photon numbers n, n'. The computational effort was much greater than for the previous cases; the number of channels employed to achieve convergence reached a maximum of 64, for the interval $(-10, +5)$ and the dimension of the $\mathbf{H}^f$-matrix was about 3200.

Table III. $2s$ resonance parameters at different intensities. $\omega = 0.65$ au, $l_{\max} = 6$; $\Delta = \mathrm{Re}E_R - E_{1s,0} - E_P$, $\Gamma = -2\mathrm{Im}E_R$. Numbers in parentheses are powers of ten.

		T matrix		K matrix		LOPT[a]
n	n'	Δ	Γ	Δ	Γ	$\Gamma = \sigma I$
$I = 10^{10}$ W/cm^2, $E_P = 1.1 \times 10^{-7}$ au						
4	3	8.9087(−8)	2.6638(−7)	8.9086(−8)	2.6638(−7)	2.6505(−7)
$I = 10^{14}$ W/cm^2, $E_P = 1.1 \times 10^{-3}$ au						
2	1	7.3548(−4)	2.6494(−3)	7.3589(−4)	2.6499(−3)	2.6505(−3)
2	2	6.9306(−4)	2.6499(−3)	6.9348(−4)	2.6503(−3)	
3	1	7.3416(−4)	2.6496(−3)	7.3459(−4)	2.6500(−3)	
3	2	6.9173(−4)	2.6501(−3)	6.9216(−4)	2.6505(−3)	
4	1	7.3383(−4)	2.6496(−3)	7.3425(−4)	2.6503(−3)	
4	2	6.9140(−4)	2.6501(−3)	6.9181(−4)	2.6508(−3)	
4	3	6.9040(−4)	2.6501(−3)	6.9081(−4)	2.6508(−3)	
$\Delta = 3.60(-4)$,[b] $\Gamma = 2.56(-3)$[b]						
$I = 2 \times 10^{16}$ W/cm^2, $E_P = 2.2 \times 10^{-1}$ au						
3	2	1.8963(−1)	1.2830(−1)	1.9430(−1)	1.3748(−1)	5.2842(−1)
4	2	1.9185(−1)	1.3083(−1)	1.9748(−1)	1.4091(−1)	
4	3	1.9105(−1)	1.3404(−1)	1.9615(−1)	1.4533(−1)	
$\Delta = 1.95(-1)$[b], $\Gamma = 1.4(-1)$[b]						

[a] σ value is derived from an analytical formula [48].
[b] Dörr *et al.* [45].

Table IV. $1s$ resonance parameters (au): dependence upon maximum l value, $l_{\max}$ (in parentheses), and photon interval $(-n, n')$ around $\bar{n}$, $I = 10^{14}$ W/cm^2, $\omega = 0.184$ au, $E_P = 3.87 \times 10^{-3}$ au; $\Delta = \mathrm{Re}E_R - E_{1s,0} - E_P$, $\Gamma = -2\mathrm{Im}E_R$.

		$-\Delta \times 10^2$			$\Gamma \times 10^3$			$\lvert T^{(+)}_{1s,0}/T^{(+)}_{2s,-2}\rvert^2$		
n	n'	(3)	(5)	(7)	(3)	(5)	(7)	(3)	(5)	(7)
4	3	0.6117	1.5449	1.8452	3.5093	1.3268	0.9545	0.8760	6.2940	10.7304
5	2	0.4306	1.3206	1.6067	4.9136	1.8707	1.3525	0.4874	4.1208	7.0825
6	3	0.8563	1.7802	2.0806	4.4657	1.8535	1.3756	0.9898	4.9718	7.8121
7	2	0.5424	1.4282	1.7143	5.3761	2.1428	1.5718	0.5559	3.7507	6.2015
7	4	1.0449	1.9887	2.2959	4.3386	1.8537	1.3867	1.2017	5.1944	7.9625
8	3	0.9172	1.8400	2.1408	4.6436	1.9639	1.4667	1.0091	4.7315	7.3198
8	5	1.1460	2.0985	2.4092	4.2626	1.8503	1.3920	1.3116	5.2842	7.9734
9	6	1.1954	2.1539	2.4670	4.2335	1.8500	1.3955	1.3604	5.3100	7.9506
10	5	1.1648	2.1176	2.4285	4.2996	1.8762	1.4146	1.3141	5.2076	7.8218

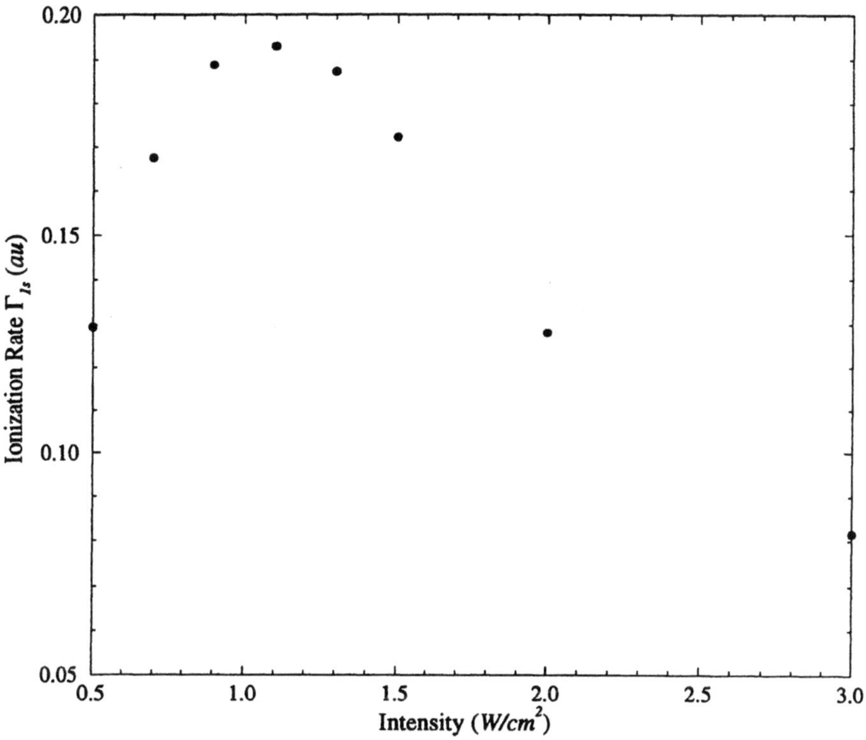

Figure 3. H ground state ionization (ω = 0.65 au): dependence of total ionization rate Γ_{1s} upon field intensity.

Table V. $2s$ resonance parameters (au): dependence upon maximum l value, l_{max} (in parentheses), and photon interval $(-n, n')$ around $\bar{n}$, $I = 10^{14}$ W/cm^2, $\omega = 0.184$ au, $E_P = 3.87 \times 10^{-3}$ au; $\Delta = \mathrm{Re}E_R - E_{2s,-2} - E_P, \Gamma = -2\mathrm{Im}E_R$.

		$\Delta \times 10^2$			$\Gamma \times 10^3$			$\lvert T^{(+)}_{1s,0}/T^{(+)}_{2s,-2}\rvert^2$		
n	n'	(3)	(5)	(7)	(3)	(5)	(7)	(3)	(5)	(7)
4	3	8.4301	5.7616	4.9681	1.3937	1.7478	1.8304	0.1729	0.0317	0.0166
5	2	7.9587	4.7124	3.9641	1.4080	1.8561	1.9539	0.2613	0.0465	0.0248
6	3	7.5954	4.2702	3.4614	1.4899	1.9002	1.9864	0.2137	0.0451	0.0256
7	2	7.6742	4.1645	3.3390	1.4317	1.8997	1.9935	0.2845	0.0542	0.0300
7	4	7.4448	4.0629	3.2236	1.5165	1.9726	2.0005	0.2009	0.0455	0.0264
8	3	7.4597	4.0144	3.1673	1.4985	1.9155	2.0038	0.2227	0.0486	0.0279
8	5	7.3800	3.9680	3.1117	1.5279	1.9204	2.0055	0.1964	0.0458	0.0270
9	6	7.3493	3.9194	3.0537	1.5323	1.9222	2.0071	0.1946	0.0462	0.0274
10	5	7.3490	3.9060	3.0388	1.5287	1.9221	2.0076	0.1986	0.0469	0.0277

The position and the width of the $1s$ and of the $2s$ resonances for various values of l and n, n' are presented in Tables IV and V. The entries correspond only to the cases such that $l_{\max}$ and $n + n'$ are odd, when the number of channels arising from allowed l, ν, coupling is balanced whatever parity of $l + \nu$. The poor results obtained using a basis set comprising functions with low values of l maximum emphasize the particular coupling due to the interaction operator employed. As one can see by looking at the selection rules, there is no limit to l values of states coupled by $\hat{h}_a(t)$, and an expansion in power series of $|\mathbf{a}(t)|$ demonstrates that the coupling strength depends upon the field intensity and upon the differences between the l values of coupled states.

At the intensity considered, LOPT provides a total decay width for the $1s$ state $\Gamma_{1s}^{\text{LOPT}} = 4.5 \times 10^{-3}$ au [45], too high with respect to that derived from calculations where higher-order processes are taken in consideration. Our calculations converge to $\Gamma_{1s} = 1.4 \times 10^{-3}$ au, perfect agreement with literature [45, 46].

It is important to note that the convergence is faster for the $2s$ state, with respect to $1s$; this is not unexpected if it is considered that its ionization potential (0.125 au) is low enough to allow one-photon ionization ($\omega = 0.184$ au), which appears to be by far the dominant process at the intensity considered.

In all the cases, the matching between $\hat{T}^{(+)}$ and $\hat{K}$ results, in view of the different numerical procedures involved, may be considered satisfactory.

References

1. D. Dill and J. L. Dehmer: *J. Chem. Phys.* **61**, 692 (1974).
2. P. W. Langhoff: in B. Dalton, S. Grimes, J. Vary, and S. Williams (eds.), *Theory and Application of Moment Methods in Many-Fermions Systems*, p. 191, Plenum, New York (1980).
3. R. R. Lucchese and V. McKoy:: *Phys. Rev. A.* **21**, 112 (1980). R. R. Lucchese, G. Raseev, and V. McKoy: *Phys. Rev. A* **25**, 2572 (1982).
4. L. A. Collins and B. I. Schneider: *Phys. Rev. A* **29**, 1695 (1984).
5. I. Cacelli, V. Carravetta, A. Rizzo, and R. Moccia: *Phys. Rep.* **205**, 283 (1991).
6. R. Moccia and P. Spizzo: *Phys. Rev. A* **43**, 2199 (1991).
7. I. Cacelli, V. Carravetta, and R. Moccia: *Chem. Phys.* **184**, 213 (1994).
8. R. Moccia and P. Spizzo: *J. Phys. B: At. Mol. Opt. Phys.* **18**, 3537 (1985); **18**, 3555 (1985).
9. R. Moccia and P. Spizzo: *J. Phys. B: At. Mol. Opt. Phys.* **21**, 1121 (1988).
10. R. Moccia and P. Spizzo: *J. Phys. B: At. Mol. Opt. Phys.* **21**, 1133 (1988); **21**, 1145 (1988).
11. J. R. Moccia and P. Spizzo: *Phys. Rev. A* **39**, 3855 (1988).
12. U. Fano: *Phys. Rev.* **124**, 1866 (1961).
13. R. G. Newton: *Scattering Theory of Waves and Particles*, Springer-Verlag, New York (1982); G. Staszewska and P. G. Truhlar: *J. Chem. Phys. Phys.* **86**, 2793 (1987).
14. S. Mengali and R. Moccia: *J. Phys. B: At. Mol. Opt. Phys.* **29**, 1597 (1996).
15. S. Mengali and R. Moccia: *J. Phys. B: At. Mol. Opt. Phys.* **29**, 1613 (1996).
16. I. Cacelli, V. Carravetta, and R. Moccia: *Chem. Phys.* **90**, 313 (1984).
17. S. Bashkin and J. O. Stoner: *Atomic Energy Levels and Grotrian Diagrams*, North-Holland, Amsterdam (1978).
18. U. Fano and J. W. Cooper: *Phys. Rev.* **137**, A1364 (1965).
19. D. J. Bradley, C. H. Dugan, P. Ewart, and A. F. Purdie: *Phys. Rev. A* **13**, 1416 (1976).
20. R. E. Bonanno, C. W. Clark, and T. B. Lucatorto: *Phys. Rev. A.* **34**, 2082 (1986).
21. Y. L. Shao, C. Fotakis, and D. Charalambidis: *Phys. Rev. A.* **48**, 3636 (1993).

22. V. I. Lengyel, V. T. Navrotsky, E. P. Sabad and O. I. Zatsarinny: *J. Phys. B: At. Mol. Opt. Phys.* **17**, 465 (1984).
23. C. Mendoza and C. J. Zeippen: *Astron. Astrophys.* **179**, 346 (1987).
24. N. T. Chang: *Phys. Rev. A* **34**, 4554 (1986).
25. N. T. Chang: *Phys. Rev. A* **36**, 5468 (1987).
26. N. E. Karapanagioti, D. Charalambidis, C. J. G. J. Uiterwaal, and C. Fotakis: *Phys. Rev. A* **53**, 2587 (1996).
27. V. Pejcev, D. Rassi, and K. J. Ross: *J. Phys. B: At. Mol. Opt. Phys.* **13**, L305 (1980).
28. M. A. Baig and J. P. Connerade: *Proc. R. Soc. A* **364**, 353 (1978).
29. D. Rassi, V. Pejcev, T. W. Ottley, and K. J. Ross: *J. Phys. B: At. Mol. Opt. Phys.* **10**, 2913 (1977).
30. F. H. H. Faisal: *Theory of Multiphoton Processes*, Plenum Press, New York, London (1987).
31. R. M. Potvliege and R. Shakeshaft: *Phys. Rev. A* **36**, 5678 (1987); **38**, 1098 (1988); **38**, 4597 (1988); **38**, 6190 (1988); **40**, 3061 (1989); **41**, 1609 (1990).
32. H. P. Breuer and M. Holthaus: *Z. Phys. D. – Atoms, Molecules and Clusters* **11**, 1 (1989); H. P. Breuer, K. Dietz, and M. Holthaus: *J. Phys. B: At. Mol. Opt. Phys.* **22**, 3187 (1989); H. P. Breuer, K. Dietz, and M. Holthaus: *J. Phys. B: At. Mol. Opt. Phys.* **24**, 1343 (1991).
33. K. Unnikrishnan: *Phys. Rev. A* **46**, 7281 (1992).
34. T. Millack: *J. Phys. B: At. Mol. Opt. Phys.* **23**, 1693 (1990); T. Millack, V. Véniard and J. Henkel: *Phys. Lett. A* **176**, 433 (1993).
35. M. Plummer and J. F. McCann: *J. Phys. B: At. Mol. Opt. Phys.* **28**, 4073 (1995).
36. S.-I. Chu and J. Cooper: *Phys. Rev. A* **32**, 2769 (1985).
37. G. Yao and S.-I. Chu: *Phys. Rev. A* **45**, 6735 (1992).
38. J. H. Shirley: *Phys. Rev.* **138**, B979 (1965).
39. H. Sambe: *Phys. Rev. A* **7**, 2203 (1973).
40. V. N. L. Manakov and L. P. Rapoport: *Phys. Rep.* **141**, 319 (1986).
41. A. Maquet, S-I Chu, and W. P. Reinhardt: *Phys. Rev. A.* **27**, 2946 (1983).
42. T. Mercouris, Y. Komminos, S. Dionissopoulos and C. A. Nicolaides: *Phys. Rev. A* **50**, 4109 (1994).
43. W. C. Henneberger: *Phys. Rev. Lett.* **21**, 838 (1968).
44. W. Pauli and M. Fierz: *Nuovo Cimento* **15**, 167 (1938).
45. M. Dörr, M. Terao-Dunseath, J. Purvis, C. J. Noble, P. G. Burke, and J. Joachain: *J. Phys. B: At. Mol. Opt. Phys.* **25**, 2809 (1992).
46. M. Dörr, P. G. Burke, C. J. Joachain, C. J. Noble, J. Purvis, and M. Terao-Dunseath: *J. Phys. B: At. Mol. Opt. Phys.* **26**, L275 (1993).
47. B. R. Junker: *At. Mol. Phys.* **18**, 207 (1982); W. P. Reinhardt: *Ann. Rev. Phys. Chem.* **33**, 223 (1982).
48. H. A. Bethe and E. E. Salpeter: *Quantum Mechanics of One and Two Electrons Atoms*, Plenum (1977).
49. M. Pont, N. Walet, M. Gavrila, and C. W. McCurdy: *Phys. Rev. Lett.* **61**, 939 (1988); M. Pont, N. Walet, and M. Gavrila: *Phys. Rev. A* **41**, 477 (1992); M. Gavrila and J. Z. Kaminsky: *Phys. Rev. Lett.* **52**, 613 (1984).
50. M. Pont and M. Gavrila: *Phys. Rev. Lett* **65**, 2362 (1990); M. Pont: *Phys. Rev. A* **44**, 2141 (1991); M. Pont: *Phys. Rev. A* **44**, 2152 (1991).
51. X. Tang and S. Basile: *Phys. Rev. A.* **44**, R1454 (1991): K. Burnett, P. L. Knight, and V. C. Reed: *Phys. Rev. Lett.* **66**, 301 (1991); K. C. Kulander and J. K. Krause: *Phys. Rev. Lett.* **66**, 2601 (1991); Q. Su and J. Javanainer: *Phys. Rev. Lett.* **64**, 862 (1990).

R. McWeeny et al. (eds.), Quantum Systems in Chemistry and Physics, 373–395.

Revisiting the Off-Center Impurity Problem: Reorientational Rates of Hindered Rotators*

P. PETROVA
Institute of Nuclear Research and Nuclear Energy, Bulgarian Academy of Sciences, 72 Tsarigradskio Chaussee, 1784 Sofia, Bulgaria.

M. IVANOVITCH and M. GEORGIEV**
Institute of Solid State Physics, Bulgarian Academy of Sciences, 72 Tsarigradskio Chaussee, 1784 Sofia, Bulgaria.

M. MLADENOVA
Department of Condensed Matter Physics, Faculty of Physics, University of Sofia, 5 James Bourchier Boulevard, 1126 Sofia, Bulgaria.

G. BALDACCHINI and R. M. MONTEREALI
ENEA C.R.E. Frascati, C.P. 65, Via Enrico Fermi 27, 00044 Frascati (Roma), Italy.

U. M. GRASSANO
Dipartimento di Fisica, Universita "Tor Vergata", Via della Ricerca Scientifica 1, 00133 Rome, Italy.

A. SCACCO
Istituto di Fisica, Universita "La Sapienza", Piazzale Aldo Moro 2, 00185 Rome, Italy.

(Received and accepted: 17 June 1997)

Abstract. The off-center displacements of impurity ions in polar crystals are regarded as a *Pseudo-Jahn–Teller effect.* They are shown to arise from first-order vibronic mixing, while third-order mixing creates barriers for reorientating an off-center impurity. The rotational equation in two dimensions is *Mathieu's equation,* with known eigenstates and eigenvalues. Rotational bands are found, allowing for reorientational transitions. Intraband transition probabilities in terms of Mathieu's eigenfunctions and eigenvalues are derived and compared with those based on harmonic-oscillator eigenstates and eigenenergies. We check the reorientational rates derived by both means against paraelectric resonance data.

Key words: Vibronic oscillator, reaction-rate theory, reorientational rate, configurational-tunneling probability, statistical ensemble.

1. Introduction

Substitutional cationic impurities such as Li^+, Ag^+ aud Cu^+ in the alkali halides are known to displace under certain conditions at fraction of an Ångstrom unit away from their normal lattice sites. Classically, the off-center displacements

* Dedicated to Professor S. G. Christov on the occasion of his 85th birthday.
** Author for correspondence.

occur because of the balance of polarization and repulsive overlap interactions beween a smaller size impurity and the surrounding host ions [1]. From a quantum-mechanical viewpoint, the symmetry lowering linked with an off-center displacement results from the vibronic mixing of opposite-parity electronic states at the impurity by an odd-parity vibration via the *pseudo-Jahn–Teller Effect* [2, 3].

The off-center impurity ions in alkali halides also provide examples of hindered rotators in solids. These ought to overcome local periodic potentials as they jump around their normal lattice sites. *Schrödinger's equation* of a hindered in-plane rotator is *Mathieu's equation*, familiar from the theory of wave propagation in periodic structures [4]. Due to mathematical complexities, various approximations are often used for the eigenstates and eigenvalues: at low reorientational barriers, the hindered rotation obviously has many features in common with a free rotor, while at energies near the potential minima at the reorientational sites with finite barriers the rotational states are close to those of a harmonic oscillator. Only recently Mathieu eigenfunctions and eigenvalues have been incorporated for discovering the exact rotational energy bands and band-like states [5].

Another problem related to hindered rotation is posed by the reorientational rate in the multiwell pattern. From a pragmatic viewpoint, it is tempting to apply the well-bottom harmonic-oscillator solutions to deriviug a rate that holds good approximately at ground state energies [6–8]. However, the accuracy of harmonic-oscillator-based calculations at higher vibronic energies is sometimes grossly overestimated. Recently, rate calculations using exact Mathieu eigenstates and eigenvalues have been done and tested against the experiment [5]. Both bottom-well and Mathieu-based calculations predict finite zero-point rates which persist in a range before the gradual transition to an Arrhenius temperature dependence. In concommitance, temperature-independent branches are readily observed at lower temperatures in the relaxation rates of off-centered Ag^+ but not of Li^+ where one Arrhenius portion is appareutly preceded by another finite, though lower, temperature slope. This discrepancy suggests that relaxation rates in an ensemble of uncoupled vibronic oscillators may be inadequate when applied to the off-centered Li^+ impurities, known to undergo considerable pair interactions [9]. Apart from its general scientific value, the interest in Li^+ rotation in alkali halides is also due to it being an essential ingredient of the fundamental color center laser.

In what follows we consider the reorientational problem of an off-centered impurity in a polar crystal. We show that while off-center displacements occur to first-order in the electron-vibrational mixing interaction, rotational barriers hindering the free rotation around the central site occur as a third-order mixing effect. We approach the reorientational relaxation by means of *Christov's reaction-rate theory* [10, 11], and apply Mathieu eigenstates and eigenvalues to deriving a rate. These rates are compared with ones derived by means of *Hermite polynomials* which are nearly as good at ground-state energies. We also minimize the free energy of a system of coupled vibronic oscillators in relevance to the temperature dependences of the relaxation rate. The predicted rate vs. temperature dependences are compared with available experiniental data on the relaxation of Li^+ off-center impurity dipoles in KCl [12].

2. Off-Center Ions

2.1. PSEUDO-JAHN–TELLER HAMILTONIAN

A Hamiltonian accounting for the off-centeredness of impurity ions in crystals has been derived elsewhere [3, 13, 14]. We add up higher-order terms to the electron-vibrational interaction, giving

$$H = \sum_{\tau} E_{\tau} a_{\tau}^{+} a_{\tau} + (1/2) \sum_{\tau} \sum_{i} [P_i^2/M_i + M_i \omega_i^2 Q_i^2]_{\tau} + \\ + \sum_{\alpha\beta} \left[\sum_i b_i(\mathbf{r}_e) Q_i + \sum_{ij} c_{ij}(\mathbf{r}_e) Q_i Q_j + \sum_{ijk} d_{ijk}(\mathbf{r}_e) Q_i Q_j Q_k \right]_{\alpha\beta} a_{\beta}^{+} a_{\alpha}, \tag{1}$$

where a_α and a_β are appropriate electron-annihilation operators.

To tackle the impurity-lattice problem by *Born and Oppenheimer's* theorem we define an *adiabatic Hamiltonian* $H_{AD} \equiv H - \sum_l [P_l^2/2M_l]_\tau$ thereby discarding the nuclear kinetic energy operator. Once eigenvalues $E_{AD}(Q_i)$ and eigenstates $v_{AD}(Q_i)$ of H_{AD} are deduced, the vibronic problem is solved as the electronic eigenvalues are used for potential energies of the nuclear motion at the impurity

$$\left\{ \sum_l [P_l^2/2M_l] + E_{AD}(Q_i) \right\} u_{\text{vib}}(Q_i) = E_{\text{vib}} u_{\text{vib}}(Q_i). \tag{2}$$

Equation (2) yields vibrational and rotational states and energies.

2.2. THREE-DIMENSIONAL (3-D) ROTATOR

There is accumulating evidence that substitutional impurities in alkali halides are driven off-center by the vibronic mixing of nearly-degenerate electronic states at the impurity ion by ungerade vibrations which render the nornial lattice site configuration unstable. For a Li^+ impurity ion, we assume a_{1g} and t_{1u} electron states mixed by a T_{1u} local vibration. We retain the band-off-diagonal terms ($\alpha \neq \beta$) to construct a mixing Hamiltonian:

$$H_{\text{eL}} = \sum_{\alpha\beta_i} \left[\sum_i b_i(\mathbf{r}_e) Q_i + \sum_{ij} c_{ij}(\mathbf{r}_e) Q_i Q_j + \\ + \sum_{ijk} d_{ijk}(\mathbf{r}_e) Q_i Q_j Q_k \right]_{\alpha\beta_i} (|\alpha\rangle\langle\beta_i| + |\beta_i\rangle\langle\alpha|), \tag{3}$$

where $\alpha = a_{1g}$ and $\beta_i = t_{1ui}$ $(i,j,k = x,y,z)$. The eigenvalues and eigenstates of the adiabatic Hamiltonian H_{AD} with H_{eL} from (3) at $c_{ij} = d_{ijk} = 0$ have been derived and discussed elsewhere [14]. We extend these results, solving for *Schrödinger's equation* $H_{\text{AD}}|n, Q_l\rangle = E_{\text{AD}}|n, Q_l\rangle$ via the linear combination $|n, Q_l\rangle = A_\alpha(Q_l)|a_{1g}\rangle + \sum_i A_{\beta i}(Q_l)|t_{1ui}\rangle$. We set $c_{ij} = 0$, since α and β_i are opposite-parity states, as Hamiltonian (1) should conserve parity, aud solve the secular equation to get energy eigenvalues

$$E_\pm(Q_l) = (1/2)\sum_i K_i Q_i^2 \pm \left[\sum_i (b_i Q_i + \sum_{jk} d_{ijk} Q_i Q_j Q_k)^2 + (E_{\alpha\beta}/2)^2\right]^{1/2} \tag{4}$$

where $K_i = M_i\omega_i^2$ and $E_{\alpha\beta} = E_\beta - E_\alpha$ is the interlevel energy gap. Using (4) the vibronic Hamiltonian in (2) is

$$\begin{aligned} H_{\text{vib}} = &-(\hbar^2/2M)\sum_i(\partial^2/\partial Q_i^2) + (1/2)\sum_i K_i Q_i^2 \pm \\ &\pm\left[\sum_i\left(b_i Q_i + \sum_{jk} d_{ijk} Q_i Q_j Q_k\right)^2 + (E_{\alpha\beta}/2)^2\right]^{1/2}. \end{aligned} \tag{5}$$

The adiabatic potential energy surface is mainly controlled by b_l, while mixed $b_i d_{ijk}$ terms create sites of hindered rotation [13].

To illustrate the foregoing statement we choose a T_{1u}-symmetry d-tensor: $d_{ijj} = d_b$, $d_{iii} = d_c$, $d_{ijk} = 0$ otherwise [3, 13]. The first-order tensor b_i is also reduced to a single component and so is the spring constant: $b_i = b$, $M_i\omega_i^2 = M\omega^2$. Introducing a radial coordinate $Q = \sqrt{(\sum_i Q_i^2)}$ neglecting small terms under the root sixth order in Q_i we get:

$$\begin{aligned} E_\pm(Q_l) = &(1/2)M\omega^2 Q^2 \pm \left\{(bQ)^2 + 2b\left[(d_c - d_b)\sum_i Q_i^4 + d_b Q^4\right]\right. \\ &\left.+(E_{\alpha\beta}/2)^2\right\}^{1/2}. \end{aligned} \tag{6}$$

Neglecting all d-terms for a while we get

$$E_\pm(Q)_{d=0} = (1/2)M\omega^2 Q^2 \pm \{(bQ)^2 + (E_{\alpha\beta}/2)^2\}^{1/2} \tag{7}$$

and then minimize in Q to obtain

$$Q_0 = \{(2E_{\text{JT}}/K)[1 - (E_{\alpha\beta}/4E_{\text{JT}})^2]\}^{1/2}, \tag{8}$$

which is the off-centered radius for $4E_{\mathrm{JT}} \geq E_{\alpha\beta}$. Hereafter $E_{\mathrm{JT}} = b^2/2M\omega^2$ stands for *Jahn–Teller's energy* of the off-centering process. Inserting $Q = Q_0$ into Equation (5) we get the lower-sign *vibronic Hamiltonian* for free rotation upon the off-center sphere with $\sum_i Q_i^2 = Q_0^2$:

$$H_{\mathrm{vib}}(d=0) = -(\hbar^2/2I)\Delta_{\Theta,\Phi} - E_{\mathrm{JT}}[1 + (E_{\alpha\beta}/4E_{\mathrm{JT}})^2], \tag{9}$$

where we have introduced spherical coordinates (Q_0, Θ, Φ) for Q_i, as well as the inertial moment of the off-center entity $I = MQ_0^2$.

Rehabilitating the hindering terms, the d-dependent correction to Equation (5) being small, Equation (6) converts to

$$\begin{aligned} E_{\pm}(Q_0) &= \pm\left[(d_c - d_b)\sum_i Q_i^4 + d_b Q_0^4\right](M\omega^2/b) + \\ &\quad + E_{\mathrm{JT}}[(1 \pm 2) - (E_{\alpha\beta}/4E_{\mathrm{JT}})^2], \end{aligned} \tag{10}$$

which defines the potential energy controlling the hindered rotation upon a reorientational sphere. Indeed from (5) we get

$$\begin{aligned} H_{\mathrm{vib}}(d \leq 0) &= -(\hbar^2/2I)\Delta_{\Theta,\Phi} \pm (M\omega^2/b)Q_0^4\{(d_c - d_b)[(\cos\Phi\sin\Theta)^4 + \\ &\quad +(\sin\Phi\sin\Theta)^4 + (\cos\Theta)^4] + d_b\} + E_{\mathrm{JT}}[(1 \pm 2) - \\ &\quad -(E_{\alpha\beta}/4E_{\mathrm{JT}})^2]. \end{aligned} \tag{11}$$

We see that the rotation is hindered by fourth power terms in Q_i which arise from the asymmetry between d_c and d_b.

Reorientational sites are the minima of $E_L(Q_0)$ with respect to Q_i, while the maxima are saddle points in between. From Equation (10), the spatial curvature of the angle-dependent part of vibronic potential along Q_i gives rise to the following *effective force constants* $K_{\mathrm{eff}} = (1/3)\sum_{i=x,y,z}(\partial^2 E_{\pm}/\partial Q_i^2)$:

$$K_{\mathrm{min}} = +4(I\omega^2)[(d_b - d_c)/b] \tag{12}$$

at the bottom of a reorientational well and

$$K_{\mathrm{max}} = -4(I\omega^2)[(d_b - d_c)/b] \tag{13}$$

at a saddle point between two wells. Introducing $K = M\omega^2 \equiv M\omega_{\mathrm{bare}}^2$, $K_{\mathrm{min}} = M\omega_{\mathrm{renII}}^2$, we define a *renormalized rotational frequency*

$$\omega_{\mathrm{renII}} = \omega_{\mathrm{bare}}[4(d_b - d_c)/b]^{1/2}Q_0 = \omega_{\mathrm{renI}}[8E_{\mathrm{JT}}(d_b - d_c)/bK]^{1/2}. \tag{14}$$

$$\omega_{\mathrm{renI}} = \omega_{\mathrm{bare}}[1 - (E_{\alpha\beta}/4E_{\mathrm{JT}})^2]^{1/2} \tag{15}$$

is the *renormalized off-centering frequency* [14] with $\omega_{\text{bare}} \equiv \omega$ standing for the *bare vibrational frequency*.

2.3. TWO-DIMENSIONAL (2-D) ROTATOR

We next simplify the eigenvalue problem by freezing-in one of the vibrational coordinates: By setting $\Theta = \pi/2$, we consider the hindered rotation of an off-centered impurity in the equatorial plane. Thereby we convert the original symmetry from octahedral O_h (T_{1u} symmetry-breaking mode) pertinent to an isolated impurity to axial C_{4v} (E_u symmetry-breaking mode) pertinent to an impurity confined to a plane. For example, an O_h to C_{4v} symmetry lowering results from the immobilization of a vibrating halogen pair along the Q_z-axis as an F center seats in the [001] site [14]. The 2-D rotational Hamiltonian reads

$$\begin{aligned} H_{\text{vib}}(2D) = & -(\hbar^2/2I)(\partial^2/\partial\Phi^2) + (I\omega^2/b)\{-(d_b - d_c)[(1/4) \\ & \times(3 + \cos 4\Phi)] + d_b\}Q_0^2 + \\ & +(1/2)[(1 \pm 2)(b^2/M\omega^2) - M\omega^2 E_{\alpha\beta}^2/4b^2]. \end{aligned} \quad (16)$$

There are eight extrema at $\Phi = n(\pi/4)$ with $n = 0$, 1 to 7. For $d_b - d_c > 0$, four of the extrema are maxima at $n = 0, 2, 4, 6$ and four are minima at $n = 1, 3, 5, 7$, for the lower-sign (–) branch. The minima are metastable rotational sites, while the maxima form barriers in-between. For the upper-sign (+) branch, the extrema are similar, though the minima and maxima interchange each other.

The azimuthal dependence of the vibronic potential energy in Equation (10) is depicted in Figure 1(a,b) showing both the off-center to on-center maximum and reorientational barriers. We also verified the off-center displacements and cos(4Φ) rotational potentials by extended Hückel calculations [15], as in Figures 2 and 3, respectively.

The *reorientational barrier* E_{BII}, viz. the energy difference between a maximum and a minimum on E_- (respectively E_+), is

$$E_{\text{BII}} = I\omega_{\text{renII}}^2/8,$$

while the *energy splitting* of the two branches, the gap between a minimum on the upper branch and a maximum on the lower branch, is

$$E_{12} = 4E_{\text{BII}}[d_c/(d_b - d_c)] + E_{\text{JT}}.$$

3. Mathieu's Equation

Sorting out the constant terms in Equation (1), we rewrite the 2-D equation

$$-(\hbar^2/2I)(\partial^2 u/\partial\Phi^2) + 2B_\pm \cos(4\Phi)u + (C_\pm - E)u = 0 \quad (17)$$

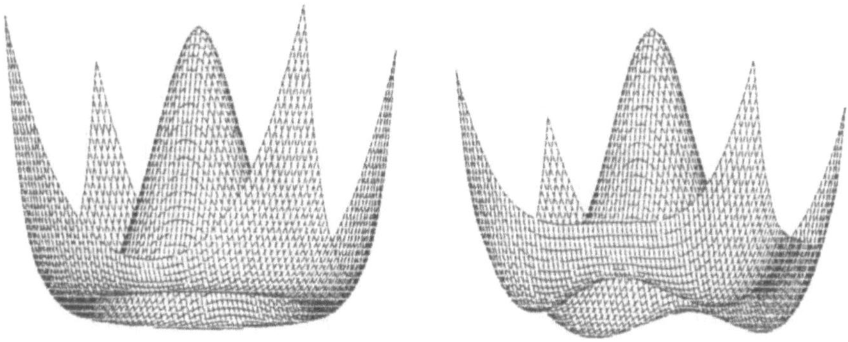

Figure 1. Relief of the lower adiabatic potential energy surface (APES) as calculated using Equation (6) at $Q_z = 0$ with KCl parameters estimated from the coupling constants model of Ref. [5]. Part (a) shows the APES resulting from first-order electron-mode mixing (left panel), while in part (b) third-order mixing terms are also incorporated (right panel).

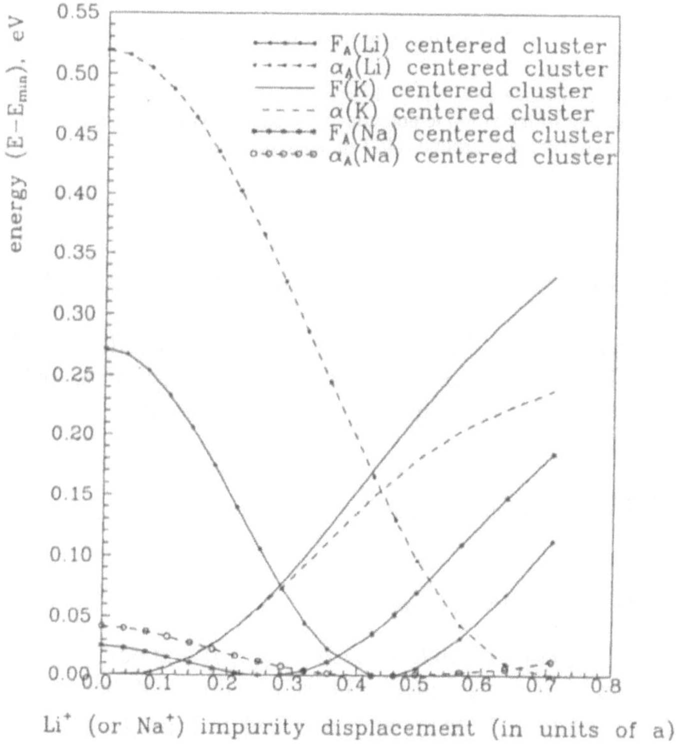

Figure 2. Extended Hückel calculations of off-center displacements along $\langle 110 \rangle$ of Li^+ and Na^+ impurities in KCl: Calculated total energies E for both F_A(Li)- or F_A(Na)- and α_A(Li)- or α_A(Na)- centered clusters against the impurity displacement. Lines without symbols are potential energies of impurity-free clusters. $a = 3.147$ Å.

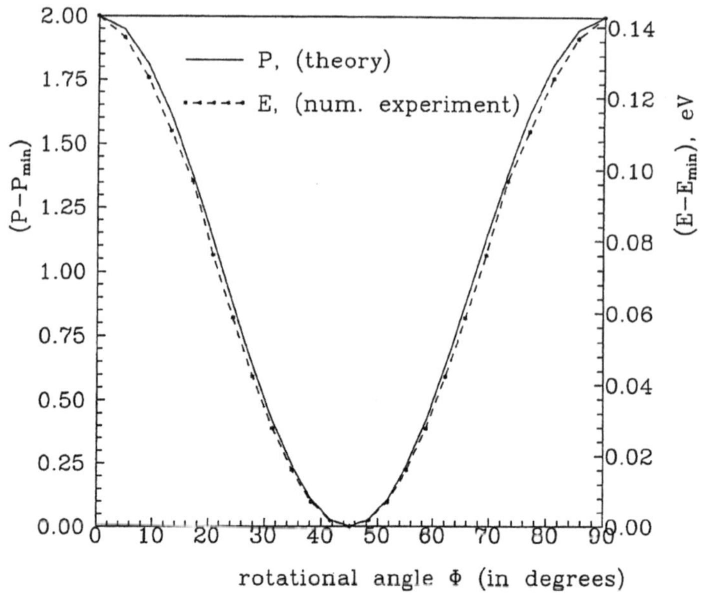

Figure 3. Extended Hückel calculations of the reorientation of off-centered Li^+ ion: Comparison of the angular part of the vibronic potential $P = \cos 4\Phi$ with the total electronic energy E against the rotational angle Φ around $\langle 001\rangle$ in plane parallel to (001).

$$B_\pm = \mp(E_{\mathrm{BII}}/4),$$

$$C_\pm = \mp(1/2)E_{\mathrm{BII}}[(3d_c + d_b)/(d_b - d_c)] + E_{\mathrm{JT}}[(1 \pm 2) - (E_{\alpha\beta}/4E_{\mathrm{JT}})^2],$$

and $u = u(\Phi)$. Equation (17) is Mathieu's equation and its solutions are Mathieu functions [16–20]. The conventional form of Mathieu's equation is: $\mathrm{d}^2Y/\mathrm{d}z^2 + (a - 2q\cos 2z)Y = 0$, the relationship between different notations being $z = 2\Phi$, $Y = u$,

$$q = 2(B_\pm I/\hbar^2) = \mp(2E_{\mathrm{BII}}/\hbar\omega_{\mathrm{renII}})^2. \tag{18}$$

3.1. Mathieu Eigenstates and Eigenvalues

There are two types of periodic solution, even $ce_m(z,q)$ and odd $se_m(z,q)$ with eigenvalues $a_m(q)$ and $b_m(q)$ respectively. These functions will be normalized so that $(1/\pi) \times \int_0^{2\pi} Y_m^2(x)\,\mathrm{d}x = 1$. At $|q| \ll 1$, $ce_0(z,q) \simeq 1$, $ce_m(z,q) \simeq \cos(mz)$, $se_m(z,q) \simeq \sin(mz)$; $a_m \simeq m^2$, $b_m \simeq m^2$.

$$E_{am,bm}(q) = (\hbar^2/2I)a_m(q), b_m(q) + C_\pm. \tag{19}$$

The eigenvalue expansions in q are known at small and large $|q|$.

3.2. Rotational Bands

Mathieu's eigenvalue Equations (17) define allowed energy bands corresponding to $q < 0$ (upper-sign branch) and $q > 0$ (lower sign branch), respectively, as in Figure 4 (cf. [4]). These bands display the general features known from textbooks on band theory, in particular they widen as m is increased at fixed q. When $q = (2E_{\rm BII}/\hbar\omega_{\rm renII})^{1/2}$ is increased; however, the allowed bands narrow tending to turn into single levels in the limit of too high barriers. The borderlines of the allowed regions in the energy-versus-q plane are described by Mathieu's parameters $a_m(q)$ or $b_m(q)$, or, alternatively, by their corresponding periodic eigenfunctions $ce_m(z,q)$ or $se_m(z,q)$ at consecutive integral $m-1$, m $(m = 1,2,3,4,\ldots)$, while each respective line in the interior is generated by pairs of Mathieu functions $ce_m(z,q)$ and $se_m(z,q)$ at an intermediate nonintegral m. Mathieu's eigenfunctions $ce_m(z,q)$ and $se_m(z,q)$ each have m zeros in the $0 < z < \pi/2$ interval, and either a basic function or its derivative is vanishing at $z = \pi/2$ (cf. [16]). Within an allowed energy band, Mathieu's eigenstates and eigenvalues are functions of the wavenumber $k = \pi m$ where m runs from 0 to 1 for the first Brilloum zone [4].

We distinguish between two cases corresponding to the upper- and lower-sign branches in Equation (10), respectively: In the-former 'negative q' case, the allowed bands are borderlined as follows: (a_0, a_1), (b_1, b_2), (a_2, a_3), (b_3, b_4), etc., while in the latter 'positive q' case the border pairs are (a_0, b_1), (a_1, b_2), (a_2, b_3), (a_3, b_4), etc. When q is large, each pair of borderlities approach one another and we can attach to the bands e.g. linear combinatious of borderline eigenstates of the form, respectively: $(1/2)(ce_0+ce_1)$, $(1/2)(se_1+se_2)$, $(1/2)(ce_2+ce_3)$, $(1/2)(se_3+se_4)$, etc. for the 'negative q', and of the form: $(1/2)(ce_0+se_1)$, $(1/2)(ce_1+se_2)$, $(1/2)(ce_2+se_3)$, $(1/2)(ce_3+se_4)$, etc. for the 'positive q'. Each linear combination is again an eigenstate whose energy falls within a 'squeezed band'. It is obvious that the allowed bands in the 'negative q' case are of definite alternating parities, even, odd, etc. in the increasing m order, while the 'positive q' bands are mixed parity.

3.3. Approximate Well-Bottom Solution

The vibronic Equation (17) converts to a harmonic-oscillator type at $|\Phi| < \pi/4$ if we introduce $\Phi' = \pi/4 - \Phi$ and expand $\cos(4\Phi') \sim 1 - (1/2!)(4\Phi')^2$ around a well bottom at $\Phi = \pi/4$. Introducing the dimensionless coordinate $X = (I\omega_{\rm renII}/\hbar)^{1/2}\Phi$, the lower sign equation turns into

$$\begin{aligned} Eu = & -(\hbar\omega_{\rm renII}/2)(\partial^2 u/\partial^2 X) + (\hbar\omega_{\rm renII}/2)[X - (I\omega_{\rm renII}/\hbar)^{1/2}(\pi/4)]^2 u - \\ & -\{E_{\rm BII}(d_c+d_b)/(d_b-d_c) + E_{\rm JT}[1+(E_{\alpha\beta}/4E_{\rm JT})^2]\} \end{aligned} \quad (20)$$

with eigenvalues and eigenstates, respectively,

$$E_n = (n+1/2)\hbar\omega_{\rm renII} - E_{\rm BII}(d_b+d_c)/(d_b-d_c) + E_{\rm JT}[1+(E_{\alpha\beta}/4E_{\rm JT})^2] \quad (21)$$

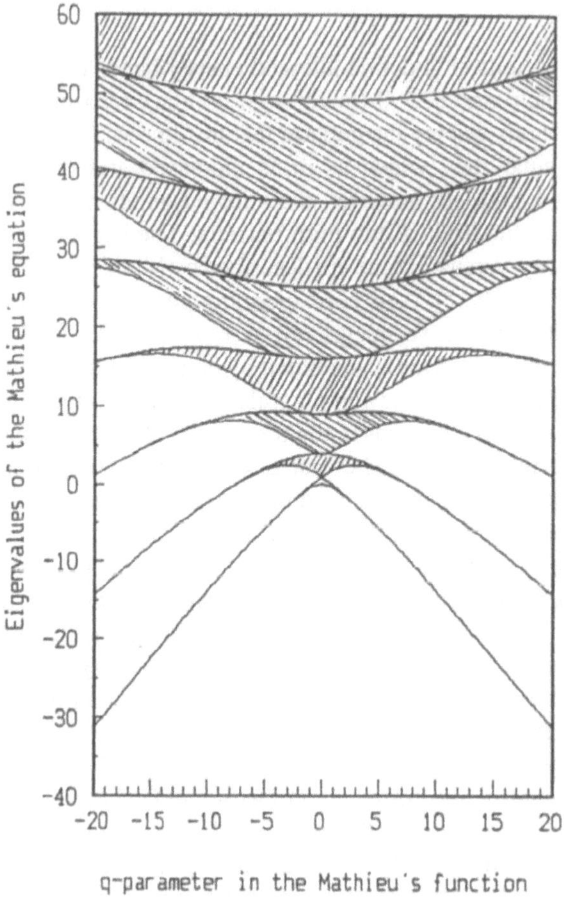

Figure 4. Allowed rotational energy hands (hatched) corresponding to $q < 0$ (upper-sign branch) and $q > 0$ (lower sign branch), calculated by Mathieu's eigenvalue expansions.

$$u_n(\Phi) = [(I\omega_{\text{renII}}/\hbar)^{1/2}/\pi^{1/2}2^n n!]^{1/2} H_n(X)\exp(-X^2/2) \tag{22}$$

where $H_n(X)$ stands for the Hermite polynomial of nth order. The harmonic-oscillator eigenvalues approximate for the rotational energy spectrum near the vibronic ground state. Another related quantity is the *lattice reorganizational energy* which defines as one expended for creating neighboring reorientational sites:

$$E_{\text{RII}} \equiv E_+(\Phi = -\pi/4) - E_+(\Phi = +\pi/4) = \pi^2 E_{\text{BII}}. \tag{23}$$

4. Reaction Rate

We next turn to the relaxation time, inversely proportional to the relaxation rate of an off-centered impurity. The qualitity is characteristic of the system as it relaxes to thermal equilibrium following an initial disturbance which has overpopulated some reorientational sites at the expense of others [21]. We calculate the relaxation rate by *Christov's Reaction-Rate Theory* [10]. For this purpose we rewrite the two-site rate expression in terms of exact Mathieu eigenvalues. We calculate configurational tunneling probabilities using the exact eigenfunctions. Whereas impurity tunneling reorientation in the harmonic approximation has been described previously [6], our Mathieu-based analysis is novel.

4.1. GENERAL TWO-SITE RATE

Whereas the reorientational motion of an off-centered impurity is actually controlled by a multiwell potential surface along the *relaxation coordinate*, a two-site analysis may be applicable leading to $t_{\rm rel} = (gk_{12})^{-1}$ where k_{12} is the two-site relaxation rate and g is an effective number of equivalent wells depending on the symmetry of dipole, lattice, and external electric field.

The *reaction-rate approach* (RRA) is based on an occurrence-probability formula incorporating both classical and quantal effects [10, 11]:

$$k_{12}(T) = k(T)(k_BT/h)(Z_1^{\#}/Z_1)\exp(-E_B/k_BT). \tag{24}$$

$$k(T) = \sum_n W_n(E_n)\exp(-\epsilon_n/k_BT)\delta(\epsilon_n/k_BT) \tag{25}$$

is a correction factor to the rate. $k(T)$ accounts for nonadiabaticity and quantal effects. Z_1 is the complete partition function of the initial state, assumed to be in thermal equilibrium, $Z_1^{\#}$ is the initial-state partition function of the *nonreactive (accepting) modes* obtained by excluding the *relaxation-promoting mode* from the domain of all the modes entering into Z_1. E_B is the barrier height between the two neighboring reorientational sites at 1 and 2 along the relaxation coordinate. The motion along the latter is quantized with n, E_n being an allowed energy level; $\epsilon_n = E_n - E_B$ is the excess energy relative to the barrier peak, while $\delta\epsilon_n = E_{n+1} - E_n$ is the separation between subsequent energy levels.

Calculating the quantum correction $k(T)$ often reduces to an uni-dimensional problem which has been solved when the relaxation coordinate is separable dynamically from the domain of all mode coordinates. $W_n(E_n)$, the transition probability at energy E_n, is assumed to factorize in concert with Condon's approximation [10]:

$$W_n(E_n) = W_{Ln}(E_n)W_{en}(E_n) \tag{26}$$

into W_L, the probability for configurational rearrangement, and W_e, the probability for a change of the electronic state at the crossover configuration $\{Q_c\}$ between initial and final electronic states ($|a_{1g}\rangle$ and $|t_{1u}\rangle$), under an energy conservation condition. Transitions with $W_e = 1$ are called '*adiabatic*', those with $W_e < 1$ are '*not adiabatic*', while those with $W_e \ll 1$ are '*nonadiabatic*'. We shall mostly deal with the adiabatic reorientational rate assuming $W_e(E_n) = 1$ at all rotational levels.

Assuming harmonicity of lattice vibrations, the *promoting-mode* contribution factorized out of Z_1 by virtue of separability gives

$$Z_1^{\#}/Z_1 = 2\sinh(h\nu_r/2k_BT), \tag{27}$$

where ν_r (ω_r) is the effective vibrational (angular) frequency along the relaxation coordinate. Inserting into (24) one obtains

$$k_{12}(T) = k(T)(2k_BT/h\nu_r)\sinh(h\nu_r/2k_BT)\nu_r\exp(-E_B/k_BT) \tag{28}$$

in the harmonic mode approximation. It is noteworthy that Equation (28) predicts a non-vanishing zero-point rate of magnitude

$$k_{12}(0) = \nu_r W_{eo}(E_0)W_{L0}(E_0) \tag{29}$$

at $E_0 = (1/2)h\nu_r$.

In view of the strong anharmonicity of hindered rotation, however, Equations (27) and (28) should be changed. For a hindered rotation, the eigenvalue spectrum being

$$E_{a/b,n} = (\hbar^2/2I)a_n, b_n = (\hbar^2/2I)[n^2 + c_n(q)],$$

where $c_m(q)$ is the correction either to a_m or to b_m due to the hindering potential, we get

$$\begin{aligned} Z_1^{\#}/Z_1 &= 1/\sum_{m=0}^{\infty}\exp(-\hbar^2[m^2 + c_m(q)]/2Ik_BT) = \exp(\hbar^2 c_0(q)/2Ik_BT) \div \\ &\div \left\{1 + \sum_{m=1}^{\infty}\exp(-\hbar^2[m^2 + c_m(q) - c_0(q)]/2Ik_BT\right\} \end{aligned} \tag{30}$$

summing over the eigenspectrum of the particular rotating mode if separable from the domain of all coupled modes.

The relaxation rate obtains by combining Equations (24) and (30):

$$\begin{aligned} k_{12}(T) &= k(T)(k_BT/h\nu_r)\exp(\hbar^2 c_0(q)/2Ik_BT) \div \\ &\div \left\{1 + \sum_{m=1}^{\infty}\exp(-\hbar^2[m^2 + c_m(q) - c_0(q)]/2Ik_BT)\right\} \\ &\times \nu_r\exp(-E_B/k_BT). \end{aligned} \tag{31}$$

In so far as $c_m(q)$ are small corrections to the basic m^2 terms in a_m and b_m a zero-point rate of magnitude

$$k_{12}(0) = (\hbar/4\pi I)[1 + c_1(q) - c_0(q)]W_{e0}(E_0)W_{L0}(E_0) \tag{32}$$

is predicted at $E_0 = (\hbar^2/2I)c_0(q)$.

4.2. Configurational Tunneling Probability

The transition probability along the augular coordinate Φ is

$$W_{if}(E_n) = 4\pi^2|V_{fi}|^2\sigma_i(E_n)\sigma_f(E_n), \tag{33}$$

[10] where the matrix element V_{fi} is calculated using initial and final state wave-functions u_i and u_f, respectively, as

$$V_{fi} = (-\hbar^2/2I)[u_f^*(\mathrm{d}u_i/\mathrm{d}\Phi) - u_i(\mathrm{d}u_f^*/\mathrm{d}\Phi)]|_{\Phi=\Phi_c}. \tag{34}$$

Here $\sigma_i(E_n)$ and $\sigma_f(E_n)$ are the corresponding densities of initial and final states (DOS).

4.2.1. *Exact Solution*

The exact rotational eigenstates are described by Mathieu's functions, e.g. $u_i(\Phi) = Y_i(\Phi + \pi/4)$, $u_f(\Phi) = Y_f(\Phi - \pi/4)$, while DOS are

$$\sigma(E_{a/b,n}) = \mathrm{d}n/\mathrm{d}E_{a/b,n} = (2I/\hbar^2)(\mathrm{d}n/\mathrm{d}a_n, b_n), \tag{35}$$

since the rotational eigenspectrum is $E_{a/b,n} = (\hbar^2/2I)a_n, b_n$. Now using $Y_{i/f}(\Phi, q)$, Equation (34) at the saddle point $\Phi = 0$ reads:

$$\begin{aligned} V_{if} &= (-\hbar^2/2I)\{Y_f(-\pi/4, q)[\mathrm{d}Y_i(\Phi, q)/\mathrm{d}\Phi]|_{\Phi=\pi/4} - \\ &\quad -Y_i(\pi/4, q)[\mathrm{d}Y_f(\Phi, q)/\mathrm{d}\Phi]|_{\Phi=-\pi/4}\} \end{aligned} \tag{36}$$

inasmuch as for Φ and q real $Y_{i/f}(\Phi, q)$ is a real function of Φ.

Series expansions of Mathieu functions $ce_m(z, q)$ and $se_m(z, q)$ are available. Using expansions in q, we calculate finite-valued saddle-point functions at $z = \pm\pi/2$ ($\Phi = \pm\pi/4$). We see that each of the periodic functions $ce_m(z, q)$ and $se_m(z, q)$ is either itself vanishing or has a vanishing derivative at $z = \pi/2$. For this reason $ce_m(z, q)$ and $se_m(z, q)$ are not the appropriate transitional eigenstates but such may be constructed as linear combinations.

We see that (36) type intraband transitions are non-vanishing if these occur in bands of the upper-sign branch of (10), such as by linear combinations

$$u(z, q) = (1/2)[ce_{m-1}(z, q) + ce_m(z, q)] \quad \text{for } m \text{ odd},$$

$$u(z,q) = (1/2)[se_{m-1}(z,q) + se_m(z,q)] \quad \text{for } m \text{ even},$$

whereas lower-sign branch bands, e.g. by linear combinations

$$u(z,q) = (1/2)[ce_{m-1}(z,q) + se_m(z,q)] \quad \text{for all } m,$$

do not promote intraband transitions.

However, the above linear combination eigenstates corresponding to energy eigenvalues in the middle of the allowed bands, they do not adequately account for the interior. To improve the method we make the following proposition: we attach an integer n to label a band where n is odd for (a_{n-1}, a_n) and even for $(b_{n-1} b_n)$, and let m be a running number $0 \leq m \leq 1$. In so far as no eigenfunctions $ce_m(z,q)$ and $se_m(z,q)$ of intraband states at noninteger m are known we form such by linear combinations of band-edge states:

$$\begin{aligned} ce^n_m(z,q) &= (1-m)ce_{n-1}(z,q) + mce_n(z,q) \quad (n \text{ odd}) \\ se^n_m(z,q) &= (1-m)se_{n-1}(z,q) + mse_n(z,q) \quad (n \text{ even}) \end{aligned} \tag{37}$$

with intraband eigenvalues

$$E^n_m(q) = (1-m)E_{n-1} + mE_n = (\hbar^2/2I)a^n_m(q),$$

$$a^n_m(q) = (1-m)a_{n-1}(q) + ma_n(q) \quad (n = 1, 3, 5, \ldots, \text{odd})$$

$$E^n_m(q) = (1-m)E_{n-1} + mE_n = (\hbar^2/2I)b^n_m(q),$$

$$b^n_m(q) = (1-m)b_{n-1}(q) + mb_n(q) \quad (n = 2, 4, 6, \ldots, \text{even}), \tag{38}$$

respectively. We redefine the transition probabilities (33) and (34)

$$W_{Ln}(E^n_m) = (2\pi)^2 |V^n_m(q)|^2 (\mathrm{d}m/\mathrm{d}E^n_m)^2,$$

so as to incorporate

$$\begin{aligned} V^n_m(q) &= -(\hbar^2/2I\pi)4(1-m)ce_{n-1}(z,q)[m\mathrm{d}ce_n(z,q)/\mathrm{d}z]|_{z=\pi/2} \quad (n \text{ odd}) \\ V^n_m(q) &= +(\hbar^2/2I\pi)4(1-m)se_{n-1}(z,q)[m\mathrm{d}se_n(z,q)/\mathrm{d}z]|_{z=\pi/2} \quad (n \text{ even}). \end{aligned} \tag{39}$$

The resulting probabilities (33) are maximal in the middle of a band at $m = 1/2$, $(W^{\max}_{Ln})$, and vanish at its edges at $m = 0$ and $m = 1$:

$$W_{Ln}(E^n_m) = 64 \times \begin{cases} |(1-m)ce_{n-1}(z,q)[m\mathrm{d}ce_n(z,q)/\mathrm{d}z]|^2_{z=\pi/2}(\mathrm{d}m/\mathrm{d}a^n_m)^2 \\ |(1-m)se_{n-1}(z,q)[m\mathrm{d}se_n(z,q)/\mathrm{d}z]|^2_{z=\pi/2}(\mathrm{d}m/\mathrm{d}b^n_m)^2 \end{cases}. \tag{40}$$

4.2.2. *Well-Bottom Solution*

A configurational-tunneling probability has otherwise been calculated quantum-mechanically for parabolic wells with $E_{12} \ll E_B$ using harmonic-oscillator wavefunctions to derive V_{fi} by means of Equation (34), though in other cases quasi-classical techniques have been applied to get a result. For underbarrier transitions at $E_n \ll E_B$ and isoenergetic parabolic wells whose bottoms lie at the same energy as in the reorientation under consideration, *Christov's quantum-mechanical expressions* has been derived [11]:

$$W_L(E_n) = \pi[F_{nn}(q_0, q_c)/(2^n n!)]^2 \exp(-E_R/h\nu_r) \tag{41}$$

where $n = n_1 = n_2$ is the vibronic quantum number in the initial (final) electronic state, $q_0 = q_2 - q_1$ is the interwell separation along the relaxation coordinate, and

$$\begin{aligned} F_{nn}(q_0, q_c) \; &= \; q_0 H_n(q_c) H_n(q_c - q_0) - 2n H_{n-1}(q_c) H_{n-1}(q_c - q_0) + \\ &\quad +2n H_n(q_c) H_{n-1}(q_c - q_0) \end{aligned} \tag{42}$$

with $H_n(q)$ standing for Hermite's polynomials. For overbarrier transitions at $E_n \gg E_B$ one sets $W_L = 1$. One should apply specific techiuques to deal with transitions near the barrier top $E_n \sim E_B$. We recall that $q = (M\omega_{\text{ren}}/\hbar)^{1/2}Q$ or $q = (I\omega_{\text{ren}}/\hbar)^{1/2}\Phi$ stand for the scaled vibrational or rotational coordinates. The well-bottom relaxation rate is obtained by insertiug in Equations (25) and (28).

4.3. Relaxation Rates

4.3.1. *Mathieu-Based Rate*

We shall next use the intraband transition probabilities to derive a practical expression for a relaxation rate pertinent to a system of allowed bands. Integrating over a rotational energy band and summing up for all the bands, we get a transition rate:

$$k_{12}(T) = (Z_1^{\#}/Z_1) \sum_{n=1}^{\infty} \int_{E_{n-1}}^{E_n} W_{en}(E_m^n) W_{LNn}(E_m^n) \exp(-E_m^n/k_B T)\, \mathrm{d}E_m^n/h, \tag{43}$$

where W_{LNn} are the normalized configurational probabilities (40),

$$\begin{aligned} &W_{LNn}(E_m^n) \\ &= 64N \times \begin{cases} |(1-m)ce_{n-1}(z,q)[mdce_n(z,q)/\mathrm{d}z]|^2_{z=\pi/2} (\mathrm{d}m/\mathrm{d}a_m^n)^2 \\ |(1-m)se_{n-1}(z,q)[mdse_n(z,q)/\mathrm{d}z]|^2_{z=\pi/2} (\mathrm{d}m/\mathrm{d}b_m^n)^2 \end{cases}, \end{aligned} \tag{44}$$

where the normalization factor is defined by

$$N^{-1} = 2\sum_{n=1}^{\infty}\int_0^1 W_{Ln}(E_m^n)\,\mathrm{d}m = 128\sum_{n=1}^{\infty}\int_0^1 \mathrm{d}m[m(1-m)]^2 \times$$
$$\times\left\{\begin{array}{l} |ce_{n-1}(z,q)[dce_n(z,q)/\mathrm{d}z]|^2_{z=\pi/2}(\mathrm{d}m/\mathrm{d}a_m^n)^2 \\ |se_{n-1}(z,q)[dse_n(z,q)/\mathrm{d}z]|^2_{z=\pi/2}(\mathrm{d}m/\mathrm{d}b_m^n)^2 \end{array}\right. . \tag{45}$$

The DOS of a hindered rotator will be derived from Equation (35): $\mathrm{d}m/\mathrm{d}a_m(q) = 1/[a_n(q) - a_{n-1}(q)]$, $\mathrm{d}m/\mathrm{d}b_m(q) = 1/[b_n(q) - b_{n-1}(q)]$. We extend the definition of E_m^n to negative m:

$$E_m^n = E_{n-1} + m(E_n - E_{n-1}) \quad (0 < m < 1)$$

$$E_m^n = E_n + m(E_n - E_{n-1}) \quad (-1 < m < 0)$$

and formulate the rate accordingly

$$k_{12}(T) = (16\hbar/\pi I)N(Z_1^{\#}/Z_1)\times$$
$$\times\sum_{n=1}^{\infty}\left\{\exp(-E_{n-1}/k_BT)\int_0^1 \mathrm{d}mW_{en}(E_m^n)\exp(-m[E_n - E_{n-1}]/k_BT)\times\right.$$
$$\times[(1-m)m]^2 + \exp(-E_n/k_BT)\int_{-1}^0 \mathrm{d}mW_{en}(E_m^n)\times$$
$$\left.\times\exp(-m[E_n - E_{n-1}]/k_BT)[(1-m)m]^2\right\}\times$$
$$\times\left\{\begin{array}{l} [a_n(q) - a_{n-1}(q)]^{-1}|ce_{n-1}(z,q)[\mathrm{d}ce_n(z,q)/\mathrm{d}z]|^2_{z=\pi/2} \\ [b_n(q) - b_{n-1}(q)]^{-1}|se_{n-1}(z,q)[\mathrm{d}se_n(z,q)/\mathrm{d}z]|^2_{z=\pi/2} \end{array}\right. . \tag{46}$$

For $W_{en}(E_m^n) = 1$ in an adiabatic process, the integration over m can be carried out:

$$k_{12}(T) = (16\hbar/\pi I)N(Z_1^{\#}/Z_1)\times$$
$$\times\sum_{n=1}^{\infty}(2/g_n)\{(1/g_n^2)[1 + (6/g_n) + (12/g_n^2)]\exp(-E_n/k_BT) +$$
$$[2 - (6/g_n) + (13/g_n^2) - (18/g_n^3) + (12/g_n^4)]\exp(-E_{n-1}/k_BT)\}\times$$
$$\times\left\{\begin{array}{l} [a_n(q) - a_{n-1}(q)]^{-1}|ce_{n-1}(z,q)[\mathrm{d}ce_n(z,q)/\mathrm{d}z]|^2_{z=\pi/2} \\ [b_n(q) - b_{n-1}(q)]^{-1}|se_{n-1}(z,q)[\mathrm{d}se_n(z,q)/\mathrm{d}z]|^2_{z=\pi/2} \end{array}\right. , \tag{47}$$

where $g_n = (E_n - E_{n-1})/k_B T$. The partition function derives from

$$(Z_1^{\#}/Z_1)^{-1} = \sum_{n=1}^{\infty} \int_{E_{n-1}}^{E_n} \exp(-E_m^n(q)/k_B T)(\mathrm{d}m/\mathrm{d}E_m^n)\,\mathrm{d}E_m^n$$
$$= \sum_{n=1}^{+\infty}(2/g_n)[1 - \exp(-g_n)]\exp(-E_{n-1}/k_B T). \tag{48}$$

At low temperature, $k_{12}(T) \simeq (Z_1^{\#}/Z_1)^{-1}(4/g_1)\exp(-E_0/k_B T)$, while $(Z_1^{\#}/Z_1)^{-1} \simeq (2/g_1)\exp(-E_0/k_B T)$. We get a zero-point rate

$$k_{12}(0) = (32\hbar/\pi I)N[a_1(q) - a_0(q)]^{-1}|ce_0(z,q)[\mathrm{d}ce_1(z,q)/\mathrm{d}z]|_{z=\pi/2}^2 \tag{49}$$

which accounts for the contribution of the lowest allowed band. N is derived through double integration over the $m > 0$ states:

$$N^{-1} = (128/30)\sum_{n=1}^{\infty}\left\{\begin{array}{l}|ce_{n-1}(z,q)[\mathrm{d}ce_n(z,q)/\mathrm{d}z]|_{z=\pi/2}^2(a_n - a_{n-1})^{-2}\\ |se_{n-1}(z,q)[\mathrm{d}se_n(z,q)/\mathrm{d}z]|_{z=\pi/2}^2(b_n - b_{n-1})^{-2}.\end{array}\right. \tag{50}$$

4.3.2. *Well-Bottom Rate*

It should be stressed that the harmonic oscillator-based rate formulae apply to a system of discrete or nearly discrete energy levels such as the rotational bands at large q. At smaller gaps $E_{12} \ll E_B$, the harmonic rate obtains by combining Equations (25), (28), (41), and (42). At larger gaps $E_{12} \geq E_B$, quasiclassical techniques may be advised if no quantum-mechanical formulae are available for the configurational terms. But a quantum-mechanical formula should always be applied to the relaxation from the ground-state vibronic level. Accordingly, RRA predicts a constant zero-point harmonic-oscillator rate at low temperature given by

$$k_{12}(0) = W_{e0}(E_0)(E_{\mathrm{RII}}/\hbar)\exp(-E_{\mathrm{RII}}/\hbar\omega_{\mathrm{renII}}) \tag{51}$$

at $E_0 = (1/2)\hbar\omega_{\mathrm{renII}}$, to be compared with the Mathieu rate (49).

5. Comparison with Experiment

One of the few experimental temperature dependences of relaxation times by Li^+ impurity dipoles available so far come from paraelectric resonance [22] KCl data [12]. Generally, the observed dependence is composed of two branches with thermal slopes of roughly 7 meV below 5 K and 70 meV above 12 K. It is tempting to associate the latter slope with the statistical drop of $Q(T)$ and $\omega_{\mathrm{renI}}(T)$ with the temperature T, such as discussed in Section 6. The statistical signature may result from siguificant pair interactions within the Li^+ ensemble, as evidenced by recent dielectric susceptibility data [9]. Alternatively, low temperature relaxation rate branches proportional to T have been attributed to one-phonon assisted transitions [21], not accounted for by the present rate theory which conserves the phonon occupation number. However, horizontal low temperature branches have been observed in optical experiments with other off-center ions, such as Ag^+ in RbBr [7, 8] and F^- in NaBr, KI, RbI [23].

Another plausible possibility is that the lower-temperature branch reflects the thermal activation to a second rotational band. If so, Li^+ rotation above 12 K occurs in the lowest-excited band which is odd-parity, while the ground-state band is even-parity. This interpretation is supported by the infrared spectrum of $KCl:Li^+$ where a band, presumably rotational, occurs near 47 $cm^{-1} \simeq 6$ meV and shifts in proportion to 2/3 as the Li^+ motion turns from a 3-D character in uncoloured crystals to a quasi-2D character in crystals with F_A centers [14]. The rotational optical bands associated with off-center impurities in alkali halides will be dealt with in a subsequent work.

The experimental Li^+ paraelectric data are shown in Figure 5 [12]. It also shows our Mathieu-based relaxation times as computed using Equation (47) with integrals over m calculated numerically. The computed temperature dependences exhibit a constant lower-temperature branch followed smoothly by an Arrhenius portion at higher temperatures. An alternative harmonic rate calculation also shown therein has a similar appearance. A different harmonic rate fit to the paralectric data has been reported earlier [23]. The well-bottom calculations are seen to be generally in concert with the Mathieu-based ones which is not unexpected, since the experiments have been made at low temperature ($k_BT \ll h\nu_\tau$) where only the ground state energy level matters.

6. Free Energy of Hindered Rotators

Equation (34), as well as alternative rate formulae based on well-bottom harmonic-oscillator solutions, assumes temperature-independent vibrational frequencies. Regarding the manifold of off-centered ions in a crystal as a statistical ensemble of anharmonic oscillators coupled weakly through pair interactions, the free energy of the system is $F = -k_BT \ln Z$. For brevity, we use the vibronic eigenvalues $E_\pm(Q)$ of Equation (10) at $d \equiv 0$ accounting for the main effect of radial off-centered displacements [2]:

$$\begin{aligned} F &= -k_BT \ln\{2[1+\cosh([(E_{\alpha\beta}/2)^2+(bQ)^2]^{1/2}/k_BT)]\} \\ &\quad +(1/2)KQ^2+U(R,Q), \end{aligned} \tag{52}$$

where Q is the radial coordinate, $K = M\omega_{\text{bare}}^2$, $U(R,Q)$ is the pair interaction at R. For a weak dipole–dipole coupling of vibronic oscillators [14],

$$U(R,Q) \equiv u(R)v(Q) = [R^2(\mathbf{p}_1 \cdot \mathbf{p}_2) - 3(\mathbf{p}_1 \cdot \mathbf{R})(\mathbf{p}_2 \cdot \mathbf{R})]/kR^5 \times$$

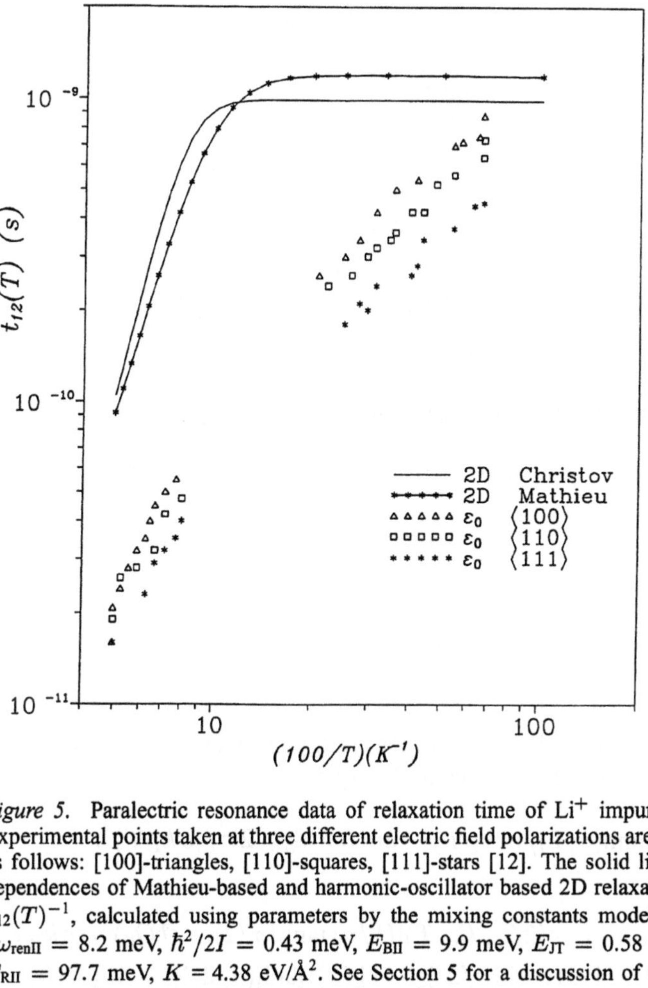

Figure 5. Paralectric resonance data of relaxation time of Li^+ impurity dipoles in KCl. Experimental points taken at three different electric field polarizations are marked by symbols as follows: [100]-triangles, [110]-squares, [111]-stars [12]. The solid lines are temperature dependences of Mathieu-based and harmonic-oscillator based 2D relaxation times $t_{12}(T) = k_{12}(T)^{-1}$, calculated using parameters by the mixing constants model of [5] as follows: $\hbar\omega_{\mathrm{renII}} = 8.2$ meV, $\hbar^2/2I = 0.43$ meV, $E_{\mathrm{BII}} = 9.9$ meV, $E_{\mathrm{JT}} = 0.58$ eV, $E_{\alpha\beta} = 1.41$ eV, $E_{\mathrm{RII}} = 97.7$ meV, $K = 4.38$ eV/Å^2. See Section 5 for a discussion of the low-temperature discrepancy between present theories and experiment.

$$\times 4b_1Q_1b_2Q_2/\{[(2b_1Q_1)^2 + (E_{\alpha\beta}/2)^2][(2b_2Q_2)^2 + (E_{\alpha\beta}/2)^2]\} \tag{53}$$

b_i and Q_i are, respectively the electron-mode mixing constant and vibrational coordinate associated with the ith oscillator. We consider two cases, in-phase vibrations ($Q_1 = Q_2 = Q$) and out-of-phase vibrations ($Q_1 = -Q_2 = Q$), and setting $\epsilon = Q_2/Q_1$ we have $\epsilon = +1$ and $\epsilon = -1$, respectively. We focus on the dipole–dipole attraction of two out-of-phase vibronic oscillators. With $\epsilon = -1$ turning over the sign to $u(R)$, we regard $u(R)$ as positively definite. We seek the extrema of (52) in order to redefine the curvatures and thereby the renormalized vibrational frequency $\omega_{\mathrm{renI}}(T)$ of the off-center oscillators' ensemble.

Introducing for brevity $\sigma = [(E_{\alpha\beta}/2)^{1/2} + (bQ)^2]^{1/2}/k_BT$, we get

$$\begin{aligned} F' &\equiv \partial F/\partial Q = KQ - \tanh(\sigma/2)(b^2/k_BT)(Q/\sigma) \\ &+2u(R)b^2(E_{\alpha\beta}/2)^2 Q/(\sigma k_BT)^4 \end{aligned}$$

$$\begin{aligned} F'' &\equiv \partial^2 F/\partial^2 Q = K\{1 - (b^2/Kk_BT)[\sigma^{-1}\tanh(\sigma/2) + \\ &+(bQ/\sigma k_BT)^2[1/2(\cosh(\sigma/2))^2 - (1/\sigma)\tanh(\sigma/2)]]\} - \\ &-2u(R)b^2(E_{\alpha\beta}/2)^2[3(bQ)^4 + 2(bQ)^2(E_{\alpha\beta}/2)^2 \\ &-(E_{\alpha\beta}/2)^4]/(\sigma k_BT)^8. \end{aligned} \tag{54}$$

The extrema all require $F' = 0$. A central maximum at $Q = 0$ yields

$$F''(0) = K\{1 - (2b^2/KE_{\alpha\beta})[\tanh(E_{\alpha\beta}/4k_BT) - 2u(R)/(E_{\alpha\beta}/2)]\} \tag{55}$$

which vanishes at $T = T_c$, the critical temperature for on-center to off-center conversion,

$$\begin{aligned} k_BT_c &= (E_{\alpha\beta}/4)\tanh^{-1}\{(E_{\alpha\beta}/4E_{\mathrm{JT}})[1 + (2u(R)/(E_{\alpha\beta}/2)) \\ &\times(4E_{\mathrm{JT}}/E_{\alpha\beta})]\} \end{aligned} \tag{56}$$

where $E_{\mathrm{JT}} = b^2/2K$ is the Jahn–Teller energy associated with the off-centered displacement. For the side minima at $Q = Q_0$:

$$1/2E_{\mathrm{JT}} = \tanh(\sigma/2)/(\sigma k_BT) - 2u(R)(E_{\alpha\beta}/2)^2/(\sigma k_BT)^4.$$

$$\begin{aligned} F''(Q_0) &= K\{1 - (2E_{\mathrm{JT}}/k_BT)[(1/\sigma)\tanh(\sigma/2) + (bQ_0)^2/(\sigma k_BT)^2 \times \\ &\times[1/2(\cosh(\sigma/2))^2 - (1/\sigma)(\tanh(\sigma/2))]] - 4u(R)E_{\mathrm{JT}} \\ &(E_{\alpha\beta}/2)^2 \times [3(bQ)^4 + 2(bQ)^2(E_{\alpha\beta}/2)^2 \\ &-(E_{\alpha\beta}/2)^4]/(\sigma k_BT)^8\}. \end{aligned} \tag{57}$$

The former of Equations (57) obtains from $F' = 0$ and is solved to give $Q(T)$. Inserting into the latter yields $\omega_{\mathrm{renI}}(T) = \omega_{\mathrm{bare}}\{\cdots\}^{1/2}$, the terms under curled brackets in (57) being the renormalization factor. Note that $\omega_{\mathrm{renI}}(0)$ agrees with (15). We see that both the vibrational frequency $\omega_{\mathrm{renI}}(T)$ and the inertial moment $I(T) = M \times [Q(T)]^2$ of the 'statistical rotator' are temperature dependent. A couple of numerical solutions to Equations (57) for $Q(T)$ and $\omega_{\mathrm{renI}}(T)$ at different pair potentials $u(R)$ are solved as shown in Figure 6.

For $u = 0$ at low temperature $2k_BT \ll [(E_{\alpha\beta}/2)^2 + (bQ_0)^2]^{1/2}$, the former Equation (57) yields the familiar result $(bQ_0)^2 + (E_{\alpha\beta}/2)^2 = (2E_{\mathrm{JT}})^2$ of an isolated off-center oscillator. The second one gives $F'' = K[1 - (E_{\alpha\beta}/4E_{\mathrm{JT}})^2]$, as expected

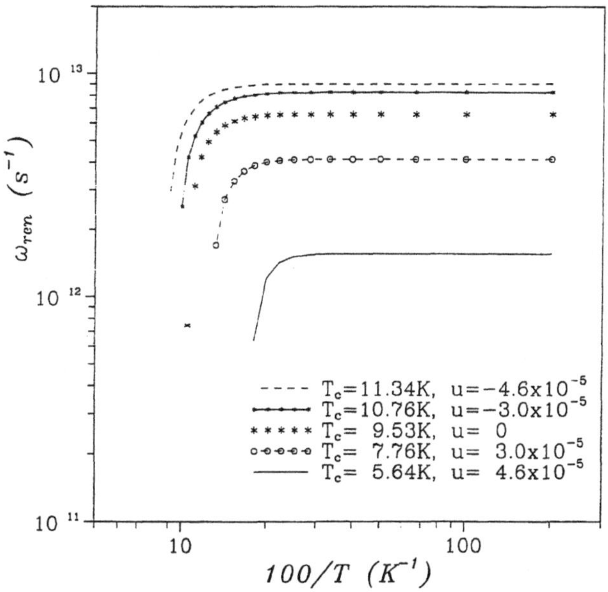

Figure 6. Temperature dependence of the renormalized vibrational frequency $\omega_{\mathrm{renI}}(T)$ in an ensemble of vibronic oscillators coupled by dipole–dipole attraction. $u = u(R)$ is the pair interaction function of Equation (53). The coupled vibrations in a pair are in-phase ($Q_1 = Q_2$) at $u < 0$ and out-of-phase ($Q_1 = -Q_2$) at $u > 0$. The parameters used are: b = 0.06 eV/Å; E_{JT} = 1.8 × 10^{-3} eV; $E_{\alpha\beta}$ = 7 × 10^{-3} eV; K = 1 eV/Å^2. $|u| = 10^{-5}$ eV corresponds to a pair separation $R \simeq 10$ Å at $|p| = 0.3$ eÅ.

on that basis. In as much as Equation (49) for the low-temperature rate $k_{12}(0)$ does not contain any ω_{renI}-dependent quantities other than $q = (2E_{\mathrm{BII}}/\hbar\omega_{\mathrm{renII}})^2 \sim (Q^2\omega_{\mathrm{renII}})^2 \sim (Q^2\omega_{\mathrm{renI}})^2$, taking due account of the statistical behavior $Q(T)$ and $\omega_{\mathrm{renI}}(T)$ signifies $q = q(T)$. Since $Q(T)$ and $\omega_{\mathrm{renI}}(T)$ are ultimate at $T = 0$ and drop continuously as T grows, then $q(T)$ will be expected to drop even more rapidly. From Equations (49) and (50) we now see that $k_{12}(0)$ increases as a result of the broadening of the lowest rotational band due to the decreasing $q(T)$. Physically the statistical 2-D rotator becomes increasingly less hindered as the temperature is raised because of the related decrease of rotational barriers $E_{\mathrm{BII}}(T) = I(T)/8 \times [\omega_{\mathrm{renII}}(T)]^2$. Ultimately at $T = T_c$ the rotational barriers vanish and the off-center impurity is free to rotate. Similar arguments apply to the harmonic bottom-well rate (51) where the temperature course is dominated by $\exp(-E_{\mathrm{RII}}/\hbar\omega_{\mathrm{renII}})$ leading to the increase of $k_{12}(0)$ as T is raised below T_c.

7. Conclusion

We have presented a theory for the occurrence and reorientation of off-center impurity ions in crystals based on the pseudo-Jahn–Teller effect. Our approach to the relaxation times of off-center dipoles combines *Christov's general reaction-rate theory* with the periodic solutions and allowed rotational bands of Mathieu's eigenvalue equation. Both configurational transition probabilities and reorientational rates are derived. These are compared with rates based on the usual well-bottom harmonic approximation. The statistical signature of Li^+ hindered rotators controlled by dipole-dipole pair interactions seems imprinted on the experimental relaxation time versus temperature dependence.

Acknowledgements

We are grateful to Dr. L. Mihailov and Dr. D. Batovsky for computational help in masterminding Figures 1 and 4, respectively. This work was supported by a grant under TN-344 of the National Foundation for Scientific Research.

References

1. R. Smoluchowski: *Proc. XV Colloque Ampère*, p. 120 (1968).
2. I. B. Bersuker: *The Jahn–Teller Effect and Vibronic Interactions in Modern Chemistry*, Academic Press, New York (1986). Russian translation: *Effekt Yana–Tellera i Vibronnie Vzaimodeystviya v Sovremennoi Khimii*, Nauka, Moskva (1988).
3. M. D. Glinchuk: in *The Dynamical Jahn–Teller Effect in Localized Systems*, Yu. E. Perlin and M. Wagner (eds.), Elsevier, Amsterdam, p. 823 (1984).
4. L. Brillouin and M. Parodi: *Propagation des Ondes dans les Milieux Periodiques*, Dunod, Paris (1956). Russian: L. Brillynen i M. Parodi, *Rasprostranenie voln v periodicheskikh strukturakh*, IIL, Moskva, p. 266 (1959).
5. G. Baldacchini *et al.*: *Preprint ENEA* (1996); M. Georgiev *et al.*: *Bulgarian Chem. Communications* (1996) (accepted); P. Petrova *et al.*: *Proc. ICDIM'96*, Wake Forest University, Winston-Salem NC (1996).
6. M. Gomez, S. P. Bowen, and J. A. Krumhansl: *Phys. Rev.* **153**, 1009 (1967).
7. A. Gochev: *Solid State Commun.* **49**, 1181 (1984).
8. M. Georgiev and A. Gochev: *Phys. Rev.* **B 31**, 5456 (1985).
9. A. Wuerger, R. Weis, M. Gaukler, and C. Enss: *Europhys. Letters* **33**(7), 533 (1996).
10. S. G. Christov: *Collision Theory and Statistical Theory of Chemical Reactions*. Lecture Notes in Chemistry, Vol 80, Springer, Berlin, pp. 49, 61 (1980).
11. S. G. Christov: *Phys. Rev.* **B 26**, 6918 (1982).
12. R. Osswald and H. C. Wolf: *Phys. Stat. Sol.* **B 50**, K93 (1972).
13. M. D. Glinchuk, M. Deigen, and A. Karmazin: *Fiz. Tverdogo Tela* **15**, 2048 (1973).
14. G. Baldacchini, U. M. Grassano, A. Scacco, F. Somma, M. Staikova, and M. Georgiev: *Nouvo Cim.* **13 D**, 1399 (1991).
15. R. Mealli and D. M. Proserpio: *J. Chem. Educ.* **67**, 399 (1990).
16. M. Abramowitz and I. A. Stegun (eds.): *Handbook of Mathematical Functions with Formulas, Graphs and Mathematical Tables*, NBS Math. Series, (1964). Russian: Nauka, Moscow (1979).
17. E. Janke, F. Emde, and F. Loesch: *Tafeln Hoeherer Funktionen*, Teubner, Stuttgart (1960). Russian: Nauka, Moscow (1964).
18. I. S. Gradsteyn and I. M. Ryzhik: *Tablitsy Integralov, summ, ryadov i proizvedenii*, GIFML, Moskva (1963). English: I. S. Gradshteyn, and I. M. Ryzhik: *Table of Integrals, Series, and Products*, Corrected and Enlarged Edition, Academic Press, Orlando (1965).
19. L. S. Bark, N. I. Dmitrieva, L. N. Zakhar'ev, and A. A. Lemanskii: *Tablitsi sobstvennykh znachenii uravneniya Mat'e*, Computing Center of the USSR Academy of Sciences, Moscow (1970).
20. M. J. O. Strutt: *Lamesche-Mathieusche und Verwandte Funktionen in Physik und Technik*, Russian Translation: GNTIU, Kharkov-Kiev (1935).
21. F. Luty: *J. Phys. (Paris)* **34**, C9–49 (1973).

22. F. Luty: *J. Phys. (Paris)* **28**, C4–120 (1967).
23. A. Diaz-Gongora, C. Medrano P., J. L. Boldu O., R. J. Gleason, and M. Georgiev: *Revista Mexicana de Fisica* **32**, 475 (1986).

Index

absorption 245
additivity rules 7
algebraic approximation 323
astrophysical plasmas 51

Brillouin–Wigner 283

calculations
  periodic Hartree–Fock 245
  VB 343
CASVB 67
cluster(s)
  dynamics 185
  multireference coupled 283
  van der Waals 219
color-centers 231
continuum spectrum of Mg 349
core-valence separation 309
correlation effects 161

Debye relaxation 149
defects in condensed matter 219

Floquet-type Hamiltonian 349
formic acid dimer 161
functional(s)
  correlation 27
  exchange 27
fusion plasmas 51

group functions 7

hard or soft acid or base 245
hydrogen atom in intense laser field 349

interaction forces 185

layered perovskites 263
LiCl dissociation 7

method(s)
  relativistic quantum defect orbital 51
  sampling 169

non-orthogonal Slater determinants 343

open-shell atom 309
oxide
  magnesium 231
  metal 245

partition technique 231
phase segregation 263
photoinduced phenomena 263
photoionization cross section 349
polarons and bipolarons 263
potential
  double-well 149
  optimized effective 27
predissociation
  rotational 185
  vibrational 185, 219
processes
  electron scattering 8
  local chemical 231
probability
  configurational tunneling 373
  transition 51

rate
  reaction 149
  reorientational 373

sampling initial conditions 219
Schrödinger equation 149
separability 7
sets
  distributed 323
  even-tempered basis 323
  finite basis 323
  Gaussian basis 323
single-root 283
spectra
  molecular 87
  optical 263
spectral lines 51
statistical ensemble 373
structure(s)
  molecular 87
  quantum 185
surface acidity 245
system(s)
  ionized 51
  tunneling 149

R. McWeeny et al. (eds.), Quantum Systems in Chemistry and Physics, 397–398.

theory
  density functional 27, 309
  perturbation 283
  reaction-rate 373
  valence bond 323
transfer
  collisional energy 169
  proton 161
transformation(s)
  local-scaling 309
  non-unitary orbital 67

unimolecular reactions 169

vibronic oscillator 373

wavefunction(s)
  factorization 7
  generalized multistructural 87
  localized 87
  valence-bond 67

TOPICS IN MOLECULAR ORGANIZATION AND ENGINEERING

Honorary Chief Editor: W. N. Lipscomb, Harvard, U.S.A.
Executive Editor: Jean Maruani, Paris, France

1. J. Maruani (ed.): *Molecules in Physics, Chemistry, and Biology.* Vol. 1: General Introduction to Molecular Sciences. 1988 ISBN 90-277-2596-9
2. J. Maruani (ed.): *Molecules in Physics, Chemistry, and Biology.* Vol. 2: Physical Aspects of Molecular Systems. 1988 ISBN 90-277-2597-1
3. J. Maruani (ed.): *Molecules in Physics, Chemistry, and Biology.* Vol. 3: Electronic Structure and Chemical Reactivity. 1989 ISBN 90-277-2598-5
4. J. Maruani (ed.): *Molecules in Physics, Chemistry, and Biology.* Vol. 4: Molecular Phenomena in Biological Sciences. 1989 ISBN 90-277-2599-3
5. E. Schoffeniels and D. Margineanu: *Molecular Basis and Thermodynamics of Bioelectrogenesis.* 1990 ISBN 0-7923-0975-8
6. A. Lund and M. Shiotani (eds.): *Radical Ionic Systems.* Properties in Condensed Phases. 1991 ISBN 0-7923-0988-X
7. P.I. Lazarev (ed.): *Molecular Electronics.* Materials and Methods. 1991 ISBN 0-7923-1196-5
8. E. Rizzarelli and T. Theophanides (eds.): *Chemistry and Properties of Biomolecular Systems.* 1991 ISBN 0-7923-1393-3
9. L.A. Montero and Y.G. Smeyers (eds.): *Trends in Applied Theoretical Chemistry.* 1992 ISBN 0-7923-1745-9
10. M.T. Pope and A. Müller (eds.): *Polyoxometalates: From Platonic Solids to Anti-Retroviral Activity.* 1994 ISBN 0-7923-2421-8
11. N. Russo, J. Anastassopoulou and G. Barone (eds.): *Properties and Chemistry of Biomolecular Systems.* 1994 ISBN 0-7923-2666-0
12. Y.G. Smeyers (ed.): *Structure and Dynamics of Non-Rigid Molecular Systems.* 1994 ISBN 0-7923-2774-8
13. A. Lund and C.J. Rhodes (eds.): *Radicals on Surfaces.* 1995 ISBN 0-7923-3108-7
14. Y. Ellinger and M. Defranceschi (eds.): *Strategies and Applications in Quantum Chemistry.* From Molecular Astrophysics to Molecular Engineering. 1996 ISBN 0-7923-3837-5
15. B. Silvi and P. D'Arco (eds.): *Modelling of Minerals and Silicated Materials.* 1997 ISBN 0-7923-4333-6
16. R. McWeeny, J. Maruani, Y.G. Smeyers and S. Wilson (eds.): *Quantum systems in Chemistry and Physics.* Trends in methods and applications. 1997 ISBN 0-7923-4699-8

MIX
Papier aus verantwortungsvollen Quellen
Paper from responsible sources
FSC® C105338
FSC
www.fsc.org

If you have any concerns about our products,
you can contact us on
ProductSafety@springernature.com

In case Publisher is established outside the EU,
the EU authorized representative is:
Springer Nature Customer Service Center GmbH
Europaplatz 3, 69115 Heidelberg, Germany

Printed by Libri Plureos GmbH
in Hamburg, Germany